International Conference on Innovations in Mechanical Sciences (ICIMS 2021)

IOP Conference Series: Materials Science and Engineering
Volume 1132

Kerala, India
19 - 20 February 2021

Editors:

Kiran Joy Irimpan
Jobin Varghese
Sudhy S. Paniker
P.K. Suresh
Biju P. Augustine

ISBN: 978-1-7138-3410-6
ISSN: 1757-8981

Printed from e-media with permission by:

Curran Associates, Inc.
57 Morehouse Lane
Red Hook, NY 12571

Some format issues inherent in the e-media version may also appear in this print version.

Institute of Physics
Dirac House, Temple Back
Bristol BS1 6BE UK

Phone: 44 1 17 929 7481
Fax: 44 1 17 920 0979

techtracking@iop.org

Additional copies of this publication are available from:

Curran Associates, Inc.
57 Morehouse Lane
Red Hook, NY 12571 USA
Phone: 845-758-0400
Fax: 845-758-2633
Email: curran@proceedings.com
Web: www.proceedings.com

TABLE OF CONTENTS

IOP Conf. Series: Materials Science and Engineering 1132 (2021) 011001 doi:10.1088/1757-899X/1132/1/011001

Preface: International Conference on Innovations in Mechanical Sciences (ICIMS'21)

The International Conference on Innovations in Mechanical Sciences (ICIMS'21) was organized by department of mechanical engineering in association with ASME student section ASIET and was held at Adi Shankara Institute of Engineering and Technology (ASIET), Kalady, Ernakulam, India during 19 – 20 February 2021.

ICIMS'21 aims to provide research aspirants form academic institutions, industries and research organizations, with a platform to share and showcase their research findings. The conference was hosted completely through google meet online platform due to the pandemic situation. Technical papers were invited in fourteen different tracks spanning from thermal engineering to allied interdisciplinary fields. Out of 68 research articles received, 45 were presented after double blind peer review. All the decisions pertinent to acceptance of articles were done by the editorial board based on the reviewer's comments. Final acceptance was made only after ensuring that the authors have incorporated the comments made by the reviewers. The acceptance ratio came out to be 70 percentage.

ICIMS'21 was first of its kind at our department and is intended to support and motivate the students and staff of our institute to do research and explore the new avenues in mechanical sciences and other.

The inaugural ceremony was initiated by extending welcome to Prof. Sabu Thomas, Honorable Vice Chancellor, MG University (a reputed state government owned university), Advocate Sri. K. Anand, Managing trustee, ASIET and Professor Jaisankar C P, Chief operations officer, ASIET and Patrons of ICIMS'21, Dr. Jacob George C, Administrator and Dr. Suresh Kumar V, Principal, ASIET and advisors for ICIMS'21, Prof. K.T. Subramanian, Head of the department and convener for ICIMS'21, and Prof. Eldose K K, Dean campus affairs and Co-convener for ICIMS'21. Inaugural address was delivered by Prof. Sabu Thomas followed by speeches delivered by Advocate Sri. K.Anand, Professor Jaisankar, Dr. Suresh kumar, Prof. K. T. Subramanian. Inaugural session was concluded with vote of thanks by Prof. Eldhose K. K, Co-convener of ICIMS'21.

The technical sessions of ICIMS'21 started with a keynote session delivered by Prof. Sabu Thomas, on the topic "Sustainable Nano materials" followed by seven keynote lectures delivered by eminent researchers around the globe. Each keynote was delivered before the commencement of each track by eminent researchers as listed. Dr. Oisik Das, Lulea University of Technology, Sweden – "*Turning biocomposites black by biochar*", Dr. C. Rajendran, Professor, Indian Institute of Technology, Madras – "Mathematical programming models for prediction classification and clustering ", Dr. Julian Gosliga, Research Associate, University of Sheffield, UK "Use of Graphs and Hypergraphs for Population-Based Structural Health Monitoring", Dr. Harikrishnan J., Principal Scientist, Roche Sequencing Solutions, CA, USA – "*Microfluidic Toolbox for Field-based Diagnostics*", Dr. Visakha Raja, GKN Aerospace, UK – "Engineering design research applied to commercial aero-engine products: some examples", Bony Thomas, Research Scholar, Lulea University of Technology, Sweden – "Energy storage materials from forest wastes: conversion of lingo-cellulosic materials to supercapacitor electrodes".

The editorial board members include: Dr. Biju B., Professor, M.A. College of Engineering, Kerala as Editor- in – chief, Dr. Kiran Joy Irimpan, Dr. Jobin Varghese, Dr. Sudhy S Paniker, Associate Professors, ASIET, Dr. P. K Suresh, Professor, ASIET and Dr. Biju Augustine P, Professor, RIT Kottayam as associate editors for ICIMS'21

Technical review panel includes: Dr. Satyanand Abraham, National Institute of Technology Warangal; Dr. Akhil VM, National Institute of Engineering, Mysore; Dr. Anand Raj Hariharan, Adi Shankara Institute of Engineering and Technology; Dr. A P Sudheer, National Institute of Technology Calicut; Dr. Ashish Alex Sam, Vellore Institute of Technology; Dr. Biju CV, Jyothi Engineering College; Prof.

IOP Conf. Series: Materials Science and Engineering 1132 (2021) 011001 doi:10.1088/1757-899X/1132/1/011001

Eldose KK, Adi Shankara Institute of Engineering and Technology; Dr. Ganesh Malayath, Indian Institute of Technology Kharagpur; Dr. Jigar Sura, Amity University Pune; Dr. Jilse Sebastian, St. Josephs College of Engineering and Technology; Dr. Jinu Paul, National Institute of Technology Calicut; Dr. Jobin Varghese, Adi Shankara Institute of Engineering and Technology; Dr. Jomy Joseph, National Institute of Technology Calicut; Dr. Joseph Babu, Rajagiri School of Engineering and Technology, Ernakulam; Dr. Krishnadas Narayanan Nampoothiri, Indian Institute of Technology Madras, Dr. Kiran Joy Irimpan, Adi Shankara Institute of Engineering and Technology; Dr. Kiran Raj M, University of Waterloo; Dr. Jayachandran KN, Indian Institute of Technology Bombay; Dr. K Sajun Prasad, Ohio State University; Dr. Rupesh S, PES College of Engineering; Dr. Manu S Natesan, Adi Shankara Institute of Engineering and Technology; Dr. Mubarak M, MEA Engineering College; Dr. Praveen KM, Muthoot Institute of Technology & Science; Dr. Raghu Kumar C, Goverment Engineering College Kozhikode; Dr. Rahul Antony, SAINTGITS College of Engineering; Dr. Rofin TM, National Institute of Technology Karnataka, Suratkal; Dr. Sebin Jose, Indian Institute of Technology Kharagpur; Dr. Shainath Kalamkar, Indian Institute of Technology Kanpur; Dr. Sreehari VM, SASTRA Deemed University; Dr. Sudhy S Panicker, Adi Shankara Institute of Engineering and Technology, Dr. Sujith Kumar S, Indian Institute of Technology Bombay; Dr. Suresh PK, Adi Shankara Institute of Engineering and Technology; Dr. Swaroop Hangal, Indian Institute of Technology Bombay,;Dr. Syam S, Indian Institute of Technology Kharagpur; Dr. Thomas A Vempiliyath, Indian Institute of Technology Kharagpur; Dr. Vineesh KP, National Institute of Technology Calicut, Dr. Ashmi M, National Institute of Technology Calicut

International advisory committee comprises of Dr. Prahlad Vadakkepat, Professor, NUS, Singapore; Dr. Oisik Das, Luleå University of Technology, Sweden; Dr. S. Jayanarayanan, Scientist, NASA, USA; Dr. S. B. Weerakoon, University of Peradeniya, Srilanka; Dr.-Ing. Hariharan Venkitachalam, Manager, ZF group, Germany; Dr. Harikrishnan J., Principal Scientist, Roche Sequencing Solutions, CA, USA.

International advisory committee comprises of Dr. Sabu Thomas, Vice Chancellor, Mahatma Gandhi University, Kerala; Dr. S. Ayoob, Pro. Vice Chancellor, KTU, Kerala; Dr. R. Sasikumar, Director, CAPE, Kerala; Dr. (Prof.) C. Rajendran, IIT Madras; Dr. (Prof.) Chandramouli P., IIT Madras; Dr. M Mathirajan, Chief Research Scientist, IISc; Dr. (Prof.) R. Shridharan, NIT, Calicut; Dr. (Prof.) P K Rajendrakumar, NIT Calicut; Dr. (Prof.) M. Duraiselvam, NIT Tiruchirappalli; Dr. (Prof.) A Kumaraswamy, DIAT, Pune; Dr. (Prof.) Sabu Sebastian, Scientist G, NPOL; Dr. (Prof.) Ganesh Natarajan, IIT Palakkad; Dr. (Prof.) M.C. Santhosh Kumar, NIT Tiruchirappalli; Dr. (Prof.) Balagopal G Menon, IIT Kharagpur; Dr. (Prof.) Chethan Mistry, IIT Kharagpur; Dr. (Prof.) A. P. Sudheer, NIT Calicut; Dr. (Prof.) Satyanand Abraham, NIT Warangal; Dr. (Prof.) Rajesh P. Nair, CUSAT

Organizing committee for ICIMS'21 comprises of Prof. K. T. Subramanian, head of Mechanical Engineering Department and convener; Prof. Eldose K.K., Dean, Campus Affairs and Enrichment Activities and Co-convener for ICIMS'21; Dr. Kiran Joy Irimpan, Dr. Jobin Varghese, Dr. Sudhy S Paniker, Associate Professors, ASIET and Shri. Leo Francis, Assistant Professor, ASIET and coordinators for ICIMS'21.

Members of organizing committee include: Mr. Tony K P, Mr. Muhammed Shiyas, Mr. Sreenivasan S. N., Mr. Sandeep O. S. and Akhil Shankar, Assistant professors, ASIET.

The technical sessions of the conference ICIMS'21 is concluded with a validatory function. Welcome speech for the validatory function was delivered by Dr. Kiran Joy, Coordinator and associate editor for ICIMS'21. The overall sessions of the conference were evaluated in a nut shell through a speech by Dr. Biju baby, Editor – in – Chief followed by short speech by Prof. K.T Subramanian. The validatory function was concluded with a vote of thank delivered by Dr. Jobin Varghese, Coordinator and associate editor for ICIMS'21.

IOP Conf. Series: Materials Science and Engineering 1132 (2021) 011002 doi:10.1088/1757-899X/1132/1/011002

Peer review declaration

All papers published in this volume of *IOP Conference Series: Materials Science and Engineering* have been peer reviewed through processes administered by the Editors. Reviews were conducted by expert referees to the professional and scientific standards expected of a proceedings journal published by IOP Publishing.

- **Type of peer review: Double-blind**

  A double – blind peer review was performed for each article before accepting the article for the international conference ICIMS'21. We were having a repository of 45 reviewers and each reviewer had reviewed a maximum of three papers for the conference. A single paper was reviewed by two reviewers who holds their Ph D from reputed university in INDIA and abroad. Reviewing was performed in a very strict manner and only articles advised to accept by the reviewers are considered to present in the conference.

- **Conference submission management system:** Created with google form
- **Number of submissions received:**
  We have altogether received 68 articles
- **Number of submissions sent for review:**
  63 articles were sent for the review before taking a decision on its acceptance.
- **Number of submissions accepted:**
  We have only accepted 48 articles out of 68 articles received. 5 substandard articles were rejected by the editor itself and 15 articles were rejected as per the advice of the reviewers.
- **Acceptance Rate (Number of Submissions Accepted / Number of Submissions Received X 100):**
- Acceptance rate = (48/68) *100 = 70.59
- **Average number of reviews per paper:** 6 Reviews per paper
- **Total number of reviewers involved:** 45 Reviewers
- **Any additional info on review process:** A maximum of 3 papers are reviewed by each reviewer. However, 60% of reviewers have reviewed only 2 papers. All papers were reviewed by experts in the particular field that paper belongs to.
- **Contact person for queries:**
  Dr. Jobin Varghese
  Associate Professor
  Department of Mechanical Engineering
  Adi Shankara Instiute of Engineering and Technology Kalady
  Ernakulam district, Kerala
  Jobin.me@adishankara.ac.in

Investigations on Joining of High Density Poly Ethylene Sheets using Resistance Welding Technique

K. Pradeep[1], Sudhy S Panicker[2], K Panneerselvam[1*] and P. Asokan[1]

[1]Department of Production Engineering, National Institute of Technology, Tiruchirappalli 620015, Tamil Nadu, India

[2]Department of Mechanical Engineering, Adi Shankara Institute of Engineering and Technology, Kalady 683574, Kerala, India

*kps@nitt.edu

Abstract. As a result of growing potential for high performance applications, parts made of HDPE materials is gaining greater interest for industry. Hence, the studies on joining of HDPE has got much significance. In this work, resistance welding of HDPE sheets was performed focusing on the governing set of parameters, current, pressure and time. Design of experiment (DOE) is done using full factorial design. The Analysis of Variance (ANOVA) was done with shear strength of weld as main parameter with the assistance of MINITAB software. Optimum parameters were identified in the study were medium level parameters produced good strength and better finished welded samples. Microstructure taken through Scanning electron microscopy (SEM) was supportive for this observation as there was a uniform flow of HDPE material without any void around the implant SS mesh for the best weld condition. The Finite element (FE) modeling of the resistance welding process was also performed using ANSYS to predict the thermal distributions occurring during the process depending on different levels of current and time inputs.

Keywords: HDPE, resistance welding, ANOVA, FE modeling

1. Introduction

High-density polyethylene (HDPE) is a thermoplastic prepared from petroleum products. HDPE is a commonly used thermoplastic material which has a wide variety of application in several industries. HDPE has become a leading polymeric material for numerous components due to several beneficial properties above metals such as light-weight, very good chemical and corrosion resistance, ease of joining, wide availability, higher shelf life, shorter processing times, and excellent reusability [1–6]. Gas and water distribution pipes, chemical-resistant piping systems, coax cable inner insulators (dielectric insulating spacer), geothermal heat transfer piping systems, etc. are some applications of HDPE products. Also, HDPE material is used in manufacturing several medical-related appliances. As this material has these many environment friendly applications, the studying of joining HDPE material structures is essential [7].

Numerous methods of adhesive bonding, mechanical fastening and fusion bonding are applied for joining the HDPE structures. The major issues like stress localisation and increase in weight during

IOP Conf. Series: Materials Science and Engineering 1132 (2021) 012001 doi:10.1088/1757-899X/1132/1/012001

mechanical fastening may be avoided by using adhesive bonding technique. However, adhesive bonding requires good pre-joining surface preparation and produces ambiguous joining quality. Moreover, adhesive bonding is also very sensitive to the chemically inert thermoplastics [8]. Fusion bonding is a method where the interface temperature at the joint crosses a definite value, i.e. T_g= glass transition temperature for amorphous polymers or T_m=melting temperature for semi-crystalline polymers. Further details of different conventional and non-conventional methods adopted to perform fusion welding of thermoplastics could be found in Jafrey and Panneerselvam [7,9]. Among these methods, resistance welding, also called as electrical-resistance fusion or electro-fusion or resistive implant welding, is an interesting technique for welding thermoplastics [1,3,10].

In resistance welding, an electrically resistive implant is placed in between the bonding surfaces of the base material to generate the desired heat input at the joint [1,11,12]. The resistive implant, when electric current is passed across it, generates heat energy E according to Joule's Law, $E=I^2Rt$, where R= is the resistance of the implant, I=current and t=elapsed time. Resistance welding is a rather simple technique and involves simple tooling and minimal surface preparation. When the energy supplied exceeds a threshold, the laminates temperature near to bonding surfaces starts to rise and with increase in welding time or current, the heat penetrates into inner layers of the material. It is advantageous to keep this heat affected zone (HAZ) closer to the bonding surface to avoid probable fibre disruption in the laminates. Resistance welding is usually applied for joining comparatively large and complex parts that does not require close-loop weld joints. This technique is effectively applied in the aerospace industry, especially by Fokker Special Products, the Netherlands [10,13]. Hence, the understanding of thermal bonding or welding behaviours of HDPE through resistance welding method is of high significance [1,2,8]. Recently, this technique was used by Xiong et al. [14] for hybrid joining of titanium alloy/thermoplastic composites using a stainless steel (SS) mesh implant. It was seen that rarely any study discussed a parametric feasibility on resistance welding of HDPE sheets under different levels of current, pressure and time used for joining.

Finite element analysis (FEA) is usually used in novel product design conceptualisation, and refinement of existing production line. FEA has become a solution to the task of predicting the behaviour of materials under various loading conditions. FE methods gives flexibility to designers by highlighting problem areas in the material and to understand the basis of deformations occurring within the material. Hence, it enables accurate prediction of structural deformations and temperature distributions happening within the investigated sample which are usually difficult to measure during actual physical experimentations [3]. Hence, this method aids in superior manufacturing technologies and to reduce the costs accrued during actual building and testing of samples. It was found that no previous work reported the investigations on FE simulation of resistance welding of HDPE sheets.

In this work, joining of HDPE sheets were performed through resistance welding process giving different levels of important input parameters, i.e. current, pressure and time. The analysis of variance (ANOVA) was done with the aid of MINITAB software. The full factorial combination approach was done to generate the experimental set of data. Mechanical characterization through tensile test, and metallurgical characterization using scanning electron microscopy (SEM) to assess better weld conditions. Finally, FE simulation was performed using ANSYS software to understand the temperature distribution occurring within the weld region.

2. Materials and methods

2.1. Specimen preparation

The HDPE sheets of dimension 150mm x 20mm x 4mm are produced using a semi-automatic injection molding machine shown in Fig. 1. The HDPE granules used for this purpose were Relene polyethylene grade M60075 from Reliance polymers. The standard properties of this polymer material were, melt flow 5to8 gm/10 minute, tensile strength 18 MPa, melting point 130-138°C, and continuous use temperature 180°C. Prefixed quantity of pure HDPE granules was fed to the heater and the heater temperature was adjusted to 200°C so that easy flow of molten material is possible. Separate levers were provided to control the opening and closing of die and plunger movement.

2.2. Resistance welding setup

The schematic of the resistance welding arrangement is shown in Fig. 2a [1]. Actual dimensions of HDPE sheets and the heating element mesh used in the present study is shown in Fig. 2b. It consists of following standard components: HDPE sheets with sandwiched heating element, insulating slabs, tool for applying pressure, electric power supply and wires, clamping device, ampere meter and voltmeter. Fabric type stainless steel (SS) fabric type mesh of 120μm wire diameter with 20mm width was placed as the implant material. Ageorges et al. [11] had shown that better uniform heating in the weld interface was given by fabric type mesh than unidirectional ones. Stainless steel material was selected in this study as it resists rusting and will not get oxidised even at higher temperatures [8]. TORNADO201 electric welding machine was used, and the temperature generated in the mesh was calibrated with varying currents. Temperature generated in the mesh at different time intervals for 20A, 30A, 40A, 50A currents was found out using IR pyrometer as enlisted in Table 1. The maximum temperature reached for maximum current in minimum time condition.

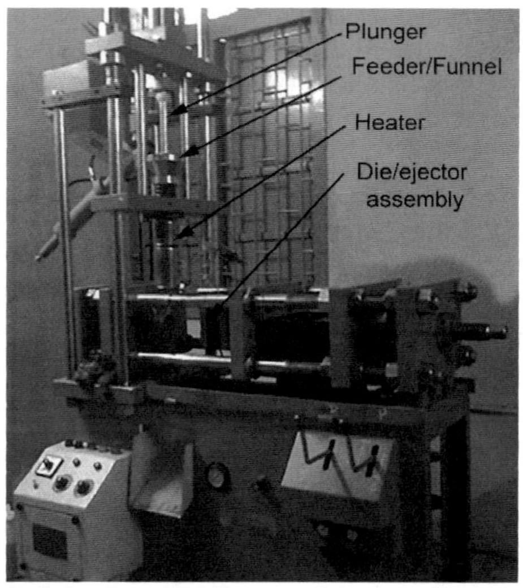

Figure 1 Semi-automatic injection molding machine used to manufacture HDPE sheets

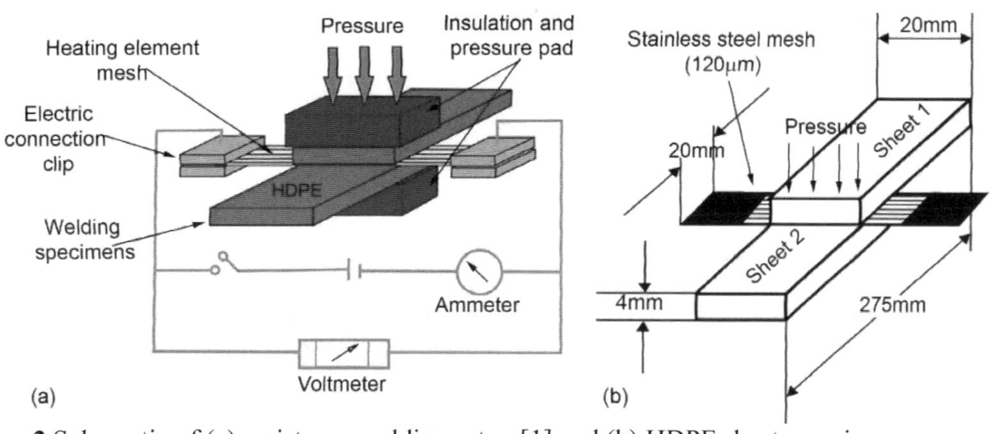

Figure 2 Schematic of (a) resistance welding setup [1] and (b) HDPE sheets specimen arrangement (all dimensions are in mm, representative figure and not to scale)

Table 1 Maximum temperature reached in different current values

Current (A)	Maximum temperature (°C)	Time taken to reach max. temperature (s)
20	214	100
30	260	90
40	324	85
50	354	75

A pressure gauge was used to accurately measure pressure applied during the process. A fixture designed for perfect holding of HDPE plates and mesh implant are made having the dimensions 90mm X 90mm. Another main function of fixture is to transfer the pressure uniformly to the welding area. It has machined groves of 3mm depth and 20mm width, in which the plates are perfectly seated so that during weld time, it is hold tight. The mesh was placed between the sheet specimens as shown in Fig. 2b. This assembly was then placed into the fabrication set-up and tightened well after properly checking the alignment of plates. The current was switched off when nominal melting was achieved, and the joint was cooled under adequate pressure. The pressure applied during the welding enabled proper contact between the laminate surfaces at the bond line and promoted molecular diffusion at the joint. After the welding, the heating element was trapped inside the joint. The critical factors considered for the present study are current (I (A)), time (T (s)) and pressure (P (kg/cm^2)).

2.3. *Design of experiments*

The full factorial approach was followed in the present study to perform number of experiments. This approach also facilitates the study of interactions between the factors. The present experiment was planned with has three factors, i.e. current, time and pressure varying in three levels as provided in Table 2. So, the total number of experiments conducted with different combinations was 27. The Design of experiments (DOE) was done with the support of MINITAB software. After making all arrangements, the current level was set to any one of the desired value. Then welding machine was switched on, measured the time to get it welded by means of a stop watch and switched it off after keeping it till the time reached as per DOE. With the welding set up and DOE, all 27 experiments are conducted, and all the welded specimens are shown in Fig. 3a. During the experiments, it was identified that the time required for melting is lower as melting starts with 20A current itself. With higher pressure, melting and joining occurs in 5s interval, and flames were formed at 12s. For very high values of current, time and pressure, liquid state HDPE spills out from the joint section as the melting occurs in a rapid rate.

Table 2 Process parameters selected with values at 3 levels

Levels	Pressure-P (kg/cm^2)	Current-I (A)	Time-T (s)
Level 1	4	20	5
Level 2	6	25	10
Level 3	8	30	15

2.4. *Finite element simulation*

In this work, FE simulation was done using the widely used ANSYS software. The basic material properties assumed for FE modelling is given in Table 3. An important problem was the lack of material data available at high temperature. At first, values of the physical parameters of HDPE changing with varying temperature (0 – 700 K) were defined. Then the interpolation method as adopted to obtain the required thermal property parameters in the unknown temperature range. The melting point of HDPE material was assumed as 470 K. Please note, the temperature mentioned for FE simulations are in Kelvin scale. The welding is a non-linear transient state process, where the physical and thermal properties of

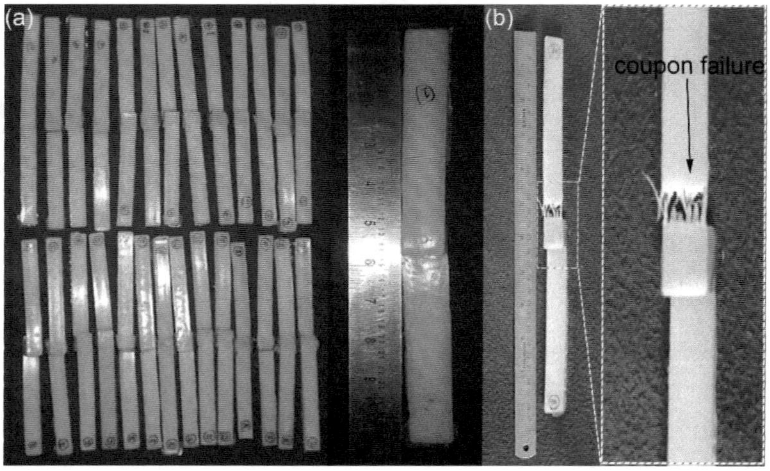

Figure 3 (a) HDPE specimens after welding and (b) Macrographs of specimens after tensile tests

Table 3 Properties of HDPE at different temperatures

Sl No.	Temperature (K)	Thermal conductivity (W/mK)	Density (kg/m^3)	Young's modulus (MPa)
1	300	.460	965	1400
2	400	.470	955	1200
3	425	.480	950	1150
4	475	.490	950	1100
5	500	.500	945	900
6	550	.510	940	800
7	600	.515	935	700
8	675	.520	935	650

the welding material rapidly changes with temperature. The controlling equation of heat conduction is given in Eq. 1.

$$\rho c\frac{\partial T}{\partial t}=\frac{\partial}{\partial x}\left(\lambda\frac{\partial T}{\partial x}\right)+\frac{\partial}{\partial y}\left(\lambda\frac{\partial T}{\partial y}\right)+\frac{\partial}{\partial z}\left(\lambda\frac{\partial T}{\partial z}\right)+Q \qquad (1)$$

where, ρ, c and λ are density, specific thermal capacity and heat conductance of material respectively. These terms were defined as functions of temperature and Q was the power of heating element. The solid 70 (Brick 8 node 70) element having 3-D thermal conduction definition was adopted for the present FE modelling. The element also compensated for mass transport during heat flow from a constant velocity field. The heat generation rates were given as element body loads at the specified nodes.

The FE model is was created as three distinct parts as shown in Fig. 4a. Two models of dimensions 150mm x 20mm x 4mm form the two work pieces to be welded together. The part in the middle portion which represents the SS mesh implant having dimensions 20mm x 20mm x 0.2mm are arranged as in experiment. Further, meshing was done such that the middle element is fine meshed and the other parts are coarse meshed, as represented in Figs. 4b,c. A free mesh, without any restriction of element shapes, and unspecified pattern was used with automatic meshing tool with a fineness level of 6/10. Heat generated in the heating element was calculated and given as input to this particular part. At the initial condition, i.e. t = 0, the work piece was given uniform initial temperature, which was the room temperature = 300K. Then, the convection heat transfer was assumed on the surfaces with surrounding temperature as 300K and convective heat transfer coefficient assumed was 25 W/m^2K. Ideal conditions

IOP Conf. Series: Materials Science and Engineering 1132 (2021) 012001 doi:10.1088/1757-899X/1132/1/012001

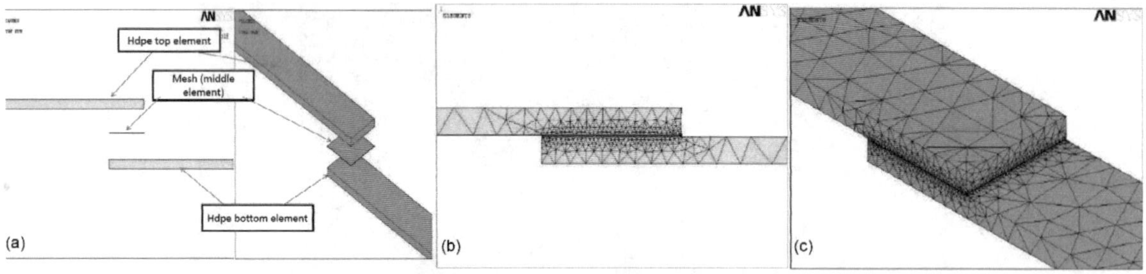

Figure 4 (a) FE model of resistance welding parts, (b), (c) view of the parts after meshing

Table 4 Calculated heat generation according to the current input

Current (I)	Heat generated (kJ)
20 A	642600
25 A	803250
30 A	963900

were selected at all levels of simulation process, where the parameters current time and pressure were given as inputs (all in SI units). As mentioned earlier, the middle part designed for mesh is acting as the heat source since heat generated in that area only which will melt the two welding parts and joining occurs. Heat generation occurring because of the current and voltage applied through stainless steel mesh calculated as follows. The SS mesh of 120microns with specified contained 120 threads. Hence, the heat generated depended only on 120 threads. As the welding machine supplied a constant output voltage of 63 V with an efficiency of 85%, the total power produced by mesh was calculated using Eq. 2.

$$P = (V \text{ x } I \text{ x } 0.85 \text{ x } 120)/(0.2e^{-3}) \quad \textbf{(2)}$$

As the welding was a transient based analysis, another important parameter was time. This parameter was directly fed into the programme as sub-step time. The heat generated for three levels of current as given in Table 4 was applied to the defined intervals.

3. Results and discussion

3.1. Shear strength testing

To verify whether the welding had achieved improvement and to determine how good the approximation matched actual performance, all 27 samples were subjected to tensile loading. This is one of the most important and generally used method to measure the ability of a material to resist breaking/shearing under tensile load. Shear strength testing of all 27 specimens were done using UTM D60 with digital load-displacement data collection facility. Table 5 enlists tensile test data such as maximum strength, maximum elongation, and maximum load conditions for all the 27 welded samples. The macrograph of the test specimen before and after the test is shown in Fig 3b. The maximum strength and elongation of each specimen was recorded, and it was found that the shear strength of the welded samples varied from 18 MPa to 24 MPa confirming that weld strength is either greater than or equal to the tensile strength specified for the base material. Fig. 5a and b shows load-displacement plot for specimen showing high and moderate strength. However, the failure happened at heat affected zone (HAZ), proving the material strength less at this region. The coupon failure mode [8,15] was observed for fractured samples, as observed in Fig. 3b with the SS mesh intact at the joint. The working parameters, i.e. current, pressure, time corresponding to these welds were 30A, 6kg/cm^2, 10s, and 25A, 8kg/cm^2 and 10s, respectively. Visual inspection of the welded samples revealed that specimens welded with medium level parameters corresponded to strength of 20 to 22 MPa showed better appearance and finishing. Breakage load was found to be the ultimate load itself, and fractured specimens showed brittle nature.

3.2. Evaluation of optimum welding parameters

The analysis of variance (ANOVA) was done using MINITAB software with factors affecting strength (current, pressure, time) as inputs. The purpose of ANOVA was to determine the factors and combination of factors that significantly affect the welding process. The regression equation obtained is shown in Eq. 3, with R^2 value of 75.86%. It should be noted that the linear model was adopted for simplicity of the problem.

$$\text{Strength} = 14.1 + 0.056\ \text{Current} + 0.559\ \text{Pressure} + 0.056\ \text{Time} \qquad \textbf{(3)}$$

Table 5 Shear strength and elongation of the welded specimens

Sl. No.	Current (A)	Pressure (kg/cm^2)	Time (s)	Strength (MPa)	Max elongation (mm)	Max Load (kN)
1	20	4	5	2.125	0.5	0.17
2	20	4	10	20.625	8.7	1.65
3	20	4	15	20.625	4.72	1.65
4	20	6	5	22.125	8.12	1.77
5	20	6	10	22.875	8.6	1.83
6	20	6	15	20.625	6.92	1.65
7	20	8	5	16.125	4	1.29
8	20	8	10	22.500	6.78	1.8
9	20	8	15	18.750	6	1.5
10	25	4	5	19.125	4.9	1.53
11	25	4	10	23.250	8.88	1.86
12	25	4	15	17.625	4.4	1.41
13	25	6	5	21.000	7.16	1.68
14	25	6	10	21.750	19.2	1.74
15	25	6	15	19.875	5.96	1.59
16	25	8	5	21.000	38.8	1.68
17	**25**	**8**	**10**	**21.125**	**0.5**	**1.69**
18	25	8	15	21.375	8.2	1.71
19	30	4	5	19.125	5.48	1.53
20	30	4	10	18.000	6	1.44
21	30	4	15	18.000	4.69	1.44
22	30	6	5	21.375	5	1.71
23	30	6	10	24.000	5.52	1.92
24	30	6	15	13.125	8.64	1.05
25	30	8	5	21.000	6.8	1.68
26	30	8	10	18.750	4.59	1.5
27	30	8	15	18.000	3.61	1.44

IOP Conf. Series: Materials Science and Engineering 1132 (2021) 012001 doi:10.1088/1757-899X/1132/1/012001

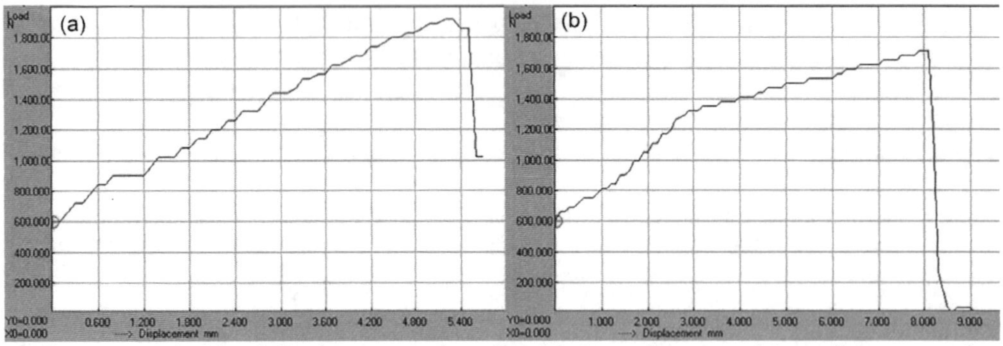

Figure 5 Representative load-displacement plot of welded specimens at (a) medium strength and (b) high strength conditions

Table 6 Analysis of Variance for Strength

Source	DF	SS	MS	F	P
Current	2	23.43	11.72	0.86	0.460
Pressure	2	47.00	23.50	1.72	0.239
Time	2	56.90	28.45	2.08	0.187
Current*Time	4	104.13	26.03	1.91	0.203
Current*Pressure	4	41.65	10.41	0.76	0.578
Pressure*Time	4	70.02	17.51	1.28	0.353
Error	8	109.22	13.65		
Total	26	452.35			

S = 3.69491 R-Sq = 75.86% R-Sq(adj) = 21.53%

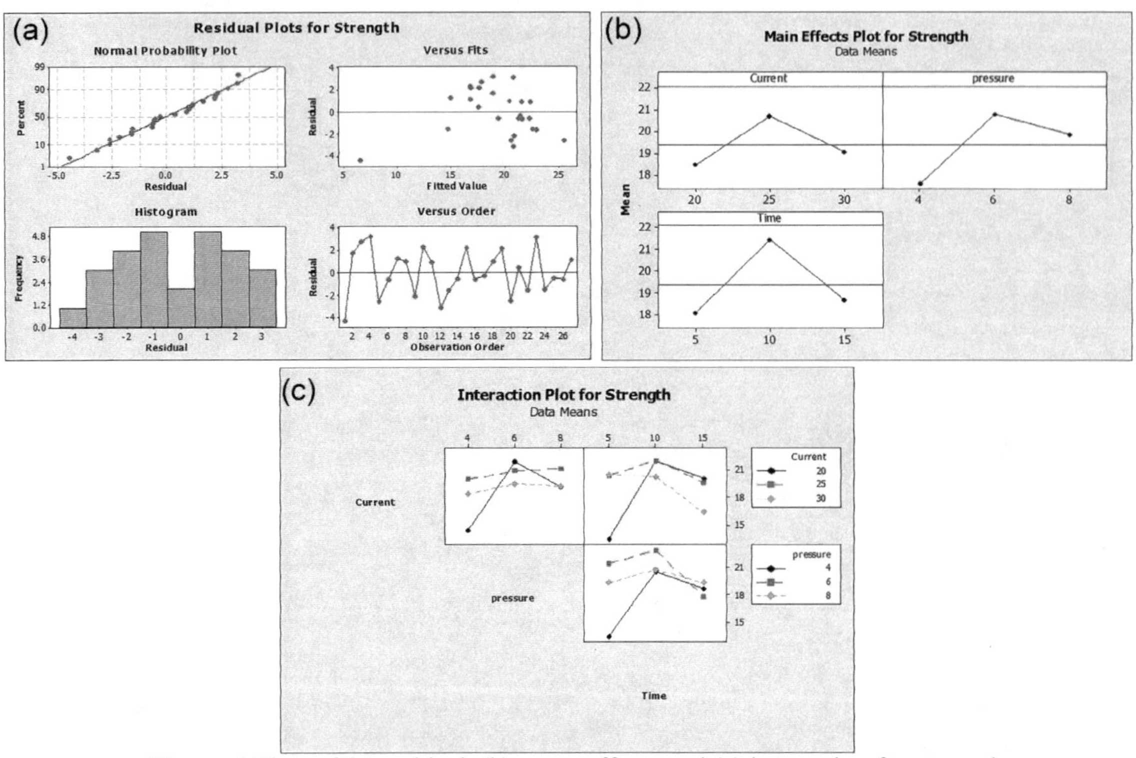

Figure 6 Plots of (a) residual, (b) mean effects and (c) interaction for strength

From the generated values using MINITAB, ANOVA for strength is as given in the Table 6. Residual plot and mean effect plot for strength generated are as also shown in Fig. 6a and b, respectively. Among the parameters, current and time were found to be more significant. From the main effects plot it was seen that pressure is not as significant as current and time as the slope gradient is comparatively less. The interaction plot (Fig. 6c) also showed that better strength was obtained for the average values of current (=25A) and time (=10s) and pressure (=8kg/cm^2). Similar conclusions was drawn from the surface plots of strength with influencing factors, i.e. strength vs time, current (Fig. 7a), strength vs pressure current (Fig. 7b), and strength vs time, pressure (Fig. 7c). Corresponding surface response plots are shown in Figs. 7d-e.

Contour plot of strength vs current, time (Fig 7d) shows strength is better in the value of current ranging from 24A to 30A, and time range of weld in 8s to 15s. Fig. 7e shows that strength is maximum in pressure range 5.2 kg/cm^2 to 7.9 kg/cm^2 and for current values between 22A to 28A. The contour plot of strength vs pressure, time in Fig. 7f implied that strength is high for pressure varying from 5 kg/cm^2 to 8 kg/cm^2 and the welding time ranging from 9s to 13s. Thus, by the analysis of parameters used for the resistance welding experiments, optimized process parameters obtained for experiment number 17 where current = 25A, pressure = 8 kg/cm^2 and time = 10s for the specimen that produced shear strength of 21.125 MPa (Table 5).

3.3. Microstructure investigation

Microstructures at the cross section of welds were taken using scanning electron microscopy (SEM). The lap region were cut using diamond cutting wheel and the regular polishing procedure was performed for SEM analysis [7,8]. SEM images of a poor joint which are having lower level parameters (current=20A, pressure=4kg/cm^2, time=5s), medium level parameters (current=25A, pressure=6kg/cm^2, time=10s) and high level parameters (current=30A, pressure=8kg/cm^2, time=15s) are presented in Fig. 8a, b, and c respectively. The defective joint shown in Fig. 8a shows large amount of voids at the interface, i.e. between the inter layers of SS mesh and the parent metal. The flow of melted base material through the heated mesh is not uniform and the joining was not intact between two parts of specimen. So, the sample showed poor strength properties. Similar observation was also reported by previous researchers [7,8,15] as well. However, the implant material (SS mesh) was completely embedded without any voids in the case of medium and high level parameters. A proper flow of molten base material was visualised as seen in Figs. 8b and c. A good adhesion of two laps was observed that confirmed a good weld. However, comparing Figs. 8b and c, the sample welded using medium level parameter showed relatively better consistency in HDPE material flow and intactness of mesh material. The welding with medium level parameters maintained the heat affected zone (HAZ) closer to the bonding surface that avoided probable fibre disruption in the laminates.

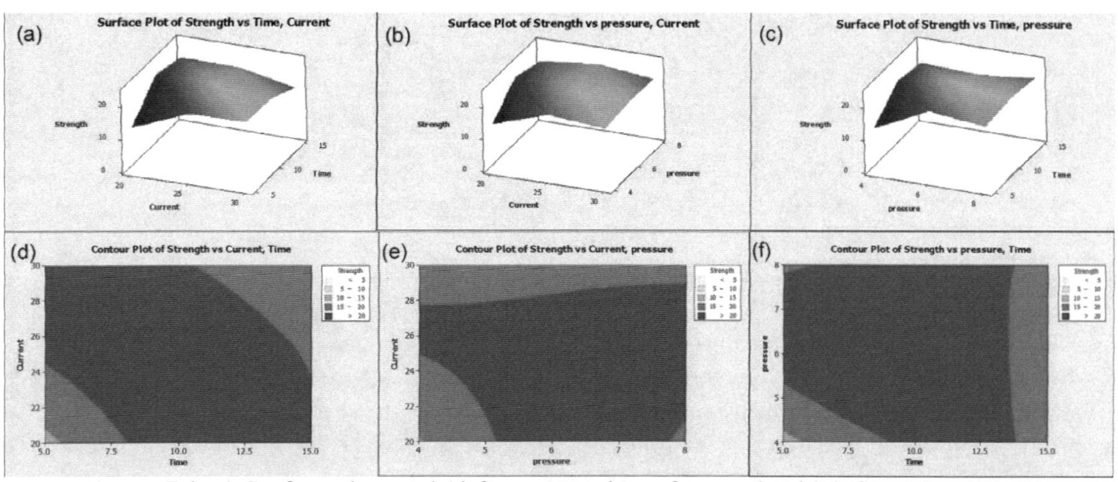

Figure 7 (a-c) Surface plots and (d-f) contour plots of strength with influencing factors

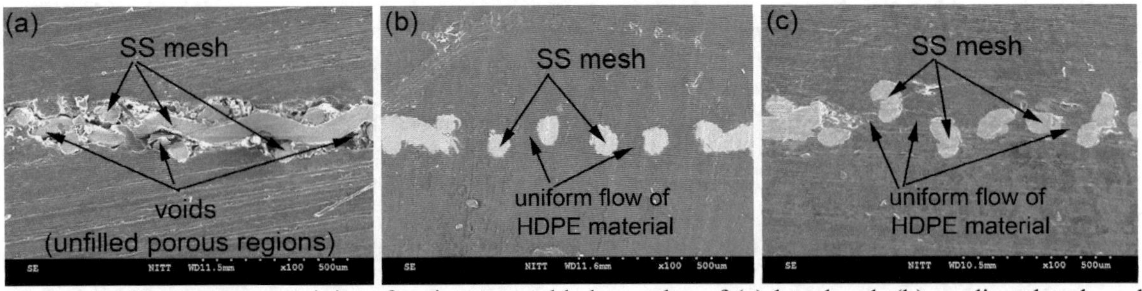

Figure 8 Microstructure at joint of resistance welded samples of (a) low level, (b) medium level, and (c) high level parameters.

3.4. Temperature distribution

The FE modelling was used to predict the temperature generation during the present resistance welding process. A clear visualisation of the heat distribution and movement of heat flux occurred inside the welded material and clear visualisation was obtained. Nodal temperature distribution was the transferring of heat from the heating element through the material. Temperature values were plotted at the middle part of the welded specimen. Maximum temperature was visualised at a little distance away from the centre. This was because of the spacing of nodes generated by the program and heat generated is applied across the node. The temperature values are listed against the depth of the material and part

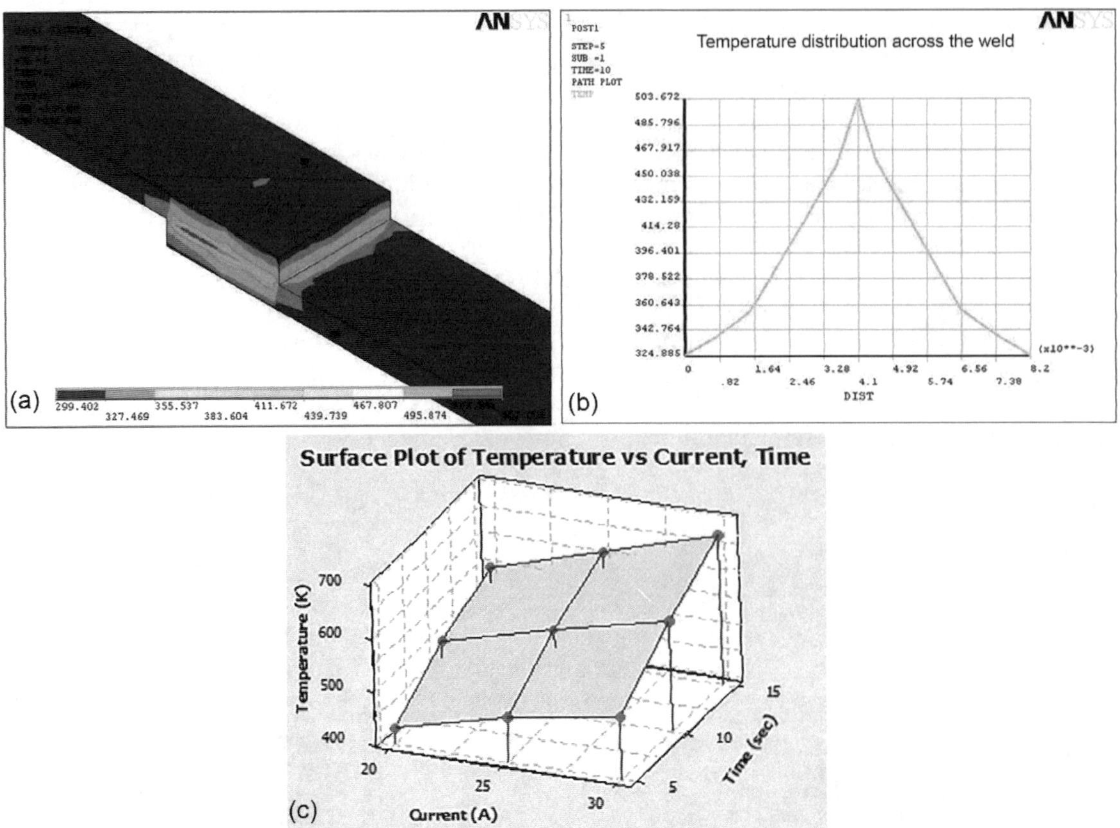

Figure 9 (a) FE simulated representation of temperature profile developed in resistance welded sample, (b) corresponding temperature distribution across the weld and (c) surface plot of Temperature v/s Current, Time

of the material above melting point. Fig. 9a shows the FE simulated image of temperature profile developed within the welded sample for the optimum welding parameter condition (i.e. for experiment 17, refer to section 3.2). It is clearly visible in Fig. 9b, that the temperature distribution was uniform across the depth at the middle section of the specimen. Similarly, FE simulations for three levels of current (20A, 25A, 30A) and for three levels of time (5s, 10s, 15s) were performed. The maximum temperature recorded for these conditions are tabulated and plotted as shown in Fig. 9c. Maximum temperature was obtained for the parameters 30A, 15s condition. It was observed that the maximum temperature obtained from simulation was in the similar range of that obtained for mesh temperature at different time intervals measured with the help of infrared pyrometer (Table 1). The differences in values occurred because of several assumptions used for FE simulations and actual values may differ according to the process conditions. The consideration of vaporization of the material at elevated temperatures were also negated while FE modeling.

4. Conclusions and future scopes

In this work, the effect of current, pressure and time on the resistance welding properties of HDPE sheets are studied. The FE simulation of HDPE resistance welding considering welding current and time was performed to predict the temperature distribution. Following are the major conclusions drawn from the study.

- Experimental setup was designed to perform lap joint resistance welding of HDPE sheets with SS mesh as implant material. DOE was done using full factorial approach, and ANOVA was carried out to obtain optimum welding parameters of current, pressure and time corresponding to shear strength data obtained through tensile testing.
- From different plots such as residual, main effect, interaction, surface and contour plots, it was concluded that the optimum resistance welding condition was for experiment number 17, where current = 25A, pressure = 8 kg/cm^2 and time = 10s for the specimen that produced shear strength of 21.125 MPa. The failed samples showed coupon failure and the fracture was brittle in nature.
- The SEM images revealed that there was a uniform flow of HDPE material without any void at the weld region showing good adhesion of both the plates with perfect implanting of SS mesh. The visual inspection and SEM images were also supportive to the ANOVA results showing that resistance welding with medium level parameters produced better finished components with reliable strength. This was because the heat affected zone (HAZ) was maintained closer to the bonding surface that avoided probable fibre disruption within the laminates.
- The temperature profile obtained for each level of process parameters were visualized through FE simulations. The heat penetration increased with increase in the value of current and time. The central area of mesh showed maximum temperature and these predicted temperature were in almost matching with the experimentally measured mesh temperature under corresponding current and time values.

Performing a coupled thermal-structural analysis is important future work for quantitative prediction of temperature-dependent variation of size and shape of the HDPE sheets. In this context, FE simulations also considering melt flow and vaporization of the HDPE material at elevated temperatures will give more accurate predictions of resistance welding process. Moreover, a future work on preparation, optimization of the parameters, mechanical and microstructural characterization, and corresponding FE analysis of resistance welding of HDPE nanocomposites is an important aspect considering the potential of applied polymer composites.

References

[1] Stavrov D and Bersee H E N 2005 *Compos. Part A Appl. Sci. Manuf.* **36** 39–54
[2] Kumar S, Kumar Y and Singh A 2014 *Int. J. Eng. Tech. Res.* **2** 89–96
[3] Ageorges C and Ye L 2001 J. *Thermoplast. Compos. Mater.* **14** 449–475
[4] Offringa A.R 1996 Compos. *Part A Appl. Sci. Manuf.* **27** 329–336
[5] Zehtabeyazdi A, Zebarjad S M, Sajjadi S A and Abolfazli Esfahani J 2007 *Express Polym. Lett.*

1 92–97
[6] Chang I Y and Lees J K 1988 *J. Thermoplast. Compos. Mater.* **1** 277–296
[7] Jafrey D D and Panneerselvam K 2018 *J. Adhes. Sci. Technol.* **32** 2350–2363
[8] Panneerselvam K, Aravindan S and Noorul Haq A 2012 *Mater. Des.* **41** 453–459
[9] Jafrey D D and Panneerselvam K 2017 P*olym. Eng. Sci.* **57** 495–504
[10] Nino G F, Ahmed T J, Bersee H E N, Beukers A 2009 *Compos. Part B Eng.* **40** 237–248
[11] Ageorges C, Ye L and Hou M 2000 *Compos. Sci. Technol.* **60** 1191–1202
[12] Villegas I F, Moser L, Yousefpour A, Mitschang P and Bersee H E N 2013 *J. Thermoplast. Compos. Mater.* **26** 1007–1024
[13] Mathijsen D 2016 *Reinf. Plast.* **60** 405–407
[14] Xiong X, Zhao P, Ren R, Zhang Z, Cui X and Ji S 2019 *Diam. Relat.Mater.* **101** 107611
[15] Hou M, Yang M, Beehag A, Mai Y and Ye L 2006 *Compos. Struct.* **47** 667–672

IOP Conf. Series: Materials Science and Engineering 1132 (2021) 012002 doi:10.1088/1757-899X/1132/1/012002

A Study on Mechanical and Durability Properties of Structural Concrete using Pumice and Sintered fly ash Aggregates

M. Sai Yaswitha Reddy*, V.K.Visweswara Rao and M. Srinivasula Reddy

Department of Civil Engineering, G. Pulla Reddy Engineering College(Autonomous), Kurnool, A.P, India.

*Email: saiyaswithareddy.m@gmail.com

Abstract. Lightweight aggregates has low density, high permeability than the conventional coarse aggregate. Most commonly available lightweight aggregate (LWA) are Pumice aggregate (PA), which is a natural aggregate, sintered fly ash aggregate (SFA), which is made by sintered processing the fly ash obtained as a by-product of coal based Thermal power plants. Among the many artificial aggregates, SFA is used as a replacement to coarse aggregate to produce lightweight aggregate concrete. This paper presents an experimental study of structural lightweight aggregate concrete (SLWAC) using combination of PA and SFA as a partial replacement for coarse aggregates. In this, the mechanical and durability properties of SLWAC using different percentages i.e., 25%, 50%, 75% and 100% of PA and the rest natural coarse aggregate, fully replacement of coarse aggregate with 50% PA and 50% SFA, are evaluated by conducting comprehensive series of tests on strength and permeability parameters. The test results showed better strength and adequate density, at coarse aggregate replacement with 50% PA and 50% SFA, to be accepted as SLWAC as per ASTM standards. When the coarse aggregate is completely replaced by pumice aggregate alone, the strength achieved is poor and is suitable for non-structural members. In this study, the mixes are designed for M25 grade concrete and the results are then compared with M25 grade of conventional concrete (CC).

1. Introduction

Concrete is a conglomeration of cement, water, fine and coarse aggregates. Concrete is the most widely used construction material in the construction industry because of its robustness, aesthetic and mouldability properties. The performance of the concrete developed largely depends on the properties of the ingredients used. In general, conventional concrete (CC) produced using natural coarse aggregate have densities in the range of 2400-2500 kg/m^3. It is well known that the density of the concrete plays an important role in defining the size of the structural members. Reducing the size of the structural members is very beneficially in reducing the dead weight. This can be achieved by replacing natural coarse aggregate with LWAs, as the coarse aggregate has a major hold in defining the concrete density. Moreover, the vast majority of conventional concrete producers utilize crushed stone as coarse aggregate and its use in concrete has triggered disparity naturally. Astonishing exploitation of natural rock masses has been creating havoc and is leading to land sliding, avalanches, etc. The influence of this on the environment has made to think toward unnaturally formed aggregates as an alternative to the natural rock aggregates and also to some of other natural sources. About 50% of concrete is filled with coarse aggregate, so instead of using conventional aggregate, usage of LWAs helps in not only avoiding destruction of natural rocks but also contributes to the reduction in overall density of concrete. Nowadays, a wide range of LWAs are available. Pumice aggregate (PA), natural resource, and Sintered fly ash aggregate (SFA), from artificial resource, have been used as a replacement of normal coarse aggregate in this study.

PA is a natural LWA formed due to the rapid cooling of liquid volcanic matter exploded. As the volcanic matter is exploded from the earth crust, it is rich in siliceous content [1]. Pumice is a very light rocky material whose cell structure consists of air pockets or air voids, the air voids forms when the gases contained in the liquid magma escapes during the cooling of magma leaving tiny air voids. The air voids are not inter-connected and the cell structure is stretched out and the cells are inter-connected to a great extent. PA has been used all across the World as a coarse aggregate in the production of LWA [2]. The major drawback with using PA as a coarse aggregate in making concrete is that, its mechanical strength properties are not sufficient enough to be used for structural purposes and can be used for specific purposes only by utilizing PA in making permeable concrete [3]. Even though, the permeable concrete with a very high permeability is useful for some applications like permeable pavements, thermal barriers, sound insulator, its incompetence in taking large loads makes its less useful for structural purposes [4]. Moreover, the water preservation behavior of PA is essentially high than typical coarse aggregate and found to have an impact on the consistency of trials during mixing and fresh concrete properties testing, which can be addressed by pre-soaking them before mixing in the concrete [5-6]. The strength properties of PA based lightweight concrete can be improved by adding SFA to the PA based lightweight concrete, by taking a cue from the research report of several researchers, where they reported that structural lightweight concrete can be developed using SFA as a replacement to coarse aggregate [7].

Utilization of SFA has several advantages apart from the improvement in the strength properties alone. It also helps in increased utilization of fly ash, an industrial by-product of coal based thermal power plants [8-9]. If fly ash is not properly disposed off; it can cause severe environmental pollution. India, China and USA being the top users of coal for power plants and their contribution to fly ash generation is also high. SFA LWA is found to be a very useful product both as a substitute to the natural coarse aggregate and to increase the utilization of fly ash for useful means, which otherwise would have been dumped futilely. SFA LWA is generally made by first mixing fly ash, clay and coke breeze in various proportions at a specified degree and sintering them at a temperature about 1200-1300 °C The sintering technique used is mainly Downdraft Sintering Technique [10]. Decreasing the self-weight of the concrete without compromising on its usefulness as a structural material becomes an essential viability of the concrete. SFA is one of such potential materials to make concrete lighter than CC [11]. High strength lightweight cement can likewise be delivered by totally supplanting the CA with sintered fly ash aggregates. In this paper, M25 grade SLWAC by supplanting CA with PA in various percentages(25%, 50%,75%,100%), SFA by half substitution of CA and a mix of both lightweight aggregates, have been created and their mechanical properties (compressive strength, splitting tensile strength) and durability properties (water absorption test) are examined and analyzed. To stay away from the water absorption issues related with LWA, it is recommended to use submerged LWAs during the concrete creation.

The main objectives of this study is to determine whether pumice stone and sintered fly ash aggregate concrete can be used as structural lightweight aggregate concrete (SLWAC), also to develop the mix design for producing M25 grade concrete using pumice (PA) and sintered fly ash aggregate (SFA) and then comparing the mechanical properties and durability properties of structural concrete with conventional/normal concrete (CC/NC).

2. Experimental Program

The test specimens have been prepared by using material as Cement (OPC 53 grade), coarse aggregate of size 12.5mm passing , pumice aggregate and Sintered Flyash Aggregate of size 10mm, 4.75mm passing river sand as fine aggregate and Water.

2.1 General

Structural concrete (SC) has been prepared with pumice and sintered fly ash aggregates and tested for compressive strength, split tensile strength and water absorption according to ASTM C642-13 [12].

2.2 Specimen Specifications

100mm x 100mm x 100 mm size cubes have been casted and subjected for testing the compressive strength. 100 mm diameter and 200 mm height cylinders, have been prepared for testing the splitting tensile strength. 100mm x 100mm x 100 mm dimensions cubes have been casted for estimating percentage of water absorption and this testing has been carried out after curing 28 days.

2.3 Properties of Material

Ordinary Portland Cement (OPC), sand (as fine aggregate), pumice aggregate (coarse aggregate), sintered fly ash (Coarse aggregate), Potable water have been used for casting cubes and cylinders in M25 mix design.

2.3.1. Cement. Ordinary Portland cement (BrilaA1 cement) 53 grade, confirming to IS:12269-1987 [13] has been used. Physical properties of Ordinary Portland Cement are given below in Table 1.

Table 1. Properties of cement

Property	Test Results
Consistency of cement	32%
Specific gravity	3.14
Initial setting time	24 mins
Final setting time	268 mins

2.3.2. Coarse Aggregate. Coarse aggregate of 10mm size was used for this investigation, aggregate taken from the local quarry confirming to IS:383:1970 [14]. The specific gravity of CA is 2.77 and bulk density is 1752.6 Kg/m^3.

2.3.3. Pumice Aggregate. The properties of pumice aggregate are shown in Table 2 and Figure 1 depicts the pumice used in the present study.

Table 2. Properties of pumice aggregate

Physical Properties	Test Results
Maximum size	10mm
Fineness modulus	2.705
Specific gravity	1.05
Bulk density	724 Kg/m^3
Water absorption	18.81%

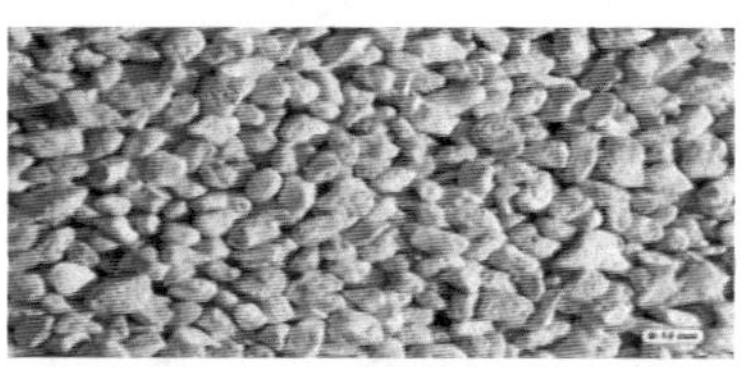

Figure 1. Pumice lightweight aggregate

2.3.4. Sintered fly ash aggregate. The properties of sintered fly ash aggregate are shown in Table 3 and the figure 2 depicts the SFA used in the present study.

Table 3. Properties of sintered fly ash aggregate

Physical Properties	Test Results
Maximum size	10mm
Fineness modulus	2.275
Specific gravity	1.785
Bulk density	903.6 Kg/m^3
Water absorption	17%

Figure 2. Sintered fly ash aggregate

2.3.5. River sand as Fine Aggregate. The locally existing sand confining to zone II was used in accordance with IS: 383-1970 [14]. The specific gravity, bulk density of this fine aggregate are 2.5,1285 Kg/m^3 respectively and the fineness modulus of sand is 3.085.

3. Mix design for SLWAC

Nadesan and Dinakar (2016) [11] have proposed a new mix design specification for SLWAC. Using the concept of mix design for SLWAC proposed by Nadesan and Dinakar, a mix design for M25 has been developed, also by using natural coarse aggregate normal concrete (NC) is developed. Compression strength, Splitting tensile strength and Water absorption analysis have been carried out for various samples.

3.1 Mix Proportions for Developed Concrete

The following four are the different mixes for M25 grade with water/cement ratio 0.65, the target mean strength of M25 grade concrete is 31.6 N/mm^2 (IS 10262-2009) and their designations and proportions are shown in table.4. The different mixes are normal/conventional concrete (NC/CC), pumice aggregate concrete with different replacements ($P_{25\%}$, $P_{50\%}$, $P_{75\%}$, $P_{100\%}$), sintered fly ash aggregate concrete ($S_{50\%}$), combination of pumice and sintered fly ash aggregate ($PS_{50\%}$).

Table 4. Proportions of M25 Mix in kg/m^3

Designation	Cement	Fine aggregate	Coarse aggregate	Pumice aggregate	Sintered flyash aggregate	Water
NC	276.9	801	1084.73	0	0	180
$P_{25\%}$	276.9	801	813.55	102.795	0	180
$P_{50\%}$	276.9	801	542.36	205.59	0	180
$P_{75\%}$	276.9	801	271.18	308.38	0	180
$P_{100\%}$	276.9	801	0	411.18	0	180
$S_{50\%}$	276.9	801	542.36	0	349.5	180
$PS_{50\%}$	276.9	801	0	205.59	349.5	180

4. Results and Discussion

4.1 Density

The air dry and oven-dry densities of all the SLWAC and CC have been estimated and given In Table5. When CA is replaced with PA (at 75%, 100%), SFA (at 50%) light weight concrete can be achieved.

Table 5. Density of concrete in kg/m^3

Designation	NC	$P_{25\%}$	$P_{50\%}$	$P_{75\%}$	$P_{100\%}$	$S_{50\%}$	$PS_{50\%}$
Air dry density	2686	2476	2354	2016	1825	2400	2033
Oven dry density	2665	2461.33	2320	1973.33	1786.66	2360	1985

4.2 Compressive Strength

The least compressive strength necessary for structural lightweight Concrete has been 17 MPa as per ASTMC-330[15] standards. For this test the specimens of size 100 mm x 100 mm x 100 mm are cast, cured for 28 days. The compressive strength is the ratio of the maximum load to the surface area of the concrete cube. Three cubes were tested for each mix ratio & the average of three specimens is taken as the compressive strength. By using 25% pumice aggregate a decrease in compressive strength is noticed, about 60% of compressive strength has been decreased on further increase of percentage of pumice aggregate (by 75%&100%). Compressive strength of NC for 28 days is outraged the target mean strength, this may depend on different factors like size of aggregate, water-cement ratio, cement strength. The compressive strength of different concrete mixtures are shown in Table 6 and maximum compressive strength of concretes for $P_{50\%}$,$S_{50\%}$ and $PS_{50\%}$ are compared with normal concrete in figure3.

Table 6. Compressive strength of concrete at 28 days

Designation	Compressive strength (MPa)
NC	50.67
$P_{25\%}$	24.67
$P_{50\%}$	34.33
$P_{75\%}$	18.67
$P_{100\%}$	16.00
$S_{50\%}$	40.33
$PS_{50\%}$	27.00

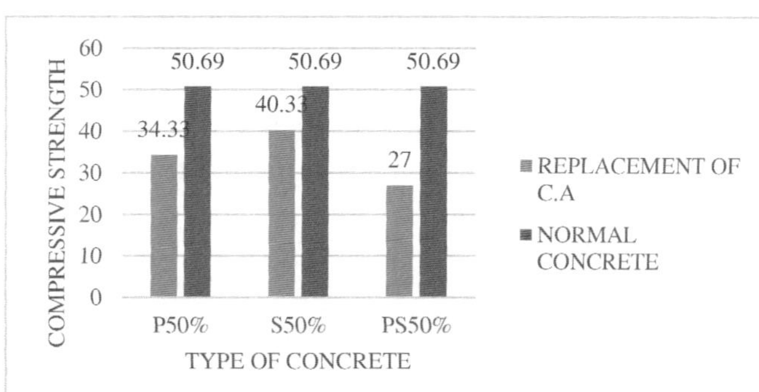

Figure 3. Compressive strength for $P_{50\%}$, $S_{50\%}$ and $PS_{50\%}$

4.3 Splitting Tensile strength

The minimum tensile-strength required for structural grade concrete is 2MPa for LWC according to ASTMC-330[15]. For this test the specimens of size 100 mm diameter (D), 200 mm length (L) are cast, cured for 28 days. The splitting tensile strength of specimens for different mixes were tested on universal testing machine and are tabulated in Table 7 and maximum compressive strength of concretes for $P_{50\%}$,$S_{50\%}$ and $PS_{50\%}$ are compared with normal concrete in Figure 4.

Table 7. Splitting tensile strength after 28 days

Designation	Splitting tensile strength (MPa)
NC	4.290
$P_{25\%}$	2.864
$P_{50\%}$	3.342
$P_{75\%}$	2.546
$P_{100\%}$	2.228
$S_{50\%}$	3.978
$PS_{50\%}$	3.024

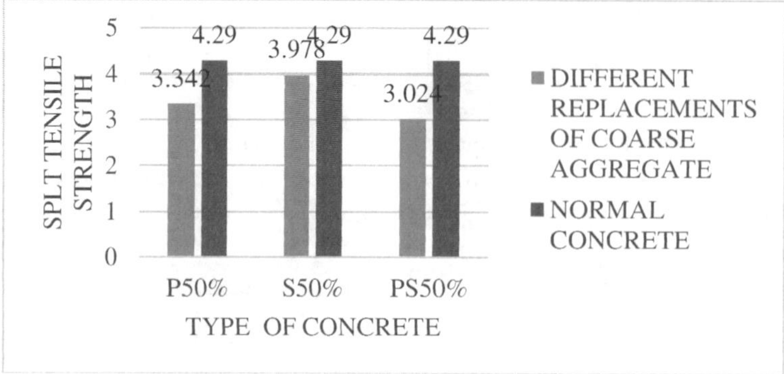

Figure 4. Splitting tensile strength at 28 days

4.4 Water Absorption

The water absorption test was performed according to ASTM C642-13 [12] which also includes the estimation of density, absorption and percentage of voids in hardened concrete. Test is conducted on three mix specimens ($P_{50\%}$, $S_{50\%}$ and $PS_{50\%}$) and compared with M25 grade CC and showed in Table 8.

Table 8. Water absorption of concrete

Designation	Weight of sample after oven dry (g)	Weight of surface dry sample after immersion (g)	Water absorption (in %)
NC	2513	2590	3.06
$P_{50\%}$	1996	2130	6.71
$S_{50\%}$	2173	2296	5.66
$PS_{50\%}$	1800	1946	8.11

5. Conclusion

Based on the results and analysis the following conclusion are made. The compressive strength of pumice aggregate and sintered fly ash aggregate concrete at 50% replacement of CA is more comparable with target mean strength of normal concrete. On further increase in percentage of PA instead of CA, a decrease in strength was noticed. Structural lightweight aggregate concrete can be achieved by using both PA &SFA. The results obtained for splitting tensile strength of 100 x 100 x 100 mm samples of SLWAC satisfies the requirements of structural applications according to ASTMC330. It has been observed that M25 grade with SLWACs has higher water absorption compared to standard aggregate concrete/NC.

References

[1] Bonewitz R, *Rocks and minerals*. 2nd edition, DK Publishing: London (2012).

[2] Khandaker M. Anwar Hossain, *Properties of Volcanic Pumice Based Cement and Lightweight Concrete*, Cement and concrete research **34** (2004).

[3] J. Rex, B. Kameshwari, *Studies on Pumice Lightweight Aggregate Concrete with Quarry Dust Using Mathematical Modeling Aid of ACO Techniques*", Advances in Materials Science and Engineering (2016) Article ID 9583757.

[4] Shadi Saadeh, Avinash Ralla, Yazan Al-Zubi, Rongzong Wu, John Harvey, *Application of fully permeable pavements as a sustainable approach for mitigation of stormwater runoff*, International Journal of Transportation Science and Technology **8(4)** (2019).

[5] Nihat Kabay, Fevziye Aköz, *Effect of prewetting methods on some fresh and hardened properties of concrete with pumice aggregate*, Cement and Concrete Composites **34(4)** (2012).

[6] C.W. Tang, *Effect of presoaking degree of lightweight aggregate on the properties of lightweight aggregate concrete*, Computers and Concrete **19** (2017).

[7] J.P. Behera, B.D. Nayak, H.S. Ray, B. Sarangi, *Lightweight concrete with sintered fly ash aggregate: a study on partial replacement to normal granite aggregate*, IE (I) Journal (**85**) (2004).

[8] R. Manikandan, K. Ramamurthy, *Physical characteristics of sintered fly ash aggregate containing clay binders*. Journal of Material Cycles and Waste Management Volume **14** (2012).

[9] Manu S. Nadesan, P. Dinakar, *Structural Concrete using Sintered Flyash Lightweight Aggregate: A review*, Construction and Building materials (2017).

[10] H.E. Exner, E. Arzt, Sintering Processes in Physical Metallurgy, Fourth Edition, Elsevier (1996).

[11] Manu S. Nadesan, P. Dinakar, *Influence of Type Binder on High-performance Sintered Flyash Lightweight Aggregate Concrete*, Construction and Building materials (2018).

[12] ASTM C642-13, Standard Test Method for Density, Absorption, and Voids in Hardened Concrete, ASTM International, West Conshohocken, PA, 2013, www.astm.org

[13] IS 12269: Standard specification for 53 grade ordinary Portland cement, BIS New Delhi, India (1987).

[14] IS 383: Standard specification for Coarse and Fine Aggregates from Natural Sources for Concrete, BIS New Delhi, India (1970).

[15] ASTM C 330: Standard Specification for Lightweight Aggregates for Structural Concretes, ASTM International, United States (2014).

IOP Conf. Series: Materials Science and Engineering 1132 (2021) 012003 doi:10.1088/1757-899X/1132/1/012003

Analysis On Crystal Formation And Physio-Tribological Properties Of Lithium Based Grease Developed From Coconut Oil (Cocos Nucifera)

Joseph Babu K

Department of Mechanical Engineering, Rajagiri School of Engineering & Technology, Ernakulam, India

josephbabuk@gmail.com

Abstract. The increase in demand for alternative base oil for lubricant has led to the study of non-depleting source like vegetable oils. Some studies have reported the development of grease from sunflower oil and castor oil, which establish the probability of use of vegetable oil as potential base oil. In the present study, an eco-friendly NLGI grade 3 grease was developed using coconut oil as base oil and lithium stearate as thickener. The inherent tribological properties of the coconut oil and grease were evaluated. Tests results of newly formulated grease procured from cone penetration, rolling stability, water wash out, grease leakage, friction and wear are compared to a commercial sample of same grade. Even though the newly formulated grease is found to be competent to the commercial sample, the crystal formation at low temperature has restricted its applications at lower temperatures. Crystal formation at low temperature is due to the agglomeration of saturated fatty acids commonly known as wax occurred due to the low cold flow characteristics of the oil (pour point of coconut oil 26.55°C). The tribological results show that the performance of the newly formulated grease is comparable to commercial greases and its application is restricted to moderate climatic conditions.

1. Introduction

The need for non-toxic, non-depleting and environmental friendly lubricants have paved way to the study of potential tribological properties of vegetable oils. In recent years, tribologists around the world have identified vegetable oils such as sunflower [1], palm [2], rice bran [3], rubber seed [4] and coconut oil [5-7] as good base oil for lubricant formulations. The green lubricants (vegetable oil) with additives were found to be competent to the commercially available oils in a few applications such as engine oils [3-6], lubricating greases [8,9] and cutting fluids [10,11].

The mechanism of vegetable oil film formation over the metal surface can be explained by its head and tail structure of fatty acid. The hydrophilic head of the triglyceride (glycerol molecule) get attracted to the metal surface whereas hydrophobic tail (fatty acids) gets suspended in the oil resulting in the formation of a lubricating film between the interacting surfaces. Vegetable oils also possess inborn tribological properties such as higher flash–fire point and superior friction-wear characterises thereby showing its domination over mineral base oil. The main disadvantage of vegetable oil is its poor oxidative stability due to the presence of unsaturated fatty acids and higher acid value due to the presence of free fatty acids.

Among vegetable oils, coconut oil has superior tribological properties in its natural state. Higher percentage of saturated fatty acids (90%), higher oxidation stability and low acid value promote the use of coconut oil as potential lubricant base oil. Coconut oil with additives has been established as good lubricant in two stroke [5], four stroke engines [6] and as a good cutting oil [10]. Hence, this work focuses on extending the application of coconut oil in the formulation of semisolid lubricants (grease). The influence of tribological properties of coconut oil in coconut oil based grease was analysed in the current study and was found to have high influence over the properties of grease.

Mechanism of grease lubricant involves the entrapment of lubricating oil inside a network of soap structure. The metallic soap acts like a surfactant resulting in the entrapment of oil inside the network structure [12, 13]. When pressure is applied over the grease, oil moves out from the network structure resulting in lubrication and while releasing pressure, surfactant restructures itself to entrap the oil molecules. Grease (NLGI Grade 3) was formulated using lithium stearate (metallic soap) as thickener and coconut oil as base oil. The tribological properties of the newly formulated grease were evaluated as per ASTM standards and were found to have satisfactory performance when compared to commercial grease sample of same grade.

Crystal formation was observed in newly formulated coconut oil based grease when preserved below 25°C. Detailed study on crystals was done using gas chromatography technique (GC). The result procured from GC confirms the crystal particles as an agglomeration of longer chain saturated fatty acids, commonly known as wax. Wax formation in the grease was due to the poor pour point property of coconut oil. Hence, application of coconut oil based grease was restricted below its liquid-solid transition temperature (31°C).

2. Materials and methods

The raw materials (chemicals and coconut oil) used for manufacturing the grease where procured from the resident market.

2.1. *Materials*

Lithium monohydrate and stearic acid used for the preparation of metallic soap was procured from Chemical house, Ernakulam, India. Magnetic and manual stirrers were used for controlled stirring operations. Deionized (DI) water with conductivity of 18.5 was used for dissolving lithium monohydrate during grease manufacturing process.

2.2. *Coconut oil*

Coconut oil is an edible vegetable oil extracted from coconut palm tree (Cocos nucifera). The flesh of coconut seed is dried and crushed to extract oil. The extracted oil is later heated to remove the moisture and preserved.

2.3. *Metal soap preparation*

Formulation of metallic soap is evolved through a chemical reaction as shown in equation 1. One mole of lithium monohydrate and one mole of stearic acid combine to form one mole of lithium stearate (soap) and two moles of water molecule. Lithium stearate metallic soap is extracted by removing the moisture content by heating.

$$LiOH.H_2O+C_{18}H_{36}O_2 = C_{18}H_{35}O_2Li+2H_2O \quad (1)$$

Initially, 100g of lithium monohydrate is dissolved in 500ml of deionized water under continuous stirring action for 30 minutes using a magnetic stirrer. Once dilution is complete, solution is poured into the molten stearic acid (678g) at 70°C under continuous stirring action using a mechanical stirrer. The Li+ ions combine with stearic acid to form lithium stearate by replacing hydrogen ions at the polar head of the fatty acids. The free hydrogen ions (H+) combine with hydroxyl ions (OH-) to from water molecules. The moisture content is removed by heating it above 100°C.

IOP Conf. Series: Materials Science and Engineering 1132 (2021) 012003 doi:10.1088/1757-899X/1132/1/012003

Heating is further continued up to 190°C (melting point of lithium stearate 180°C) to ensure complete soap formulation. The metal soap is cooled and preserved for future use.

2.4. *Grease formulation*

Different grades of grease (NLGI 000 to 6) can be manufactured by varying oil-soap ratio. For manufacturing NLGI grade 3 grease, 1500ml of coconut oil was mixed with molten lithium stearate (previously manufactured) at 190°C. The mixing was carried out under constant stirring action using a mechanical stirrer for 1hr. Once formulation was complete, the grease was cooled and milled for 3 hrs. The newly formulated grease was preserved for further analysis.

2.5. *Tribological test*

Tribological properties of coconut oil and grease were evaluated using different ASTM standards. Friction, wear, flash point of the oil was carried out as per ASTM D5183, D4172 and D92 respectively. The thermal degradation and pour point of the oil was studied using thermo gravimetric analysis (TGA) and differential scanning calorimetry (DSC). Viscosity of the oil was measured using a rheometer. The tribological properties of grease such as consistency, rolling stability, grease leakage and water wash out were evaluated as ASTM D217, D1831, D1263 and D1264 respectively.

2.6. *Crystal analysis*

The crystals formed in the grease were analysed using gas chromatography (GC) technique. Macroscopic structural image of the crystal was acquired using scanning electron microscope (SEM).

3. Results and Discussions

The properties of the lubricating oil and grease where analysed. The tribological properties of the newly formulated grease were compared to a commercial sample procured from the market. Studies were also carried out on the crystals formed in the grease when preserved at low temperatures.

3.1. *Physiochemical and tribological properties of coconut oil*

Coconut oil mainly consists of 90% saturated fatty acids, 7% mono unsaturated and 2% poly unsaturated fatty acids. The physical properties of coconut oil are listed in table 1. Coconut oil is pale yellow coloured oil with a density of 0.925 gm/cm3. The oil has very low iodine (7) and acid value (0.4) when compared to other vegetable oils [3,4]. The lower percentage of unsaturated bonds of the oil has resulted on low iodine value. It infers the ability of the oil to resist oxidation and thermal degradation. Lesser amount of free fatty acids in the oil has resulted in lower acid values. The resistance of the oil towards corrosion is partially proportional to the amount of free fatty acid present in it. The oil has comparatively higher saponification value of 255. It shows the tendency of the fatty acids to get separated from glycerol molecules to form metal soaps.

Coconut oil has low cold flow properties. Cloud point of the oil is close to 29°C and solidifies at 26.55°C. The low cold flow property of the oil is due to the higher percentage of saturated fatty acids in the oil. The oil has high flash point of 289°C and fire point of 295°C, which is one of the main advantage of using vegetable oils as base oil. The inherent dynamic viscosity of the oil is 25.1mPas at 40°C and 5.6mPas at 100°C. Coefficient of friction of the oil is 0.093, were as wear scar diameter is 0.645mm. Thermal stability of the oil was found through thermo gravimetric analysis (TGA). It was found that coconut oil is thermally stable up to 240.9°C and starts degrading on further heating as shown in figure 1.

Table 1. Physical properties of coconut oil

Physical properties of coconut oil	
Colour	Pale yellow
Density [g/cm3]	0.925
Iodine value [gl2/100 g]	7
Saponification value [mg KOH/g]	255
Acid Value [mg KOH/g]	0.4
Cloud point [°C]	29
Pour point [°C]	26.55
Flash point [°C]	289
Fire point [°C]	295
Viscosity at 40°C [mPas]	25.1
Viscosity at 100 °C [mPas]	5.6
Coefficient of friction	0.093
Wear scar diameter mm	0.606

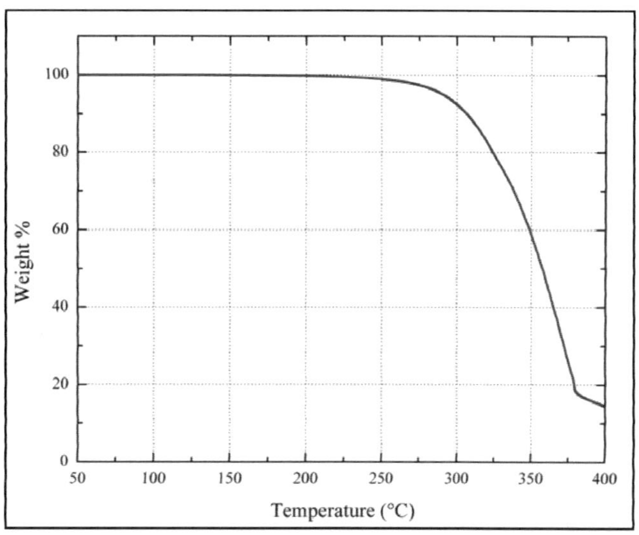

Figure 1. Thermo gravimetric analysis (TGA) conducted of coconut oil.

3.2. *Tribological properties of CO grease*

The tribological properties of the newly developed grease (CO grease) and commercial grease (COM grease) was evaluated using different test methods as shown in table 2.

Table 2. Comparison of tribological properties of CO grease with COM grease

Tribological properties	CO grease	COM grease
Unworked penetration	222	230
Penetration after 60 strokes	240	231
Rolling stability test (2 hrs)	246	231
Water wash out test (39°C)	0.31 grams lost	0.23 grams lost
Water wash out test (79°C)	0.43 grams lost	0.25 grams lost
Grease leakage test (6 hrs)	No leakage	No leakage

The unworked penetration of the newly formulated grease was found to be 222; it signifies that the newly formulated grease (CO grease) was under grade 3 (NLGI Grade 3 range 220-250). After completion of 60 stokes, a penetration difference of 7.5% (222 to 240, grade 3) was found when compared to unworked sample. The commercial grease seemed to be stable even after 60 strokes (230 to 231). Both samples (CO, COM grease) remained under same grade even after 60 strokes.

The shear stability of the grease was analysed by rolling stability tester. The stability of the grease was found satisfactory as the penetration variation of 9.7% (222 to 246) was observed when compared to unworked sample. Commercial sample was stable throughout the test (230 to 231).Both samples have passed the test as they remained in same grade even after continuous shearing operation.

The resistance of the grease towards water was analysed through water wash out test. The results infer the loss of 7.75 % (0.31g) and 10.75% (0.43g) of CO grease lost when tested at 39°C and 79°C respectively. Whereas, in the case of commercial sample 5.75% (0.23g) and 6.25% (0.25g) of grease was lost during testing. Even though the commercial sample seemed to be more resistant to water, a variation of 2% and 4.5% was only observed between CO and COM grease at lower (39°C) and elevated (79°C) temperatures respectively.

The bleeding tendency of oil from grease under running condition was analysed through grease leakage test. While testing both samples (CO, CMO grease), the grease collectors of the machine was found empty after 6 hours test. It infers the rare chances of oil to bleed under operating conditions. CO and COM grease have dropping point of 180°C (melting point of lithium 180°C) hence it did not exhibit any bleeding tendency throughout the test.

Friction-wear characteristics of coconut oil and grease samples were compared using four ball tester (1200rpm, 40kg, 75°C) as shown table 3.

Table 3. Comparison of friction-wear properties of coconut oil, coconut oil grease (CO grease) and commercial grease (COM grease).

Property	Coconut oil	CO Grease	COM Grease
Coefficient of friction	0.081	0.065	0.096
Wear scar diameter [mm]	0.606	0.712	0.892

The COF of CO grease is 19.6 % (0.065 mm) lesser than that of coconut oil (0.081) and 32.3% lesser than commercial sample (0.096). The reduction in frictional properties is due to the combined action of oil and metallic soap [14]. The newly formulated grease (CO grease) seemed to be consistently giving lesser COF than oil and commercial grease (COM) as shown in figure 2.

In the case of wear, crude coconut oil showed more resistant to wear when compared to grease samples. The capacity of the oil to dissipate heat at the contact surface and the hydrodynamic action would have resulted in lesser and uniform material removal as shown in figure 3. The anti-wear property of the newly formulated coconut oil grease is 20.2% (0.712mm) more when compared to commercial sample (0.892mm). The wear scars on the steel balls of the commercial sample have a uniform pattern and the load seemed to be equally distributed over the region when compared to CO grease in figure 4 and 5. This might be due to the presents of extreme pressure additives in the commercial sample.

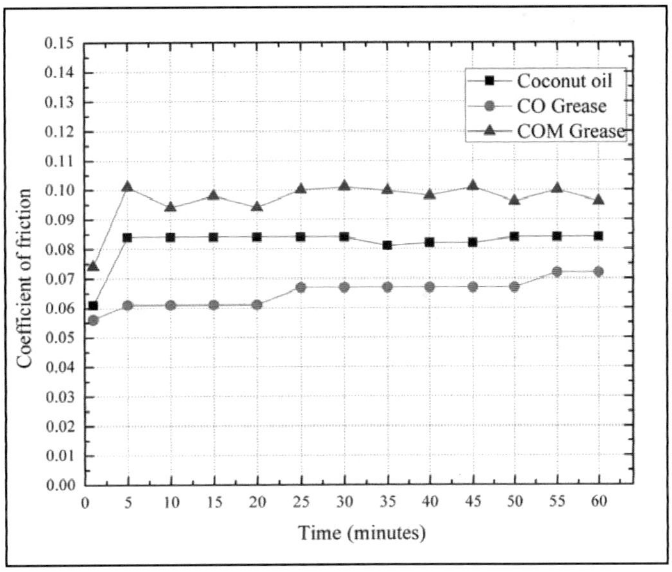

Figure 2. Comparison of COF of coconut oil, CO grease and COM grease

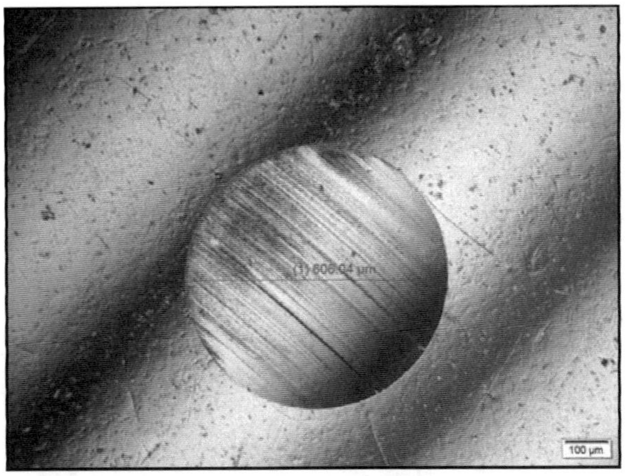

Figure 3. Microscopic image of wear scar over steel ball when tested with coconut oil.

Figure 4. Microscopic image of wear scar over steel ball when tested with CO grease.

Figure 5. Microscopic image of wear scar over steel ball when tested with COM grease.

3.3. *Crystal formation*

Crystal like formation was observed in CO grease when preserved below 25°C. Most of the crystals were spherically shaped as shown in figure 6. Detail study on the composition of crystals was done using gas chromatography technique. The results confirmed the presents of saturated fatty acids in agglomerated form as shown in figure 7. Thus the crystals found in CO grease were confirmed to be wax crystals. The structural images procured from SEM (scanning electron microscope) confirm close resemblance of crystal structures to that of paraffin wax as shown in figure 8 and 9.

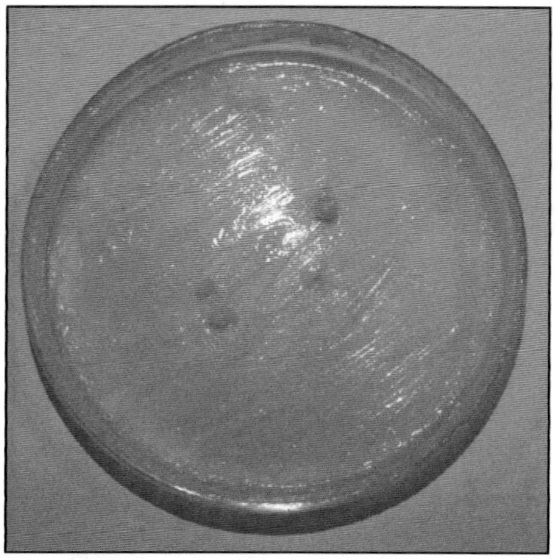

Figure 6. CO grease sample along with crystals spread over a petri dish.

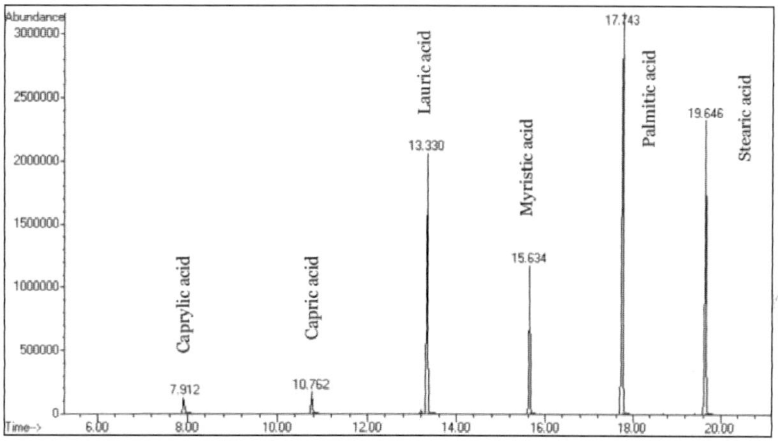

Figure 7. Gas chromatography analysis of crystals.

Figure 8. SEM image of crystals found in grease

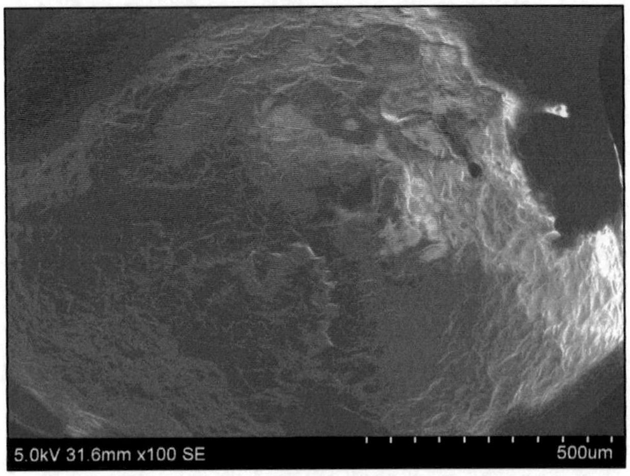

Figure 9. SEM image of paraffin wax

Wax formation is one among the main challenges faced by petroleum industries while transferring crude oil below liquid-solid transition temperature [15-17]. The liquid-solid transition temperature of coconut oil and grease was found using differential scanning calorimetry (DSC) as shown in figure 10. The structural changes in coconut oil and CO grease was observed between 9°C to 33°C and 10°C to 31°C respectively, when heated from -50°C to 50°C at a constant rate of 10°C/min [18]. The peak value of the samples was observed as 26.55°C for oil and 25.52 °C in the case of grease. The close resemblances of structural changes in the samples infer that grease exhibits similar phase change characteristics to that of its base oil. This is due to the presents of higher percentage of base oil (70%) when compared to total grease composition. Hence when coconut oil grease was preserved below 25°C, micelle structure with saturated fatty acid in the CO grease solidified to form wax crystals. Hence, the application of coconut oil grease was restricted below 31°C.

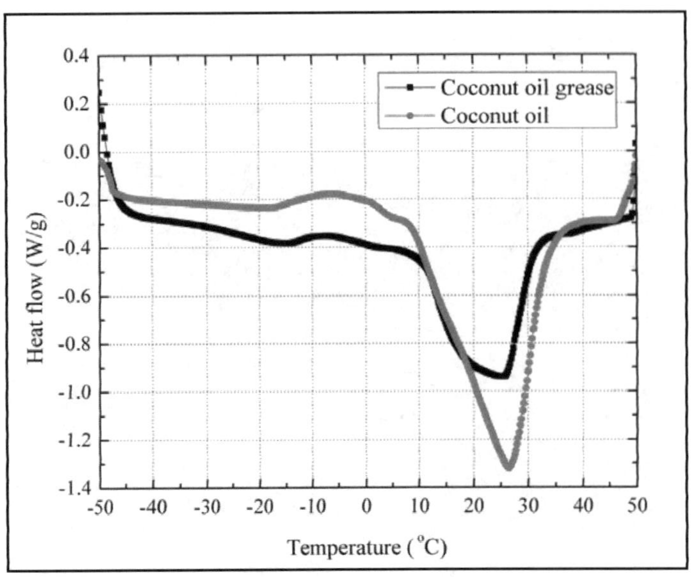

Figure 10. Differential scanning calorimetry of coconut oil.

4. Conclusions

The results procured from tribological test conducted on coconut oil based lithium stearate grease seemed to be satisfactory when compared to a commercial sample of same grade (NLGI grade 3). The coefficient of friction and wear characteristics of the newly formulated grease is seen superior than commercial grease by 32.3% and 20.2% respectively. Commercial grease is more resistant towards water when compared to newly formulated grease, a variation of 2% and 4.5% is only observed between CO and COM grease at lower (39°C) and elevated (79°C) temperatures respectively. The worked penetration and shear stability of commercial sample is better when compared to coconut oil grease (CO grease), but both samples retained their grade after testing (penetration, rolling stability), thereby satisfying the requirements of the standards.

The wax crystal formation in coconut oil grease is due to the low pour point characteristics of the oil, thereby restricting its application below 31°C. The properties of the newly formulated grease can be improved by chemical processes or by adding suitable additives.

Acknowledgments

The authors wish to acknowledge Dr. Satish Vasu Kailas, Dr. Jagadeesh K.M. and Dr. Suvin P.S., Department of Mechanical Engineering, IISc Bangalore, India and Dr. Jilse Sebastian, Mr. Eldho P.V., Department of Mechanical Engineering, NIT, Calicut, India for their help and encouragement throughout this work. Sincere thanks to the entire team of SAIF CUSAT, Chemistry Department, NIT, Calicut for their technical support.

References

[1] Fox N J, Tyrer B, Stachowiak GW. Boundary lubrication performance of free fatty acids in sunflower oil. Tribology Letters 2004; 16: 275-281.
[2] Masjuki HH, Maleque MA, Kubo A, Nonaka T. Palm oil and mineral oil based lubricants—their tribological and emission performance. Tribology International 1999; 32: 305–314.
[3] Rani S, Joy ML, Prabhakaran Nair K. Evaluation of physiochemical and tribological properties of rice bran oil – biodegradable and potential base stoke for industrial lubricants. Industrial Crops and Products 2015; 65: 328–333.
[4] Amith Aravind, Joy ML, Prabhakaran Nair K. Lubricant properties of biodegradable rubber tree seed (Hevea brasiliensis Muell. Arg) oil. Industrial Crops and Products 2015; 74: 14–19.
[5] Jayadas NH, Prabhakaran Nair K, Ajithkumar G. Tribological evaluation of coconut oil as an environment-friendly lubricant. Tribology International 2007; 40: 350–354.
[6] Jagadeesh KM, Satish VK. Experimental investigation of coconut and palm oils as lubricants in four-stroke engine. Tribology online 2011; 6: 76-82.
[7] Jayadas NH, Prabhakaran Nair K. Coconut oil as base oil for industrial lubricants – evaluation and modification of thermal, oxidative and low temperature properties. Tribology International 2006; 39: 873–878.
[8] Ponnekanti N, Prashant K. Eco-Friendly multipurpose lubricating greases from vegetable residual oils. Lubricants 2015; 3: 628-636.
[9] Dwivedi MC, Sapre S. Total vegetable-oil based greases prepared from castor oil. Journal of Synthetic Lubrication 2002; 19: 229-242.
[10] Rakesh S, Suvin PS, Divya PG, Satish VK, Dipshikha C. Eco-friendly, non-toxic cutting fluid for sustainable manufacturing and machining processes. Tribology Online 2016; 11: 4.
[11] Shashidhara YM, Jayaram SR. Vegetable oils as a potential cutting fluid—an evolution. Tribology International 2010; 43:1073–1081.
[12] Kaneta M, Ogata T, Takubo Y, Naka M. Effect of a thickener structure on grease elastohydrodynamic lubricant films. Proceedings of the Institution of Mechanical Engineers 2000; 214: 327-336.
[13] Radulescu I, Radulescu AV, Vasiliu F. The structure of lubricating greases by electron microscopy. Tribology in industry 2004; 26: 3-4.

[14] Jagadeesh KM, Satish VK. The effect of oxidation on the tribological performance of few vegetable oils. Journal of Materials Research and Technology 2012; 1: 91-95.
[15] John A, Svendsen. Mathematical modelling of wax deposition in oil pipeline systems. AIChE Journal 1993; 39: 1377-1388.
[16] Probjot Singh, Ramachandran Venkatesan, Scott Fogler H. Formation and aging of incipient thin film wax-oil gels. AIChE Journal 2000; 46: 1059-1074
[17] Zhenyu Huang, Hyun Su Lee, Michael Senra, Scott Fogler H. A fundamental model of wax deposition in subsea oil pipelines. AIChE Journal 2011; 57: 2955-2964.
[18] Govindapillai A, Jayadas NH, Bhasi M. Analysis of the pour point of coconut oil as a lubricant base stock using differential scanning calorimetry. Lubrication Science 2009; 21: 13–26.

IOP Conf. Series: Materials Science and Engineering 1132 (2021) 012004 doi:10.1088/1757-899X/1132/1/012004

Design and fabrication of six axial arm robot

K N Gunasekaran, Vasu C , Vishnu Ram N S, Tharun Athi Raj G, Vijayamoorthi R S

Department of Mechanical Engineering, Sri Krishna College of Engineering and Technology, Coimbatore,Tamil Nadu.

E-mail: gunasekarankn@skcet.ac.in

Abstract: Industrial revolution 4.0 is paradigm shift towards automation to cater the needs of ongoing demand in industrial production and to meet the requirements of exponential increase. In line with that six axial robotic arm can pave the way to complete the laborious task with in stipulated time. Six axial robotic arm it has a six axial robotic arm driven by six servomotors; a simple 6-DOF motion can be obtained. Robotic arm encompass of mechanical structure and control unit, Control unit controls whole motion of the system. It has two pieces of plastic gears, motors linkages tightened with fasteners and some sophisticated combination of laser cutted acrylic components. This fabricated model has the applications of robotic arm are pick and place spot welding, spray painting etc. The main objective of this paper is to propose a design and fabricate a six axial arm robot using servomotors which helps in the movement of links also the arm is allowed to extend freely with base fixed. The Arduino board connected to the motor to inspect and control the process

1. Introduction

The robotics is a term, which is well defined as the design, study and use of robotic systems for production and manufacturing. Robots are commonly used to execute hazardous, highly repetitive, unsafe and unpleasant jobs. They may perform various operations such as assembly, arc welding,material handling, resistance welding, loading and unloading functions of machine tools, spraying, painting etc.

There are generally two different types of robots: a service and an industrial robot. Either a service robot is one, which operates at fully or semi autonomously in performing services which are useful to the goodness of humans and equipment, eliminating manufacturing processes. Industrial robot, automatically controlled and multipurpose manipulator programmable in three or more axis. Industrial robots are designed to move material, parts, tools, or specialized devices over variable programmed motions in order to perform an innumerable range of jobs. It includes, sensors needed for the robot in performing the operations as well as sequencing or observing communication interfaces. In 2025, the world market grew by 7% with approximately 713,000 newly installed industrial robots. At the end of 2025 there were around one million industrial robots in use,compared with an estimated 50,000 service robots for industrial use. The main objective of the paper is to propose a design and fabricate six axial arm robot-using six servomotors, which helps in the movement of links also the arm is allowed to extend freely with base fixed. The Arduino board connected to the motor to inspect and control the process A servomotor is an electrical device, which can push or rotate an object with great

IOP Conf. Series: Materials Science and Engineering 1132 (2021) 012004 doi:10.1088/1757-899X/1132/1/012004

precision. In order to rotate an object with particular distance or angles, servomotors are best suitable one. It consists of modest motor, which operate by utilizing the mechanism of servo. Once DC power is supplied to the servo motor to run then it is termed as DC motor, and if AC source is utilized then it is termed as AC servomotor. The main advantages of using the servomotor is very high torque can be easily obtained even from a compact to lightweight. Due to this feature, it plays a vital role in most of the real time applications in RC Helicopters, Robotics, machine etc. Mostly servomotors are graded in kg/cm (kilogram per centimetre) and hobby servomotors are categorized at 3kg/cm or 6kg/cm or 12kg/cm. This kg/cm denotes the load that the servomotor can lift at a particular distance. For example: A 12kg/cm Servo motor would have the capacity to lift 12kg when the load is actually suspended 1cm away from the shaft of motor[2]. From this it is understood that when the distance is more the weight carrying ability is smaller. Along with this the position of servo motor is defined by electrical pulse and the circuitry is located at the side of the motor.

Servo mechanism is controlled with three parts: Controlled device, Output sensor and Feedback system. The servomotors consist of three wires where the positive and the negative is used for supply and the other is to send the signal from MCU. PWM (Pulse width Modulation) controls the servo motor using wires. From its neutral position, Servomotor can rotate 90 degree from both directions. For every 20 milliseconds (ms) the servo motor expects to see a pulse and the rotation of motor depends on the length of the pulse. The servomotor expects to see a pulse every 20 milliseconds (ms) and the length of the pulse will determine how far the motor turns [3]. For example, 1.5ms pulse would turn the motor to 90° position, as if pulse is smaller than 1.5ms shaft, which turns it to 0°, and if it is longer

Arduino is known for open source computer hardware, user community that involves in designing and manufacturing of single board micro controller kits with building digital devices of interactive objects that can sense and control objects in the real world. Using Arduino, the motors are controlled and are connected to the motors via motor drivers. In addition, a Bluetooth module is also connected to the via Bluetooth is compiled in the Arduino IDE.

In this model, 12v lead acid battery is used to supply current to all the six motors effectively. Lead and lead dioxide, the active materials on the battery's plates, react with sulphuric acid in the electrolyte to form lead sulphate. The lead sulphate first forms in a finely divided, amorphous state, and easily reverts to lead, lead dioxide and sulphuric acid when the battery recharges. During the battery recharge, cycle lead sulphate (sulfation) begins to reconvert to lead and sulphuric acid. During the recharging process as electricity flows through the water portion of the electrolyte and water, ($H2O$) is converted into its original elements, hydrogen and oxygen [4].

This system is designed in the view to improve pick and place. This achieved by ensuring perfection in all the aspects such as speed, accuracy, flexibility, safety, data storage and easy retrieval during service and maintenance of robot.[5]

The primary focus is to deliver simple and effective robots, which are cheap due to its simplicity. In the usual case, due to the human error the dc geared motors the movement of the robotic arm may not seems perfect in all the cases. But replacing that system which is complex and expensive by servo mechanism help to increase the performance and mobility of arm in desired work space.[6].

The simplicity of this project provides more flexibility to the user. This system provides the facility to the user to control the robot accordingly to the work space required. This system can be upgraded by making use of positional sensors , through which the accuracy of the arm can be maintained .

In any process human (labour) safety holds the first priority. In this work, there will be no risk in controlling the robot. This project setup allows the worker to handle the robot effectively and easily without any harm to the user.

2. Material and method

2.1. Degrees of freedom

No. of links = 6, No. of binary joints = 6, DOF = 3(6)-2(6)-0
= 18-12 Therefore, DOF = 6

2.2. Work Volume

Length of the arm at maximum extension is found through work volume.

V=0.5m

$4/3 * \pi r^3 = 0.5 \Rightarrow r = 350\text{mm}$

2.3. Length of arm

The length of the links at maximum extension is obtained by dividing the total length.
Lengths of,Link 1 + Link 2 + Link 3 + Link 4 (End Effector) = 350
As dividing as per geometry,
Link 1 = 121 mm

Link 2 = 98 mm
Link 3 = 32 mm
End effector/wrist = 99 mm

2.4. Torque of motor

Here we considered the weight of end effector and payload whole as the load for link 3.

Maximum Payload = 100 g
= 0.100 * 9.81
= 0.981 N
Gripper = Gripper weight + Motor Weight
= (0.25 * 0.981) + (0.04 * 9.81)
= 2.47 + 0.39
= 2.86 N
Therefore, total load to link 3 will be (0.98 + 2.86) = 3.84 N----- (1)

2.5. Torque of motor 3

$\tau_3 = (\text{load} * L_3) + (W_3 * L_3/2)$

= (3.84 * 0.032) + (0.09 * 9.81 * 0.032)
= 0.12288 + 0.015696
= 0.1384 N m or 13.84 N cm or 1.406 kg cm--------------------- (2)

2.6. Torque of motor2

$\tau_2 = (m * g * (L_3 + L_2)) + (W_3 * (L_2 + L_3/2)) + (W_{m3} * L_3) + (W_2 * L_2/2)$

= (3.84 * (0.032 + 0.098)) + ((0.1 * 9.81) + (0.98 + 0.015)) + (0.15 * 9.81 * 0.049)
= 0.4992 + 0.0144 + 0.01569 + 0.0225 + 0.0721
τ_2 = 0.6238 N m or 62.3 N cm or 6.35 kg cm ---------------------(3)

2.7 Torque of motor1

$\tau_1 = (\text{load} * (L_3 + L_2 + L_1)) + (W_3 * (L_1 + L_2 + L_3/2)) + (W_{m3} * (L_1 + L_2)) + (W_2 * (L_1 + L_2/2)) + (W_{m2} * L_1) + (W_1 * L_1/2)$

= (3.84 * 0.25) + (0.1 * 9.81 * (0.12 + 0.098 + 0.032/2)) + (0.05 * 9.81 * (0.012 + 0.98)) + (0.05 * 9.81 * (0.12 + 0.98/2)) + (0.05 * 9.81 * 0.12) + (0.25 * (0.12/2)
= 0.96 + 6.276642 + 0.106929 + 0.024868 + 0.5886 + 0.015
τ_1 = 1.666 N m or166.6 N cm or 16.92 kg cm --------------------- (4)

IOP Conf. Series: Materials Science and Engineering 1132 (2021) 012004 doi:10.1088/1757-899X/1132/1/012004

3. Results and Discussion

3.1 Deciding thickness of links :

The thickness of the link can be fixed from the stress acting on the link. Shear Stress , $\sigma = F/A$

$\sigma = F/2t^2$

Also, σ = Yield Stress/FOS

$A = b * t$ [By Golden ratio, $b = 1.6t$; $b = 2t$; therefore, $A = 2t^2$] Assumption, FOS = 2

Therefore, $t = \sqrt{\frac{F}{2 * 32.45 * 10^6}}$

Force on link 3 = 3.84 N

Therefore, thickness, $t = \sqrt{\frac{4{\cdot}5}{2 * 32{\cdot}66 * 10^6}}$

$= 2.62 * 10^{-4}$

t = 0.000262 m-- (5)

Therefore, it is not possible to select material of this thickness. Standardizing the thickness we can take it as 2 mm (from equation 5)approximately. Thus, the design calculations for the servo motors were done. These calculations were done keeping in mind the dimensional restrictions for this work, along with economical views. The values are well within the limit and the design is safe and accurate

3.2. The fabricated model

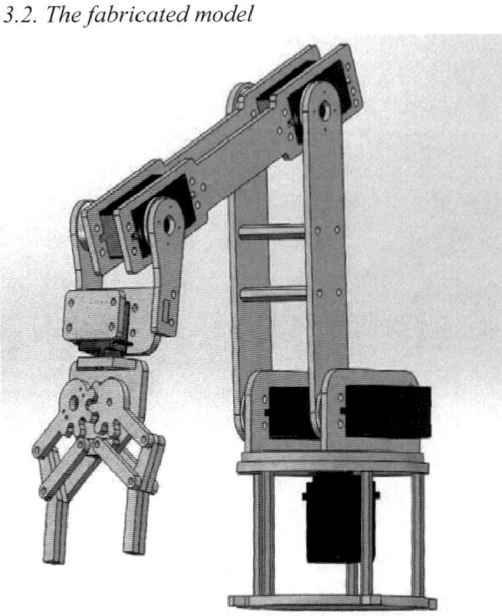

Figure 1. CAD model of six axial robot

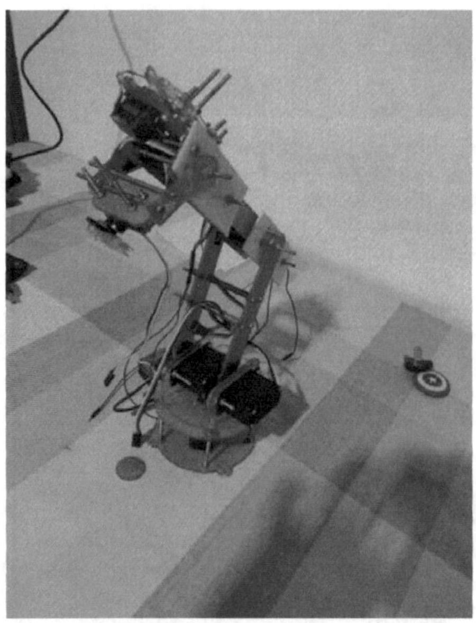

Figure 2. Prototype of Six axial robot

The Proposed model shown in Figure 1 designed using Solid Works modelling software. Based on this design, the design calculations were done, prototype is fabricated shown in Figure 2 and the machine is kept within these dimensions as per calculations

4. Conclusion

Thus the six axial arm robot designed using Solid Works Software, calculations were done and fabricated a prototype and tested model by performing different basic activities with the help of Arduino Board programming It had a control of the motor to perform clockwise and anticlockwise rotation is accomplished with in stipulated time

References

[1] Manipulating Industrial Robots—Vocabulary(2015), International Organization for Standardization Standard 8373,

[2] Rinat Galin and Roman Meshcheryakov (2019), Automation and robotics in the context of Industry 4.0: the shift to collaborative robots,*IOP Conf. Ser.: Mater. Sci. Eng.* **537** 032073

[3] Ye Lu and Kai Wang (2021),Kinematics Analysis and Trajectory Planning of Polishing Six-axis Robot, *IOP Conf. Ser.: Earth Environ. Sci.* **632** 052074

[4] R. J. Wang, J. W. Zhang, et al., (2000),The Multiple-Function Intelligent Robotic Arms, FUZZ-IEEE Journal, Korea, 20-24 August 2009, pp. 1995-2000.

[5] Ibrahim Qadri, Amgad Muneer and Suliman Mohamed Fati (2021),Automatic robotic scanning and inspection mechanism for mines using IoT, *IOP Conf. Ser.: Mater. Sci. Eng.* **1045** 012001

[6] D Sakthi, T P Vijay and Venkat Subramaniam (2021), A platform to test and automation on gesture and motion controls of play station 4 using robotic arm, *IOP Conf. Ser.: Mater. Sci. Eng.* **1070** 012105

[7] I Staretu (2021), Robotized application of assembly and soldering – case study, Robotized application of assembly and soldering, *IOP Conf. Ser.: Mater. Sci. Eng.* **1009** 012056

[8] S Y Khalapyan *et al* (2021), Interaction algorithm of the mechanisms of relative manipulation in a 6-DOF robot control system,*IOP Conf. Ser.: Mater. Sci. Eng.* **1064** 012019

[9] Industrial and Service Robots, IFR International Federation of Robotics, 2020. http://www.ifr.org/home

[10] Case Studies and Profitability of Robot Investment, The IFR Statistical Department, 2018. http://www.ifrstat.org/downloads/2008_Pressinfo_english .pdf

Breath-holding - A Strategy in Enhancing Drug Delivery to the Lungs: A Review

Balasubramanian R, Jayakumar J S * and Vikas Rajan

Department of Mechanical Engineering, Amrita Vishwa Vidyapeetham, Amritapuri, India

*E-mail: jsjayan@gmail.com

Abstract. After inhalation of pharmaceutical aerosols, holding of breath prior to exhalation is one of the simpler inhaler techniques when using pressurized metered-dose inhalers and dry powder inhalers. Many benefits are offered by breath holding in enhancing particle deposition, drug efficacy, reducing deposition of particles in the undesired sites and masking of hotspots. Many of the patients are not aware why this procedure is recommended by medical professionals and thus fail to get these benefits. If the breath hold step is successfully addressed, pulmonary drug delivery to the lungs will be more successful. This review addresses computer simulation studies, in particular computational fluid dynamics studies, on the influence of breath-hold in deposition patterns and behaviour of drug particles in the respiratory tract of humans. Along with the studies on breath holding reviewed here, we also briefly discuss the particle phase modelling tools like the discrete element method and discrete phase method.

Keywords: Breath hold; drug deposition; computational fluid dynamics (CFD); Discrete element method (DEM); Discrete phase method (DPM); gravitational sedimentation.

1. Introduction

Drug delivery to the lungs through inhalation is relatively complex due to the resistance mechanisms of the lungs to the inhaled particles and improper usage of inhaler devices [1]. Poor inhaler usage involves lack of coordination between inhalation and actuation of the device, improper positioning of the device, inhalation deprived of required force, failure to completely exhale before inhaling, and lack of post inhalation breath hold pause [1]. Breath-holding means stopping of airflow for few seconds, thus retaining the whole lung capacity after drug inhalation. Breath hold is one of the simplest strategies in inhaler usage recommended by medical professionals in order to enhance the efficiency of drug particle deposition. Breath-hold for few seconds is usually recommended after drug inhalation through pressurized metered-dose inhalers (pMDI) and dry powder inhalers (DPI) before the exhalation phase [2]. Breath holding requires the lowest rate of skill and care amongst various complex drug inhalation procedures.

In most of the experimental and computer simulation studies, the inhalation and exhalation phases are taken into account in research [2,3]. In the analysis of inhalation and exhalation processes, the effects of size, shape, electrostatic charge, and hygroscopicity of drug particles are considered. Other factors influencing drug deposition are airway morphological characteristics, inhalation rate, respiratory volume, inhalation patterns, turbulence, and shear flow impacts. Just a few researchers have suggested breath-holding manoeuvre as a method of improving the accumulation of drug aerosols inhaled. Also due to safety and regulatory restrictions, in vivo testing of particle behaviour in real airways of humans along with breath-hold pauses are limited. Imaging studies employing scanning technologies are also insufficient in resolution.

Computational fluid dynamics (CFD) makes use of computational strategies and algorithms to address fluid flow problems and the results can be analysed with in vivo and in vitro data [4]. With the development of high-performance computer software and hardware, CFD is a reliable tool that allows medical researchers to carryout analysis of drug delivery using 3-D models of human respiratory physiology. This review focuses on the contributions of computer simulations in particular CFD, for a better understanding of the impacts of breath-holding strategy in enhancing drug delivery and drug deposition efficiency. This helps in understanding why the medical recommendations advise this strategy after inhaling pharmaceutical aerosols. We also briefly address about different types of geometric models of lung airways used for CFD study. The influence of gravitational sedimentation in settling of particles during the breath hold and the role of breath holding in drug targeting are also addressed. Particle phase modelling tools like the Discrete Element Method (DEM) and Discrete Phase Method (DPM) are also briefly discussed based on the few breath-hold studies investigated here in this review.

2. Geometric Models Of Lung Airways

In 1963 Weibel brought an idealized symmetrical version (extensively called Weibel A model) of the respiratory tract represented through generations beginning from the trachea (0th generation) and ending at alveoli (twenty-third generation) for the research of fluid and particle delivery into the human lungs figure 2 [5]. Weibel version can only give the general trend of fluid and particle flow but cannot give the accurate prediction due to the absence of curvature and asymmetry. The uneven nature of the human lungs is because of presence of two left lobes and three right lobes and variations in size of the daughter airway generations. Based on the lung cast studies conducted by researchers, the geometric dimensions of the asymmetric model of lung airways and lung acinus are available [6]. However, these asymmetric models are also limited due to the fact that it is far from the realistic human lung models and they are not patient-specific representation. Patient-specific noncircular curved geometric version of real human airways can be digitally reconstructed with the help of CT scan and MRI data, which can give an accurate prediction of drug particle transport and deposition [6]. The Weibel, asymmetric and CT based models are represented in figure 1(a), 1(b) and 1(c) respectively.

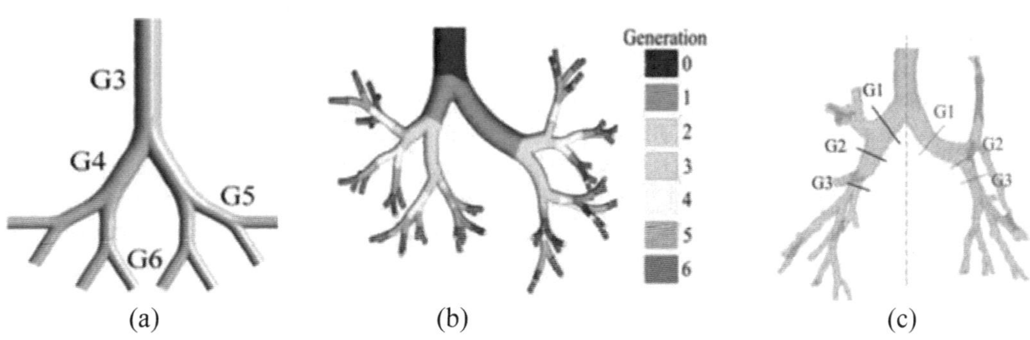

Figure 1. Different geometric models of tracheobronchial tree (a) Weibel based 3-D model of a four generations and triple bifurcation [10]; (b) asymmetric hybrid model by combination of Weibel and Horsefield models [13]; (c) realistic patient specific airway model generated from CT data [11].

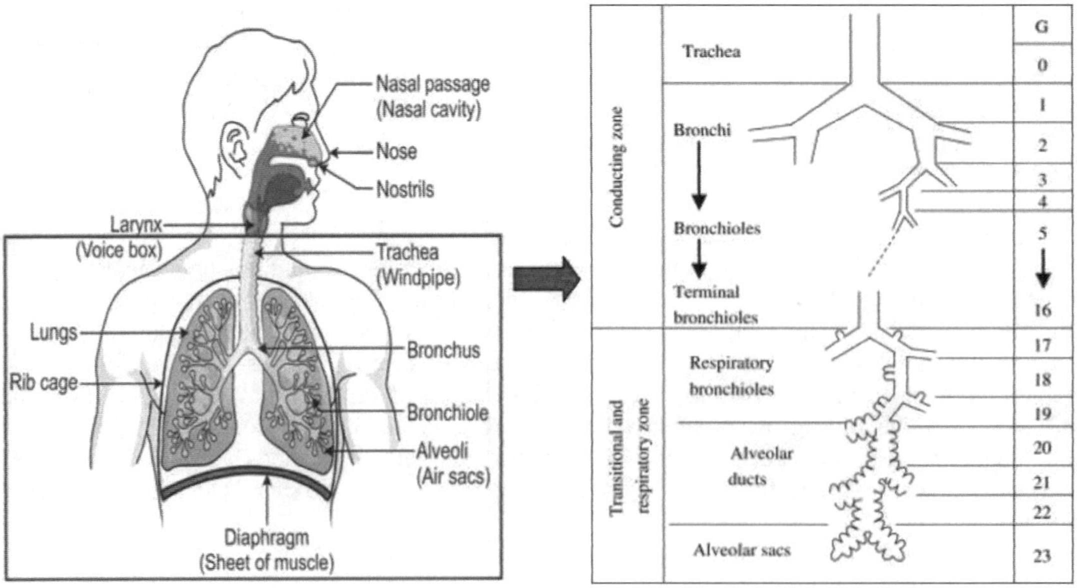

Figure 2. Representation of airway generations (G) [7].

3. Deposition During Breath Holding

In order to examine particle deposition during breath holding, Augusto *et al.* [7] used the CFD technique, and the main drawback of this research was lack of a comparative study of inhalation, exhalation, and breath holding. Inthavong *et al.* [8] conducted a CFD study of breath holding for two seconds after deep inhalation of aerosols. The results show that there was no increase in particle deposition efficiency until reaching airway generation-6. However, breath-hold is recommendable for locally targeted depositions up to the first few generations due to sedimentation impacts. Later, by adding a breath-hold with the exhalation stage, Augusto *et al.* [9] updated the previous studies and found substantial increase in particle deposition. The findings suggest that the breath-holding phase should be introduced in real-life scenarios because particles escape through the bifurcations of the airway resulting in lesser deposition efficiency when patients exhale immediately after drug inhalation. The behaviour of particle deposition with and without breath holding and comparison of particle deposition with exhalation was investigated by Kadota *et al.* [10]. It is clear from their analysis that without breath hold, particles are emitted on exhalation without being deposited in the airways. Later Kadota *et al.* [11] investigated the particle deposition in the realistic human airway models of asthma patients and healthy individuals with different inhalation patterns without including breath hold. However, the behaviour of particle deposition during terminal velocity was similar to that of a breath holding condition as in his previous study [10].

In the numerical research of Ostrovski *et al.* [12], breath holding was one of the tactics used in the pulsed aerosol bolus drug targeting integrated with the inhalation technique. Their results indicate that breath-holding manoeuvres are ideally suitable for increasing drug deposition of particles in the upper respiratory airways when the particle size is 5 to 10 μm. The findings of the CFD study of Tao *et al.* [13] on the 2D pulmonary acinus recommend that gravity only influences the local deposition fraction all through breath-holding and gravitational force is dominant for large particles (1-3 μm), Brownian force for smaller particles (0.01 μm) and medium-sized particles (0.1-0.5 μm) are marginal. This is indicated in figure 3. The research work of Khajeh *et al.* [14] shows that a breath-hold pause longer than 10s is required for the whole deposition of smaller particles (< 2 μm) inside the alveolar region. Faster deposition of large particles (>5 μm) was observed for shorter breath pause time (<10s). Many researchers observed the phenomenon of accumulation of drug particles dominated by inertia creates hot spots around airway divisions, thus significantly increasing particle accumulation in the locality of carinal ridges. These regions where the undesirable accumulation of particles takes place are known as hotspots. It is evident from the investigation of Imai *et al.* [15] that shorter breath-holding is a solution for minimizing unwanted particle accumulation in these upper airway regions. The breath-hold manoeuvre can deliver a uniform dispersal of particles and thus hot spots become insignificant [15].

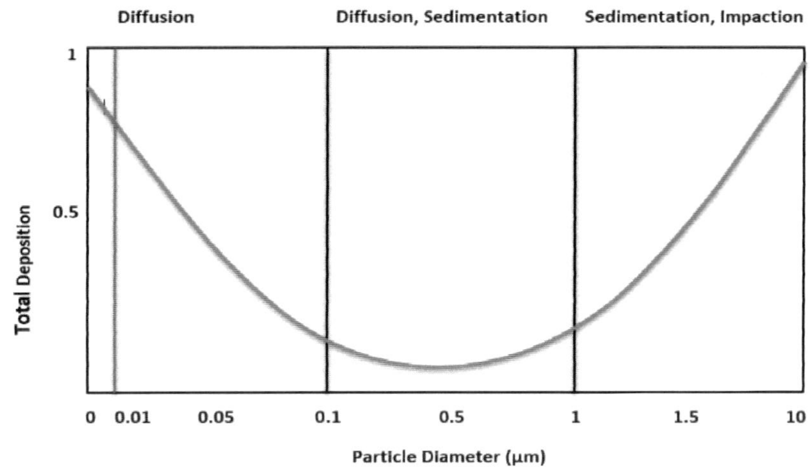

Figure 3. Drug particle transport mechanisms corresponding to particle size [16].

4. Discussions

The factors related to the CFD simulation such as the type of anatomy (realistic/ideal geometry), flow model (laminar/turbulent), particle tracking method (Lagrangian/Eulerian), number of airway generations and types of mesh are summarized in table 1.

Table 1. Particle depostion under breathholding condition (TB: Tracheobronchial)

Flow Model	Author	Year	Particle Modelling	Airway Models	Airway Generations	Mesh type
Laminar	Augusto *et al.* [7]	2013	Lagrangian Tracking	Symmetric Weibel Model TB tree	G3-G6	Hexahedral
LRN k-ω	Inthavong *et al.* [8]	2010	Lagrangian Tracking	CT scan TB tree	G0-G5	Tetrahedral
Laminar	Augusto *et al.* [9]	2016	Lagrangian Tracking	Symmetric Weibel Model TB tree	G3-G6	Hexahedral
SST k- ω	Kadota *et al.* [10]	2017	Lagrangian Tracking	CT realistic TB tree	G0-G7	Polyhedral
k-ω	Kadota *et al.* [11]	2019	Lagrangian Tracking	CT realistic TB tree	G0-G7	Polyhedral
Laminar	Ostrovski *et al.* [12]	2018	Lagrangian Tracking	Asymmetrical Hybrid Model (Weibel + Horsfield) TB tree	G0-G6	Tetrahedral + prism layer
Laminar	Tao *et al.* [13]	2018	Lagrangian Tracking	2D Pulmonary Acinar Model + TB tree	G0-G8	Not given
Laminar	Khajeh *et al.* [14]	2014	Lagrangian Tracking	Simplified Acinar Model	D2-D4	Tetrahedral

4.1 Flow Model

The complex oral cavity extending to the larynx of the respiratory airways figure 2 is ignored in most CFD studies thus airflow and particle simulation from the trachea figure 1 is initiated so the flow becomes laminar. Based on the experimental results, it is found that the flow in the generations G3-G6 is laminar [7]. So, Augusto *et al.* [7,9] particularly selected G3-G6 to avoid the turbulence effects, and also Reynolds number (Re) calculated in these four generations found to be in the range 173-669 confirming laminar flow. For low Re and low inlet flow rate, a laminar flow model was used in the studies of Ostrovski *et al.* [12] and Tao *et al.* [13]. Transitional flow and turbulent intensity are due to the laryngeal jet formed by the constriction as a result of the soft palate and glottis on the oral area and throat area respectively and additionally because of the flow instabilities on the carinal ridges which affect drug-aerosol transport and deposition [8,17]. For the correct prediction of particle deposition in the upper airways, the choice of turbulence model is essential. The minimum turbulent dispersion of particles due to the assumption of absence of the laryngeal jet, Inthanvong *et al.* [8] preferred LRN k-w model over other turbulence models. The use of the SST k- ω model [10] for the investigation of turbulent flow in realistic human airways provides better results than the k- ω model [11]. This is because of the possibility to use k- ω formulations in the proximity of the wall and k-ε formulations away from the wall [10]. For its good behaviour in adverse pressure gradients and flow separation, the SST k- ω model is advantageous than k- ω model. Although the computational cost is higher, for the accurate prediction of micron particle deposition under unsteady conditions, to solve eddies with large-scale energy and the ability to provide instantaneous velocity fluctuations, LES and DNS models can be used in the future to investigate the particle deposition under breath-hold conditions.

4.2 Investigation of Particle Dynamics

Aerosol system is a two-phase scheme comprising of a fluid phase (continuous) and a particle phase (dispersed). Lagrangian and Eulerian are the two typical approaches used for dealing with two-phase schemes [18]. Clouds of particle or individual particles are monitored as they pass across the computational domain in the Lagrangian method [19]. The Lagrangian method is also called the approach of a moving frame so that a particle's instant position can be determined as a function of time in relation to its point of entry [19]. The second law of motion of Newton governs the motion of particles meant to be propelled by force of gravity, aerodynamic forces, Coulomb forces, etc. [18]. The Lagrangian method equation of particle motion is given by

$$m_p \frac{du_p}{dt} = F_D + m_p g + F_{Br} + F_a \tag{1}$$

where m_p signifies the particle mass, u_p is the velocity of the particle, F_D denotes drag force, F_{Br} represents force due to Brownian motion, g is the acceleration due to gravity and F_a are the additional forces to be considered. The influence of gravitational effect in particle deposition during breath hold is dominant for relatively larger particles (>1μm) [20]. Therefore, the force due to Brownian motion term in (1) is ignored. This term is only necessary for submicron and nano particles, where the diffusion behaviour predominates [10, 11]. For the sake of simplicity, additional forces like electrostatic, thermophoresis and terms depending on the density ratio like the forces due to buoyancy, pressure, Basset force, Saffman lift force, and virtual mass effect may be ignored [8, 18]. Eulerian approach is common when the concentration of particles is large and the mean diameter is small [19]. Therefore, the Eulerian approach is generally not used for the investigation of particle tracking in breath hold condition [18].

4.2.1 DPM and DEM Approach of Particle Tracking. DPM method is most suitable for monodispersed spherical particles and also when the volume percentage of the particle phase is very small [9]. Therefore, the one-way coupling is taken into consideration and the fluid forces only influence the trajectories of the particles in this method, while the particle phase does not affect the fluid phase [9].

While the CFD-DEM is a useful tool for the construction of irregularly shaped particles simulating strongly coupled gas-solid flow [21] and also when the non-spherical particles are injected along with the monodispersed spherical particles like that of the aerosol bolus study of Ostrovski *et al* [12]. In general, both the CFD-DPM and DEM methods are suitable for successfully simulating drug particle delivery and deposition in humans' respiratory tracts. However, the CFD-DPM method is mostly preferred in the studies investigated here due to the fact of consideration of the uniform size of drug particles for the sake of simplicity. As compared to the lower respiratory tract such as the alveolar zone, the upper respiratory tract wall movement during respiration is very slight. Therefore, it is recommended the usage of user-defined functions to simulate the movement of the airway walls and the deposition of particles in the alveolar regions during respiratory cycles [14].

4.3 Airway Models

Due to the exclusion of the effects of the oral region, larynx and further lower branches like alveolar regions, the airway models [7-12] was limited. Due to the injection of particles from the trachea rather than introduction from the mouth, as in actual physical scenarios, relative deposition fractions were measured as an alternative to the real amount of particles reaching the targeted sites. Therefore, it is necessary to consider the whole airway model rather than considering segmental models figure 1. Most of the investigations [7,9,12] use Weibel model because of the availability of dimensions, low computational cost and easiness in generating 3-D models. However, the estimation of particle deposition during breath holding in the Weibel model are far away from the actual prediction given by the realistic CT based models [22]. This is due the lack of curvature and asymmetry [22]. So, it always advisable to use the CT based models for getting accurate prediction of particle deposition [22]. However, using algorithms artificially generated models that resemble the realistic complicated human lung structures are available. One of the finest examples is the KG models of 3 D airway tree, 4D pulmonary acinus, 4D alveolar system, and 4D lung shape developed by Kitaoka *et al.* [23]. The simulation results are close to those of patient-specific airway models are clear from the analysis of Kim *et al.* [24] using KG models. The use of the KG model is suggested as a potential area of future study since this model is computer-efficient than CT based patient-specific airway models.

4.4 Drug Deposition by Sedimentation

Sedimentation due to gravity is the main mechanism behind the transport of drug particles during the breath-hold scenario. This is more prevalent in regions of small dimensions and low-velocity areas like smaller respiratory tracts and alveolar regions figure 4. Diffusion and sedimentation dominate for the transport of the small sized particles 0.1-1μm. For large size particles (1-10μm), impaction and sedimentation are the prominent mechanism of drug transport figure 3 [20]. The significant factors that have an effect on the amount of particle deposition due to sedimentation are particle size and residence time [18]. Sedimentation becomes more pronounced as the particle size increases. To express the significance of gravitational sedimentation in the deposition of the aerosol, a characteristic length known as the settling distance is used. Settling distance is represented by the equation

$$S_s = V_t\, t \tag{2}$$

where t is the residence time, V_t is terminal velocity given by $V_t = g\,\tau$, g signifies acceleration due to gravity and τ denotes the relaxation time [18]. The aerosol deposition and drug retention within the targeted sites are also influenced by the inhalation rate. With the rise in inhalation rate, the particle residence time decreases, resulting in the reduction of the impact of sedimentation [18].

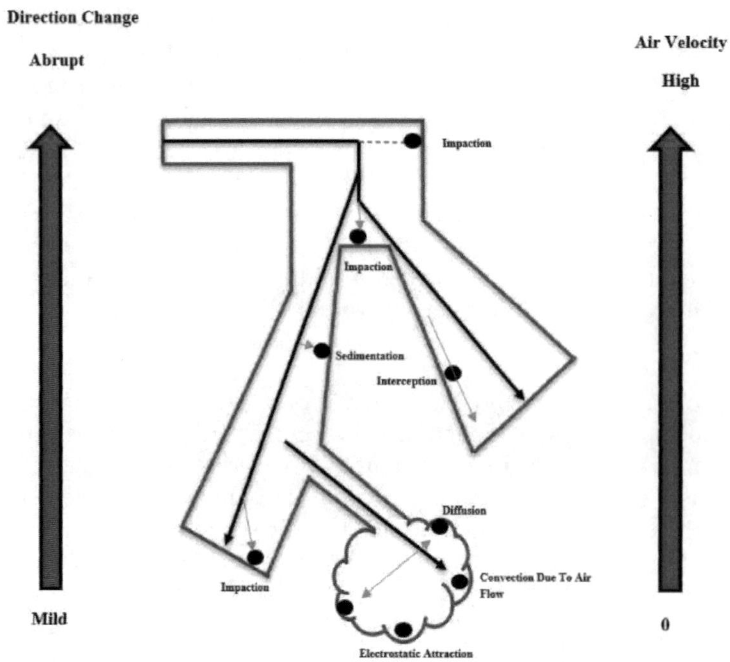

Figure 4. Aerosol deposition mechanisms in human respiratory airways [18].

5. Conclusions

Computer simulation studies using CFD have been discussed here with a focus on the impact of breath holding on increasing particle deposition and reason behind the recommendation of this strategy after inhaling pharmaceutical aerosols in clinical trials. It is clear that progress in delivery of respiratory medicine through inhalation needs to be accompanied by breath hold manoeuvre. Breath holding has got many advantages like it increases the efficiency of particle deposition in targeted site especially in intra thoracic regions (lower bronchi, alveolar regions etc.), rises the dose carried to the targeted sites which in turn rises the drug efficacy, avoids undesirable accumulation of particle in the upper airway generations. As compared to non-breath-holding conditions, they provide uniform distribution of particles and masking of hot spots thus preventing undesirable deposition of particles around airway bifurcations. Based on the breath-hold studies investigated, it is seen that the particle tracking tools like the DEM and DPM approaches are suitable for multiphase flow prediction in the human airways. However, DPM approach is commonly preferred by the researchers for the particle tracking. Laminar flow model is mostly suitable for low Reynolds number, low inlet flow rate and for generations G3-G6 Even though, the SST k- ω model is advantageous than k- ω model and LRN k- ω based on the studies investigated in this review, LES and DNS models can be used in the future to investigate the particle deposition under breath-hold conditions. Even though medical professionals recommend a breath hold of 5-10s, a future direction is to find the optimum time for breath-hold, which increases the total drug deposition efficiency of the lungs. The use of the KG model is suggested as a potential area of future study since this model is computer-efficient than CT based patient-specific airway models and Weibel model.

References

[1] Newman S P 2017 Drug delivery to the lungs: challenges and opportunities *Therapeutic Delivery* **8** 647-61
[2] Cheng Y S 2014 Mechanisms of pharmaceutical aerosol deposition in the respiratory tract *AAPS PharmSciTech* **15** 630-40
[3] Koullapis P G, Kassinos S C and Lin C L 2015 Computational fluid dynamics (CFD) simulations

of aerosol deposition in the lungs *9th International Symposium On Turbulence And Shear Flow Phenomena* 1-6

[4] Versteeg H K and Malalasekera W 2007 *An Introduction to Computational Fluid Dynamics : The Finite Volume Method* (Second ed.) England: Pearson Prentice Hall

[5] Weibel E R 1963 *Morphometry of the Human Lung* New York: Springer Verlag/Academic Press Heidelberg

[6] Clement Kleinstreuer, Zhang Z and Zheng Li 2008 Modeling airflow and particle transport/deposition in pulmonary airways *Respiratory Physiology & Neurobiology* **163** 128-38

[7] Augusto L X Lopes G C , Silveira A and José G 2013 Particle deposition during breath holding in human lung airways using computational fluid dynamics *Proceedings of the XXXIV Iberian Latin-American Congress on Computational Methods in Engineering*

[8] Inthavong K, Choi L-T, Tu Jiyuan , Songlin Dinga and Francis Thienb 2010 Micron particle deposition in a tracheobronchial airway model under different breathing conditions *Medical Engineering & Physics* **32** 1198-1212

[9] Augusto L X Lopes G C and Gonçalves J A 2016 A cfd study of deposition of pharmaceutical aerosols under different respiratory conditions *Brazilian Journal of Chemical Engineering,* **33** 549-58

[10] Kadota K , Imanaka A , Shimazaki M, Takemiya T, Kubo K , Uchiyamaa H and Yuichi Tozukaa 2017 Effects of inhalation procedure on particle behavior and deposition in the airways analyzed by numerical simulation *Journal of the Taiwan Institute of Chemical Engineers* **90** 1-7

[11] Kadota K, Inoue N , Matsunaga Y, Takemiya T, Kubo K, Imano H and Tozuka Y 2019 Numerical simulations of particle behaviour in a realistic human airway model with varying inhalation patterns. *Journal of Pharmacy and Pharmacology* 1-12

[12] Ostrovski Y, Dorfman S, Mezhericher M , Kassinos S and Sznitman J 2018 Targeted drug delivery to upper airways using a pulsed aerosol bolus and inhaled volume tracking method. *Flow Turbulence Combust* **102** 73-87

[13] Tao W , Hongyan Z and Haihang C 2018 Numerical simulation of the effect of gravitational direction on particle deposition in pulmonary acinus under the condition of breath holding. *Nanotechnology and Precision Engineering* **1** 66-72.

[14] Khajeh-Hosseini-Dalasm N and Longest P W 2015 Deposition of particles in the alveolar airways: inhalation and breath-hold with pharmaceutical aerosols *Journal of Aerosol Science* **79** 15-30

[15] Imai Y, Miki T, Ishikawa T, Aoki and T Yamaguchi T 2012 Deposition of micrometer particles in pulmonary airways during inhalation and breath holding *Journal of Biomechanics* **45** 1809-15

[16] Heyder J, 2004, Deposition of inhaled particles in the human respiratory tract and consequences for regional targeting in respiratory drug delivery *Proceedings of the American Thoracic Society* **4** 315-20

[17] Kleinstreuer C and Zhang Z 2003 Laminar-to-turbulent fluid-particle flows in a human airway model *International Journal of Multiphase Flow* **29** 271-89

[18] Rostami A A 2009 Computational modeling of aerosol deposition in respiratory tract: a review. *Inhalation Toxicology* **21** 262-90

[19] Zhang Z and Chen Q 2007 Comparison of the Eulerian and Lagrangian methods for predicting particle transport in enclosed spaces *Atmospheric Environment* **41** 5236-48

[20] Tsuda A , Henry F S and Butler J P 2013 Particle transport and deposition: basic physics of particle kinetics *Comprehensive Physiology* **3** 1437-71

[21] Chen X , Zhong W , Zhou X , Jin B and Sun B 2012 CFD–DEM simulation of particle transport and deposition in pulmonary airway *Powder Technology* **228** 309-18

[22] Nowak N, Kakade P P and Annapragada A V 2003 Computational fluid dynamics simulation of

airflow and aerosol deposition in human lungs *Annals of Biomedical Engineering* **31** 374-90
[23] Kitaoka H 2011 A 4D model generator of the human lung *Forma* **26** 19-24
[24] Kim Y, Tong Z, Chan H and Yang R 2019 CFD modelling of air and particle flows in different airway models *Journal of Aerosol Science* **134** 14-28

Nonisothermal Warm Deep Drawing Behavior of Automotive Grade Aluminum Alloy Sheets

Sudhy S Panicker [1, 2*] **and Sushanta Kumar Panda** [1]

[1] Department of Mechanical Engineering, Indian Institute of Technology Kharagpur, Kharagpur, West Bengal, India-721302

[2] Department of Mechanical Engineering, Adi Shankara Institute of Engineering and Technology, Kalady, Ernakulam, Kerala, India-683574

*email: sudhyspanicker@gmail.com

Abstract. In this work, thermomechanical FE modeling of nonisothermal warm deep drawing of automotive grade AA5754 and AA6082 aluminum alloy sheets were performed. The deformable blanks were modeled using temperature dependent Cowper-Symonds constitutive equation along with Barlat-89 yield criterion. The predicted cup heights, earing, % thinning and surface strain distributions were successfully validated within acceptable error of 5%. The development of earing profile, cup thinning pattern and surface strains along 0°, 45° and 90° from rolling direction (RD) were plotted using the validated data. The ear profile developed in AA5754 cups was prominent compared to AA6082 due to relatively higher anisotropy at elevated temperatures. Thickening was more prominent along 45° to RD for AA5754, whereas thinning was higher in this direction for AA6082 material. The strain states developed at cup corner along plane strain condition was lower than the FLD_0 values of materials. Thus there was uniform surface strains at cup corner indicating the improved ability of material to withstand more deformations. Comparatively, AA5754 exhibited relatively better formability compared to AA6082 material.

Keywords: FE modeling, warm deep drawing, aluminum alloys, anisotropy, strain distribution

1. Introduction

Lightweighting is a key method under consideration in automotive sector to cope with the stringent emission rules [1]. Aluminum alloys have potential application in this regard owing to its high strength-to-weight ratio and excellent corrosion resistance. Typically, AA5754 and AA6082 series alloy sheets are used due to their moderate strength properties, and decent weldability [2]. Deep drawing process is an important process followed to produce both shallow and deep cup shaped components used for different automotive components such as automotive door panels, fender, valve enclosures, round enclosure, slender cups, spring seat, filters, fuel tanks, wiper motor housing, cartridge case, lids, containers, etc. [3,4]. As aluminum alloy sheets displays poor formability, techniques such as warm forming has to be adopted to improve the drawing limit window of aluminum alloy sheets. This process is considered because of the easiness in modification of an existing setup to accommodate heating arrangements with reduced installation investments while stamping complex components [3,5].

Moreover, the deep drawing process with differential temperature gradient, i.e. nonisothermal warm deep drawing, was reported to produce significant improvement in part depth [3,6].

The formability studies using computer aided finite element (FE) methods are necessary to predict the warm forming behavior implementing temperature dependent material properties, anisotropy, strain hardening, and strain rate sensitivity [7]. The FE analysis helps to significantly reduce the material wastage during expensive experimental trial and error tooling tryouts [8,9]. The proper identification of a constitutive model to incorporate temperature dependent stress-strain behavior and yield function is vital in accurate formability predictions using FE methods [8]. Generally, forming limit diagrams (FLD) are used as the failure criteria to define the necking/failure strains in sheet metal forming processes [10,11].

In this regard, a comparative study into the temperature-dependent deformation behavior of automotive grade thin aluminum alloy sheets under nonisothermal warm deep drawing conditions is important. Hence, in this study, thermomechanical FE modelling of this nonisothermal process for such as AA5754 and AA6082 was developed using temperature-dependent Cowper-Symonds coefficients and Barlat89 yield parameters. The FE predicted data was first validated with experimental measurements, and correspondingly, the deformation behavior was studied in terms of earing, thinning and surface strain development.

2. Methodology

2.1. Material selection and characterization

Two automotive grade aluminum alloy sheets selected in the present work were the strain hardened AA5754 and precipitate hardened AA6082 of 1mm thickness. These aluminum alloy sheets are usually used in manufacturing thin-walled components including panels and hoods of lightweight automotive cars. The AA5754 material is a strain hardened alloy having moderate strength and elongation, but produces undesirable stretcher marks while deforming. Hence, these sheets are used for inner panels. The AA6082 is a heat-treatable alloy that attains strength with heat treatment post forming, and hence applicable for exterior panels.

The uniaxial tensile testing was performed at various temperatures ranging from 30°C-250°C as detailed in previous work of the authors [5,12,13]. Tensile tests were performed under the strain rates of $1\text{x}10^{-3}$ s^{-1}, $1\text{x}10^{-1}$ s^{-1} and $1.6\text{x}10^{-1}$ s^{-1} to evaluate the strain rate sensitivity of the materials [12,13]. Lankford anisotropy parameters (*r* values) along rolling direction (0°), 45°, and 90° at different temperatures were evaluated at all the investigated temperatures [5,13], and the respective values are enlisted in Table 1. The widely used Cowper-Symonds model (Eq.1) was used to evaluate the strain rate sensitivity of the materials at different temperatures.

$$\bar{\sigma}(\varepsilon,\dot{\varepsilon}) = K\bar{\varepsilon}^{n}\left[1+\left(\dot{\varepsilon}/C\right)^{1/P}\right] \quad \textbf{(1)}$$

Here, $\bar{\sigma}$ =effective flow stress at a particular temperature, *K* is strength coefficient, *n* is strain hardening exponent, and *C*, *p* are Cowper-Symonds strain rate sensitive parameters. These parameters evaluated at different temperatures were correlated to develop temperature dependent material flow properties. The variation in *K* value and *n* value is depicted in Fig. 1a, and the variations of *C* and *p* are also shown separately in Fig. 1b and Fig. 1c respectively. The decreasing trend of *K* and *n* values with temperature as observed from Fig. 1a showed that the materials are thermally softened with rise in temperature. The exponential reduction in *C* value in Fig. 1b demonstrated the increasing strain rate sensitivity of materials. However, *p* values showed a polynomial decrement nature with temperature (Fig. 1c).

The Barlat89 yield criterion [14] represented in Eq. 2 was used in FE model and this model was referred as *MAT_036 in LSDYNA solver.

Table 1 Experimental *r*-values (r_0, r_{45}, r_{90}), average anisotropy value ($\bar{r}$) and planar anisotropy value (Δr) evaluated for AA5754 and AA6082 sheets at 30-250°C

Material	AA5754					AA6082				
Temperature (°C)	30	100	175	200	250	30	100	175	200	250
r_0	0.71	0.74	1.02	1.13	1.23	0.60	0.64	0.69	0.78	0.93
r_{45}	0.56	0.59	0.91	1.00	1.12	0.68	0.72	0.79	0.89	0.99
r_{90}	0.79	0.81	1.12	1.17	1.28	0.56	0.61	0.66	0.75	0.91
$\bar{r}^*$	0.65	0.69	0.99	1.08	1.19	0.63	0.67	0.74	0.83	0.96
Δr^{**}	0.19	0.18	0.15	0.14	0.13	-0.09	-0.09	-0.12	-0.13	-0.07

$^*\bar{r} = \left[\dfrac{r_0 + 2r_{45} + r_{90}}{4}\right]$ $\quad$ $^{**}\Delta r = \left(\dfrac{r_0 - 2r_{45} + r_{90}}{2}\right)$

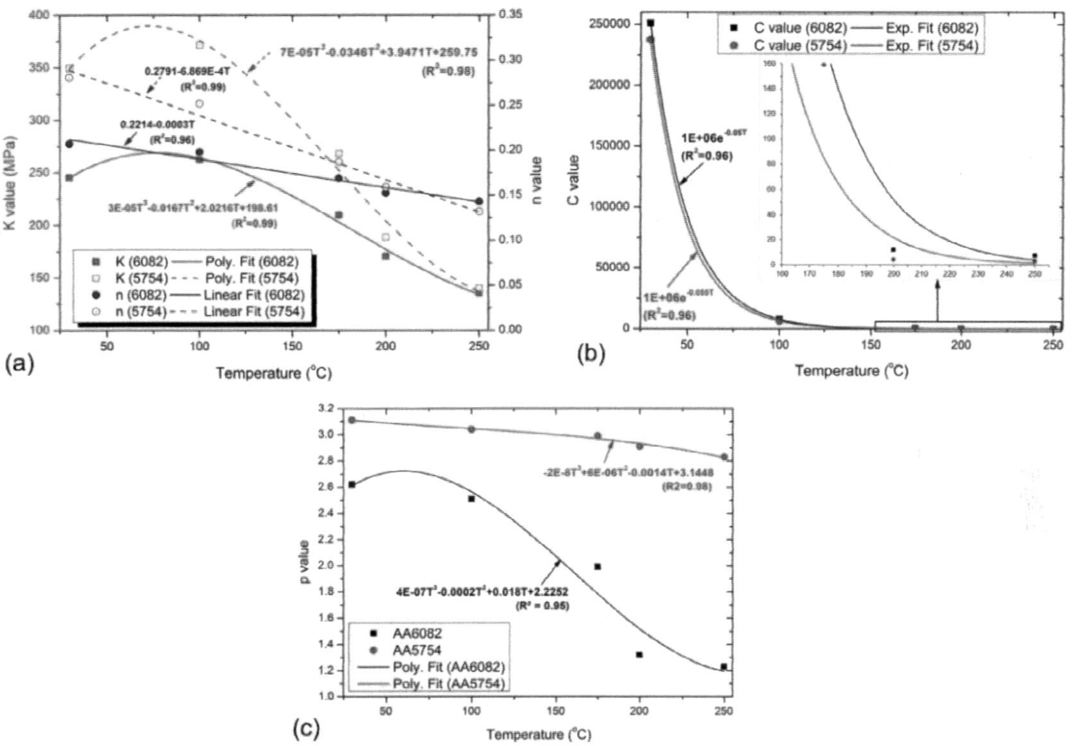

Figure 1 Temperature dependent variation with respective curve fitting equations of (a) *K*, *n* values, (b) *C* value, and (c) *p* value

$$\psi(\sigma_{xy}) = a|\beta_1 + \beta_2|^M + a|\beta_1 - \beta_2|^M + c|2\beta_2|^M = 2\bar{\sigma}^M \tag{2}$$

Here, $\bar{\sigma}$ = yield stress of the material, M=8 for aluminum alloy; β_1, β_2 =stress invariants expressed in terms of terms of $\sigma_x, \sigma_y, \sigma_{xy}$ (plane stress components in orthotropic axis) as shown in Eq. 3.

$$\beta_1 = \frac{\sigma_x + l\sigma_y}{2}, \text{ and } \beta_2 = \left[\left(\frac{\sigma_x - l\sigma_y}{2}\right)^2 + \left(q\sigma_{xy}\right)^2\right]^{1/2} \tag{3}$$

The Lankford anisotropy coefficients evaluated along three directions (r_0, r_{45}, r_{90}) were also made temperature dependent with a third order polynomial curve fitting method. Further, according to these variations in r values, the temperature dependent values of Barlat89 anisotropy coefficients (a, c, l and q) were calculated after solving a set of non-linear equation following the steps as detailed in [10,13]. The variation in Barlat89 yield locus of both the sheets at 100°C, 175°C, and 200°C are compared in Fig. 2. It was observed that the size of yield locus was decreasing, as the blank material possessed lower stress levels at elevated temperatures. The shrinkage of yield locus of was more in case of AA6082 compared to AA5754 material.

2.2. Finite element model

The nonisothermal warm deep drawing experiments were performed using cooling punch with water circulation which was maintained near to room temperature (30°C) while the dies were heated to 200°C. The schematic of the setup with dimensions is shown in Fig. 3a. The initial diameters of the blanks considered were 110 mm for AA5754 and 103 mm for AA6082 material. During warm deep drawing, flange materials were under thermally softened state at 200°C, whereas cup wall, corner and center was

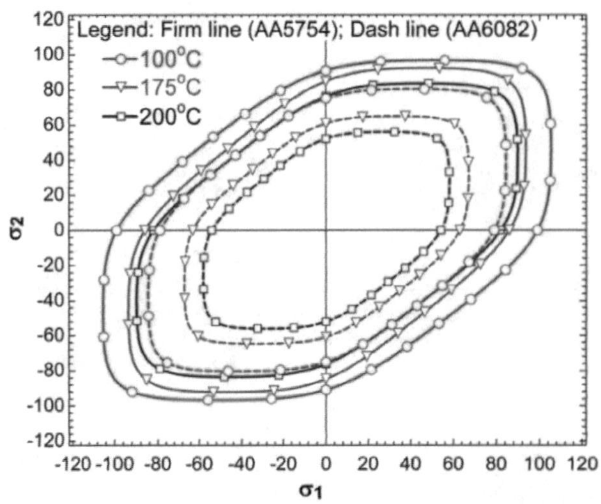

Figure 2 Variation in Barlat89 locus of AA5754 and AA6082 materials under warm temperatures

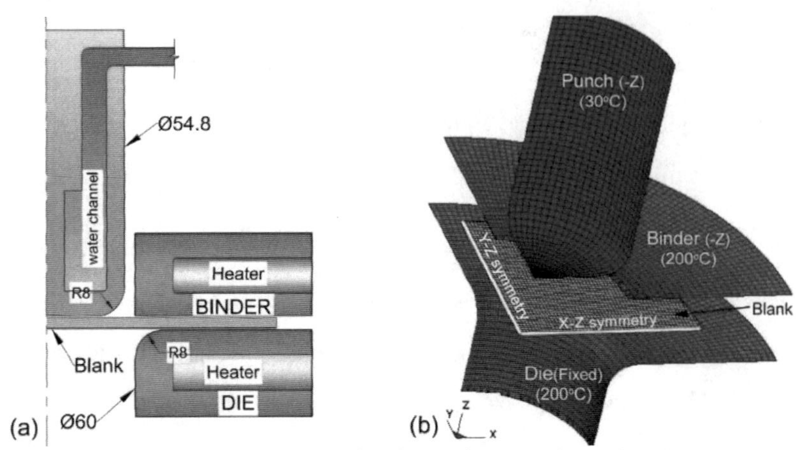

Figure 3 (a) Schematic of the experimental nonisothermal warm deep drawing process (all dimensions are in mm.) and (b) FE model developed for nonisothermal process

under cooled conditions because of direct contact with the punch at 30°C (refer Fig. 4). The FE model of the nonisothermal warm deep drawing process was done using LSDYNA, and corresponding tooling and blank is shown in Fig. 3b. The tools were assigned as rigid bodies, and deformable blanks as four node quadrilateral Belytschko-Tsay shell elements of size 1mm×1mm. The overall computational time was reduced as quarter-symmetric model was considered. Both the punch and binder were allowed to move in –Z (downward direction) and die was assigned as fixed body. The blank was placed over the die and the blank holding force (BHF) of 1.5kN was applied. The initial temperature assigned for the die and binder was 200°C and that of punch was 30°C. All the relevant thermal and structural properties of steel tools and aluminum blank were taken from [15]. The punch was ramped up to 1000mm/s during the deformation stage to utilize the time scaling method which can reduce overall computational time. Accordingly, all the rate dependent properties were also scaled in the FE model. The coefficient of friction at the die-blank and punch-blank interfaces were assumed as 0.23 and 0.33 respectively [10]. The experimentally evaluated FLDs at 200°C of respective materials [5,13] were used as the failure criterion.

2.3. Background on deformation rate

During warm deep drawing, most of the critical plastic deformation occurred over the die profile to form the cup wall region. The flange materials were always maintained within the heated die, and the selected materials were strain rate sensitive at elevated temperatures. Hence, it was essential to understand the strain rate arising in the critical deformation zone. Fig. 4 represents the heat transfer and velocity of material flow during the nonisothermal warm deep drawing process. The material element in flange region first moved inward in the radial direction and reached at the position *b*. This element was prone to plastic bending and unbending when it flowed over the die profile radius to form the cup wall in the radial position x_i. The radial draw-in velocity (v_i) of the material at the entrance of the die cavity (at distance x_i) was assumed to be almost equal to the punch velocity (v_p = 20 mm/min). Also, assuming the flange deformation to be under plane strain condition, the effective plastic strain rate of the flange element ($\dot{\varepsilon}_{flange}$) located at a distance x ($x_0 > x > x_i$) was estimated according to Eq. 4 [16].

$$\dot{\varepsilon}_{flange} = -\frac{2}{\sqrt{3}}\frac{v_x}{x} \quad \textbf{(4)}$$

The $\dot{\varepsilon}_{flange}$ of an element initially located at 3 mm inside the perimeter of the blank (i.e. at the radial distance of 52.0 mm for AA5754 and 48.5 mm for AA6082) was calculated using the above relation. It was found that $\dot{\varepsilon}_{flange}$ increased from 4.96x10^{-3} s^{-1} to 9.99x10^{-3} s^{-1} during the sliding over the die. However, the $\dot{\varepsilon}_{flange}$ of the material during drawing into the cavity over the die profile radius was almost

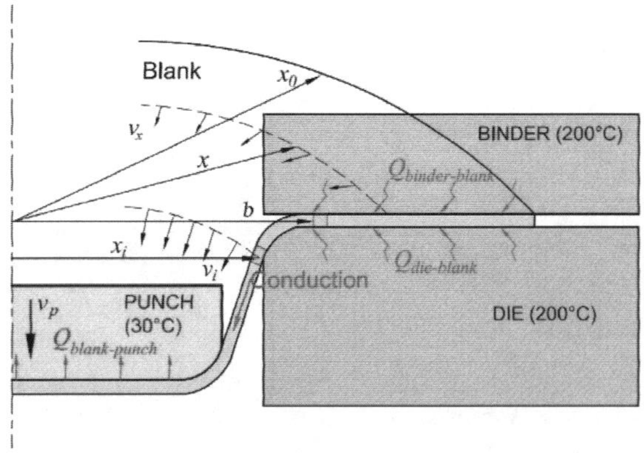

Figure 4 Schematic of the heat transfer and material flow during nonisothermal deep drawing

constant, i.e. 9.99x10^{-3} s^{-1} to 13x10^{-3} s^{-1}. Hence, the identification of Cowper-Symonds coefficients evaluated from the tensile test responses at the lowest initial strain rates (10^{-3} s^{-1} and 10^{-1} s^{-1}) could be able to provide a reasonable description of flow stress during thermomechanical FE modelling of the warm deep drawing process.

3. Results and discussion

It was found that the maximum blank diameter drawn without failure during nonisothermal warm deep drawing process of AA5754 material was 110 mm, and that of AA6082 was 103 mm. The developed FE model was then successfully validated by comparing ear profile, thickness and surface strain distributions. Fig. 5a and b respectively show the FE predicted quarter section of completely drawn cups of AA5754 and AA6082 materials. Fig. 5c shows the experimentally drawn cups of both the materials. Minor wrinkling observed was also predicted well with the developed FE model.

3.1. Ear profile development

The ear profile developed during intermediate stages at similar flange diameters along rolling direction (RD) for both the materials with die internal radius as reference is shown Fig. 6a. The radial inward flow of flange materials through the die corner radius was not uniform at different angles from RD to transverse direction (TD). This was because of higher anisotropy experienced by flange material which is in contact with the die and binder maintained at 200°C. It was observed that the evolution of ear profile was different for both materials along diagonal direction (along 45° from RD). This waviness was better visible in the flange at an instantaneous cup depth of 30 mm, and in the completely drawn cups as shown in Fig. 6b. The average cup heights (H_{avg}) was slightly under predicted within 5% error, i.e. 40.4 mm and 34.6 mm against experimental values of 41.6 mm and 35.5 mm respectively for AA5754 and AA6082 cups. The ear profile was demonstrated in terms of $\%\,\text{earing} = (h_i - h_{\min}) / h_{\min}$, where h_i is the instantaneous cup height, and $h_{\min}$ is the minimum cup height). Fig. 6c demonstrates the % earing profile developed for both the materials. It was observed that peak of ear was developed along 45° for AA5754, whereas trough was observed at similar location for AA6082. This difference in peak location was mainly because of the corresponding positive and negative planar anisotropy (Δr) values of AA5754 and AA6082 respectively (Table 1). The ear profile was more pronounced for AA5754 as it showed relatively higher anisotropy response at higher temperatures compared to AA6082 material (Table 1).

3.2. Thinning development

The measured thinning profile of deep drawn cups along RD was matching very well with the FE predictions. Fig. 7 provides the thinning distributions of both the materials from cup center through corner, wall and towards the top of the cup. There was minor stretching of materials during the initial stages of deep drawing. With further increment in punch displacement, the material bends over the

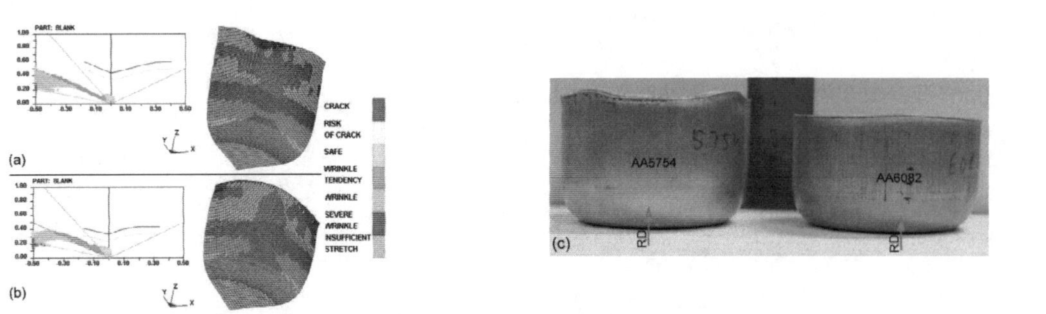

Figure 5 FE predicted deep drawn cups of (a) AA5754 and (b) AA6082 and (c) representative experimental deep drawn cups (arrow represents rolling direction (RD))

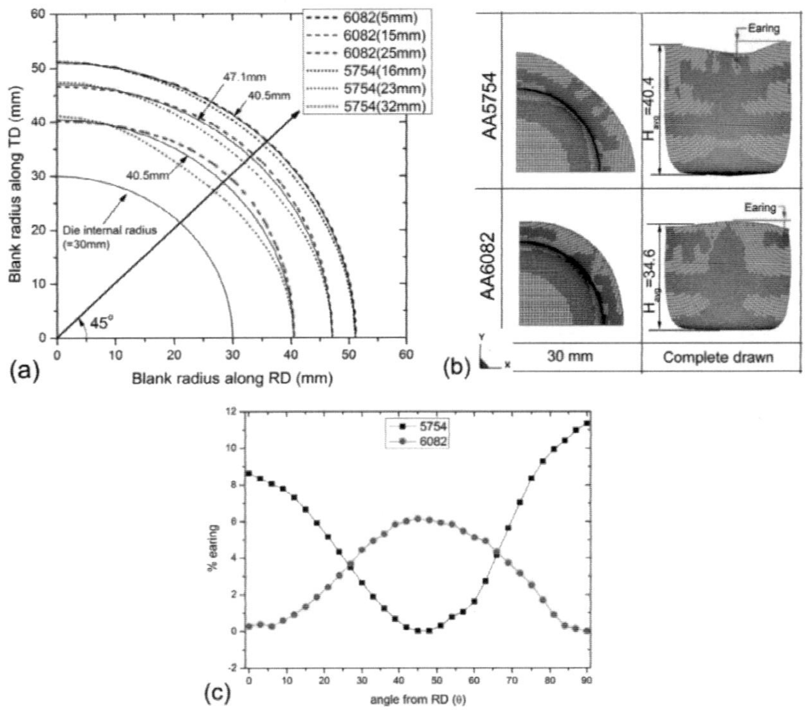

Figure 6 Comparison of earing profile development from FE simulations: (a) at similar flange radius along RD, (b) at an intermediate cup depth of 30mm (top view) and completely drawn cups, and (c) %earing developed in fully drawn cups

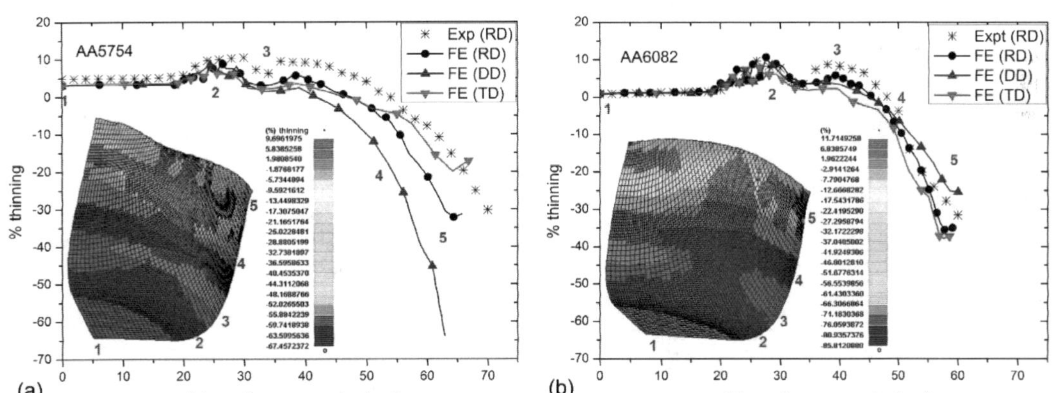

Figure 7 Thinning developed along RD, DD, and TD in completely drawn cups of (a) AA5754 and (b) AA6082 materials with the experimental data along RD

punch profile radius and bending and unbending happens simultaneously over the die profile radius. Hence, the flange sheet experiences radial tensile load. During the warm deep drawing test, the materials were always under heated conditions within the die (at 200°C) which results in lower strength property and *r* values. Generally, higher the *r* value, higher the deep drawability with enhanced resistance to thinning of sheet materials. AA5754 showed improved $\overline{r}$ at elevated temperature (Table 1). Thus more resistivity towards flow was offered along RD and TD, which resulted in increased thinning along these directions compared to that along 45° (DD). Such non-uniformity resulted in relatively more compressive hoop stresses to act towards DD. This resulted in about 110% thickening at the cup top along DD when compared to that along RD as observed from Fig. 7a. The converse was true for AA6082 materials, and thus higher thickening was observed along TD and RD compared to that along DD. However, as observed from Fig. 7b, almost similar thinning pattern was observed for AA6082 in RD and TD. The thickening along DD was lesser by around 60% than that along RD. Negative thinning demonstrated increased thickening in the cup wall region. These variations in anisotropy and thickening also attributed to the variation in ear profile development in completely drawn cups (Fig. 6). It was observed that there was no abrupt thinning at cup corner region.

3.3. Surface strain distribution

The surface strain data obtained from FE simulations along RD, DD and TD for both the cups are plotted in Fig. 8. Both major and minor true strain trend was very well predicted and matched with that of experimental strains along RD. As relatively more stretching occurred in the cup wall region along TD and RD for AA5754 and along DD for AA6082, the major true strain distributions was higher in the respective directions. The maximum of major true strains developed along cup wall of AA5754 in DD was lesser by about 90% with that of RD. The trend shown by minor true strains along DD in the cup wall was also minimum. Conversely, the maximum of major true strain evolved was about 60% higher along DD compared to RD for AA6082 cup, and the minor true strains was also higher along DD. The lower *C* value of AA5754 material represented its relatively higher strain rate sensitivity [5] (Fig. 2b). The higher strain rate sensitivity indicated higher strain redistribution capability of the material [10]. Moreover, the enhanced resistance to thinning due to higher *r* values also reflected the relative improvement in drawability of AA5754 compared to AA6082 material. The variations of strain values from general trend at cup top region was due to wrinkling and uneven thickening happened in these regions. The minor and major true strains evolved in completely drawn cup center, corner, wall and top regions captured from FE simulations consolidated for both the materials are given in Table 2. The positive major and minor true strains indicated marginal initial stretching which also resulted in minor thinning at cup center (Fig. 7). The minor true strains recorded at the cup corner were approximately

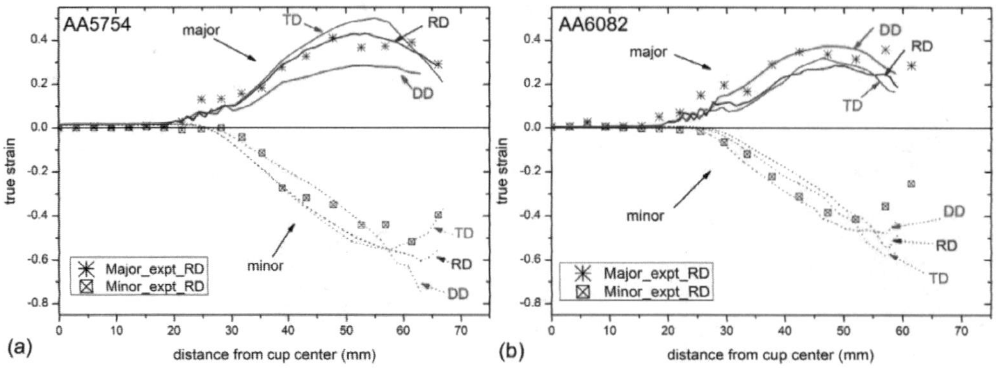

Figure 8 Surface strain developed along RD, DD, and TD in completely drawn cups of (a) AA5754 and (b) AA6082 materials with the experimental data along RD

Table 2 Minor and major true strains at different locations in completely drawn cup compared with FLD_0 at 200°C

Location on the cup	AA6082		AA5754	
	Minor	Major	Minor	Major
Center	0.004	0.006	0.014	0.017
Corner	-0.008	0.085	-0.006	0.079
Wall	-0.279	0.279	-0.383	0.368
Top	-0.480	0.230	-0.588	0.308
FLD_0 (at 200°C)	0	0.351	0	0.432

equal to zero, which signified that material deformation was under plane strain condition. Usually, the part failure happens in a deep drawing process when the necking/failure strains at cup corner exceeds the limiting strains in plane strain conditions. This particular strain state represented in forming limit diagram is represented as FLD_0. The FLD_0 of AA5754 and AA6082 sheet materials evaluated at 200°C is also given in Table 2.The major true strains at cup corner was very much lower than that at FLD_0 value, signifying the avoidance of sudden localization of thinning/necking.

4. Conclusions

In this study, thermomechanical FE modeling of nonisothermal warm deep drawing process of AA5754 and AA6082 materials was successfully developed incorporating temperature-dependent material properties. The Cowper-Symonds parameters and Barlat89 yield coefficients evaluated at different temperatures ranging from 30°C to 250°C were incorporated into the FE model. The FE predicted results showed good agreement with experimental ear profile, thinning and surface strain data. It was observed that cup height and ear profile developed in AA5754 cups was prominent due to higher planar anisotropy at elevated temperatures. The peak of earing was along the diagonal direction (DD) for AA6082 due to negative planar anisotropy. The highest thickening of approximately 110% was observed for AA5754 cups along DD at the cup top. This thickening resulted in about 60% reduction of maximum major true strain evolved at cup wall region. Conversely, the maximum thickening in cup wall regions was 60%, and the corresponding maximum major true strains evolved was 60% lower along RD for AA6082. Most of the deformations were observed to be in the tension-compression mode. Non-localization of thinning and surface strains at cup corner indicated improved ability of materials to withstand more deformation without failure. The AA5754 showed comparatively better formability due to its higher temperature dependent rate sensitivity, yielding and anisotropy responses compared to that of AA6082 material.

References

[1] Hirsch J 2011 *Mater. Trans.* **52** 818–824
[2] Dursun T and Soutis C 2014 *Mater. Des.* **56** 862–871
[3] Panicker S S, Prasad K S, Sawale G, Hazra S, Shollock B and Panda S K 2019 *Mater. Sci. Eng. A.* **768** 138445
[4] Amino H, Nakamura K and Nakagawa T 1990 *J. Mater. Process. Tech.* **23** 243–265
[5] Panicker S S and Panda S K 2019 *J. Mater. Eng. Perform.* **28** 2967–2982
[6] Palumbo G and Tricarico L 2007 *J. Mater. Process. Technol.* **184** 115–123
[7] van den Boogaard A H and Huétink J 2006 *Comput. Methods Appl. Mech. Eng.* **195** 6691–6709
[8] Ghaffari Tari D, Worswick M J, Ali U and Gharghouri M A 2014 *Int. J. Plast.* **55** 247–267
[9] Bandyopadhyay K, Panda S K, Saha P and Padmanabham G 2015 *J. Mater. Process. Technol.* **217** 48–64
[10] Panicker S S and Panda S K 2017 *J. Manuf. Sci. Eng.* **139** 031013
[11] Dhara S, Basak S, Panda S K, Hazra S, Shollock B and Dashwood R 2016 *J. Manuf. Process.* **24** 270–282
[12] Panicker S S, Prasad K S, Basak S and Panda S K 2017 *J. Mater. Eng. Perform.* **26** 3954–3969
[13] Panicker S S, Singh H G, Panda S K and Dashwood R 2015 *J. Mater. Eng. Perform.* **24** 4267–4282
[14] Barlat F and Lian J 1989 *Int. J. Plast.* **5** 51–66
[15] Abedrabbo N, Pourboghrat F and Carsley J 2007 *Int. J. Plast.* **23** 841–875
[16] Naka T and Yoshida F 1999 *J. Mater. Process. Technol.* **89–90** 19–23

IOP Conf. Series: Materials Science and Engineering 1132 (2021) 012007 doi:10.1088/1757-899X/1132/1/012007

Distribution of effectiveness and Nusselt number over a corrugated surface impinged by a row of circular jets

S Abraham[1*] and R P Vedula[2]

[1]Assistant Professor, Mechanical Engineering Department, NIT Warangal, India
[2]Professor, Mechanical Engineering Department, IIT Bombay, India

[*]E.mail: abrahams@nitw.ac.in

Abstract. The characteristics of effectiveness and heat transfer for a row of jets impinging on a corrugated surface were experimentally investigated. The Reynolds number, Re which was calculated based on the exit velocity and the hydraulic diameter of the nozzle, was 20000 and the jet-to-target spacing, L, was varied from 2 to 10 times the jet diameter, d. The spacing between adjacent jets, Sx, was kept constant at 4d. The actual angle of impingement and jet-to-target distance were different for different jets in the row due to the corrugated surface. This results in difference in the interaction of walljets of adjacent jets, which inturn influence the local distribution of effectiveness and Nusselt number

1. Introduction

Jet impingement gives high local heat transfer coefficient and can be used for the cooling of surfaces with severe thermal conditions. A row of jets can be used to reduce the drawback of sudden drop in heat transfer coefficient at locations far from the impingement region. Saripalli [1] studied flow visualization of impingement of two circular jet on a flat plate and observed a fountain due to the interaction of wall jets. Abrahams and Vedula [2] reported studies on impingement of column of circular jets on the convex surface with and reported that the distribution of Nusselt number for each jet is different due to the contact of the adjacent walljets. The flow behavior and heat transfer of obliquely impinging jets are different as more fluid diverts to the downhill side. Donovan and Murray [3] reported that the increase in the inclination decreases the turbulence generation in the downhill side.

The temperature of jet is needed to be different from the ambient for some applications and such jets are termed non-isothermal jets. The ambient gets entrained into the jet and changes the recovery temperature on the impinging surface from the jet temperature. This recovery temperature can be considered to be the adiabatic wall temperature, T_{aw}, and needed for the precise value of the heat transfer over the surface. Baughn et al. [4] and Goldstein et al. [5] studied on non-dimentional T_{aw} data, termed as effectiveness, on a flat surface impinged with circular jet. Highest effectiveness observed at the impingement region and the variations were Reynolds number independent. Goldstein and Seol [6] studied row of jets on flat surface and observed maximum effectiveness at midway between jets at large stream wise direction. Abraham and Vedula [7] studied effectiveness for circular jets row impinging at different inclinations on a convex surface and observed that the interaction of walljets reduce the entrainment in the inner regions and shows high effectiveness. Abraham et al. [8] reported effectiveness data on convex surface for slot jet impinging at an angle. The effectiveness was decreasing on the uphill side and increasing on the downhill side as the inclination increased.

The present study is on jets arranged in a single row impinging on to corrugated surface for the application of cooling of after-burner of gas turbine engines. The corrugated liners of the afterburners must withstand very high temperatures and very effective cooling methods are required to keep the liner from failing. The studies on external cooling of a corrugated surface are very rare and are mainly restricted to film cooling. Funazaki et al. [9] reported cooling of a typical corrugated liner of an after-burner using film cooling method. The increase in the film cooling fluid velocity was observed to increase the penetration of the film into the primary flow. This causes the film to lift-off from the wall and causes reduction of film effectiveness. Present investigation reports the detailed effectiveness and Nusslet number distribution on a corrugated surface impinged with circular jets arranged in a single row.

2. Experimental setup and data reduction

2.1. Experimental setup

Figure 1 shows the layout of experimental setup. A compressor supplies air to a pressure vessel. The air was then supplied to a venturimeter by adjusting the control valves to the bypass and to the test setup. The pressure drop measured across the venturimeter gives the mass flow rate. The air was the heated above the ambient temperature for non-isothermal experiments by using heaters connected in series. Heaters were made by fixing high resistance electric rods with helical fins inside metal pipes. The air then entered the plenum chamber designed with to get a uniform flow at the exit by giving a divergence to the inlet section and then a constant area section. Plexiglass was used to fabricate the plenum chamber. The jet nozzle was fabricated with acrylic tubes and was attached to the plenum chamber. K-type thermocouples were used to measure the jet temperature.

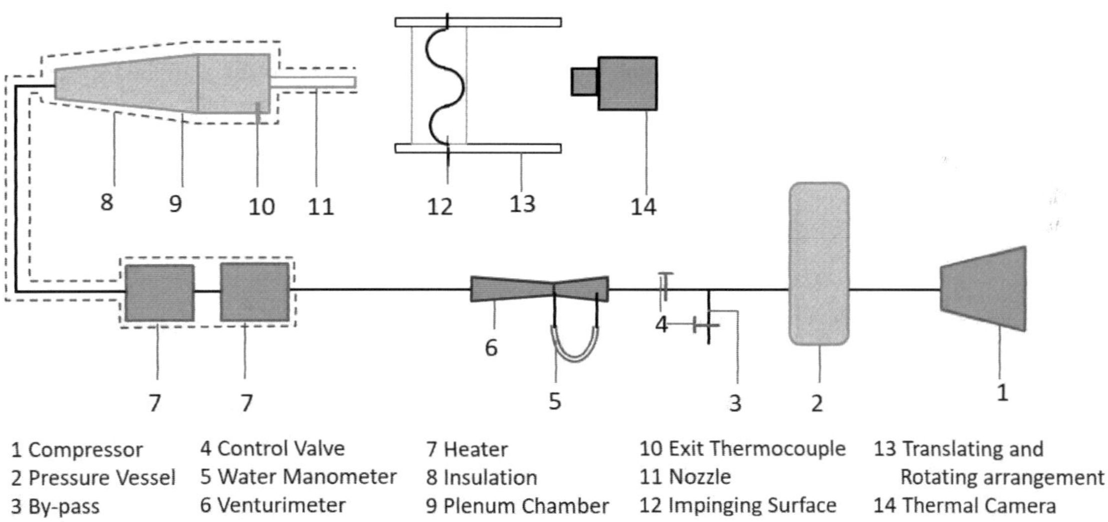

Figure 1. Schematic of experimental test rig

Figure 2 shows the representative figure of the jet impingement on corrugated surface. The present corrugated surface consists of a convex portion at the middle and two concave parts at the sides. The radius of curvature for the convex and concave parts was kept as 100mm and the pitch as 160mm, giving an amplitude of 17mm. Corrugated cut of required dimensions were made on two Teflon sheets using CNC and the SS foil was fitted between the Teflon sheets. The foil was sandwiched between a pair of copper bars at both ends for support and for electric power connection.

The row of nozzles positioned along the corrugated surface. The diameter of the jet was 10mm, and was given a developing length of 25d. The spacing between the adjacent nozzles, Sx, was given as

4d and the jets were named as shown in the figure for the ease in discussion of the results. The middle jet, jet A impinged at the apex of the convex surface and jets C and C^I impinges at the centre of the concave portion. The jets B and B^I impinge at the merging position of the convex and concave surfaces, making the jets impinging at an inclination of 23^0. The jet-to-target distance, L/d, is specified as the distance of the jet A from the apex of the convex portion, i.e. r/d=0 line. The distance along the height of the impinging surface is measured as y/d.

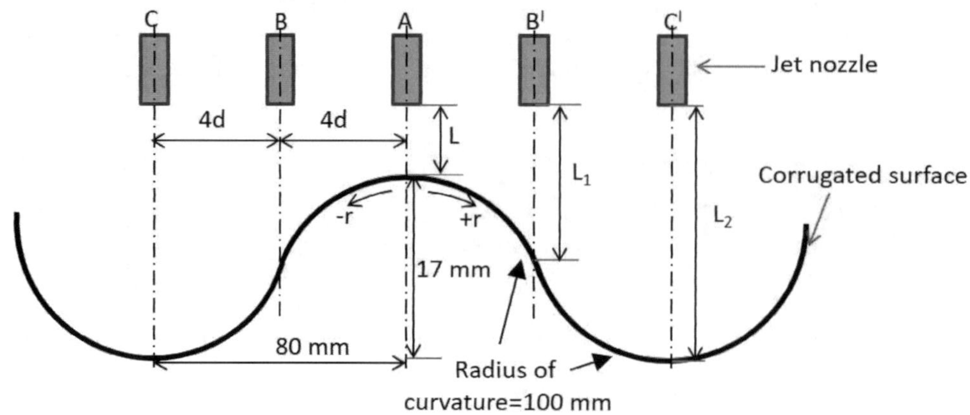

Figure 2. Schematic of jets impinging on corrugated surface

The actual jet-to-target distances for different jets vary due to the corrugated nature of the target and are shown in table 1. The spacing between the adjacent jets along the circumference (r/d direction) is different from the Sx/d value due to the curved nature of the impinging surface and is also included in Table 1

Table 1. actual jet-to-target distances for different jets

Jet	Actual jet-to-target distance	r/d position of impingement
Jet A	L	0
Jet B and B^I	L_1=L+0.85d	-4.1 and +4.1
Jet C and C^I	L_1=L+1.7d	-8.2 and +8.2

2.2. Data reduction

The data reduction methodology is reported in detail in [8] and is briefly discussed here. The impingement plate was made with 0.12 mm thick foil of Stainless steel. The plate surface opposite to the impinging surface was painted with black colour to take thermal image at steady state condition using a Infrared camera (Mikron make M7600PRO). Boundary condition as constant heat flux was obtained by applying Alternating Current to the foil. The total power given to the target surface (foil), q_{total}, was calculated from the measured values of current flowing through the foil and the voltage drop across it. A minor quantity of the heat supplied was lost to the ambient as one side of foil was open for thermal imaging. This loss in the heat, q_{loss}, was calculated with a no flow experiment and a curve between q_{loss} and the temperature difference between the ambient and the foil was obtained. The difference in total heat flux supplied and the heat loss is the net heat flux, as in equation below

IOP Conf. Series: Materials Science and Engineering 1132 (2021) 012007 doi:10.1088/1757-899X/1132/1/012007

$$q_{total} = VI \quad q''_w = \frac{q_{total} - q_{loss}}{A_{foil}} \tag{1}$$

The heat transfer coefficient was calculated based on the neat heat flux as shown in equation (2) for isothermal jet experiments (where jet temperature same as ambient). The jet diameter "d" was used as the characteristic length for the calculation of Nusselt number 'Nu' as shown in eq. (3).

$$h_{iso} = \frac{q''_w}{(T_w - T_b)} \tag{2}$$

$$Nu = \frac{h_{iso}\, d}{k} \tag{3}$$

The effectiveness was measured for non-isothermal jet experiments, and is defined as follows.

$$\eta = \frac{T_{aw} - T_{amb}}{T_j - T_{amb}} \tag{4}$$

The method for calculating T_{aw} is as follows.

$$q''_w = h_h (T_w - T_{aw}) = h_h T_w - h_h T_{aw} \tag{5}$$

In the equation (5), h_h and T_{aw} are constants, and gives equation of straight line for the plot of q''_w vs T_w. The slope at each location gives h_h and the intercept on the temperature axis gives T_{aw}. Each experiment was performed with different heat inputs and the plots were always linear. A typical case for slot jet on on a convex surface is shown in Figure 3 ([8]).

.

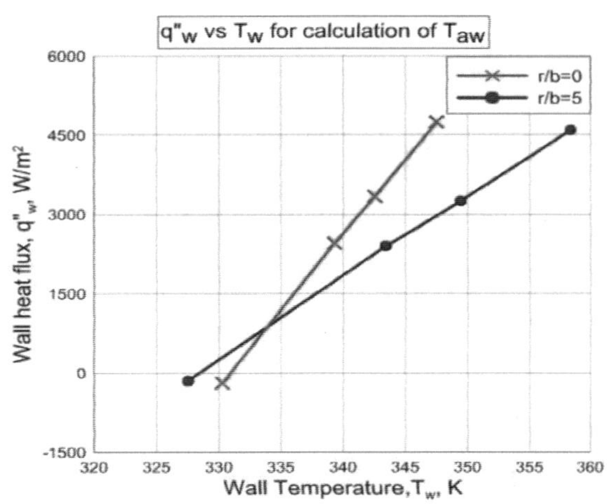

Figure 3. q''_w vs T_w for calculation of T_{aw} on convex surface ([8])

2.3. Uncertainty of the measured and calculated values

The uncertainties in the temperature measured using thermocouple and thermal camera were ±0.3^{0}C and ±.1.1^{0}C. The uncertainties of the calculated quantities were estimated using method mentioned in Coleman and Steele [10]. Minimum uncertainty for effectiveness and Nusslet number were calculated as 15% and 8% respectively.

3. Results and discussion

The studies were conducted for jets in a single row with spacing between adjacent jets equal to 4d impinging normally on the corrugated surface. Experiments were mainly conducted for a horizontal row impinging along the corrugated surface as shown in **Figure 2**. Few experiments were conducted with single jet impinging on the convex apex of the target surface to compare present results with literature. All the experiments were conducted for Re=20000.

3.1. Single jet

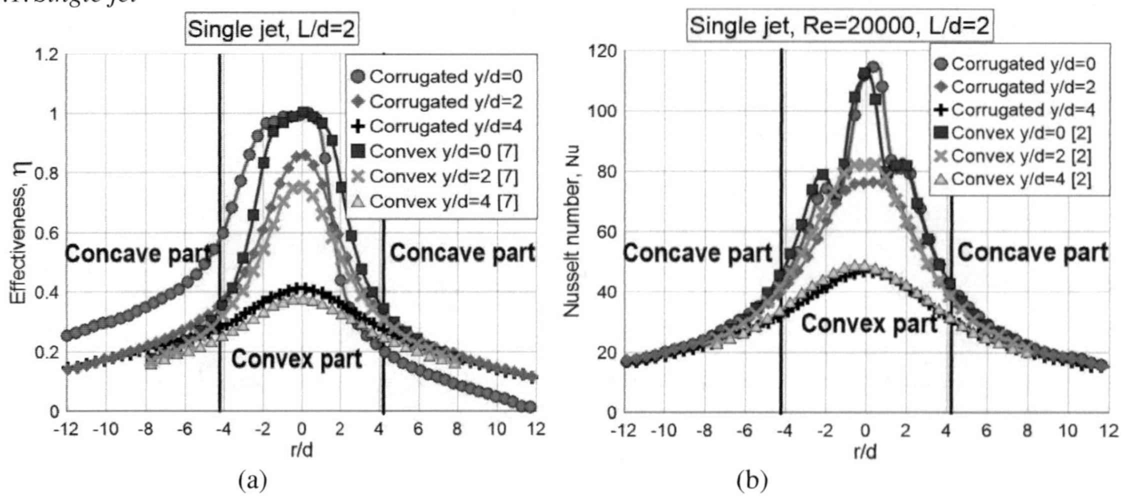

Figure 4. Comparison with literatture for single jet (a) effectiveness (b) Nusselt number

Figure 4 (a) and (b) compares the effectiveness and Nusselt number variations for a single jet respectively on current corrugated and convex surfaces ([7] for effectiveness and [2] for Nusselt number) for the distance L/d=2. Data are provided at y/d=0, 2 and 4. At r/d=±4 the corrugated surface turns from convex to concave geometry, whereas for convex target the geometry is convex for the whole r/d range. The variations on convex surface are matching in the convex portion of the corrugated surface (± 4) for both Nusselt number and effectiveness, but the values differ in the concave portion of corrugated surface. However, the differences in values are within the uncertainty limit.

3.2. Row of jets along the corrugated surface

3.2.1. Effectiveness. Figure 5 (a) gives the effectiveness distribution on a corrugated surface impinged with the row of circular jets at jet to target distance L/d=2. The impinging location of various jets are shown as dotted lies in the figure. The data is symmetric about y/d=0 and the variations of effectiveness only along the +r/d portion of the target surface are discussed. The contours of constant effectiveness near jet A retain the circular shape of the jet till r/d≈2, which is almost the midway between jets A and B^I. The impingement of jet B^I on the inclined portion causes a diversion of impinging air to the corresponding downhill side (r/d>4.1), making the contours near the impingement point depart from the round shape. The jet C^I, which impinges on the apex of concave surface, travels 1.7d distance more compares to jet A and 0.85d more to jet B^I before impinging on the target, is likely to be affected by the spent fluid from the other jets. The interaction of walljets of jets B^I and C^I results in the inclination of interaction zone towards concave portion and is characterized by high local effectiveness. The effectiveness values in the convex portion at high ±y/d positions are noticed to drop faster compared to that in the concave part due to the increased flow towards the concave part by the jets A and B^I.

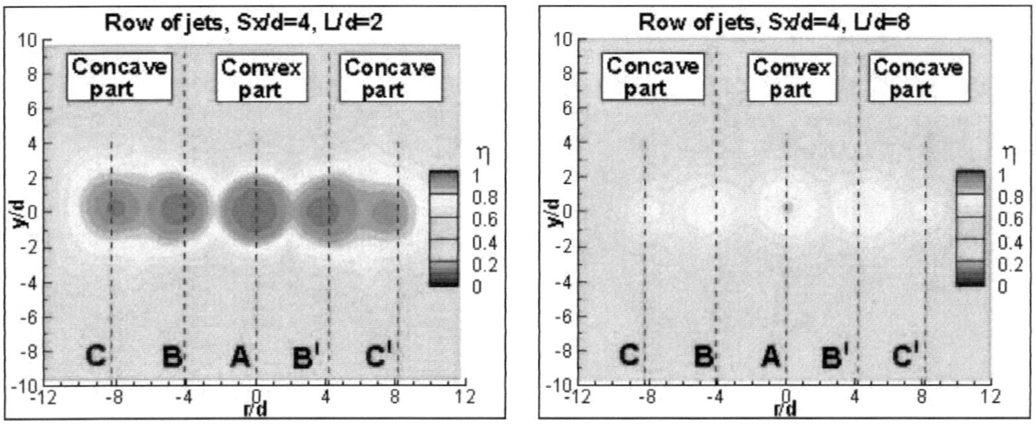

(a) (b)

Figure 5. Detailed distribution of effectiveness (a) L/d=2 (b) L/d=8

Figure 5 (b) show the effectiveness distribution on a corrugated surface at L/d=8. The general behavior of the effectiveness distribution are comparable to L/d=2. However, the values at L/d=8 are lower compared to values at a lower distance (L/d=2) because of the higher entrainment of stagnant surrounding air to the jets prior to the impingement.

Figure 6 (a) shows effectiveness data over corrugated surface at various y/d locations at L/d=2. Even though the actual jet-to-target distances are different for different jets at y/d=0 (table 1), all are within the potential core length and therefore the peak effectiveness corresponds to all the jets are almost equal. The difference in the interactions of different pairs of adjacent jets results in the difference in drops of effectiveness between adjacent jets; with higher drop between jets A and B^I compared to that between jets B^I and C^I. The drop in effectiveness for jet B^I is also noticed to be higher on the respective uphill side (r/d<+4.1), compared to the downhill side (r/d>+4.1). The high effectiveness with similar variations persists till y/d=1, after which the values drops. At y/d=2 the drop in effectiveness between the jets B^I and C^I is small due to the spread of jet B^I on the concave surface due to the inclined nature of the jet. The r/d position of the peak value appearing in the concave region increases with increase in y/d due to the inclination of the interaction zone of walljets of the jets B^I and C^I. The walljet momentum of jet B^I is higher due to the inclination of jet B^I towards the jet C^I and results in the inclination of the interaction zone. The high effectiveness along the interaction zone is due to the weaker entrainment into the boundary layer at interaction zone as reported in [7].

Figure 6 (b) shows effectiveness variation at L/d=8. At y/d=0 the peak effectiveness value is shown at r/d=0 (the apex of the convex surface) and the values are lesser than L/d=2 because of the higher entrainment of the surrounding air before impingement. The actual jet-to-target distance of all the jets are outside the potential core region at this L/d and therefore the effectiveness at the impinging point reduces progressively as ±r/d increase. The crossflows formed by the walljets of adjacent jets can be also influencing the effectiveness variations at large |r/d| positions. The variation of effectiveness at various y/d locations are comparable to that for L/d=2.

Experiments were also conducted for L/d= 4, 6 and 10 and the effectiveness distribution at different y/d locations relative to y/d=0 are similar to L/d=2 and 8. Figure 7 shows the effectiveness data for all the jet-to-target distances investigated, but only at y/d=0. The peak effectiveness values for all the jets are almost equal to unity at L/d=2. As the distance of target from jet exit (L) increases, the highest effectiveness values decreases and the drop is greater for L/d>6, which is roughly the length of potential core. The drop in effectiveness values can be observed to be higher for the jets at higher |r/d| positions because of the larger jet-to-target distance.

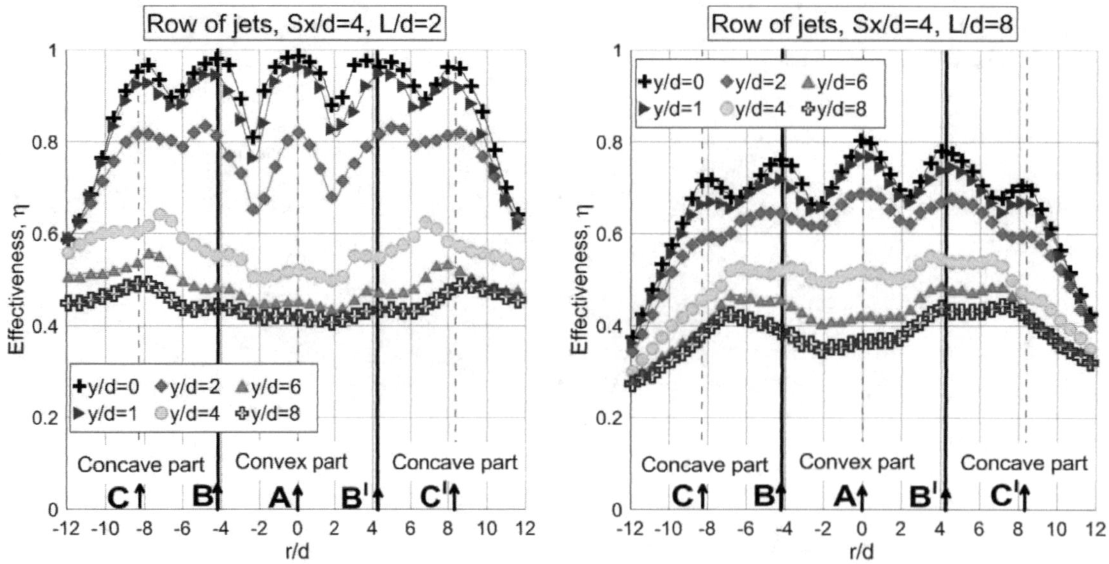

(a) (b)

Figure 6. Effectiveness variation at different y/d (a) L/d=2 (b) L/d=8

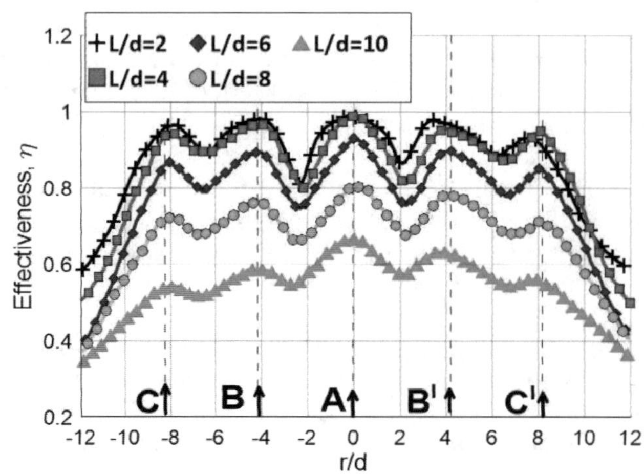

Figure 7. Effectiveness variation at y/d=0 for differnet L/d

3.2.2. Nusselt number. Figure 8 (a) show the distribution of Nusselt number for Re=20000 at L/d=2. The peak values of jets A and C^I occurs at the impinging region, whereas the peak values of jet B^I occurs at the corresponding uphill side. At low y/d positions the interaction of walljets by jets A and B^I has been observed to have minimum effect, whereas the interaction of walljets of jets B^I and C^I causes a local disturbance in variation of Nusselt number. The contours of high Nusselt number at the location near to the jet C^I are deviated from the circular shape of the impinging jet due to the influence of the walljets from the upstream position. Figure 8 (b) shows the distribution of Nusselt number at L/d=8 for Re=20000. The Nusselt number variations due to the interaction of jets are lesser than L/d=2.

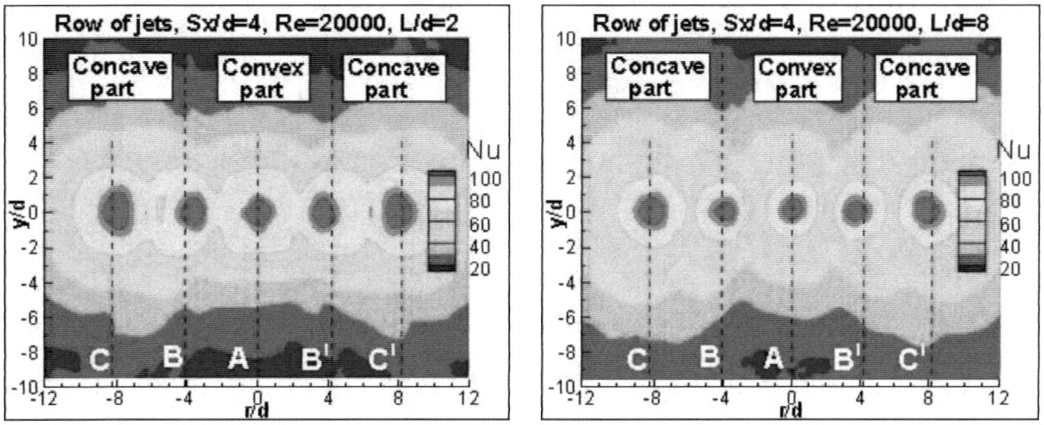

(a) (b)

Figure 8. Detailed distribution of Nusselt number (a) L/d=2 (b) L/d=8

Figure 9 (a) describes the Nusselt number data at several y/d positions at Re=20000. The peak Nusselt number of jets A and C^I occurs at the impinging point, whereas the peak of jet B^I shifts to the corresponding uphill side. The drop in Nusselt number along r/d after the peak values are similar to a normally impinging jet for jets A and C^I and similar to an obliquely impinging jet for jet B^I. The gradient of Nusselt number along y/d is also different at different jet impinging locations. At r/d=0 the Nusselt number drops rapidly along y/d direction till y/d≈1.5, after which a local peak is noticed. The y/d=2 position gives higher Nusselt number than y/d=1.5 as the walljet transforms turbulent flow, after which the values drops again. At other jet impingement positions (r/d≈4.1 and r/d≈8.2) the values drop monotonically along y/d, however, at different rate. The difference in the peak values of Nusselt number for different jets maybe due to the difference in interference of adjacent jets. However, any physical interpretation cannot be given with certainty as the difference of the peak values are within the uncertainty limit. Figure 9 (b) reports the Nusselt number data at L/d=8. The drop in Nusslet number along y/d is monotonic as the significance of walljet transition to turbulence is minimum or nill at this high L/d position

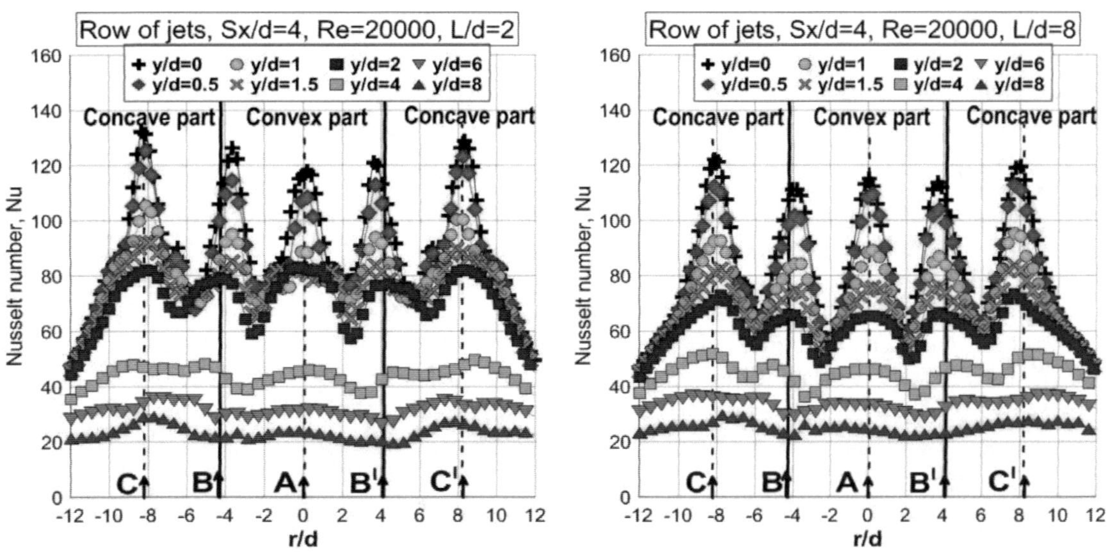

(a) (b)

Figure 9. Nusselt number variation at different y/d (a) L/d=2 (b) L/d=8

Experiments were conducted for L/d = 4, 6, and 10 also. Nusselt number behaviour for all configurations are similar to that explained for Re=20000 at L/d=2 and 8. Therefore for the other configurations the Nusselt number variations along only y/d=0 line are discussed. Figure 10 show the variation of Nusselt number for y/d=0 line for all jet-to-plate distances studied. The maximum Nusselt number at stagnation region are observed for L/d=4.

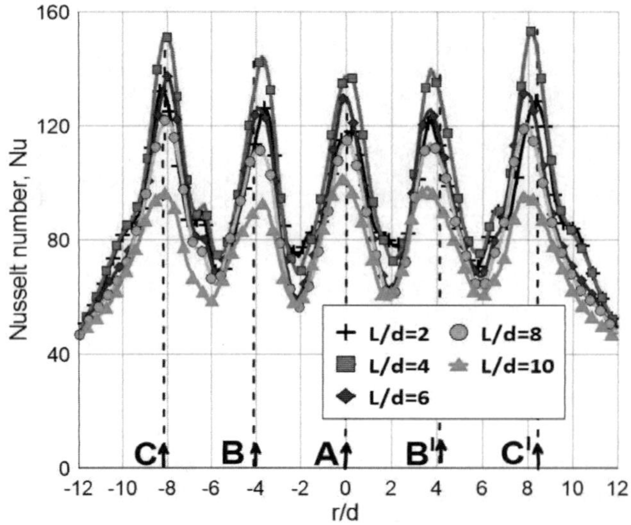

Figure 10. Nusselt number variation along r/d at y/d=0 for differnet L/d

The Nusselt number at the stagnation point can be observed to be increasing as |r/d| position of jet increases. This might be due to the deviation f spent fluid from jets ar r/d=0 and ± 4 towards the concave portion and its interaction with the impinging jets. The peak Nusselt number values can be observed to be decreasing drastically for L/d > 6. The drop in Nusslet number between the adjacent jets are monotonic at high L/d values due to the decrease in the momentum of interaction caused by higher entrainment of ambient.

4. Summary and conclusion

An experimental investigation on the detailed profile of Nusselt number and effectiveness for circular jets in a single row impinging on corrugated surface was conducted. The row of jets was oriented along the corrugated surface. Studies were done at and jet-to-plate distances (L/d) equal to 2, 4, 6, 8 and 10, and Reynolds number equal to 20000.

The actual inclination and jet-to-target distances of jets at different positions are different due to the geometry of target, which affects the Nusselt number and effectiveness variations. The effectiveness values in the convex part at high ±y/d positions are noticed to drop faster compared to that in the concave part. As jet-to-plate distance increases the peak effectiveness values decreases and drop is greater for L/d>6. The drop in effectiveness values can be observed to be higher for the jets at higher |r/d| positions caused by the larger jet-to-target distance. The difference in the impingement angle of different jets causes difference in the Nusselt number distribution between adjacent jets. The stagnation point Nusselt number is noticed to be increasing as |r/d| position of jet increases for L/d>2.

References

[1] Saripalli K.R., 1983, *AIAA Journal,* 21, 483–484.

[2] Abraham S., Vedula R.P., 2019, *Int. J. Heat and Mass Trans.*, 137, 751-764.

[3] O'Donovan T. S., Murray D.B., 2008, *Int. J. Heat and Mass Trans.,* 51, 6169–6179.

[4] Baughn J.W., Hechanova A.E., Yan X., 1991, *J. Heat Transfer,* 113, 1023–1025.

[5] Goldstein R.J., Sobolik K.A., Seol W.S., 1990, *J. Heat Transfer,* 112, 608–611.

[6] Goldstein R.J., Seol W.S., 1991, *Int. J. Heat and Mass Trans.,* 34, 2133–2147.

[7] Abraham S., Vedula R.P., 2018, *Int. J. Heat and Fluid Flow* 69, 210–223.

[8] Abraham S., Kakade A.B., Vedula R.P., *Pro. World Congress on Engineering 2015 Vol II*, July 1 - 3, 2015, London, U.K., 1273–1278.

[9] Funazaki K.S. K., Igarashi T., Koide Y., 2001, *Proc. ASME turbo expo,* June 4-7, 2001, New Orleans, USA, 1–8.

[10] Coleman H. W., Steele W.G., 1989, John Wiley, New York

IOP Conf. Series: Materials Science and Engineering 1132 (2021) 012008 doi:10.1088/1757-899X/1132/1/012008

Design of a Low-Cost Lower Limb Rehabilitation Exoskeleton System

Ullas U* and Rajendrakumar P K

Dept. of Mechanical Engineering, NIT Calicut, Kozhikode, India.

*Ullas1795@gmail.com

Abstract. Human locomotion is probably the single most crucial thing to carry out daily activities in a person's life. It is hence a necessity than an option to help people walk, when they lose that essential ability, for any reason. Rehabilitation is a methodical process, usually carried out by trained professionals in this regard. With the current lifestyle, the increase in factors leading to disabilities is on the rise, and there is a pressing need to rehabilitate the needy. With a lack of specialized medical facilities and personnel to meet the rising demands, especially in many rural parts of India, a low cost rehabilitation device is necessary. The devices that exist for this purpose remain inaccessible to the masses mainly due to their cost, availability, and ability to adapt to different patients. In many of the works from literature, it is seen that actuators play a significant role in driving the costs high. Also, the designs need considerable customization for every individual patient, which makes mass production difficult. In this work, an affordable lower limb exoskeleton system is designed. This device is adaptable to the varying anthropometry of the Indian population while being affordable by the masses. Electric linear actuators are used to keep the costs low, with a simple control system. Adjustability and modularity are maintained throughout the design as an essential criterion.

1. Introduction

Walking disabilities constitute a significant concern in the world today. In a developing country like India, it is seen that the movement disability accounts for 20.2% of all disabilities, according to the 2011 census. House to house surveys put this number at 31.4% [1]. Not being able to walk poses many problems to human beings, firstly, inability to perform day to day chores, being unable to move. Secondly, being confined to bed causes issues such as osteoporosis, pneumonia, etc. Lastly, it can lead to psychological trauma, which can drive the patients' and their caretakers' morale down [2]. Hospital-based rehabilitation in physiotherapy has its high perks due to constraints such as costs, equipment availability, trained personnel, etc. [3].

A significant contributor to movement disability is stroke [4]. WHO has published a set of guidelines called the STEPS Guidelines, followed in a good number of surveys in India and abroad. These surveys observe that India is in the middle of a stroke epidemic [5][6]. The rehabilitation services in India are not up to the needs of the country [2]. The risk factors that lead to stroke are on the rise [7]. After Covid – 19, doctors have observed a rise in stroke cases [8] [9]. However, this needs to be ascertained by scientific studies. Walking disabilities are also caused by Osteoarthritis (OA) and Cerebral Palsy (CP) [10] [11]. Aging causes primary OA, which is the most commonly seen form of OA. Other diseases or conditions can cause secondary OA. The male population has a higher risk of suffering from OA (males – 31.6% and females – 28.1%). A physically demanding work style such as that of daily wage workers, significantly lowers the chance of being affected by OA (22.2%) [11]. Abnormalities in brain development, often before birth, lead to Cerebral Palsy (CP). Perinatal asphyxia (brain damage at the time of childbirth) is seen more commonly in India compared to other countries [12]. In 15-20% of children suffering from CP, physical therapy and orthotics can substantially improve their quality of life [10].

An exoskeleton is an external support suit that takes the wearer's load and may be powered or passive. A rehabilitation exoskeleton is a device that can address some of the concerns discussed above, if not all of them. Research on the development of exoskeletons has been going on for decades, but affordability and availability to the masses is still a distant dream in India. The main contributor to the cost would be the actuator used in the devices. In the present work, a low cost, lower limb rehabilitation active exoskeleton, shown in **Figure 1**, is designed using electric linear actuators. A discussion on the same is presented in this paper.

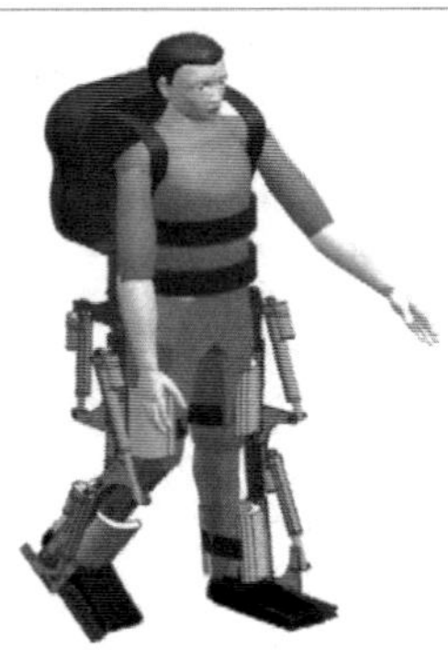

Figure 1. Exoskeleton system assembled onto a human being. The figure shows each module assembled and strapped onto the human body.

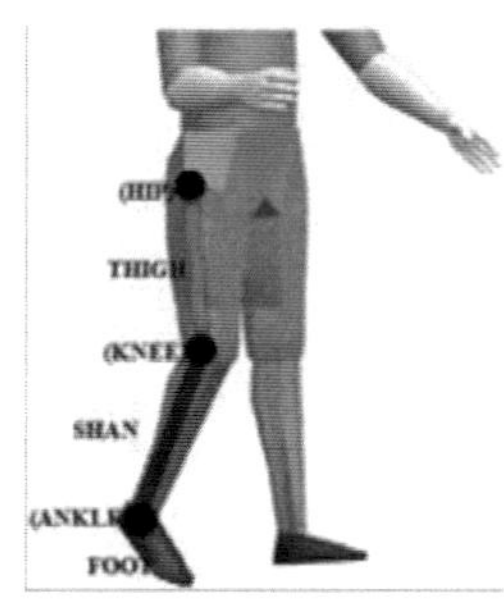

Figure 2. Parts of the lower limb of a human being, showing the limbs (thighs, shank and foot) and joints (hip, knee, and ankle)

1.1. Human lower limb and gait

The human lower limb consists of four parts, namely the hipbones, thigh, shank, and foot, connected through the hip, knee, and ankle joints, as shown in Figure 2. Each joint is unique and allows different types of motions, actuated through the muscles. These joint motions are studied along biological planes, which hypothetically divide a human body into different segments. The planes are named sagittal, coronal, and transverse planes. The movement of limbs about different joints and along different planes are tabulated in Table 1.

Table 1. Lower limb joints and their motions.

Movement	Movement	Angle from ref (°)[a]	Range of Motion (°).
Hip	Flexion	130	140[b]
	Extension	-10	
	Abduction	-40	70
	Adduction	30	
	Lateral (external) rotation	-40	70
	Medial (internal) rotation	30	
Knee	Flexion	-140	140[b]
	Extension	0	

	Lateral rotation	-30	40
	Medial rotation	10	
Ankle	Dorsiflexion	30	80[b]
	Plantarflexion	-50	
	Inversion	-35	55
	Eversion	20	
	Pronation	-30	90
	Supernation	60	

[a] Reference is a Vertical imaginary line passing through limb joints, staying perpendicular to the ground.
[b] Sagittal plane.

The walking cycle of a human being is called the 'Human Gait Cycle'. Conventionally, it starts with the human being's foot contacting the ground, to the subsequent contact with the ground during locomotion. Researchers use angular and torque sensors to measure the joint angles and torque at each joint during gait, and the data hence published is called the Clinical Gait Analysis (CGA) data [13] [14]. This data is used in the design of exoskeleton systems to know the range of motion and force/ torque outputs required from each actuator. The joint angles and torque are measured w.r.t % gait cycle. A sample plot of CGA data is sown in Figure 3 [13]:

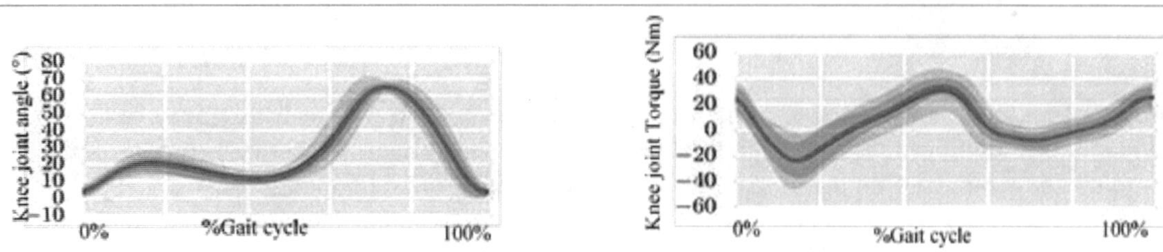

(a) (b)

Figure 3. The plots of **(a)** knee joint angle (°) against % gait cycle and of **(b)** knee joint torque (Nm) against % gait cycle provided by Moore et al. [13].

In most lower-limb exoskeleton designs, it is seen that the gait motion along the sagittal plane alone is considered, as most of the human gait motion occurs along with it. This approach simplifies the design, requiring lesser number of actuators, and hence costs much lesser. This same approach is considered here. However, it is noted that many designs do not allow limb motion out of the sagittal plane. This issue is of significant concern. Human gait varies significantly between each individual and may have out of plane motions, arresting which can cause serious injuries.

An essential requirement in the exoskeleton design is to understand human anthropometry. The limb weights and limb lengths are essential in deciding the length of linkages and for selecting actuators. The limb lengths and weights are represented as percentages of the individual's height and body weight, respectively. The approximate compilation of data on human limb weights and lengths is provided by Aydin Tözeren's work [15]. The approximate range of joint angles are obtained from the CGA data.

2. Research on rehabilitation exoskeleton systems

Research on novel exoskeleton designs has been a hot topic, whether for rehabilitation or for supporting workers' heavy loads. It is seen from literature that in the design of exoskeleton systems, the design of the structure is part of the whole design procedure. The other significant aspect is to design a control system, which is a challenge in itself. In the present work, more prominence is given to the design of the structure with a simple control system that can be upgraded later.

Chen et al. [16] designed a device, which is a modular knee-ankle-foot portable anthropometric rehabilitation exoskeleton. Series Elastic Actuators (SEA) were designed to provide mean as well as peak torques. The exoskeleton modules were designed to be used separately or together. It is seen that the SEA used increased cost as well as complexity of the device, with the motor used in SEA being considerably expensive. Wu et al. [17] developed a 3-DOF Lower Limb Rehabilitation (LLR) robot for patients' motion recovery, using a brushed DC motor with an encoder as the actuator. An adaptive sub-controller was designed to mitigate the errors induced by friction and disturbances from the patient. Emergency stop switches were provided for safety of patient. Though the link lengths were kept adjustable, the system used, utilizing screws, was discontinuous, causing misalignments between the device's joints and that of the human, resulting in discomfort. Accoto et al. [18] designed an exoskeleton called LENAR, which is non-anthropometric. Use of the non-anthropometric arrangement provides not only a large number of linkage options to select and optimize, but also helps to mitigate the joint misalignment issue. It ensures less don and doff times, and the adjustable link fits different users' anthropometry. Maxon brushless DC motors were the actuators used, with required encoders and controllers. The difficulty in this work arises due to the synthesis of non-anthropometric linkages, which are not as straightforward as their anthropometric counterparts. Hence, it is a necessity more than anything to develop an economical alternate powered exoskeleton for the needy, as physiotherapy and rehabilitation costs for post-medical treatment can push families into economic crisis [2].

3. Materials and methods

In the present work, the intention is to design an exoskeleton system that fits the requirements of the Indian population. There should be a trade-off at places to keep the costs low, but that should not hamper the functionality in any way. This approach poses inherent issues such as the gait data being unique to each individual, joint misalignments causing discomfort, inherent limitations of the modular and unified design approach, user safety, system operation without expert intervention, etc. It is hence required to consider certain requirements in the end product. The device should be a low cost, portable system, which fits different patients, with different anthropometry. The parts designed should be modular and interchangeable which will be used in uniped or biped configurations. The product should be bought off the shelf with no customisation to the individual, and should have minimum maintenance costs, if any. Don and doff time should be kept low. The passive joints are kept passive (free to move) to avoid any injury or discomfort to the wearer. Crutches will be provided with the device to avoid falls and for the patient's psychological benefit. The device will also need relevant sensors and control system, with an emergency stop button provided to the patient or the caretaker, in case of any discomfort.

The first step in designing the exoskeleton system is to select and evaluate the kinematic linkages that is to be used in the present work. The actuators that will be used was initially selected as pneumatic actuators, but due to issues discussed later, electric linear actuators is selected to be used. Interchangeable, modular links were designed and assembled to form the exoskeleton assembly in a CAD software. The model is tested for the necessary load condition and kinematical operation. A simple control system with necessary components is also designed. These topics will be discussed in detail under further headings.

3.1. Linkages of the mechanism

The linkages that is to be used in the mechanism is first selected. The listed linkages are modelled mathematically, to map the desired output, i.e. force and stroke length of actuator used to the known input. These models are evaluated mathematically, and the optimal linkage for each joint is selected.

3.1.1. Kinematics and kinetics of mechanism- Synthesis. The linkages that attach to the human limbs are either anthropometric or non- anthropometric; in a sense, it can mimic the human limbs' lengths and joints, or it may not. The latter design is challenging to synthesize, as was in the case of LENAR

IOP Conf. Series: Materials Science and Engineering 1132 (2021) 012008 doi:10.1088/1757-899X/1132/1/012008

[18]. However, it avoids discomfort to patients in case of misalignments. If a rotary actuator is to be used, the mechanism is straightforward, as the actuator is placed directly at the joint and provides a rotary motion, but these motors are expensive. The difficulty arises when the actuator used is a linear one. Mechanisms to convert the linear motion of the actuator to rotary motion at the joint, are plenty. The most optimal linkage may be selected based on the joint's specific requirements such as space constraints, torque requirements, etc. In this regard, five linkages (including one with a rotary actuator) are shown in Figure 4, along with their simplified diagrams, created using SolidWorks software.

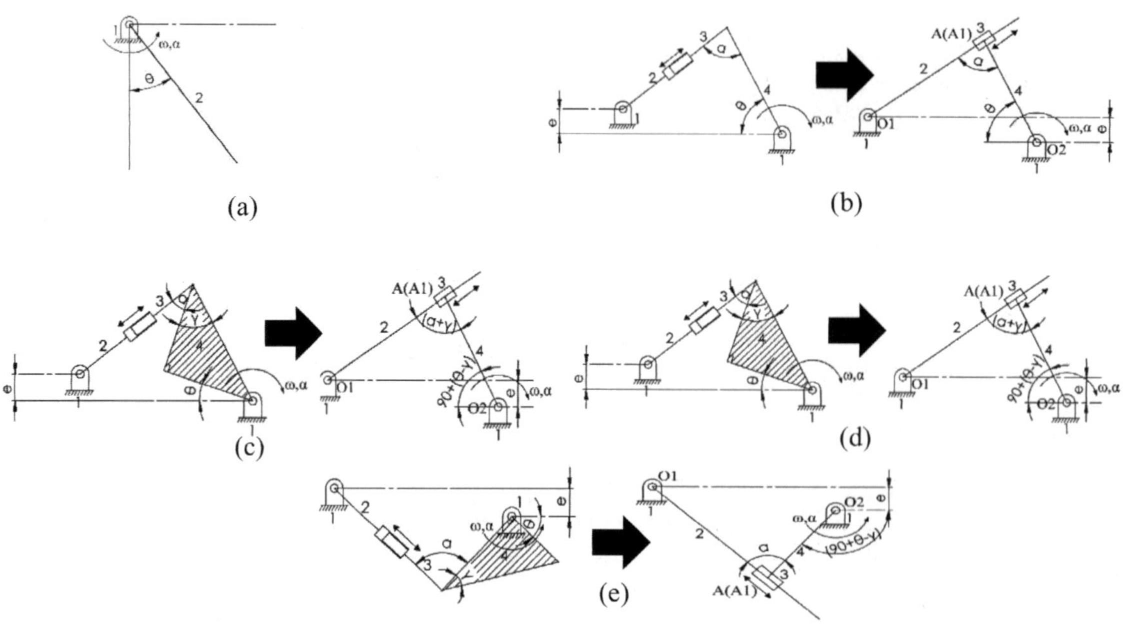

Figure 4. The kinematic linkages, along with their simplified versions of linkages for linkages **(a)**, **(b)**, **(c)**, **(d)**, **(e)**. Linkage (a) uses a rotary actuator placed in line with the biological joint. Drawings were created utilizing SolidWorks software.

The important criterion in selecting a linear actuator, be it pneumatic or electric, is to specify its force and stroke length. However, the torque and angle at the joint are the known quantities. By using principles of trigonometry, equations are derived for the force and displacement of the actuator, knowing the torque requirement and angular displacement at the joint. The stroke length is obtained indirectly. For the extreme values of joint angle θ, the values of a, which is the distance from the hinge point of the rear end of the linear actuator to the front end mounting point of its slider, is calculated, and the difference of which gives the stroke length (s) required. This data helps to select a linear actuator of required specifications. The formulae of a, F, and other vital parameters are tabulated in Table 2.

Table 2. Lower limb joints and their motion.

Kinematic linkage	a[a]	F	Other equations
a[b]	a→θ	F→τ	-NA-
b	$\sqrt{\left(b-\frac{e}{tan(\theta)}\right)^2+\left(h-\frac{e}{sin(\theta)}\right)^2-2\left(b-\frac{e}{tan(\theta)}\right)\left(h-\frac{e}{sin(\theta)}\right)cos(\theta)}$	$\frac{\tau}{hsin(\alpha)}$	$cos(\alpha)=\frac{a^2+\left(h-\frac{e}{sin(\theta)}\right)^2-\left(b-\frac{e}{tan(\theta)}\right)^2}{2a\left(h-\frac{e}{sin(\theta)}\right)}$

c	$\sqrt{((h*sin(\theta-\gamma)+b)^2+((h*cos(\theta-\gamma)-e)^2}$	$\frac{\tau}{hsin(\alpha+\gamma)}$	$cos(\alpha) = \frac{a^2+d^2-(e-(c*sin(\theta)))^2-(b-(c*cos}{2ad}$
d	$\sqrt{(b-(h*sin(\theta-\gamma))^2+(e-(h*cos(\theta-\gamma))^2}$	$\frac{\tau}{hsin(\alpha)}$	$\alpha=(\theta-\gamma)+sin^{-1}\left(\frac{(b-(h*sin(\theta-\gamma))}{a}\right)$
e	$\sqrt{(-p*sin(\delta'+\theta)+b)^2+(-p*cos(\delta'+\theta)-e)^2}$	$\frac{\tau}{sin(\beta)}$	$p=\sqrt{c^2+d^2}$ $\delta' = \tan^{-1}\left(\frac{d}{c}\right)$ $cos\beta = \frac{b^2+e^2-a^2-e^2}{2.a.p}$

[a] Stroke length, s = $a_1 - a_2$, where a_x= value of a at θ_x
[b] For linkage (a), since a rotary actuator is used, directly, the required θ and τ are obtained from the input data.

3.1.2. MATLAB Results. The optimal kinematic linkage for a joint is selected using the equations discussed in Table2, evaluated for different input conditions using MATLAB software. Here, the parameters such as the length of the limbs and other geometric dimensions are kept constant, based on human anthropometry and other physical constraints. The value of joint angle (θ) and torque (τ) are also necessary for computations. Two approaches are taken in this regard. The first is to directly use CGA data, which gives the said variables as a function of the % gait cycle. The other is to vary the joint angle based on the range of angular motion desired at each joint and evaluate the torque using a simple relation. If the limb has a mass of m, length L and for a given joint angle θ,

$$\tau = m * g * L * cos\theta \quad (1)$$

Using both the methods, each type of linkage was evaluated for the three joints, and the results obtained included plots of *F* and *a* w.r.t %gait cycle, as well as the maximum values of force and stroke lengths. However, linkage (e) was tested only for the hip joint, as it was considered out of necessity when it was found that the other linkages would not fit at the hip joint due to space constraints. The CGA data were utilized from a data set prepared by the research of Bovi et al. [14], who has compiled a data set from a vast number of trials on healthy adults and has provided the mean and standard deviation values of the parameters against percentage gait cycle. The forces and stroke lengths' peak values obtained are tabulated in Table 3, and the rows which are in bold are the linkages that were selected to be used in the final design. A very similar process was also carried out to find the velocity and acceleration required out of the actuator's slider when the joint angular velocity and angular acceleration were provided, by drawing generalized velocity triangles.

Table 3. Summary of results of *F* and *a* for different joints and linkages

		Force (N)		Stroke (mm)	
	Linkage	CGA based	Limb weight-based	CGA based	Limb weight-based
b	Hip	261.2458	81585	35.7456	38.2510
	Knee	485.3332	674.7398	105.0907	108.7687
	Ankle	3532.2	558.8533	46.5096	47.8340
c	Hip	199.4437	408.5012	107.6324	109.9089
	Knee	**224.6899**	**139.9799**	**140.3408**	**131.6203**
	Ankle	744.6044	111.4417	64.3778	70.8693

d	Hip	195.3531	601.9770	90.6298	93.3078
	Knee	252.7633	175.9442	145.2216	140.4404
	Ankle	**1096.4**	**163.1545**	**38.7812**	**43.2377**
e [a]	**Hip**	**82.6601**	**171.5064**	**90.5703**	**92.5924**

[a] Linkage (e) evaluated only for the hip joint.
* Rows in bold are the linkages selected to be used in final design.

3.2. Exoskeleton actuators.

Actuators play a vital role in the exoskeleton system, as they provide the force required for locomotion. It was noted from literature that selection of an economical actuator is vital in keeping the costs low. In this regard, two types of actuators are discussed below.

3.2.1. Pneumatic cylinders. For reasons mentioned above, it is preferred to select a linear actuator and not a rotary actuator, even if it means the added challenge of controlling the angular rotation resulting from the linear motion from the actuator. The two linear actuators that were of interest were the pneumatic cylinders and the electric linear actuators. Hydraulic system, though better in terms of force output obtained, presents challenges such as cost, complexity, and weight. Pneumatic cylinder was the actuator of choice initially in the design procedure. To select a pneumatic cylinder, the parameters to be selected are the stroke and the bore. Stroke required is directly taken for each linkage arrangement, referring to Table 3 and bore can be calculated or referred from the catalog (of a commercially available pneumatic cylinder) for the required force and for a suitable pressure input. The pneumatic cylinder hence considered was a 25mm bore and 160mm stroke pneumatic cylinder. By knowing the speed required from the pneumatic cylinder, the required flow rate was calculated. For multiple cylinders connected to an air supply tank, the capacity required for a reasonable operation period can then be determined.

As shown in Figure 5, the primary pneumatic circuit consists of an air tank (reservoir), which connects to a 4/3 DCV (directional control valve). The other port of the DCV connects to an exhaust. The outputs of the DCV connects to the two ports of the pneumatic cylinder. Throttle valves are used to control the speed of actuation. The DCV is operated by a solenoid, which in turn is controlled through Arduino. Some inherent problems to note here are the high speed of actuation, which is very difficult to precisely control, higher pressure required for actuation, and noisy operation.

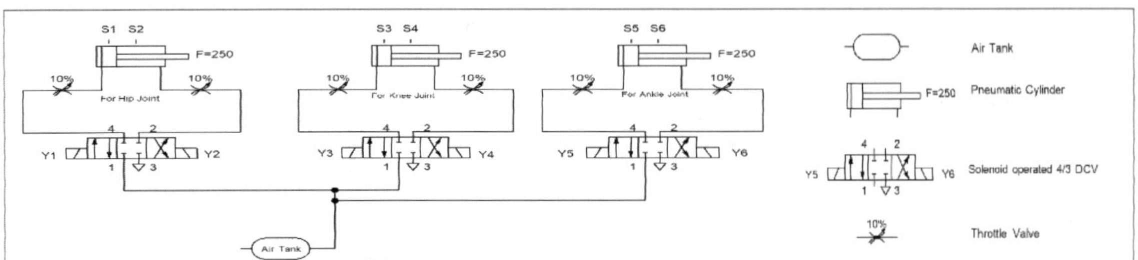

(a) (b)

Figure 5. (a) The pneumatic circuit for one leg, and **(b)** symbols used in the circuit. The circuit consists of three pneumatic cylinders, each operating a limb on one leg, either the thigh, shank, or the foot. A similar circuit will be used for the other leg, with three more cylinders. Illustration created using Festo Fluidsim software.

It was calculated that the cylinder's capacity required to operate the exoskeleton for one hour is around 100 liters, which is quite impractical, if not impossible. It is possible to use a mid-sized

IOP Conf. Series: Materials Science and Engineering 1132 (2021) 012008 doi:10.1088/1757-899X/1132/1/012008

compressor to keep the tank filled and operate the exoskeleton continuously, but those compressors are not portable. A low flow rate, small compressor can fill the reservoir of 25ltr in about 10 min, but keeping it filled at high pressure is not possible as these compressors run only for about 10 min at a time. These issues and impracticalities, along with the ones discussed earlier, makes pneumatic actuators a less feasible option.

3.2.2. Electric linear actuators. Electrical linear actuators are used in many industrial applications. They essentially are identical to the pneumatic cylinders when it comes to installation in the exoskeleton system and hence requires a minimum redesign of the device structurally, if any. They are directly powered from a 12V/24V DC power supply and can operate for long durations. These actuators comprise of a DC motor powering a lead screw with a gear train in between. A linear actuator providing around 500N force and a stroke of 150mm is selected for the present application. An issue with the linear actuators in general, is the speeds of actuation, as they are usually slow (nearly 7mm/s). However, models with faster actuation speeds are available at the cost of peak force while still meeting the force requirements of this design. The cost is higher than that of pneumatic cylinders but it is justified since the costs of reservoir, compressor, valves, and tubing are not involved. Hence, an electric linear actuator fits the requirements of the present design satisfactorily. A pneumatic cylinder and a linear actuator are shown in Figure 6. The linear actuator shown here is a Visesh Transmissions - LAG 01, with a load capacity of 50 kg (500N) and a stroke length of 150mm.

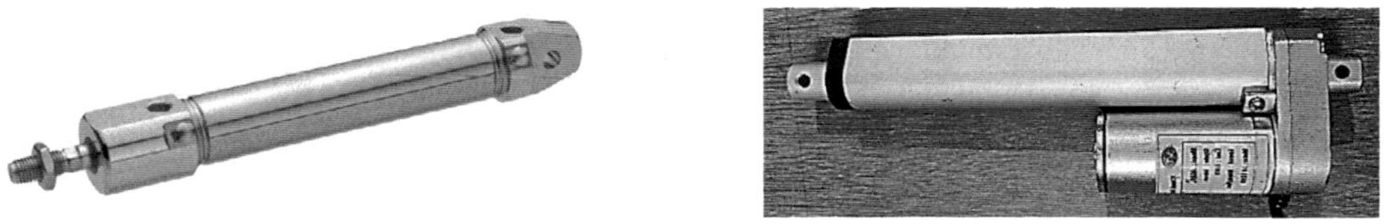

(a) (b)

Figure 6. (a) Pneumatic cylinder and **(b)** Electric linear actuator, LAG 01 from Visesh Transmissions.

3.3. Model of the rehabilitation exoskeleton

Bearing in mind all the discussions so far, the exoskeleton model is virtually built using a CAD software, Unigraphics NX 10. Aluminum fencing section (30*30, Al 6063) and aluminum box tube (1.5", square, 3mm wall) will be used to form a length-adjustable link. Aluminum fencing sections provide the advantage of easy and rapid fabrication and modifications. Mild steel and aluminum sheet metal will also be used. Modularity and interchangeability are maintained throughout the model. For the joints between links, rod end bearings (used in automobiles, like go-karts) are used to allow out of plane rotations.

3.3.1. Limb modules of the exoskeleton. A linkage that fits parallel to the human limb, with all the parts and actuators required to operate it, is called a limb module. It is so designed as to have minimal differences between each module and to keep the modules interchangeable. Each module of the exoskeleton essentially consists of a length-adjustable linkage, with mounting points for the rod end bearing. An L-shaped extension is attached to the linkage, in different configurations, based on which limb it is used on. There would be a provision to attach the rear end of the actuator onto the linkage. The actuator's moving end is attached to the L-shaped extension, with rotation allowed at both ends of the actuator. The limb module is shown in Figure 7(a). Aluminum is chosen as the material for all parts due to its lightweight and ease of fabrication where fencing sections are used. The module was subjected to structural analysis using Ansys structural, based on the load conditions expected on it, and was found to be safe. The Von-Mises stress on the module is shown in Figure 7(b).

IOP Conf. Series: Materials Science and Engineering 1132 (2021) 012008 doi:10.1088/1757-899X/1132/1/012008

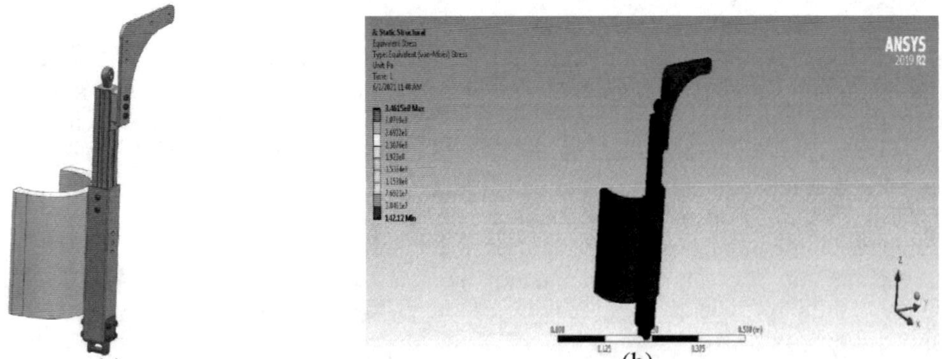

(a) (b)

Figure 7. (a) Limb module of exoskeleton and **(b)** Analysis of limb module of exoskeleton under expected load conditions, shows no failure.

Various concepts have been considered for the design of the limb module. Different parameters were changed between each concept to select the optimal module configuration, the parameters being the type of actuator, hinge, and method of adjusting the link length. A total of ten such concepts were considered. They were systematically evaluated, and a concept that uses a linear actuator, connected with a simple hinge, and having a length-adjustable link, made with an aluminum fencing section sliding in an aluminum box tube has been selected as the most optimal design, as shown in Figure 7(a).

3.3.2. Exoskeleton assembly. In the assembly of the exoskeleton system, as mentioned earlier, kinematic linkage (e) is used at the hip joint, linkage (c) at the knee joint, and (d) at the ankle joint, demanding different arrangements of the L-shaped extension. An adjustable back support system is designed to support the patient's as well as battery's load and provide attachment points to the limb modules at hip joint. The back support system and the limb modules are attached securely at respective locations using belts with adjustable tension. A bag pack is provided, which houses the battery and the control circuitry. The assembled exoskeleton system is shown in Figure 8.

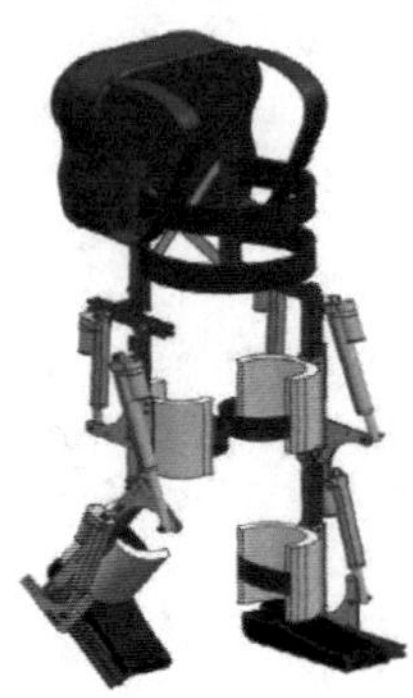

Figure 8. Exoskeleton system assembly. The limb modules, in the configurations discussed, are assembled into two sets, one for each leg, and are assembled onto the back support.

To validate if the mechanism can perform the required kinematical task of walking, Unigraphics NX 10 also provides a 'motion' workspace, where such a kinematic analysis is carried out, with the prior knowledge of CGA data and the corresponding value of a of linear actuators. The system provided a reasonable walking motion, as shown in Figure 9. The product is to be fabricated using the above mentioned materials. It was aimed throughout the design process to keep the fabrication process simple.

3.4. Control system and power supply

3.4.1. DH modelling. One of the crucial requirements of a sound control system is to accurately position the end effector, in this case, the foot, at any point in its workspace. This positioning is

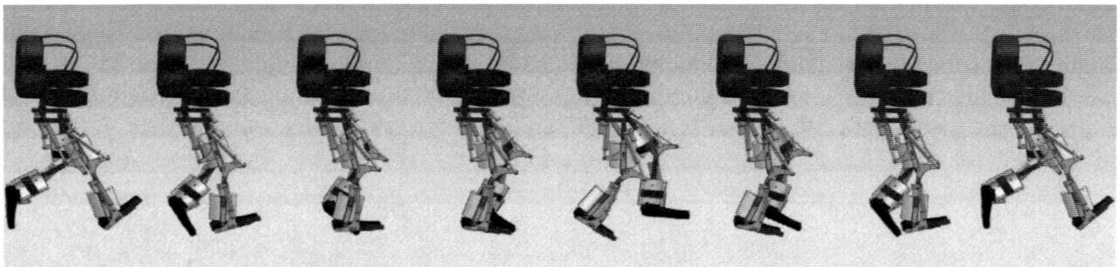

Figure 9. A walking sequence of the exoskeleton, simulated in NX 10 Motion Simulation.

accomplished by Denavit–Hartenberg (DH) modelling. The simplified model shown in Figure 10(a) is used to obtain a DH model, and a parameter table as shown in Figure 10(b) and 10(c) respectively, and the transformation matrix ${}^{0}_{3}T$ is obtained. This result is used to obtain the position matrix of point P in terms of the joint angles and using formulae from Table 2 in terms of actuator extension as well.

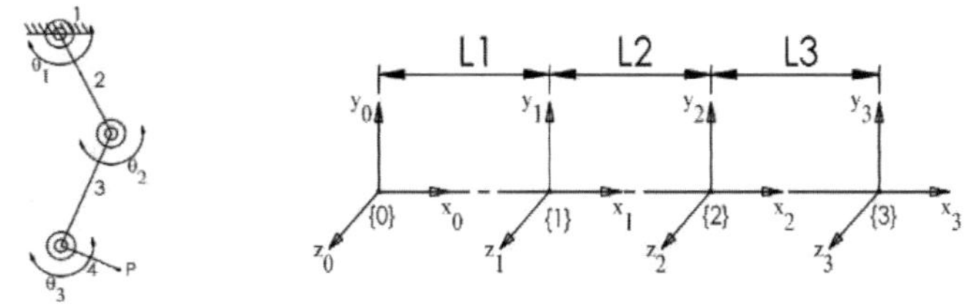

DH Parameters

Link	θ	d	a	α
1	θ_1	0	L_1	0
2	θ_2	0	L_2	0
3	θ_3	0	L_3	0

(a) (b) (c)

Figure 10. (a) A simplified model of a human leg, **(b)**Denavit–Hartenberg (DH) model, and **(c)**DH parameter table. Drawings created in SolidWorks software.

$$P = \begin{bmatrix} P_x \\ P_y \\ P_z \end{bmatrix} = \begin{matrix} L_3 \cos(\theta_1 + \theta_2 + \theta_3) + L_2 \cos(\theta_1 + \theta_2) + L_1 \cos\theta_1 \\ L_3 \sin(\theta_1 + \theta_2 + \theta_3) + L_2 \sin(\theta_1 + \theta_2) + L_1 \sin\theta_1 \\ 0 \end{matrix} \tag{2}$$

3.4.2. Power supply. For an application where the range of walking is minimal, such as while walking on a treadmill or testing the prototype, a Switch-Mode Power Supply (SMPS) of the required specifications may be used. According to the linear-actuator specifications, for a 24V power supply, current I = 3A. Since six actuators are needed to operate all the joints, and a current divider circuit is used, I = 6*3= 18A. A 24V, 18Ah battery is used, and an 80% efficient circuit is assumed. Time of operation of the actuator powered by the battery of capacity Ah ampere-hours, drawing current I, can be shown to be:

$$T = \frac{0.8\ X\ Ah}{I} \tag{3}$$

Substituting the known values, T comes out to be 48 minutes, which is a good duration for the device. The battery weighs around 10kg (5kg x 2), supported by the back support as discussed. Four 9Ah sealed lead-acid batteries may also be used, weighing a couple of kilograms less in total.

3.4.3. Control system. A simple, effective control system is to be designed for the device. The system should independently operate each actuator and verify if the desired angle is reached at the joint, which is set by the biological limits of that joint. The electric linear actuator is essentially a motor powering a lead screw through a gearbox. The speed can hence be varied by using Pulse Width Modulation (PWM). A potentiometer or an absolute rotary actuator is used at the joint to verify if the required angle is

IOP Conf. Series: Materials Science and Engineering 1132 (2021) 012008 doi:10.1088/1757-899X/1132/1/012008

reached. The motor power is cut off / reversed at this point. An Arduino Uno board is used to carry out the control, interfacing with the motor through a motor controller such as BTS7960 or through a simple relay board. However, the relay board cannot provide PWM. Load cells placed at the actuator front end measures the force output from the actuator. An emergency stop button is provided to the patient, which retracts the device to the neutral position and cuts off the power supply in the event of any injury or discomfort to the patient. Electromyography (EMG) sensors provide muscle activity information to the control system. A simple flowchart showing the operation of the control system is shown in Figure 11. A simple plywood mock-up to understand the system's operations, using Arduino, potentiometer, and a relay board was built and tested for different angular ranges. The mock-up is as shown in Figure 12.

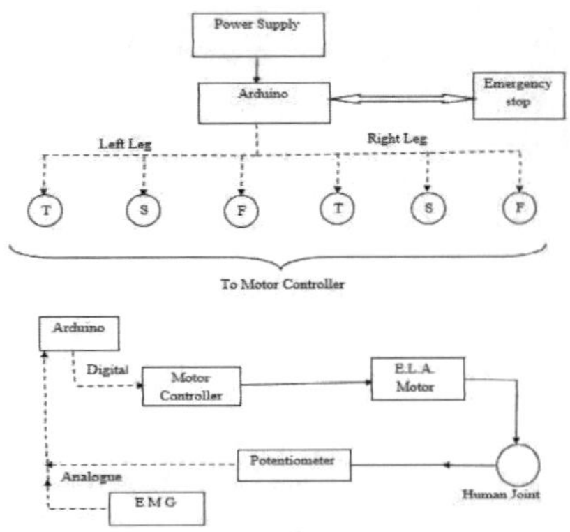

Figure 11. Simple control system for the exoskeleton system. Here, T, S, F, ELA, and EMG stand respectively for Thigh, Shank, Foot, Electrical Linear Actuator, and Electromyography.

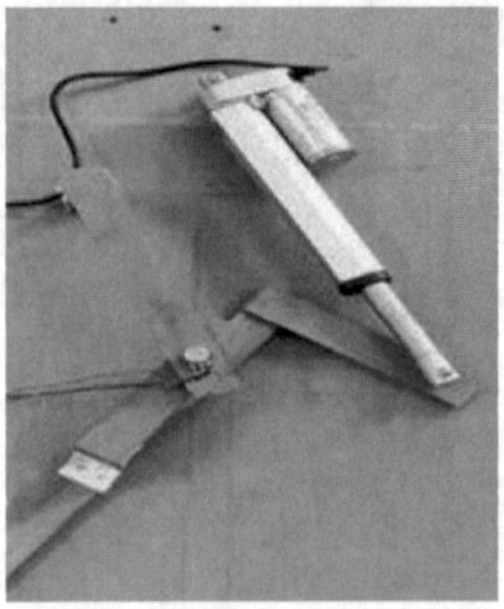

Figure 12. Wooden mock-up with an electric linear actuator and a potentiometer. The actuator is controlled by Arduino through a two-channel relay board. Also provided is an emergency stop switch.

4. Results and discussions

A rehabilitation exoskeleton system that is economical in comparison to existing products is designed, considering the needs of the Indian population. The product is modular, adjustable, with interchangeable parts, and uses electrical linear actuators. Further, it is intended to fabricate this design and test it in lab conditions and on human subjects. EMG sensors and PWM needs to be integrated into the control system. The modular and adjustable design poses few inherent limitations. It may not be possible to comfortably fit the device to each individual due to gait and anthropometry variations, which can cause joint misalignments. The motion of limb out of sagittal plane is allowed by the use of rod end bearings, but there is a limitation to that even then. Usage of linear actuators to produce an angular motion at the joint is not the most efficient way to produce and control such a motion, and can cause difficulties.

The design's nature provides opportunities for modifications in the future, such as changing the actuators, designing a much more robust control system, etc. One area to explore is to use pneumatic cylinders with compressed carbon-dioxide, where large quantities of the gas can be stored in a small container and hence it can operate the system for longer durations. The limitations have to be mitigated, and design modifications are to be explored in the future. Through this work, it is intended to make the exoskeleton technology accessible to those who need it the most.

5. Conclusion

An economic lower limb rehabilitation exoskeleton system was designed, considering the varying needs of the Indian population. Concepts of modularity and interchangeability was maintained throughout the design. Different possible kinematic linkages were identified, and their force and stroke requirements for given angular motion and torque at the joint were mathematically modeled and evaluated using

IOP Conf. Series: Materials Science and Engineering 1132 (2021) 012008 doi:10.1088/1757-899X/1132/1/012008

MATLAB software. Based on the understanding of the kinematics of different linkages, and the space constraints at each joint, an exoskeleton system model was built in Unigraphics NX 10. It was found that the electrical linear actuators were better than their pneumatic counterparts. The requirements of the battery, DH modeling, and a simple control system were also briefly discussed.

References

[1] Dandona R, Pandey A, George S,Kumar G A and Dandona L 2019 India's disability estimates: limitations and way forward *PLoS ONE* **14** 9.

[2] Sureshkumar K, Murthy G V S, Prost A, Subbulakshmy N, Pant H, Naveen C, Goenka S and Kuper H 2016 Rehabilitation needs of stroke survivors after discharge from hospital in India *Arch. Phys Med and Rehab* **97** 9.

[3] Kumar S G, Roy G and Kar S S 2012 Disability and rehabilitation services in India: issues and challenges *J. Family Med. and Prim. Care* **1** 1.

[4] Rossi S, Patanè F, Sette F D, and Cappa P 2014 Wake-up: a wearable ankle knee exoskeleton *5th IEEE RAS & EMBS Int. Conf on Biomedical Robotics and Biomechatronics* (São Paulo: IEEE).

[5] Pandian J D and Sudhan P 2013 Stroke epidemiology and stroke care services in India *J. Stroke* **15** 3 pp 128-34.

[6] Sridharan S E, Unnikrishnan J P, Sukumaran S, Sylaja P N, Nayak S D, Sarma P S and Radhakrishnan K 2009 Incidence, types, risk factors, and outcome of stroke in a developing country: the Trivandrum stroke registry *Stroke* **40** 4 Pp 1212–18.

[7] Nagaraja D, Gururaj G, Girish N, Panda S, Roy A K, Sarma G R K and Srinivasa R 2009 Feasibility study of stroke surveillance: data from Bangalore India *Indian J. Med. Research* **130** 4 Pp 396-403.

[8] Pandve H T and Patil S P 2020 Post covid syndrome: a challenge to address *J. Epidemiol. Public Health Rev.* **5** 4.

[9] Suraksha P Oct 17, 2020, Covid-19 'stroke' deals a cognitive blow *Deccan Herald* (https://www.deccanherald.com/science-and-environment/covid-19-stroke-deals-a-cognitive-blow-903046.html.)

[10] Jindal P, Macdermid J C, Rosenbaum P, Direzze B, Narayan A and Nayak S L 2019 Treatment and rehabilitation of children with cerebral palsy in India: a scoping review *Developmental med. & child neurology* **61** 9 pp 1050-60.

[11] Pal C P, Singh P, Chaturvedi S, Pruthi K K and Vij A 2016 Epidemiology of knee osteoarthritis in India and related factors *Indian J. Orthopaedics* **50** 5 Pp 518-22.

[12] Bhati P, Sharma S, Jain R, Rath B, Beri S, Gupta V K and Aneja S 2019 Cerebral palsy in north Indian children: clinico-etiological profile and comorbidities *J. Pediatric Neurosciences* **14** 1 pp 30-35.

[13] Moore J K, Hnat S K and van den Bogert A J 2015 An elaborate data set on human gait and the effect of mechanical perturbations *PeerJ* (eCollection - 10.7717/peerj.918).

[14] Bovi G, Rabuffetti M, Mazzoleni P and Ferrarin M 2011 A multiple-task gait analysis approach: kinematic, kinetic and emg reference data for healthy young and adult subjects *Gait & Posture* **33** 1 pp 6-13.

[15] Tözeren A 1999 Human Body Dynamics: Classical Mechanics and Human Movement (New york: Springer publications) p 302.

[16] Chen G, Qi P, Guo Z and Yu H 2016 Mechanical design and evaluation of a compact portable knee–ankle–foot robot for gait rehabilitation *Mechanism and Machine Theory* **103** pp 51–64.

[17] Wu J, Gao J, Song R, Li R, Li Y and Jiang L 2016 The design and control of a 3dof lower limb rehabilitation robot *Mechatronics* **33** pp 13-22.

[18] Accoto D, Sergi F, Tagliamonte N L, Carpino G, Sudano A and Guglielmelli E 2014 A lower limb non-anthropomorphic wearable robot *IEEE Robotics and Automation Magazine* **21** 4 pp 15-55

IOP Conf. Series: Materials Science and Engineering 1132 (2021) 012009 doi:10.1088/1757-899X/1132/1/012009

The Role of Boundary Layer Theory in Soccer Ball Dynamics

Abhijith Namboodiri Kapothanillath, Sidharth Pradeep, Shriram Sathishkumar and K Balaji*

Department of Mechanical Engineering, Amrita School of Engineering, Amrita Vishwa Vidyapeetham, Coimbatore, India

*k_balaji@cb.amrita.edu

Abstract. The boundary layer theory plays a major role in the development of fluid mechanics applications especially in the sports ball dynamics. Whenever an object travels through the air, it experiences aerodynamic forces acting on it. Sport balls, uses these forces to deviate from its expected path of flight. Thus, creating an advantage which can be exploited by the playing members to improve their skills. The present study focuses on numerical simulation of aerodynamic flow over a soccer ball and estimated the drag coefficient. In order to a get more precise results, the effect of gravity is included to study the dynamics of flow. The variation of drag coefficient is plotted as a function of Reynolds number and wind speed. The computational result had a good agreement with existing empirical data.

1 Introduction

Fluid dynamics is the science of movement of liquids and gases. Studying it allows us to understand them and the variety of concepts that come with it. Aerodynamics is a subset of fluid dynamics, dealing with the flow of air. Soccer is one of the most played and loved games in this modern era. In this paper, we discuss the interaction of the ball in flight with the flowing air. The drag coefficient (C_D) for different velocities is calculated and validated with experimental values. The aerodynamic properties of a football are a topic of research, which is lagging in its development, as the research carried out in this topic is few.

Experimental analysis done on soccer ball aerodynamics, [5], using a wind tunnel experiment, compared the coefficient of drag for different Reynolds numbers on both rotating and non-rotating balls. A hand-stitched soccer ball and two thermally bonded soccer balls were used for the experiments. A C_D value of 0.43 in the sub-critical region and ~ 0.15 in the super-critical region for a non-rotating soccer ball was obtained. And the critical Reynolds value was in the range of ~ 2.2–3.0 × 10^5. A drag crisis occurring in between the speeds of 12 - 15 m/s was also identified.

The effects of surface roughness on a stationary circular cylinder were discussed and the flow around it was numerically simulated [14]. By running different cases, the effect of surface roughness on C_D value was identified. The drag was very high, around 2.5 times higher than that of a smooth sphere. The effect of lifts was also understood.

The aerodynamic properties of 3 footballs, especially Jabulani, Teamgeist and Fevernova balls of different panel numbers were identified [1]. The C_D values were determined to be 0.15, 0.19 and 0.21 respectively. The Jabulani ball showed a late transition compared to the other two balls. The 32 panel Fevernova ball had an early transition.

The aerodynamics performances of 32 panel (with seams and stitch) and 14 panel (seamless) balls were evaluated [2]. The balls were tested at speeds of 20 km/h to 130 km/h. Their C_D values was determined and compared. The 32 panel Nike was found to be better than the Adidas, while it had more panels and an early transition to a turbulent boundary layer, which delayed its separation (100 degrees from stagnation). For the Adidas, due to a lower number of panels, the flow was mostly laminar. The ball acted as a smooth sphere, hence the boundary layer separated early at 90 degrees.

The wake vortex interactions of a curved plate were experimentally evaluated [8]. By changing the chord length to diameter ratio and the angle of flow the generation of wakes was observed for each case. It was inferred from this paper that on a soccer ball the wake region can be deferred due to different velocities. Sometimes, the direction of wind in practical cases (flow angle) would impact the formation of the wake region drastically. The varying of the diameter and sphericity of the ball would also impact the generation of wake.

Recent study on fluid dynamics of soccer ball reported that the number of panels on the ball, their orientation and the changes in the panel shape have a major impact on the aerodynamic characteristic of the ball [9]. Changing the texture of the ball surface would change the aerodynamics of the ball. This was validated with the help of a wind tunnel experiment.

Numerically investigation the aerodynamic characteristic of a square section cylinder at low Reynold numbers was done [7]. The paper has shown C_D values for different corner ratios and concluded that the most effective configuration is of a 0.2 corner ratio which gives a percentage drag reduction of 21.49% along with a very high Strouhal number (0.17).

The rotational and transverse oscillation of a square plate and conducted numerical simulations was discussed [12]. The results were validated using mean drag coefficient for both cases from literature. The generated pressure contours were explained and the correlation between frontal area, vortex patterns and drag coefficient were identified. The influence of Re on C_D was also observed. Thus, this paper looked into the major differences when an object starts to rotate.

An experiment to find out the differences between 4 modern soccer balls and how these differences affected the trajectory of the ball was done [13]. Wind tunnel tests were used to carry out experiments and analyse the four balls. The Adidas Terrapass had optimal sphericity, thereby reducing its form drag. The Nike t90 omni epl had a good design and a panel geometry of hexagons, thereby reducing its drag at low speeds due to an early transition. The Puma v108 had 24 panels and an improved sphericity. The Adidas Terrapass replica is the replica of a Terrapass ball. It was observed that among the four balls, the Puma was better at lower velocities, showing lower C_D values due to an early transition (low critical Reynolds number) and the Terrapass had a late transition, which makes it good at higher velocities (high critical Reynolds number).

A wake is formed downwind of any object exposed to flowing air. The study of this wake can help us to understand various aerodynamic effects such as the drag. In the computational analysis of flow past a curved surface, it was found that by adjusting and introducing curve radius and the orientation of the surface, a C_D reduction of 58% was achieved [3]. The drag produced on a cylinder due to the presence of a notch was computationally analysed and its flow visualisation was studied [4]. The flow was simulated for various angles of the notch with respect to the flow of air. The C_D values were observed to increase till 30° and then decrease till 45° and then finally increase again till 140°.

Most of the research works were experimental and these papers were referred to understand the behaviour of the ball during its flight. But, setting up an experimental set up is not an easy task. Formulating a computational setup is easier. It is observed here, how the coefficient of drag changes with the structure of the ball, surface roughness and the varying velocity and was also able to observe the transition of fluid flow from laminar to turbulent (during critical Reynolds number) for different types of soccer balls.

The two basic equation that are solved are the continuity and the Navier-Stokes equations. The continuity equation is derived from the statement that the mass entering the system is equal to the mass leaving the system. Thus, the continuity equation would be equation 1. As the velocity of air flows never exceeds Mach 0.3, the flow is considered to be incompressible. We also assume that density doesn't change with time. Hence the continuity equation would eventually become equation 2.

$$\frac{\partial \rho}{\partial t} + \nabla \cdot (\rho \mathbf{u}) = 0 \quad (1)$$

$$\frac{\partial u}{\partial x} + \frac{\partial v}{\partial y} + \frac{\partial w}{\partial z} = 0 \quad (2)$$

The next equation, Navier – Stokes equation, is actually a set of partial differential equations. This combines the conservation of mass and conservation of momentum equations with Newton's second law of motion. It is a very sophisticated equation combining the density, pressure, temperature, viscosity, kinetic and potential energies of a fluid. The NS equation is given in equation 3.

$$\frac{\partial}{\partial t}(\rho \boldsymbol{u}) + \nabla \cdot (\rho \boldsymbol{u} \times \boldsymbol{u}) = -\nabla p + \mu \nabla^2 \boldsymbol{u} + \rho g \quad (3)$$

The Coefficient of Drag (C_D)is a dimensionless quantity which quantifies the resistance to fluid flow (drag) over a surface. This value is dependent on two types of drag – form or pressure drag and the skin friction drag. The drag coefficient and the drag force can be related by the equation 4.

$$C_D = \frac{2F_d}{\rho u^2 A} \quad (4)$$

The boundary layer is a region of fluid very close to the surface where the effects of viscosity are significant. The flow properties are determined by the boundary layer thickness (δ) which is the distance from the solid object's surface to the fluid whose velocity is 99% of the free-stream velocity. The thickness depends on the viscosity, velocity and the flow profile (laminar or turbulent) – essentially the Reynold's number $\left[Re = \frac{\rho v D}{\mu} = \frac{vD}{\nu}\right]$. A fluid flow with a very small Reynold's number will have a large boundary layer thickness compared to the flow with a large Reynold's number, which will have a small boundary layer. Also, a turbulent flow has a larger boundary layer thickness than a laminar flow. The boundary layer theory bridges the gap between the Euler equation (slip) and the Navier-Stokes equation (no slip condition). Using the Order of Magnitude analysis for the Navier Stokes equations: Continuity equation $\left[\frac{\partial u}{\partial x} + \frac{\partial v}{\partial y} = 0\right]$ and the Momentum equation $\left[\rho\left(\frac{\partial v}{\partial t} + (v \cdot \nabla)v\right) = -\nabla p + \mu \nabla^2 v\right]$, the Boundary layer equation and the equation for boundary layer thickness can be determined.

Boundary Layer Equation: $$u\frac{\partial u}{\partial x} + v\frac{\partial u}{\partial y} = U_\infty \frac{\partial U_\infty}{\partial x} + \nu \frac{\partial^2 u}{\partial y^2} \quad (5)$$

In this paper, our computational drag coefficient, C_D and drag force was compared with the experimental results obtained [2]. The experimental setup consisted of a ball being attached to sting with the help of adhesives. The sting had a JR-3 multi axis load cell (a 6 degree of freedom - force and torque sensor). RMIT Industrial Wind Tunnel was used. A computer was used to simultaneously interpolate the 3 forces (lift, drag and side) and their respective moments.

2 Methodology

The computational approach is used to solve for the physical equations necessary to calculate the aerodynamics of the flow across the football.

2.1 Geometry

Specification:

Ball Diameter : 200mm
Number of Panels : 32
Stitches around its surface

Figure 1 Football model
Src. Raj Gupta, GrabCAD

2.2 Pre-processing

The computations for the flow over a football was done through Ansys Fluent 2020 R2. The geometry for the football was taken from GrabCAD (figure 1). The control volume was extruded over the ball in Autodesk Inventor. The control volume was a cube of dimension 1m. Then Combine was used to

IOP Conf. Series: Materials Science and Engineering 1132 (2021) 012009 doi:10.1088/1757-899X/1132/1/012009

subtract the football from the control volume. This geometry file was imported into Ansys Mechanical to generate the mesh. Due to computational limits, we limited the size of the mesh to 200 mm. This gave the total number of cells to 464821 elements. Figure 2 shows an isometric view of the mesh domain. The figure 3 gives a more closer sectional view of the mesh near the football.

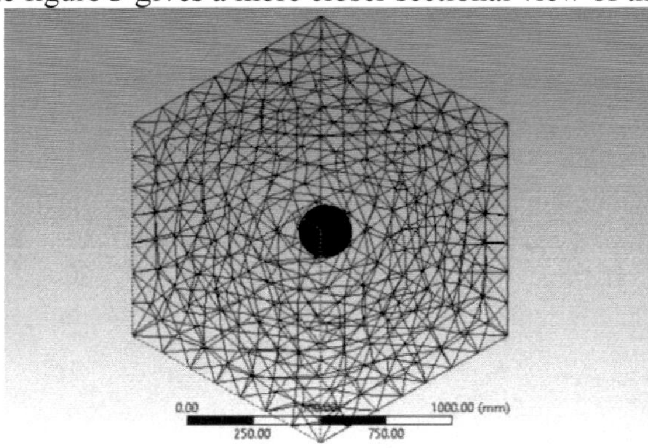

Figure 2 Control Volume Mesh

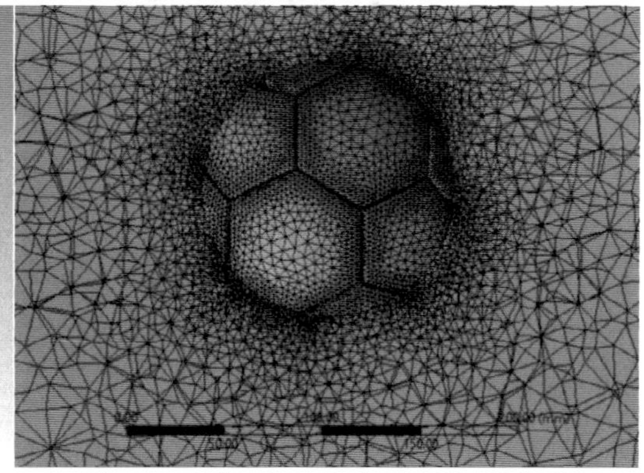

Figure 3 Sectional view of the Mesh

2.3 Setup
After establishing and checking for the mesh quality, now we have to start defining the conditions and set up other solver models which will define the method of solving the physical governing equations. Here, we have used a pressure-based solver as the flow is incompressible and transient state solver was employed to see how the flow field evolves with respect to time. A transient state (un-steady) simulation was done since an unsteady simulation best demonstrates the transition of laminar to turbulent flows and the characteristics of turbulent flows. Hence, transient state simulations were done instead of a pseudo-transient case in a steady state solver. We have chosen a standard k-omega model with standard wall treatment function to account for turbulence in the flow and as already discussed earlier, at high Reynolds number and due to seam orientation, the boundary layer separation reaches past the critical regions and results in turbulent flow after the flow separation point. This model also provides better results for observing free-shear layers and wake regions. Using this model will help in predicting the motion of an object by studying its wake region.

2.4 Boundary Conditions
Boundary conditions are what defines our problem. These conditions are given by us on the exterior of the mesh surfaces as well as to the interior region. This helps the solver to understand what problem we are trying to solve. A slight mistake in the boundary conditions could lead to very different solutions, solutions that may be in its entirety - wrong. The frequently used boundary conditions in Ansys Fluent are inlet, outlet, wall, symmetry and axis among others. Our inlet condition was given to the face facing positive Z-axis and the outlet facing the negative Z-axis. The other four surfaces were given as walls with specified shear as 0. The inlet velocities simulated were 7, 14, 25 and 36 m/s. The turbulence percentage was 5% with a turbulent viscosity ratio of 2. The outlet was given a pressure of 0 Pa. The internal pressure was maintained as 101325 Pa. The boundary conditions used are tabulated in table 1. The figure 4 gives a labelled view of the control volume. These velocities were chosen from the velocities over which the experimental studies were done. The transition point in the experiment was observed between 14 and 20 m/s. The experiment began at 5.55 m/s till 36.11 m/s. Hence the velocities 7, 14, 25 and 36 m/s were chosen.

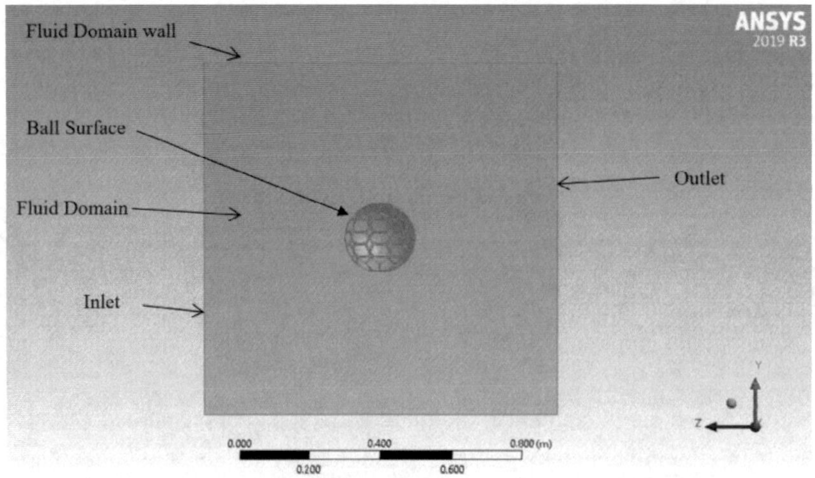

Figure 4 Mesh

Table 1 Boundary Conditions

Solver Type	Transient Pressure based
Viscous model	Standard K -omega
Inlet Air Velocity	20-130 km/h (Along Z axis)
Outlet Wall	1atm
Outer Wall	Specifies shear = 0
Ball Surface	Stationary wall
Roughness Coefficient	0.5
Roughness Height	0 m
Gravity	-9.81 m/s^2 (y - axis)
Turbulence intensity	5%
Turbulent viscosity ratio	2
Control volume dimensions	1000 x 1000 x 1000 (mm)

2.5 Solution

After the initial steps, the mesh cells are initialised using hybrid initialisation. In the reference values, the projected area of the ball, its diameter and the velocity were given as reference values in order to aid in the calculation of drag values. The equation was solved by Coupled model, with first order upwinding schemes for solving momentum, turbulent kinetic energy, turbulent dissipation and energy equations. The solution was done for a timestep size of 0.05 s for 10 timesteps. A timestep size of 0.05 s was chosen after experimenting with different timestep size values, taking into account the computational limitations (time taken for simulation). The number of iterations for each timestep was capped at 200 iterations. The residual values for the continuity equation, x, y and Z velocities were given as 0.0001. A report was set up to calculate the Coefficient of Drag and the Drag force along the negative Z axis. The values were also plotted against the time step. The latest time step values were taken into consideration for the results. The boundary layer was fully resolved at higher velocities, since the layer became fully turbulent, thereby delaying the separation. Near the boundary layer, the mesh is uniform.

3 Results

The drag coefficient C_D and drag force acting on the football over different speeds were computed. Values were obtained for 7, 14, 25 and 36m/s. These were validated with the experimental data obtained from [2]. The findings were presented in the graphs. (figures 5 and 6)

IOP Conf. Series: Materials Science and Engineering 1132 (2021) 012009 doi:10.1088/1757-899X/1132/1/012009

The obtained data closely resemble the experimental data. From the graphs (figures 5 and 6), the transition from laminar to turbulent was observed at Re ≈ 200000, which is close to the experimental value ≈ 200000. While computing, the C_D value was found to be 0.245 at the transition state. The drag coefficient was found to be maximum at Re ≈ 100000 which is 0.35 and was minimum at Re ≈ 50000 with 0.224. The Coefficient of drag decreases as the Reynolds number increases because when it becomes more turbulent the pressure drag will get reduced and at the transition period there will be an abrupt reduction in C_D. But after that there will be a slight increase due to the viscous drag.

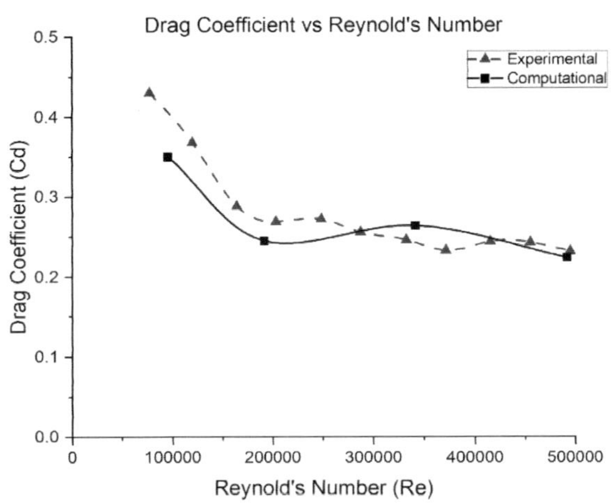

Figure 5 Drag Coefficient vs Reynold's Number

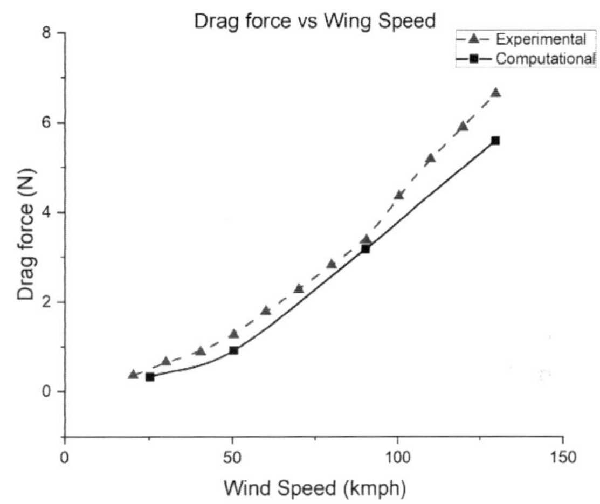

Figure 6 Drag force vs Wind speed

The computed drag force is closely following the experimental values. This can be explained by the drag force formula where the drag force is proportional to the square of velocity. The maximum variation of the drag coefficient (figure 5) is 0.05 (≈ 12%). It is seen around Re = 100000. The maximum variation of the drag force (figure 6) was seen at 130 kmph (≈ 1 N) (≈ 15% change).

From Figures 7 (a), (b), (c), (d) it can be seen that the wake area following the ball is reducing in size as the velocity increases. This is due to the increased turbulence, which delays the flow separation over the ball. This reduces the pressure drag. For 14, 25 and 36 m/s speeds even though the pressure drag is decreasing the viscous drag is increasing which further increases the drag force. The wake for the first three conditions is travelling towards the -y axis. Due to the pressure difference between both ends of the ball, the ball experiences a force along +y direction. This can be seen by its wake travelling towards the -y axis. For the fourth condition (36 m/s), the wake is travelling towards +y direction. This indicated that the ball is experiencing a force in the -y direction.

Pressure reaches maximum at the stagnation point and as it reaches the aerodynamic shoulder pressure becomes minimum and velocity becomes maximum. And as the free stream velocity increases, the stagnation pressure also increases. This can be observed from figure 7 (Velocity contour plots) and figure 8 (Pressure contour plots).

The separation points for each velocity can be observed in the figures 9 (a), (b), (c) & (d). The separation points are progressing inwards towards the wake. For the first 3 cases the separation point is at the top half of the ball. For the last case (figure 9-d), it is at the bottom half. This inversion which occurs at the fourth case can be explained. In the first three cases, the pressure distribution causes the ball to move upwards (We can observe the asymmetric wake progressing downwards).

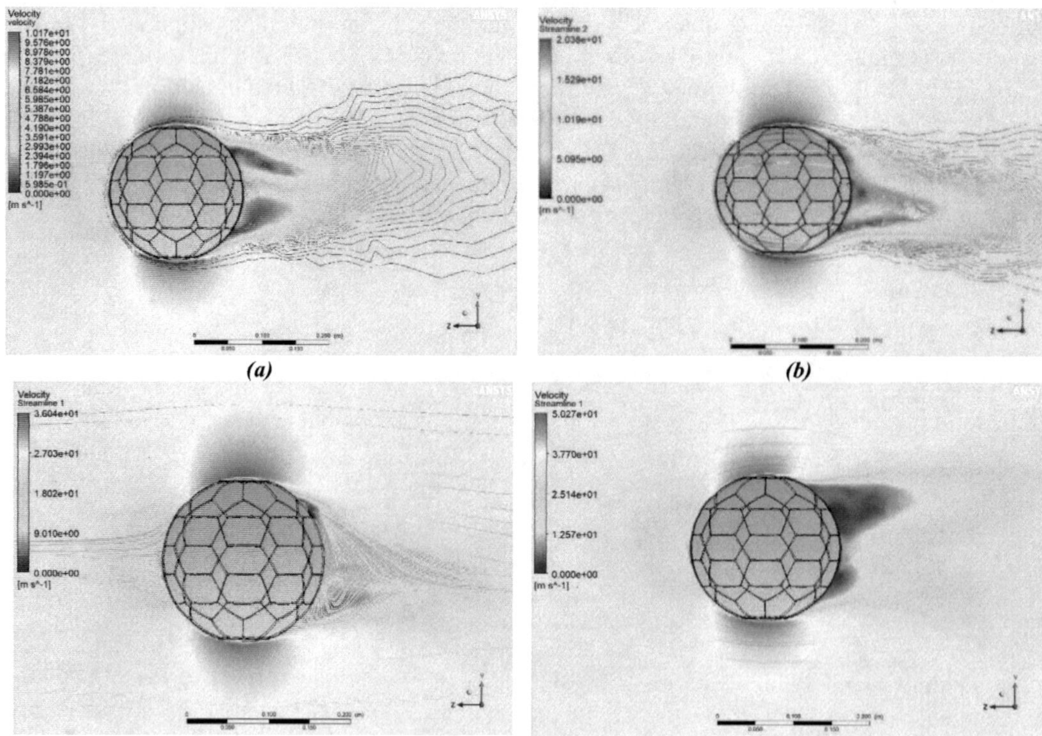

(a) *(b)*

(c) *(d)*

Figure 7 (a), (b), (c) & (d) - Velocity contours for free-stream velocities of 7, 14, 25 & 36 m/s

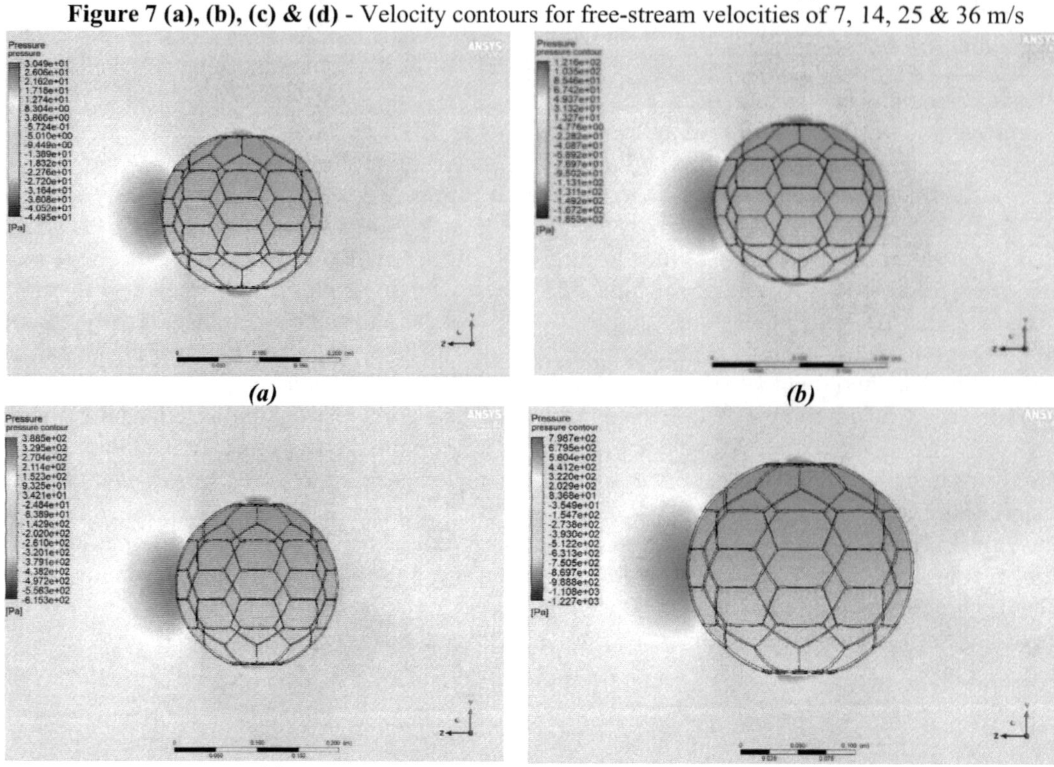

(a) *(b)*

(c) **(d)**

Figure 8 (a), (b), (c) & (d) - Pressure contours for free-stream velocities of 7, 14, 25 & 36 m/s

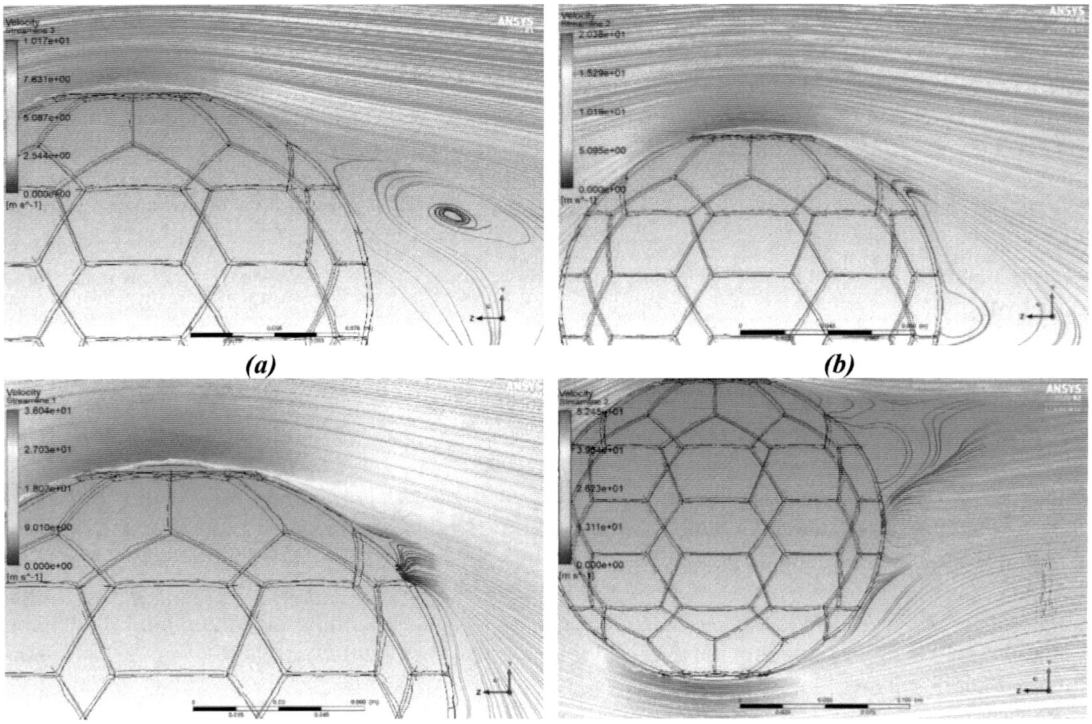

(c) (d)

Figure 9 (a), (b), (c) & (d) – Separation point for free-stream velocities of 7, 14, 25 & 36 m/s

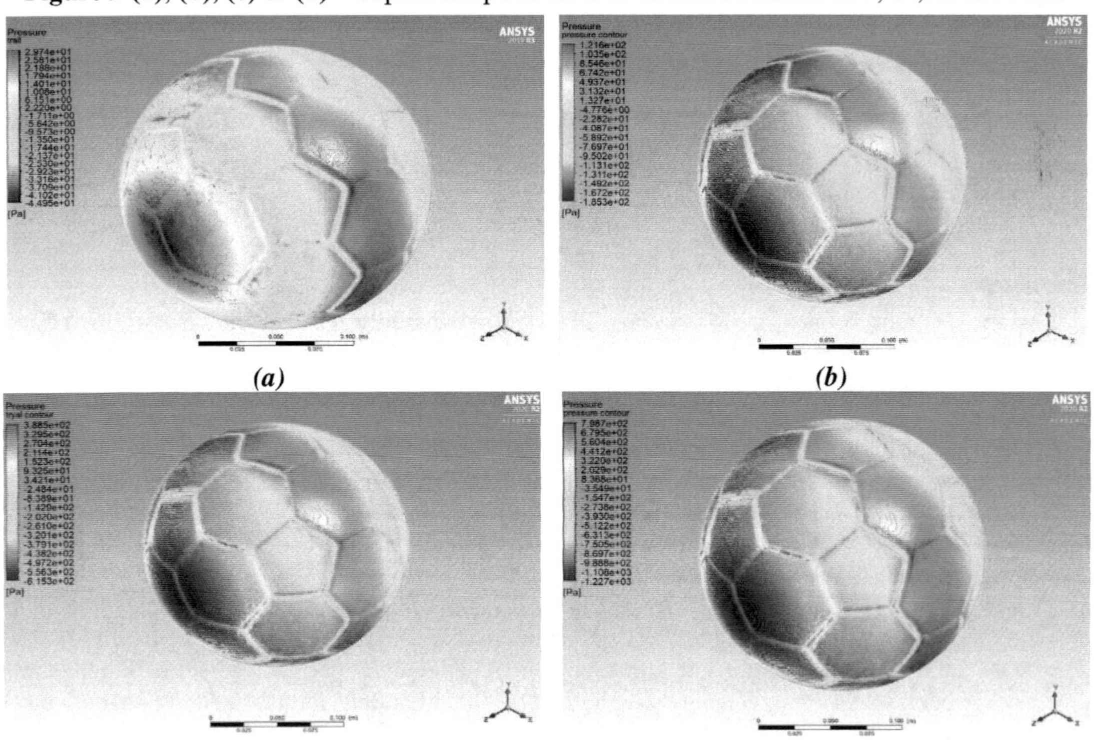

(c) *(d)*

Figure 10 (a), (b), (c) & (d) – Pressure distribution on the surface of the football for free-stream velocities of 7, 14, 25 & 36 m/s

In the fourth case (fig. 9-d), the pressure distribution inverts, causing the ball to move downwards and the asymmetric wake progressing upwards. This phenomenon is similar to the contrast swing achieved in cricket balls. Contrast swing is a type of swing which occurs when the seam of a ball is parallel to the direction of flight and the swing is achieved purely based on the difference in surface roughness between the two sides [10]. The orientation of the panels in the football, can cause difference in the overall surface roughness, causing the ball to swing.

The figure 10 shows the distribution of pressure over the surface of the soccer ball. The increase in stagnation pressure at the frontal point of contact of the ball with air can be seen with increasing wind speed as expected. The stagnation pressure is seen to increases from 29.74 Pa to 798.7 Pa for 7 m/s to 36 m/s of wind.

4 Conclusion

The computed results are validated using the experimental data. The Aerodynamic behaviour of the soccer ball was attempted to be understood. The drag coefficient is decreasing with increasing Reynolds number. The drag force is increasing as the velocity increases. With increases in fluid velocity the boundary layer separation point is delayed. The element size was maintained at 200 mm in order to make it computationally feasible as it was a large mesh. Better refinement of the mesh and access to high power computing we can expect to obtain better and more accurate results. Soccer is a very competitive sport today in Europe. Computational studies on the soccer ball can be used to study the dynamics of the ball. These studies can help a player to fully utilize these aerodynamic effects to his fullest advantage. Further study into these aerodynamic effects on the ball can help us better understand this process and implement it in the games.

4 Future study

The following modifications can be studied to further improve the aerodynamic characteristics of the football:

- Footballs with more than 32 panels.
- Curved edges between panels.
- At least 20 small dimples in a panel as in golf balls - this helps in early transition to turbulent flow, leading to delayed boundary layer separation.
- Increasing the sphericity of the ball.
- Rough material can be used for stitching to create turbulence.

References

[1] Alam, F., Chowdury, H., Moria H., Fuss, F.K., Khan, I., Aldawi, F., Subic, A. (2011). Aerodynamics of contemporary FIFA soccer balls. (Elsevier, Ed.) *Procedia Engineering, **13***, 188-93. doi:10.1016/j.proeng.2011.05.071

[2] Alam, F., Chowdhury, H., Moria, H., & Fuss, F. K. (2010). A Comparative Study of Football Aerodynamics. *Procedia Engineering.* **2**, pp. 2443-8. Elsevier Ltd.

[3] Anil, A., Arunkumar, K., Ajith kumar, R., Hariprasad, C. M., & Mani, T. (2019). Flow around curved plates at low subcritical reynolds number: Investigation of wake characteristics. In *Lecture Notes in Mechanical Engineering* (pp. 811-9). Pleiades Publishing. doi:10.1007/978-981-13-6416-7_76

[4] Arunkumar, K., & Sarath, R. (2021). Investigation of Flow Structures Around a Circular Cylinder with Notch: A Flow Visualization Analysis. In P. Pant, S. K. Mishra, & P. C. Mishra (Ed.), *Lecture Notes in Mechanical Engineering* (pp. 181-90). Bhubaneswar: Springer Science and Business Media Deutschland GmbH. doi:10.1007/978-981-15-7779-6_16

[5] Asai, T., & Seo, K. (2013). Aerodynamic drag of modern soccer balls. *Springerplus,* **2**(171). doi:10.1186/2193-1801-2-171

[6] Asai, T., Seo, K., Kobayashi, O., & Sakashita, R. (2007). Fundamental aerodynamics of the soccer ball. *Sports engineering,* **10**(2), 101-9. doi:10.1007/BF02844207

[7] Chandar, S. V., Sarath, R. S., & Ajith Kumar, R. (2017). Aerodynamic characteristics of a square section cylinder: Effect of corner arc. *2nd International Conference for Convergence in Technology,* (pp. 814-9). Pune: Institute of Electrical and Electronics Engineers Inc. doi:10.1109/I2CT.2017.8226241

[8] D. Murali, K. H. Balu Haridas, K. Arunkumar and R. A. Kumar, (2017) "Flow past a curved plate: Wake vortex characteristics," *2nd International Conference for Convergence in Technology,* Mumbai, 2017, pp. 318-324, doi: 10.1109/I2CT.2017.8226144.

[9] Hong, S., & Asai, T. (2017). Aerodynamic effects of dimples on soccer ball surfaces. *Heliyon,* ***3***(10). doi:https://doi.org/10.1016/j.heliyon.2017.e00432

[10] Mehta, R. D. (2008). Sport Ball Aerodynamics. In H. Nørstrud, *Sport Aerodynamics* (Vol. **506**, pp. 229-331). Vienna: CISM International Centre for Mechanical Sciences, Springer. doi:https://doi.org/10.1007/978-3-211-89297-8_12

[11] Nair, A. S., Kumar, S. A., Kumar, K. A., & Kumar, R. A. (2020). Aerodynamic Characteristics of a Square Cylinder: Effect of Dissimilar Leading Edges. In S. Revankar, S. S., & D. Sahu (Ed.), *Lecture Notes in Mechaincal Engineering* (pp. 189-97). Bhubaneshwar: Springer Science and Business Media Deutschland GmbH. doi:10.1007/978-981-15-7831-1_18

[12] Narayanan, B. A., Lakshmanan, G., Mohammad, A., & Kishore, V. R. (2020). Laminar Flow over a Square Cylinder Undergoing Combined Rotational and Transverse Oscillations. *Journal of Applied Fluid Mechanics,* ***14***(1), 259-73. doi:10.47176/jafm.14.01.30706

[13] Oggiano, L., & Sætran, L. (2010). Aerodynamics of modern soccer balls. *Procedia Engineering*, **2**, 2473-9. Elsevier Ltd.

[14] S. Iyer, K, A., R, A., and Dr. Srikrishnan A. R., (2017). Combined Effect of Surface Roughness and Wake Splitter Plate on the Aerodynamic Characteristics of a Circular Cylinder, *IOP Conference Series: Materials Science and Engineering*, vol. **234**, p. 012007.

IOP Conf. Series: Materials Science and Engineering 1132 (2021) 012010 doi:10.1088/1757-899X/1132/1/012010

Design, Analysis, Manufacturing and Testing of a SCARA Robot with Pneumatic Gripper for the Poultry Industry

Amal Prakash*, **Anikesh Rajendran, Pranav V K and Pramod Sreedharan**

Department of Mechanical Engineering, Amrita Vishwa Vidyapeetham, Amritapuri, India

*amalp@am.amrita.edu

Abstract. Soft robotics is replacing conventional robots in fields requiring interaction with fragile objects. Namakkal district of Tamil Nadu is responsible for over 90% of egg exports in India. Most of the eggs are processed and palletized manually which causes a lot of money on human labor. This paper discusses the design of a SCARA robot which uses asymmetric flexible pneumatic bellow actuator (AFPBA) to palletize the eggs to reduce the processing time and human labor. The design, analysis, manufacturing and testing details have been provided. The manufactured prototype was able to successfully lift 200grams without slippage.

1. Introduction

India is the third biggest producer of eggs in the world with almost 95 billion eggs being produced annually. That means that on a given day an average of 260.3 million eggs are processed. Almost 90% of the egg exports are done by the Namakkal district of Tamil Nadu. On conducting a study of different egg distributors in this district, it was seen that most of the expense in the industry was because of the need for a great amount of human labor to process the eggs. To prevent any kind of spoiled goods reaching the consumers, strict quality control is placed. Eggs need to be candled to check for any anomalies and the weight graded and palletized before produce gets to the market. Eggs being fragile objects, these processes are manually carried out by humans to prevent any damage. Given that a single person process approximately 5000 eggs a day it becomes quite difficult to carefully inspect each egg. Sometimes hairline cracks maybe overlooked which results in bad quality eggs to be exported. SCARA robots [1] can be employed in these scenarios to palletize the eggs for export. This reduces the time for processing and helps increase the export amount. But conventional hard robots fail to do these tasks. Eggs being fragile objects would break if any excess force was applied by the end effector. This situation calls for the development of a customized robot with a soft gripper to handle the eggs [2,3,4]. A SCARA robot can be used to perform the pick and place action for the palletization of eggs. This would increase the speed of production and reduce the amount of human labor required to these tasks. The works done by Ganesha Udupa et al. in developing an AFPBA made from hyper elastic materials [5-11] has been studied and modified to make a gripper that provided firm grasping and a good amount of bending.

2. Market analysis

In order to find out if the cost of implementation and maintenance of the SCARA is profitable compared to using manual labour, a study was conducted in the Nammakkal district of Tamil nadu. During the study with RDS egg distributors, one of the major distributors in the district, it was found that nearly 75000 eggs were processed at any given day. It requires 10-15 labourers a day to palletize these eggs. Given that the cost of each labour is approximately 750 rupees ($10). This gives us an approximate labour cost of 337,500 rupees ($4585). The cost taken to procure the components and build the robot and manipulators are 200,000 rupees ($2718).As the robot is made using cheap and commercially available components and has a low amount of moving parts, the maintenance of the robot is very low.

3. Design of the robot and actuator

3.1. Pneumatic actuator

For the manufacture of the pneumatic actuator, both nitrile rubber (Table 1, Figure 1a) and silicone rubber were good candidates as they provided good deflection as shown by the previous works done on asymmetric flexible pneumatic bellow actuator. The actuator has a semi-circular cross-section with a length of 10cm and eccentricity of 0.87. The eccentricity was value was selected as 0.87 because the gripper gave the maximum stable deflection at that value (Table 2). Values above 0.87 were unstable and hence were not used. The semi-circular corrugations were provided for maximum deflection. Molding was then used to manufacture the actuator (Figure 1b).

Table 1. Properties of acrylonitrile rubber.

Property	Value
Density	1000 kg/m^3
Shear Modulus	2.9*10^6 N/m^2
Poisson Ratio	0.49
Ultimate tensile strength	13.787 MPa

Table 2. Eccentricity vs deflection angle for semi-circular corrugation at 5 bar.

Eccentricity	Bending angle
0.11	0.0000
0.42	6.7489
0.76	9.1115
0.83	7.7372
0.85	7.5705
0.87	26.7550
0.92	12.9560
0.96	7.9325
0.97	6.5324
0.99	7.1692
1.01	23.0800
1.45	27.9730
1.47	35.8920
1.87	18.7260
1.96	14.5200
1.99	10.5870
2.04	6.2815
2.14	2.8724

(a) (b)
Figure 1. Asymmetric flexible actuator made from nitrile rubber(a) and mould used(b).

3.2. SCARA robotic manipulator

The selective compliance provided by SCARA robot makes it a good choice for packaging applications. They are easy to design, prototype and maintain. Even though parallel robots provide higher speed, the eggs will be prone to slippage. Hence SCARA robot has been used for building the robot.

Solidworks 2018 was used for drafting the prototype. The manipulator provides three degrees of freedom without the end effector. 3d printed components using PLA (polylactic acid) was used to build the prototype(Figure 2a and 2b).

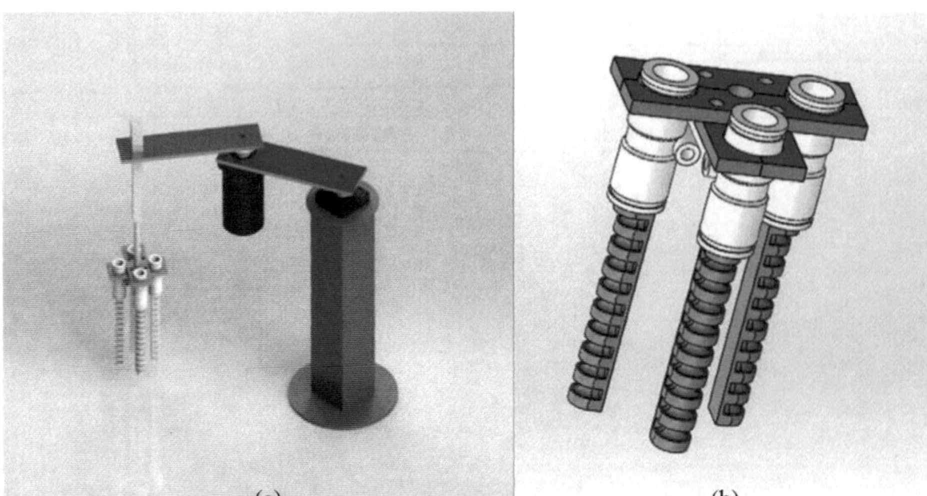
(a) (b)
Figure 2. Design of SCARA (a) and end effector (b).

4. Analysis of Actuator

4.1. Analysis of gripper

The gripper selection of the gripper configuration was based on previous work done in the design. Ansys was used for the analysis of the gripper. Two different prototypes were manufactured and tested to find the optimal material that provided a good amount of bending and withstood the pressure supplied. Even though both materials provided a decent bending, Silicone rubber tends to burst at 2 bar pressure. Hence actuator made from nitrile rubber was used to do the bending test (Figure 3). The

IOP Conf. Series: Materials Science and Engineering 1132 (2021) 012010 doi:10.1088/1757-899X/1132/1/012010

amount of deflection given by the actuator is shown in Table 3. The results almost coincide with the analysis results from previous works.

Table 3. Pressure versus deflection values for pneumatic actuator.

Pressure (bar)	Deflection in degrees
1	5
2	12
3	20
4	24
5	32

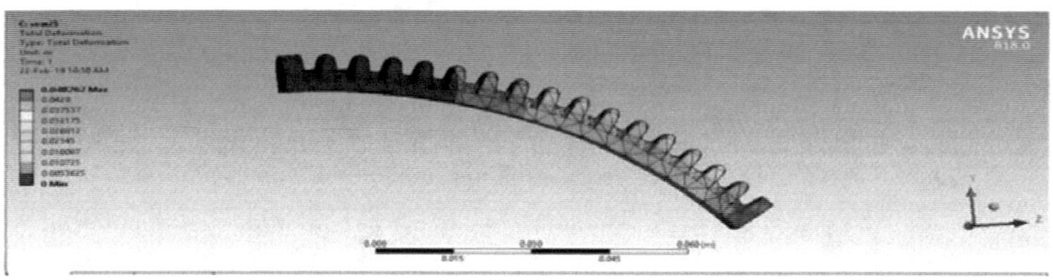

Figure 3. Structural analysis of pneumatic flexible bellow actuator at 5 bar.

4.2. Analysis of SCARA

Adams solver was used to conduct the motion study of the manipulator (Figure 4). Iterative finite element solver in Solidworks 2018(FFEPlus) was used to perform the structural analysis of the manipulator designs. The prototype was successfully manufactured and functioned perfectly. The end effector consisted of 4 AFPBA connected using a 3d printed mount. After the manufacture and assembly (Figure 5), the manipulator was able to lift weights up to 200 grams.

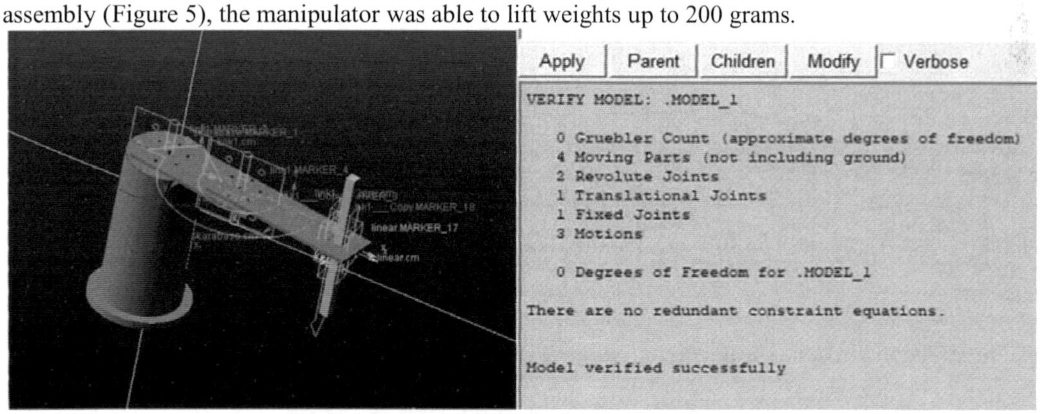

Figure 4. Dynamic analysis of SCARA robot.

Figure 5. Assembly of SCARA robot with pneumatic flexible bellow actuator.

5. Control System

Control of a soft robot is a difficult task as the equations that used for the kinematics of rigid bodies fail to apply in the case of a soft robot. Hence a control system is required to provide optimum pressure to lift objects. The Control system and the electronic components used have been specified below(Figure 6).

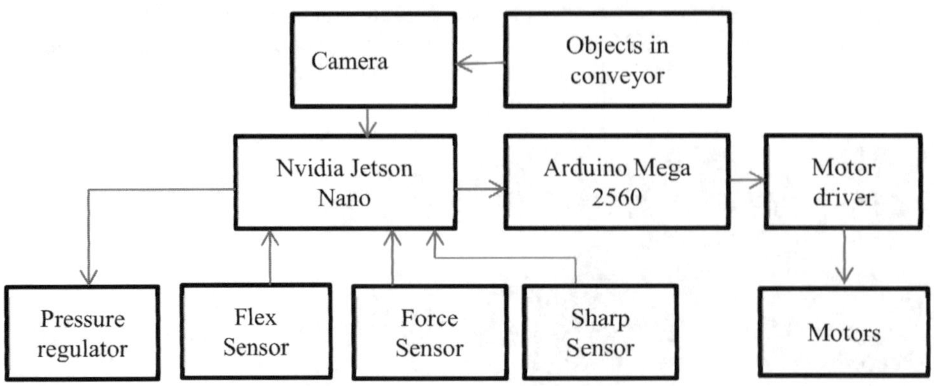

Figure 6. Control system of SCARA.

5.1. Nvidia Jetson Nano

Nvidia Jetson Nano has been used as the main processor for egg detection. As a lot of eggs needs to be palletized in a short amount of time, the image processing requires to be done as fast as possible and therefore it was necessary to use a small and powerful processor to do the task accurately and effectively. It also solves the inverse kinematics of the manipulator using trigonometry after locating the position of the egg.

5.2. Camera

A raspberry pi camera is used and mounted on the end effector to locate and centralize the gripper on top of the egg to be picked. The identification of the location is done by colour tracking the egg and then actuating the SCARA to move till the centroid of the actuator coincides with the centre of the image.

5.3. Sharp, Flex and Force sensors

Sharp sensor is used to measure the proximity of the end effector to the egg. This sensor was employed as the camera could not be used to obtain the depth information. The sensor enables the SCARA to position itself for firm grasping. It also prevents accidentally ramming the gripper onto the egg. Flex sensor was used to study the deflection given by the AFPBA at different pressures. This was required to control the pressure output by the regulator to achieve optimum bending. Force sensor is employed to prevent slippage of the grasped object

5.4. Pressure regulator

Pressure regulator is used to control the pressure output to the actuators so that the desired amount of bending is achieved. In the prototype we have used 12V solenoid pressure regulators

5.5. Arduino mega

Arduino Mega 2560 has been used for controlling the motors and the pressure regulator. It takes the angle of each motor and the pressure to be applied as the input and actuates the motors and regulators

5.6. Motors, Drivers and linear actuator

The prototype used rhino servo motor which came with its own motor driver. It had a rated torque of 120kgcm at 10rpm. The linear actuator used is a 6V L12-P linear actuator that provides a stroke length of 100mm.

6. Conclusion

Nitrile rubber has been selected as the optimal choice for manufacturing the gripper. Prototype of the gripper was 3d printed and assembled. The deflection of the end effector with change in pressure is measured. A control system was created to control the end effector force, prevent slippage and study the effectiveness of the actuator. The actuator was able to lift weights of up to 200 grams safely without slippage. The prototype proved to be effective in accomplishing the task. Further testing and improvement required before implementing the design in the industry

7. Acknowledgements

We would like to thank all faculties of Mechatronics and Intelligent Systems Research lab (MISR) and mechanical department for helping us in developing the prototype. Also, we would like to thank Mata Amritanandamayi Devi for her blessings, without which we would have never made it this far. We also thank all our friends, colleagues and well-wishers who helped and motivated us to complete the research.

References

[1] M. Shariatee, A. Akbarzadeh, A. Mousavi and S. Alimardani, Design of an economical SCARA robot for industrial applications, *2014 Second RSI/ISM International Conference on Robotics and Mechatronics (ICRoM)*, Tehran, Iran, 2014, pp. 534-539.

[2] I. Gaiser, R. Wiegand, O. Ivlev, A. Andres, H. Breitwieser, S. Schulz and G. Bretthauer (October 17th 2012). Compliant Robotics and Automation with Flexible Fluidic Actuators and Inflatable Structures, *Smart Actuation and Sensing Systems - Recent Advances and Future Challenges*, Giovanni Berselli, Rocco Vertechy and Gabriele Vassura, IntechOpen.

[3] J. M. Hollerbach, I. Hunter, and J. Ballantyne, 1991, A comparative analysis of actuator technologies for robotics, *Robotics Review* **2** pp 299-342

[4] AlineDe Greef, Pierre Lambert and Alain Delchambre, October 2009 Towards flexible medical instruments: Review of flexible fluidic actuators, *Precision Engineering* **33** Issue 4 pp 311-

IOP Conf. Series: Materials Science and Engineering 1132 (2021) 012010 doi:10.1088/1757-899X/1132/1/012010

21

[5] Pramod Sreedharan, Ganesh sundaram and Ganesh Udupa, 2016, Fluidic deflection of asymmetric flexible pipes and its control for robotic applications, Aug 29-31 *NAFEMS Int Conf on Engineering Modeling, Analysis, Simulation and 3D Printing*,

[6] I. Vaani, S. Pal, V. Indu and P. Sreedharan, "Optimal Force Control and Analysis of Soft Pneumatic Actuator," *2019 2nd Int Conf on Intelligent Computing, Instrumentation and Control Technologies (ICICICT)*, Kannur, India, 2019, pp. 501-506

[7] Ganesha Udupa, Pramod Sreedharan, P. S. Dinesh, and D. Kim, 2014 Asymmetric bellow flexible pneumatic actuator for miniature robotic soft gripper, *Journal of Robotics*, **2014**.

[8] Mata Amritanandamayi Devi, Ganesha Udupa, and Pramod Sreedharan, 2018 A novel underactuated multi-fingered soft robotic hand for prosthetic application, *Robotics and Autonomous Systems*, **100,** pp. 267-77.

[9] R Praneeth, Kiran, N V Sai, Aditya , Anjai K, Pramod S and Udupa G, 2018 Design, Analysis, Manufacturing and Testing of Artificial Hand with Fingers made of Rubber Actuators, in *Materials Today: Proceedings*, **5** pp. 25236-25244.

[10] M. Amritanandamayi Devi, Ganesha Udupa and Pramod Sreedharan, 2017 Anti-Bourdon tube pressure gauge, *Measurement*, **101** pp. 190-99

[11] Ganesha Udupa, Pramod Sreedharan and Aditya K, 2010 Robotic gripper driven by flexible microactuator based on an innovative technique, *Proceedings of IEEE Workshop on Advanced Robotics and its Social Impacts*, ARSO, pp. 111-16

IOP Conf. Series: Materials Science and Engineering 1132 (2021) 012011 doi:10.1088/1757-899X/1132/1/012011

Design Optimization and Analysis of External Corrugation and Geometry of Asymmetrical Bellow Flexible Pneumatic Actuator

Pranav V K*, Amal Prakash, Anikesh Rajendran and Pramod Sreedharan

Department of Mechanical Engineering, Amrita Vishwa Vidyapeetham, Amritapuri, India

*pranavvkumar21@gmail.com

Abstract. Pneumatic soft actuators are a highly researched topic in soft robotics. The paper discusses the design and analysis of an Asymmetric Flexible Pneumatic bellow Actuator (AFPBA). This kind of actuator is cheap, easy to produce, reliable and durable. Soft grippers commonly use this actuator to pick fragile objects. The objective is to design an actuator which produced the maximum bending and study the effect of external corrugation and eccentricity on the actuator to select the optimal design. Optimum eccentricity has been determined by two factors – bending and stability. The model has also been tested at different pressures. Nitrile rubber was used as the material to design and analyze the actuator. Finite element analysis was employed to conduct the structural analysis of the actuator. In conclusion, an actuator with a semicircular cross-section and semi-circular corrugation for the bellows gave the maximum bending. For the particular design selected with an eccentricity of 0.87 provided the maximum bending of 30.764 degrees at working pressure (5 bar).

1. Introduction

Soft robotics is a field of robotics which has been at the spotlight in recent years. Robots are used by humans to do tasks they find dull, dirty and dangerous. These environments require the robot to pick or interact with fragile objects and bodies that deform on application of force. The field of soft robotics aims at making special grippers and manipulators from highly compliant materials to aid the robots with these tasks. Soft actuators or robots are usually actuated using electric fields, shape memory alloys (thermally actuated) or pneumatic actuation. A soft robot can interact with complex environments and objects more effectively than a hard robot [1, 2] such as picking up fragile objects like eggs and assistance in surgeries [3]. The tradeoff here is that soft robots are harder to control due to their high compliance.

The benefits to using a pneumatic soft actuator are that they are usually safe, cost effective and easy to maintain. Asymmetric pneumatic actuators are tubes made of compliant material in which the tube chamber is slightly shifted from the center. When pressure is supplied into the chamber, a bending moment is created due to the force acting on the wall due to the pressure and the difference between the center of pressure and centroid of the actuator. This difference is known as eccentricity. This actuator works in a way opposite to how a bourdon tube works [6]. Providing bellows to the tube increases the amount of bending and force which makes them a good candidate for use in soft grippers [4 - 7]. This paper describes how the design configurations like the cross section of the actuator, the corrugation of the bellows and eccentricity affect the deflection angle and how to select the optimal design to be used in soft grippers.

2. Material selection

Acrylonitrile butadiene rubber (Table 1) also known as nitrile rubber was selected as the material for actuator fabrication. The selection was based on factors such as availability, abrasion resistance, tear resistance and ease of manufacture. Although nitrile rubber has poor temperature resistance which limits its use in high temperature zones, it was found suitable to be for building the prototype [6-8].

Table 1. Properties of acrylonitrile rubber.

PROPERTY	VALUE
Density	1000 kg/m^3
Shear Modulus	2.9 MPa
Poisson Ratio	0.49
Ultimate tensile strength	13.787 MPa

3. Design of actuator

The actuator is an asymmetric tube that functions in a way opposite to how a bourdon tube works. The axis of the chamber is slightly shifted to one side of the tube. This shift, known as eccentricity, along with the force exerted on the walls due to the air pressure in the chamber creates a bending moment which causes the tube to bend [6]. The objective of the design should be to maximize bending and stability all the while providing a good surface for better grasping. For this purpose we have to take into account the cross section of the tube, the corrugation of the bellows and the eccentricity of the chamber. A decent amount of wall thickness should be provided to prevent bursting of the actuator. Also, since the tube is to be connected to the air supply, it should be able to fit into a pneumatic connector. The use of an extension rod without bellows helps to connect the actuator to the connector.

3.1. Cross section determination

Initially, 3 different cross sections were taken into consideration for the actuator – circular, semi-circular(Figure 1) and square(Figure 2). Square cross section provided less amount of bending compared to circular cross sections. But the lack of a good gripping surface makes circular cross section a poor choice. In the end, half square semi-circular cross section was selected as it provided good deflection as well as a good gripping surface [9].

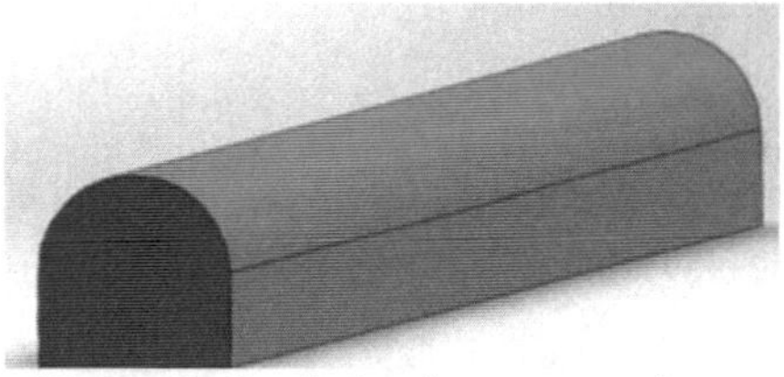

Figure 1. Semi-circular cross section.

Figure 2. Square cross section.

3.2. Corrugation shape determination

Four different corrugation designs were considered for analysis – trapezoidal, semi-circular, triangular and square (Figure 3, 4, 5 and 6). These four designs would be tested to select the optimal design which provides the best bending and stability.

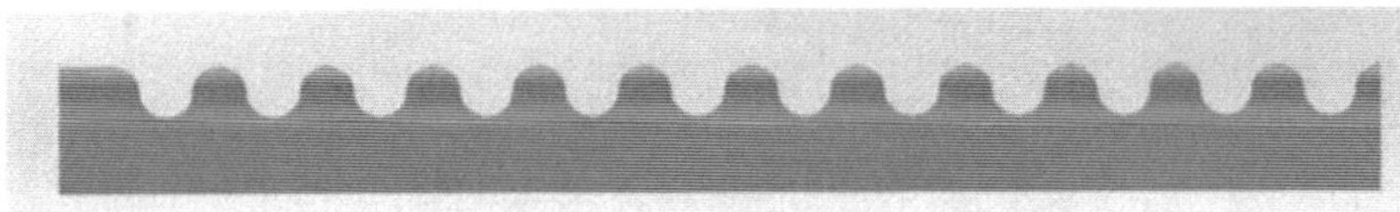

Figure 3. Semi-circular corrugation.

Figure 4. Rectangular corrugation.

Figure 5. Trapezoidal corrugation.

Figure 6. Triangular corrugation.

3.3. Pneumatic chamber and extension rod

The pneumatic chamber passes through one end of the actuator and has a length of 97mm. This length provides a wall thickness of 3mm. An extension rod of 15mm length and 12mm diameter is additionally added to the design to connect the pneumatic connector (Figure 7a). The final design has a total length of 115mm including the extension rod and 12 corrugations. Air at pressure of 5 bar is supplied into the pneumatic chamber (Figure 7b) which exerts a force on the walls of the chamber. This force and the eccentricity provided causes a bending moment in the actuator.

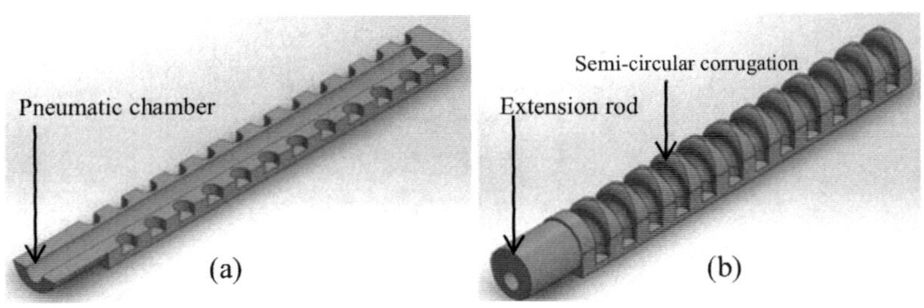

Figure 7. Sectional view of pneumatic chamber(a) and finalized design with extension rod(b).

4. Analysis of actuator.

Finite element analysis using Ansys 18.0 has been used for the analysis of the gripper [8] with nitrile rubber selected as the actuator material and medium level span angle center meshing. This divides the mesh in curved boundaries till each element span an angle of 24 to 75 degrees. The initial analysis is done to select the optimal corrugation for maximum bending. Since one end of the actuator will be connected to a pneumatic connector, we keep one end fixed. A pressure of 5 bar is given to the pneumatic chamber to simulate the actuator at working pressure. All models were subjected to 5 bar pressure with one end fixed (Figure 8, 9, 10 and 11).

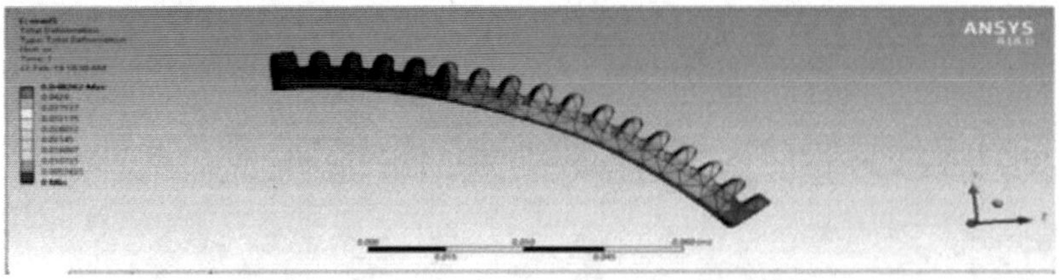

Figure 8. Structural analysis of semi-circular corrugation.

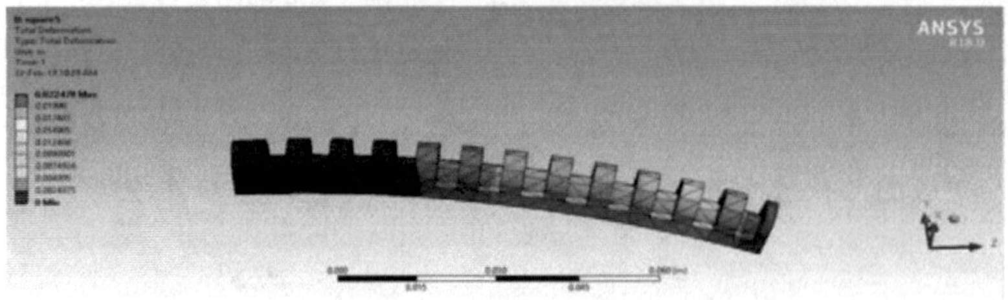

Figure 9. Structural analysis of rectangular corrugation.

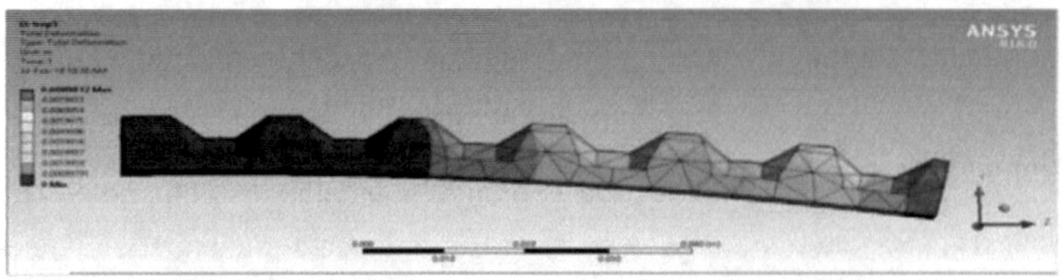

Figure 10. Structural analysis of trapezoidal corrugation.

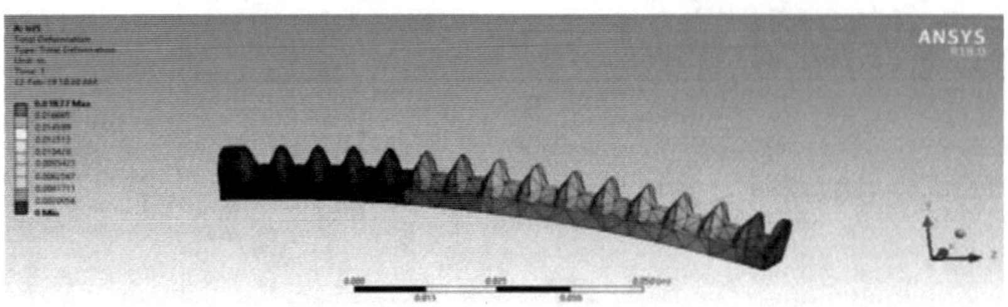

Figure 11. Structural analysis of triangular corrugation.

From the graph showing the deflection of all four corrugations at different pressures, we can see that semi-circular corrugation provides the maximum bending (Figure 12). Hence, semi-circular corrugation was selected to conduct other analyses.

The selection of optimal eccentricity was crucial in obtaining better bending. The analysis of the design with different eccentricities is done and their maximum deflection is noted. The eccentricity values between 0.11 and 2.14 are considered for the analysis. This was done because any value of eccentricity below 0.11 provided a very low amount of bending and any values above 2.14 caused failure of the actuator at 5 bar. The graph of the eccentricity versus deflection angle is plotted (Figure 13).

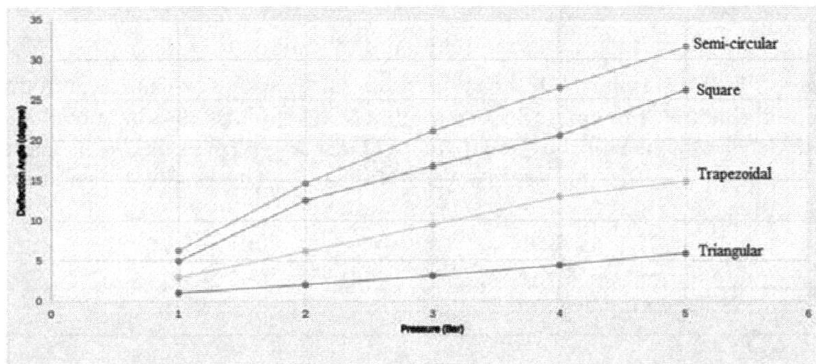

Figure 12. Graph showing deflection angle vs pressure for square, trapezoidal, triangular and semi-circular corrugation.

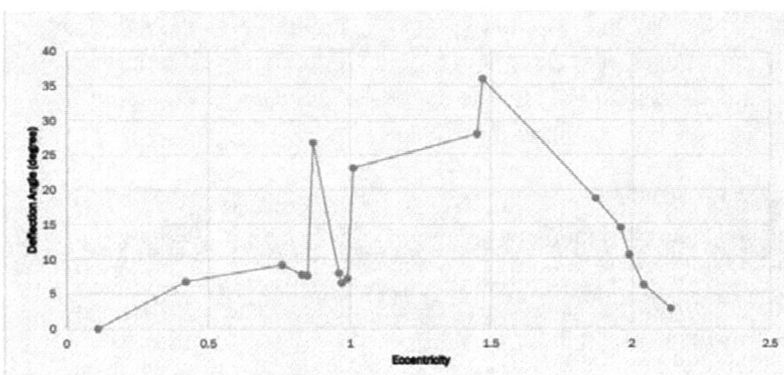

Figure 13. Graph showing deflection angle vs eccentricity for semi-circular corrugation.

The models were also tested at different pressures to check stability and although the eccentricities greater than 1 gave better deflection, they also led to instabilities in the model at higher pressures. Therefore 0.87 is selected as the optimal peak. The best model is then tested at different pressures to obtain the deflection angle and deflection length and their graphs are plotted (Figure 14).

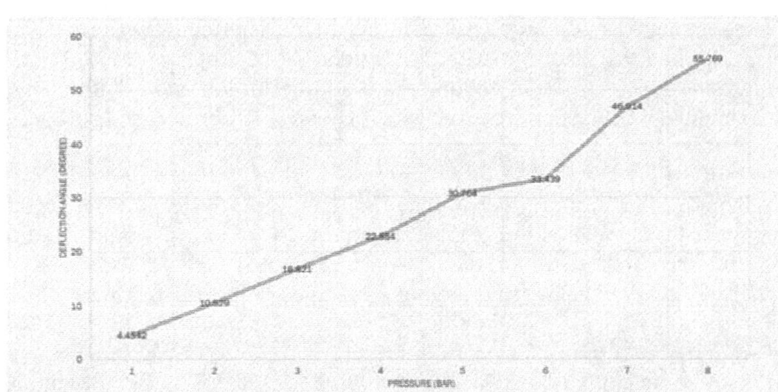

Figure 14. Deflection angle vs pressure graph of the final design.

5. Results

Semi-circular corrugation provided optimal bending and stability and was used to decide on the eccentricity required for maximum stable bending. Optimal eccentricity value was found to be 0.87. The best model produced a maximum deflection angle of 55.769 degrees. The model can withstand a pressure of up to 8 bar after which the model fails. At a working pressure of 5 bar, the deflection length is 0.0267cm and the deflection angle is 30.764 degrees. The final dimensions of the optimal model selected are given in Table 2.

Table 2. Dimensions of final design.

Parameters	Value
Total length	115mm
Diameter of actuator chamber	4.5mm
Total height	12mm
Length of actuator	100mm
Number of corrugations	12mm
Corrugation Diameter	4mm
Width	12mm
Crest to Crest diameter	8mm
Actuator chamber length(Without extension)	97mm
Length of extension rod	15mm
Diameter of extension rod	12mm
Minimum wall thickness	2mm
Maximum wall thickness	4.5mm
Cross-sectional semicircle diameter	12mm

6. Conclusion

Different designs were tested systematically to decide on the optimal model to be used as an end effector. Semi-circular cross section was chosen as the best cross section out of the ones in consideration. Semi-circular corrugations proved to be the most effective external corrugation as it gave the maximum bending and stability. After further analysis optimal eccentricity is decided. With the finalized design, the variation in deflection length and deflection angle with change in pressure along with the maximum deflection for the given specifications is measured and plotted.

7. Acknowledgements

We would like to thank all faculties of Mechatronics and Intelligent Systems Research lab (MISR) and mechanical department for helping us in completing the design. Also, we would like to thank Amma for her blessings, without which we would have never made it this far. We also thank all our friends, colleagues and well-wishers who helped and motivated us to complete the research.

References

[1] I. Gaiser, R. Wiegand, O. Ivlev, A. Andres, H. Breitwieser, S. Schulz and G. Bretthauer (October 17th 2012). Compliant Robotics and Automation with Flexible Fluidic Actuators and Inflatable Structures, *Smart Actuation and Sensing Systems - Recent Advances and Future Challenges*, Giovanni Berselli, Rocco Vertechy and Gabriele Vassura, IntechOpen.

[2] J. M. Hollerbach, I. Hunter, and J. Ballantyne, 1991, A comparative analysis of actuator technologies for robotics, *Robotics Review* **2** pp 299-342

[3] AlineDe Greef, Pierre Lambert and Alain Delchambre, October 2009 Towards flexible medical instruments: Review of flexible fluidic actuators, *Precision Engineering* **33** Issue 4 pp 311-21

[4] Pramod Sreedharan, Ganesh sundaram, Ganesh Udupa, 2016, Fluidic deflection of asymmetric flexible pipes and its control for robotic applications, *NAFEMS International Conference on Engineering Modeling, Analysis, Simulation and 3D Printing*, Aug 29-31

[5] I. Vaani, S. Pal, V. Indu and P. Sreedharan, Optimal Force Control and Analysis of Soft Pneumatic Actuator, *2019 2nd Int Conf on Intelligent Computing, Instrumentation and Control Technologies (ICICICT)*, Kannur, India, 2019, pp. 501-506

[6] Ganesha Udupa, Pramod Sreedharan, P. S. Dinesh, and D. Kim, Asymmetric bellow flexible pneumatic actuator for miniature robotic soft gripper, *Journal of Robotics*, **vol. 2014, 2014**.

[7] Mata Amritanandamayi Devi, Ganesha Udupa, and Pramod Sreedharan, 2018 A novel underactuated multi-fingered soft robotic hand for prosthetic application, *Robotics and Autonomous Systems*, **100,** pp. 267-77.

[8] R Praneeth, Kiran, N V Sai, Aditya , Anjai K, Pramod S and Udupa G, 2018 Design, Analysis, Manufacturing and Testing of Artificial Hand with Fingers made of Rubber Actuators, in *Materials Today: Proceedings*, **5** pp. 25236-25244.

[9] Philip Moseley, Juan Manuel Florez, Harshal Arun Sonar, Gunjan Agarwal, William Curtin and Jamie Paik, June 2016, Modelling, Design, and Development of Soft Pneumatic Actuators with Finite Element Method, *Advanced engineering materials*, **18** Issue 6 pp 978-988.

Numerical Investigation on Laminar Cross-flow Around a Top Surface Heated Square Cylinder

Deepak. K*, Anandhavishnu. V, Sujith Stephen and S. Ajith Kumar
Department of Mechanical Engineering, Amrita Vishwa Vidyapeetham, Amritapuri, India
E-mail: *deepakajay77@gmail.com

Abstract. This numerical study attempts to discuss about the effects in vortex shedding and aerodynamic parameters due to heating the top surface of a square cylinder for various Ri values ($0 \leq Ri \leq 1$) at $Pr = 0.71$ and $Re = 100$. Numerical calculations are carried out using a finite volume method based on the PISO (Pressure-Implicit with Splitting of Operators) algorithm. We have used ANSYS Fluent for the numerical simulation. The results in this study are promising and we found that the surface heating would rather changes or modifies the flow field and the vortex shedding phenomenon behind the square cylinder. Also it is observed that heating plays a critical role in the integral quantities such as aerodynamic coefficients, Strouhal number etc.

1. Introduction

Fluid flow around the square cylinder has brought attention of researchers due to its applicability and scope in both academics and industrial fields. Researchers noticed the formation of vortices and its shedding for various Reynolds numbers (Re) [1] [2]. The Reynolds number is a non-dimensional quantity defined as the ratio between inertial force and viscous force.

$$Re = \frac{\rho v L}{\mu} \tag{1}$$

ρ represents the density, v represents the velocity, L represents the characteristic length, and μ represents the viscosity.

Enormous studies have been reported which focus on vortex shedding and associated wake dynamics. Heating the cylinder surface adds more complexity to the wake dynamics and the vortex shedding because of buoyancy effects. Understanding this complexity has received lot of scope and are used in many applications such as cooling towers, design of bridges, designing electric components, etc.

The dependency of fluid flow and aerodynamic coefficients with Re for flow past a square cylinder is reported in Bai and Alam [1], Chatterjee and Mondal [3], Sharma and Eswaran [4]. It is established that the surface heating substantially influences the vortex shedding pattern and the shedding frequency [5]. Bhattacharyya and Mahapatra [6] studied a heated cylinder at a low Reynolds number. They found that while heating, the centreline symmetry of the cylinder wake was lost. A mean downward lift was experienced by the cylinder when the buoyancy effect is considered. The non-dimensional shedding frequency, and Strouhal number (St) were found increasing with Richardson number (Ri). The Strouhal number is defined as follows.

Nomenclature

C_D	Coefficient of drag
C_L	Coefficient of lift
f	Frequency of vortex shedding $[1/s]$
g	Acceleration due to gravity $[m/s^2]$
L	Characteristic length $[m]$
p	Pressure $[Pa]$
p^*	Non-dimensional pressure
Pr	Prandtl number
Re	Reynolds number
Ri	Richardson number
St	Strouhal number
T_{cold}	Temperature of cold wall $[K]$
T_{hot}	Temperature of hot wall $[K]$
T_{ref}	Reference temperature $[K]$
U	Free stream velocity $[m/s]$
u^*	Non-dimensional velocity in the x direction
V	Characteristic velocity $[m/s]$
v	Velocity $[m/s]$
v^*	Non-dimensional velocity in the y direction
x^*	Non-dimensional x coordinate
y^*	Non-dimensional y coordinate

Greek Symbols

α	Thermal diffusivity $[m^2/s]$
β	Coefficient of thermal expansion $[1/K]$
μ	Dynamic viscosity $[PaS]$
ρ	Density $[kg/m^3]$
τ	Non-dimensional time
Θ	Non-dimensional temperature

$$St = \frac{fL}{U} \tag{2}$$

f represents the frequency of vortex shedding, L represents the characteristic length, and U represents the free stream velocity.

Richardson number is a non-dimensional quantity defined as the ratio between buoyant force and the inertia force.

$$Ri = \frac{g\beta(T_{hot} - T_{ref})L}{V^2} \tag{3}$$

g represents the acceleration due to gravity, T_{hot} represents the temperature of hot wall, T_{ref} represents the reference temperature, β represents the coefficient of thermal expansion, L represents the characteristic length, and V represents the characteristic velocity.

Dulhani *et al* [7] had done a numerical investigation by changing the angle of incidence (α) of flow from $0°$ to $45°$ for $Re = 100$ and $Pr = 0.71$. Prandtl number (Pr) is a non-dimensional quantity defined as the ratio between momentum diffusivity and thermal diffusivity. They found that as the angle of incidence increases, the lateral wake width of the square cylinder decreases thus heat transfer rate increases. Chatterjee and Mondal [3] investigated the effect of buoyancy and heat transfer accross a square cylinder subjected to upward laminar flow for $50 \leq Re \leq 150$. A decreasing trend was observed for St while cooling the cylinder below the free stream temperature for all Re studied. Dhiman *et al* [8] studied the effects of blockage ratio (B) in the mixed convection from a heated square cylinder in a horizontal flow for $1 \leq Re \leq 30$ and $0 \leq Ri \leq 1$ for blockage ratio 0.125, and 0.25. They observed that coefficient of drag increases with an increase in blockage ratio for distinct values of Re and Ri.

From the literature it observed that heating the square cylinder increases the drag coefficient (C_D) and St. However, the effect of heating the top surface of a square cylinder is not clearly investigated in the literature. The present study attempts to explain the effect of buoyancy in

IOP Conf. Series: Materials Science and Engineering 1132 (2021) 012012 doi:10.1088/1757-899X/1132/1/012012

vortex shedding and aerodynamic parameters while heating the top surface of a square cylinder. We have created a two-dimensional rectangular flow domain with a square cylinder in mixed convection regime which is a combination of natural and forced convection. If the heat transfer occures due to forced fluid flow over the surface by external means it is called a forced convection. In contrast, if the heat transfer occures due to motion of fluid caused by buoyancy forces it is called natural convection. In this work, we are conducting a numerical study on a top surface heated cylinder for five different Ri, ranges from 0 to 1 at $Re = 100$.

2. Numerical Methodology

2.1. Geometrical Modelling and Domain Discretisation

A domain independent study is conducted in order to identify the optimum domain geometry. Computations are carried out for distinct domain geometries, and the C_D, St values are noted, as shown in Table 1. The optimum domain size was found out to be $50L$ and $25L$.

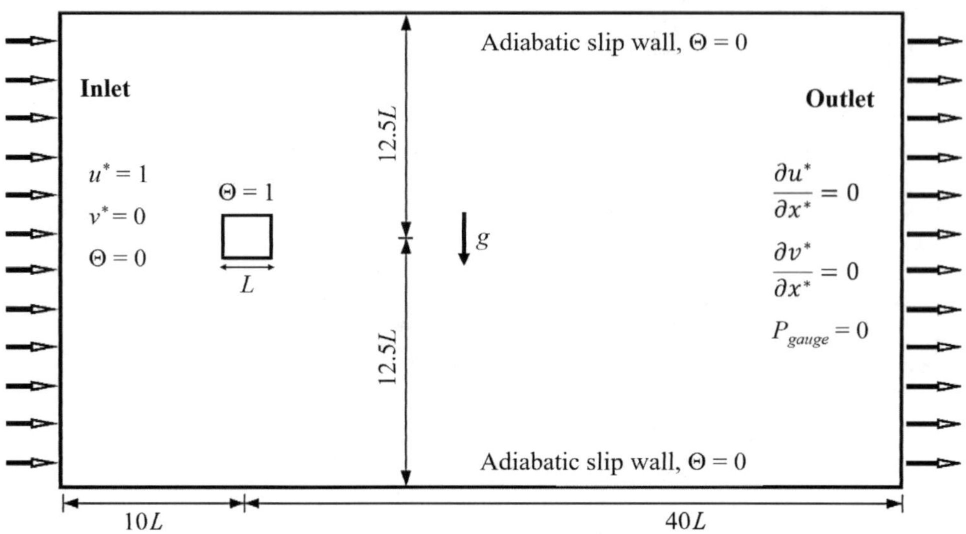

Figure 1: Representation of domain geometry with boundary conditions.

Table 1: Domain independence study at $Ri = 0$, $Pr = 0.71$ and $Re = 100$. The percentage deviation of the calculated values from the literature is noted in brackets.

Length	Width	Elements	C_D (% Error)	St (% Error)
$35L$	$10L$	15,254	1.608	0.154
$40L$	$15L$	23,498	1.533 (4.66%)	0.148 (3.89%)
$45L$	$20L$	33,514	1.512 (1.37%)	0.147 (0.68%)
$50L$	**$25L$**	**45,144**	**1.505 (0.46%)**	**0.146 (0.68%)**
$55L$	$30L$	58,286	1.503 (0.13%)	0.146 (0.00%)

Following this a grid independent study is conducted in order to obtain the optimum grid size required. A two-dimensional model was created in ANSYS SpaceClaim. Unstructured quadrilateral grid is created inside the domain using ANSYS Workbench. Figure 1 represents the domain geometry with boundary conditions. The domain variables are non-dimensionalized with the length of the square cylinder. The top surface of the square cylinder is heated and is highlighted in the figure.

Figure 2 shows the square adapted quad mesh. Calculations are done for four different grid structures. The C_D and St values are compared and the percentage deviations are noted in Table 2. The grid having 45,144 elements was found to be optimum.

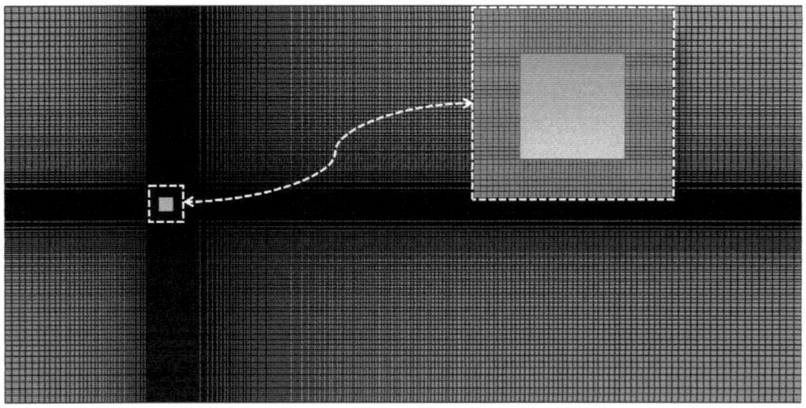

Figure 2: Mesh obtained after the grid independence study

Table 2: Grid Independence Study at $Ri = 0$, $Pr = 0.71$ and $Re = 100$. The percentage deviation of the calculated values from the literature is noted in brackets.

Elements	C_D (% Error)	St % Error)
25,088	1.530	0.145
35,567	1.511 (1.24%)	0.145 (0.00%)
45,144	**1.505 (0.39%)**	**0.146 (0.69%)**
55,159	1.501 (0.26%)	0.146 (0.00%)

2.2. Mathematical Modelling

The conservation of mass, momentum and energy equations for incompressible fluid flow under the Boussinesq approximation are given by,

$$\frac{\partial u^*}{\partial x^*} + \frac{\partial v^*}{\partial y^*} = 0 \tag{4}$$

$$\frac{\partial u^*}{\partial \tau} + u^* \frac{\partial u^*}{\partial x^*} + v^* \frac{\partial u^*}{\partial y^*} = -\frac{\partial p^*}{\partial x^*} + \frac{1}{Re}\left(\frac{\partial^2 u^*}{\partial x^{*2}} + \frac{\partial^2 u^*}{\partial y^{*2}}\right) \tag{5}$$

$$\frac{\partial v^*}{\partial \tau} + u^* \frac{\partial v^*}{\partial x^*} + v^* \frac{\partial v^*}{\partial y^*} = -\frac{\partial p^*}{\partial y^*} + \frac{1}{Re}\left(\frac{\partial^2 v^*}{\partial x^{*2}} + \frac{\partial^2 v^*}{\partial y^{*2}}\right) + Ri \cdot \Theta \tag{6}$$

$$\frac{\partial \Theta}{\partial \tau} + u^* \frac{\partial \Theta}{\partial x^*} + v^* \frac{\partial \Theta}{\partial y^*} = \frac{1}{Re \cdot Pr}\left(\frac{\partial^2 \Theta}{\partial x^{*2}} + \frac{\partial^2 \Theta}{\partial y^{*2}}\right) \tag{7}$$

u^* and v^* represent the non-dimensional velocities in the x and y directions, Θ and p^* represent the non-dimensional temperature and pressure.

$$\Theta = \frac{T_{ref} - T_{cold}}{T_{hot} - T_{cold}}, \quad p^* = \frac{pL^2}{\rho \alpha^2} \tag{8}$$

α represents the thermal diffusivity, T_{cold} and T_{hot} represent the temperatures at cold and hot walls, and p represents the pressure.

The equations govern fluid flow for the two-dimensional laminar flow are discretize and solved using a CFD tool, ANSYS 20.1. PISO algorithm is used and 10^{-6} is assigned as the convergence criteria.

2.3. Boundary Conditions

The boundary conditions of our present study are as follows.

- At the inlet, the boundary conditions are as follows.

$$u^* = 1, \quad v^* = 0, \quad \Theta = 0 \tag{9}$$

- At the exit, the boundary conditions are as follows.

$$\frac{\partial u^*}{\partial x^*} = 0, \quad v^* = 0, \quad \Theta = 0 \tag{10}$$

- The top and bottom boundary walls are considered as adiabatic slip walls and are mathematically expressed as follows.

$$v^* = 0, \quad \frac{\partial u^*}{\partial y^*} = 0, \quad \Theta = 0 \tag{11}$$

- The cylinder surface is considered as no-slip and is mathematically expressed as follows.

$$u^* = 0, \quad v^* = 0, \quad \Theta = 1 \tag{12}$$

3. Code Validation

The present study is validated with the flow around a non-heated and heated cylinder at $Re = 100$. The C_D and St values obtained from the present study are compared with literature.

Table 3: Comparison of C_D and St at Ri=0, Pr=0.71 and Re=100 with literature. The percentage deviation of the calculated values from the literature is noted in brackets.

Literature	C_D (% Error)	St (% Error)
Sohankar et al. [9]	1.500 (0.33%)	0.150 (2.67%)
Sahu et al. [10]	1.490 (1.01%)	0.149 (2.01%)
Sharma and Eswaran [4]	1.490 (1.01%)	0.150 (2.67%)
Bai and Alam [1]	1.500 (0.33%)	0.143 (2.09%)
Singh et al. [2]	1.510 (0.33%)	0.147(0.68%)
Sen et al. [11]	1.520 (0.99%)	0.145 (0.69%)
Present Study	**1.505**	**0.146**

Table 3 represents the comparison of present value of C_D and St at $Ri = 0$, $Pr = 0.71$ and $Re = 100$ with Sohankar *et al* [9], Sahu *et al* [10], Sharma and Eswaran [4], Bai and Alam [1], Singh *et al* [2], and Sen *et al* [11] with maximum deviation less than 1.01% and 2.67% respectively, which shows a fair agreement with the literature.

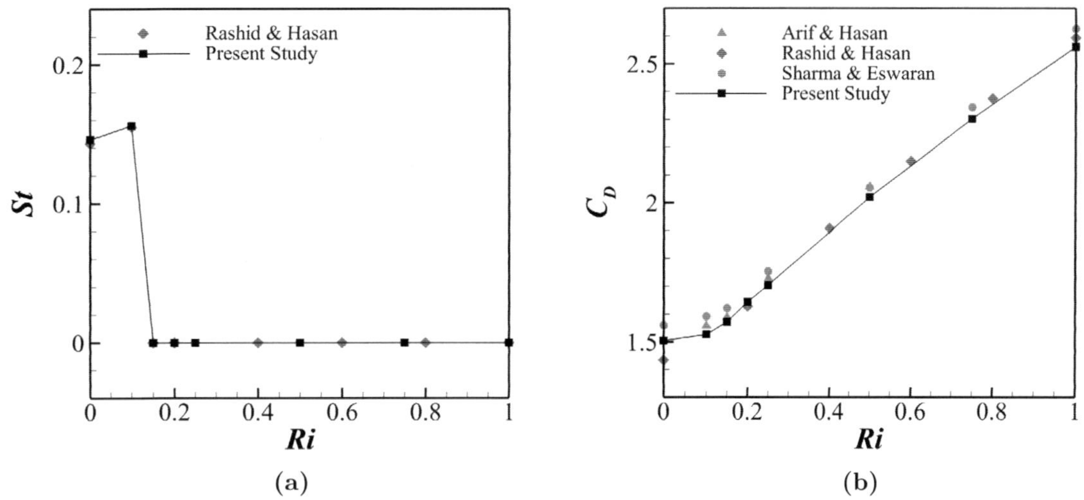

Figure 3: Comparison of (a) St and (b) C_D for various Ri at $Pr = 0.71$ and $Re = 100$.

Figure 3 (a) shows the comparison of St at $0 \leq Ri \leq 1$, $Pr = 0.71$ and $Re = 100$. The deviation is less than 0.64%, which shows a good agreement with Rashid and Hasan [5]. Figure 3 (b) shows the comparison of C_D at $0 \leq Ri \leq 1$, $Pr = 0.71$ and $Re = 100$ with Arif and Hasan [12], Rashid and Hasan [5] and Sharma and Eswaran [13]. The maximum deviation is less than 4.951%, which shows a fair agreement with the literature.

4. Results and Discussion

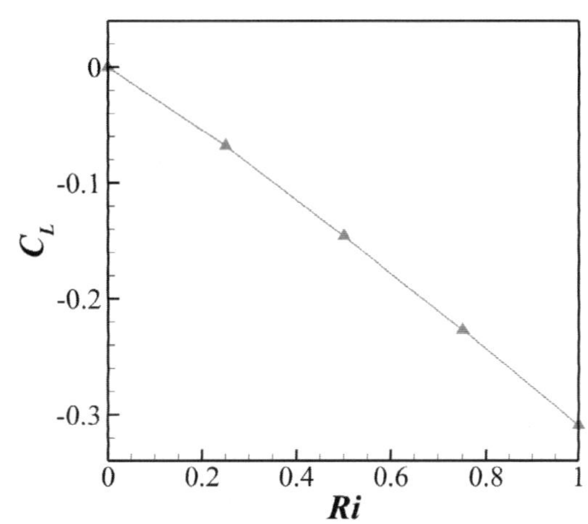

Figure 4: Variation of C_L for top surface heated cylinder at various Ri, $Pr = 0.71$ and $Re = 100$.

Studies have been conducted by heating the top surface of the square cylinder for $0 \leq Ri \leq 1$ at $Pr = 0.71$ and $Re = 100$. It was found that the pressure difference between the upper and lower surface is increasing with the increase in Ri, which causes a negative lift on the cylinder, as shown in Figure 4.

Figure 5: Vortex structure for various Ri at $Pr = 0.71$ and $Re = 100$. Shedding is symmetric to the horizontal axis for $Ri = 0$ and looses its symmetry for other Ri values.

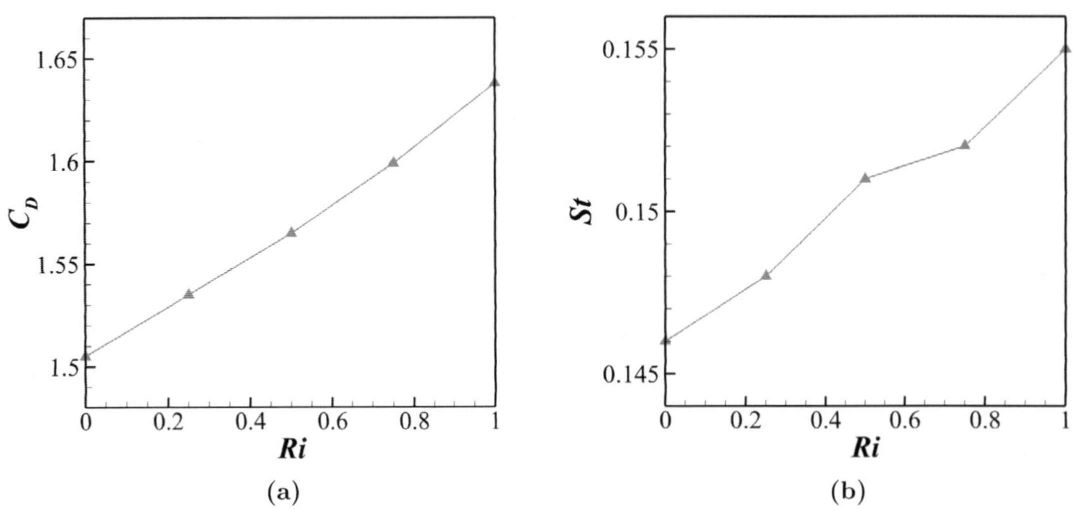

Figure 6: Variation of (a) C_D and (b) St for top surface heated cylinder at various Ri, $Pr = 0.71$ and $Re = 100$.

While heating the cylinder surface, the fluid that passes over that surface will get heated, and it becomes less dense. As a result of this, the fluid will move upwards along the direction of buoyancy. Figure 5 shows that for $Ri = 0$, the shedding is symmetric about the horizontal axis of the flow. However, when the top surface is heated, the vortex shedding became asymmetric.

When the top surface is heated, the vortices move upward, causing a widening of wake width. As a result, a reduction of pressure occurs near the base point, which increases the pressure difference between the stagnation point and base point, causing an increment in C_D, as shown in Figure 6 (a).

While heating the top surface, the vortices which sheds from the top surface of the cylinder gets dissipated earlier than vortices from the bottom surface due to unstable density stratification caused by heating. This acceleration in shedding from the top surfaces causes an increment in shedding frequency, as observed in Figure 6 (b).

5. Conclusion

A numerical investigation is conducted on flow over top surface heated square cylinder for $0 \leq Ri \leq 1$ at $Pr = 0.71$ and $Re = 100$. Variation of drag coefficient (C_D), lift coefficient (C_L), and St with Ri is presented. Major findings are,

- An asymmetry in vortex shedding found while heating top surface of the square cylinder.
- Heating the top surface of square cylinder increased the wake width and caused an increment in drag.
- An increment in shedding frequency was observed while heating due to unstable density stratification.
- The coefficient of the lift is found to be negative while heating due to modified pressure distribution.

References

[1] Bai H and Alam M M 2018 *American Institute of Physics* **30**
[2] Singh A P, De A K, Carpenter V K, Eswaran V and Muralidhar K 2009 *International Journal for Numerical Methods in Fluids* **61** 658–682
[3] Chatterjee D and Mondal B 2012 *Heat Transfer Engineering* **33**(12) 1063–1074
[4] Sharma A and Eswaran V 2004 *Numerical Heat Transfer* **45**(3) 247–269
[5] Rashid A and Hasan N 2011 *World Academy of Science, Engineering and Technology*
[6] Bhattacharyya S and Mahapatra S 2005 *International Journal for Heat and Mass Transfer* **41** 824–833
[7] Dulhani J P, Sarkar S and Dala A 2014 *International Journal of Heat and Mass Transfer* **74** 319–332
[8] Dhiman A K, Sharma N and Kumar S 2012 *Brazilian Journal of Chemical Engineering* **29** 253–264
[9] Sohankar A, Davidson L and Norberg C 1995 *Twelfth Australasian Fluid Mechanics Conference* 517–520
[10] Sahu A K, Chhabra R P and Eswaran V 2009 *Journal of Non-Newtonian Fluid Mechanics* **160** 157–167
[11] Sen S, Mittal S and Biswas G 2011 *International Journal for Numerical Methods in Fluids* **67** 1160–1174
[12] Arif M R and Hasan N 2020 *International Journal of Heat and Mass Transfer* **161**
[13] Sharma A and Eswaran V 2004 *Numerical Heat Transfer* **45** 601–624

Development of an Efficient Framework for the Distribution of Essential Relief Materials during Floods by Clustering Technique

Sudeesh S [1,2], Rameshkumar K [1*], Anbuudayasankar S P [1]

[1] Department of Mechanical Engineering, Amrita School of Engineering, Coimbatore – 641112, Amrita Vishwa Vidyapeetham, India
[2] Department of Mechanical Engineering, Rajadhani Institute of Science and Technology, Palakkad – 678613, India

*k_rameshkumar@cb.amrita.edu

Abstract. An efficient logistic network is extremely vital to distribute essential items quickly to the region wherever the natural calamities have occurred. This paper focuses on developing an efficient framework for the distribution of essential relief materials to the people in Kuttanad Taluk of Alappuzha District in Kerala, India during the monsoon season. The existing relief distribution network in the flood-affected region has been studied in detail and an improved distribution logistics has been suggested in this study. K-means clustering algorithm has been used to identify the location for the relief material distribution centres in the flood-affected region. The proposed approach reduces the efforts needed to coordinate the relief activities by optimally identifying no. of distribution centres and locations of distribution centres with an objective of reducing the total supply chain transportation distance for the supply of essential items during the flood. The proposed clustering approach reduces the traveling distance of the people during the flood to receive the relief material to an extent of 88%.

1. Introduction

A natural disaster is a notable hostile incident ensuing due to floods, landslides, volcanic eruptions, tremors, and different natural science processes. Such disasters can be harmful to life or destroy the properties. These types of disasters typically leave some financial loss to the entire community. The severity of the disaster depends on the character of the disaster, preparedness, and handiness of the infrastructure for reaching out to the affected region for the relief work. Flooding and landslides are common during monsoon season in most parts of the state in Kerala, India, where the rains falls are very heavy.

It is hard to stop disasters due to calamities like flood, earthquake, or tsunami but with essential preparedness, planning, and a fast retort can alleviate the damages of the properties and griefs of the affected individual. During such situation of any disaster, there requires a support from an external agency, either for evacuation or transferal to take back things into routine. Overall recovery without any aid looks impossible or the implications and losses of the event would be abundant larger. Housing assessment and planning in a pre-disaster stage are crucial elements of preparedness and planning. While the reorganization of affected individuals to a fairly harmless provisional housing and relief item distribution is the greatest puzzling part in the post-disaster condition. Housing apportionment and relief item distribution throughout a flood are complex optimization problems due to the contribution of several locations, severe restraints, and distinct viable area [1]. Transportation models aid as vital decision-making gears for serving faster attention to a disaster state. By the assistance of a classification-based procedure, major transportation model objective functions and restrictions that are typically used during the primary relief operations like casualty transportation and evacuation and distribution, can be understood [2].

In disaster relief actions, logistics are expected to assist the organization and employment of response processes in order to confirm their suitability and effectiveness. Mobilizing staff, equipment and belongings of humanitarian support organizations, and vacating or relocation of those affected by the disaster needs a logistics scheme to maximize the feat. In disaster management, preparedness and response actions are associated. Stochastic planning model is an appropriate tool for preparedness for of its capacity to handle with ambiguity built on probabilistic situations of disasters and their penalties [3].

Disaster management is usually outlined as a method with many phases involving planning, mitigation, detection, response, and recovery [4]. Various stages of disaster relief operations, preparedness for the relief events, necessity of information technology in humanitarian logistics were studied by Wassenhove [5], Altay and Green [6]. Kovács and Spens conferred variations between business logistics and humanitarian logistics. Their paper points out the importance of planning and carrying out logistics operations in the complicated areas of disaster relief [7]. The humanitarian operation is having diverse stages, each of which has its own objectives [8]. Suresh *et al.* explored the idea of agility in humanitarian operations on occasion of the cart festivals in India. In their study interpretive structural modelling approach is applied to find the interrelationships among the features which impact the agility [9].

Facility location studies were used for defining the site of various amenities and infrastructures including central and native warehouses, enduring relief amenities, like major hospitals, and positioned relief items and automobiles [10], [11]. Caunhye *et al.* performed an analysis on optimization models in emergency logistics, categorizing the results conferring to the process conducted relating facility location, relief distribution and casualty transfer and model kinds, choices, purposes and restraints [12]. Tzeng *et al.* established a fuzzy multi-objective linear programming model for the relief distribution arrangements. Their model studies purposes like, minimization of total costs, curtailing the total travel time and, make best use of the satisfaction through the planning phase using a forecast for the commodity demand [13]. Jalali *et al.* highlighted the principles of non-operational defense and developed a multi-objective mathematical model looking for an apt place for setting up hospitals, temporary care centres, shelters and store rooms [14]. Forero and Rodriguez concluded that relief procedures can be addressed as a multi-mode resource-constrain multi-project scheduling problem. A theoretical outline at a tactical level was created by them and the Colombian relief acts for an earthquake response was shown as a demonstrative case [15]. Gupta *et al.* considered disaster management strategies and methods and they also evaluated the effect of flash floods Leh, India on human life, health risks, and upcoming insinuations of a natural disaster [16]. Sreedharan *et al.* did a detailed review of academic articles pertain to humanitarian logistics and proposed a MAPA model, to accomplish challenges and to address the exact needs of relief procedures [17].

Özdamar L and Demir proposed a work for the aid of relief distribution and evacuation activity. They proposed a clustering algorithm called a hierarchical cluster and route procedure to manage vehicle routing [18]. Lima *et al.* suggested a methodology to find and categorize provinces by the category and regularity of calamities. The information on the clusters permit to categorize the relief stuffs required to meet each cluster. In their approach, data history of disasters is used as the input for forming the clusters using a computing tool [19]. Gutierrez and Mutuc offered a mathematical model for the recognition of optimal position of a provisional or permanent capacity in a certain topographical area under consideration. Principally, that model is useful to Humanitarian Supply Chain where it tries to find the best position for temporary relief hub so as to optimize the delivery of relief items to the informally spread evacuation centres [20].

This paper focus on a specific region where the flooding is happening frequently during the monsoon period. It is highly essential for the people located in the flood affected areas to be relocated to the elevated locations until the danger of the flooding subsides. It is important that whenever the flood warnings are issued, the people in that region quickly collect their relief items from the distribution centres and move to the safer places in the elevated region for cooking and stay back until the normally returns. Since the flood affected region is quite large, many centres for cooking have to be located in

IOP Conf. Series: Materials Science and Engineering 1132 (2021) 012013 doi:10.1088/1757-899X/1132/1/012013

that region. The locations for cooking and stay are provided such that they are easily approachable by the people in the flood affected region and the locations should not get affected by the severe flood.

2. Problem statement and objectives

A Grama Panchayath named ‘Kainakary’ in ‘Kuttanad’ taluk, ‘Alappuzha’ district of state ‘Kerala’ in India, spreading over 38.6 sq km is severely affected by floods during the south–east monsoon period. It is essential to help the affected people by supplying essential items for preparing food in shelters called gruel centres. These centres are operated on elevated locations in the flood-affected regions. In mild to moderate flood situations, when water start entering their houses, people in the flood-affected region are reluctant to vacate their homes and rather than moving to the rescue camps operated by the Government, they prefer to go to gruel centres. They can't cook in their home during these flood conditions due to water lodging at their homes. For receiving the groceries, they will be visiting the place called Maveli stores/Ration stores. After receiving the essential items from Maveli stores/Ration stores, affected people will go to gruel centres to cook their food. If the flood is very severe, the gruel centres becomes non-operational and the affected people will be shifted to rescue camps and they stay there until the normalcy returns.

During the floods, the ‘gruel center’ serves as a temporary shelter, and as well as a cooking place. At present, there are ‘74’ temporary gruel centres which are distributed randomly in the flood-affected region. The locations of the gruel centres were identified by the local people with the government support for supplying the food items and as well as for cooking the food. These are located on higher grounds and people rely on these places every time flood occurs. At the time of the disaster, the people at the disaster zones move to the gruel centres to prepare food using the food items supplied by the government agencies. Gruel center mapping in the Kainakary Grama Panchayath was done by making use of Google maps in smartphones. The geo location of the 74 gruel centres existed in the Kainakary Grama Panchayath of the Kuttanad taluk during the July – August floods of 2018 was identified and recorded. These locations were served as gruel centres in every flood situation. The locations of the gruel centres mapped in the Google map are shown in figure 1.

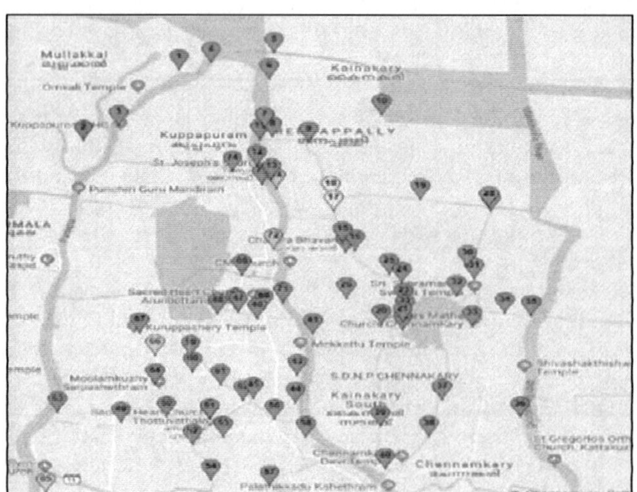

Figure 1. Location of gruel centres mapped using Google map

These gruel centres even though serve the purpose, it is very difficult for the government officials to coordinate the distribution of the food items effectively and also many of these gruel centres are not optimally utilized. Many of the gruel centres are also very close to the affected region and some are far off. These gruel centres are also affected during the severe flood conditions. So, it is essential to establish the optimal number of permanent gruel centres which can be effectively utilized during the flood conditions. Also, these centers should be in elevated locations and easily accessible by the affected people during the flood. Normal Supply chain operations are disrupted during floods and government agencies need a proper framework that will operate efficiently during floods for the supply of essential items. It is proposed to locate and construct permanent gruel centres in an optimal manner that will be used to store food items and also facilitates cooking. Clustering approach is used in this study to identify the optimum number of gruel centres, which are easily accessible by the people in the neighborhood and the located gruel centres should not get affected during the severe flood conditions.

3. Existing distribution network

At present, the commodities are primarily stored in a warehouse at the district headquarters. Then the commodities are transferred to the supply sources namely Ration stores and Maveli stores. Ration and Maveli stores are essential item distribution centres available at the Grama Panchayath level. From the Maveli stores or Ration stores, essential food items are distributed to the gruel centres during the flood for the affected people. The people in the flood-affected region will go to these stores to collect the food items for cooking in the gruel centres. Facilities are arranged at the gruel centres for cooking. Existing supply chain for the distribution of essential relief materials during the flood is shown in Figure 2.

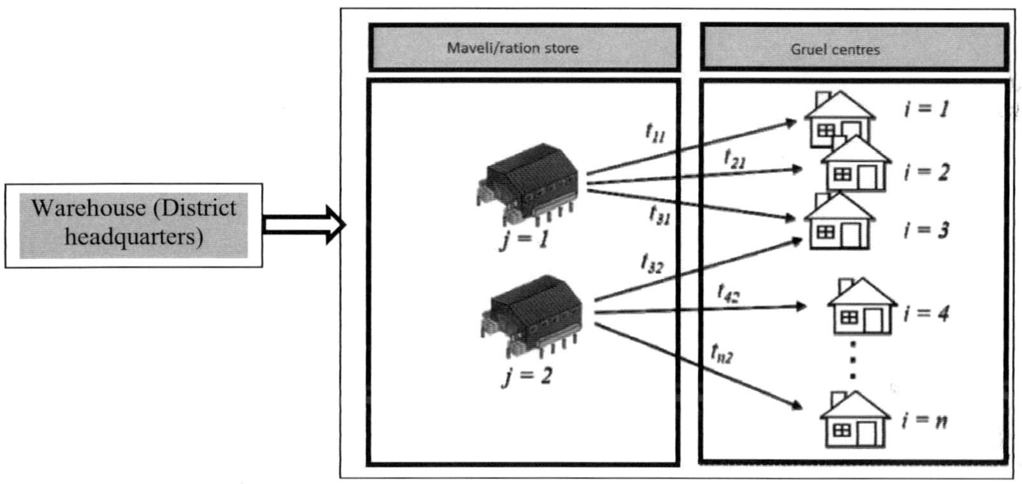

Figure 2. Existing supply chain distribution of essential relief materials during floods [21]

4. Permanent gruel centre allocation

The existing model requires a high level of coordination between the source (j) and destination (i). Many of the gruel centres in the exiting set-up are under-utilized and few of the gruel centres are affected during the severe flood conditions. A modified system is proposed for the humanitarian logistics operation during floods. Replacing all the temporary gruel centres to a few permanent gruel centres will reduce the transportation cost as well as the efforts required to coordinate the supply of essential relief materials during the flood. Permanent gruel centres can be a roofed concrete building at a reasonable height which ensures that flood wont affect its working. Further studies need to be conducted to establish permanent centres in the elevated region. Government agencies are also keen to develop such a system.

A typical model proposed for improving supply logistics is shown in Figure 3. The entire region of the flood-affected taluk will be divided into 'n' regions. The location of permanent gruel centres will be identified in that region based on the distance the affected people have to travel, distance from Maveli/Ration stores, and suitability of the location.

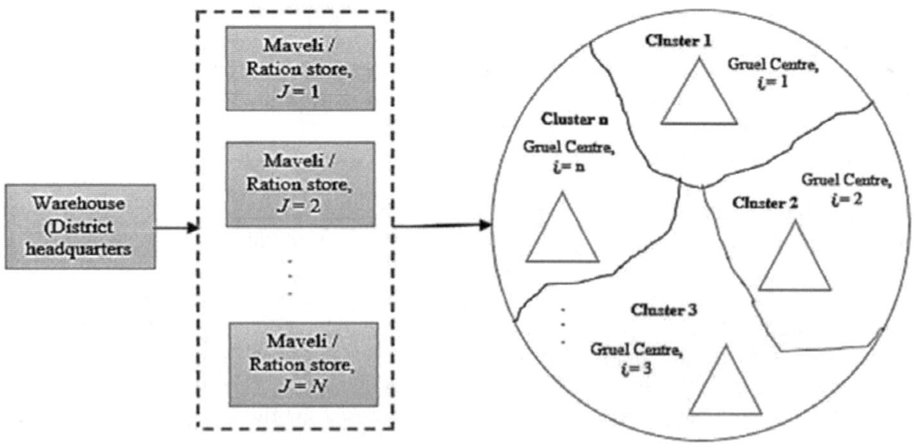

Figure 3. Proposed framework for the location of gruel centres

4.1 Grouping of Gruel centres using the k-means clustering algorithm

The *k*-means is a well-known partitioning algorithm [22]. Objects are grouped as belonging to one of '*k*' groups and the value of '*k*' is chosen as a priori. Cluster membership is determined by calculating the centroid for each group (the multidimensional version of the mean) and assigning each object to the group with the closest centroid. This approach minimizes the overall within-cluster dispersion by iterative reallocation of cluster members.

In a general sense, a k-partitioning algorithm takes as input a set '*S*' of objects and an integer '*k*', and outputs a partition of *S* into subsets $S_1, S_2, \ldots, S_k$. It uses the sum of squares as the optimization criterion. Let x^i_r be the rth element of S_i, $|S_i|$ be the number of elements in S_i, and $d(x^i_r, x^i_s)$ be the distance between x^i_r and x^i_s. The sum of squares criterion is defined by the cost function shown in equation (1).

$$c(S_i)=\sum_{r=1}^{|S_i|}\sum_{x=1}^{|S_i|}\left(d(x^i_r, x^i_s)\right)^2 \quad (1)$$

The algorithm, *k-means* works by calculating the centroid of each cluster S_i, denoted by x^{-i}, and optimizing the cost function given in equation (2)

$$c(S_i) = \sum_{r=1}^{|S_i|}\left(d(x^{-i},\ x^i_r)\right)^2 \quad (2)$$

The goal of the algorithm is to minimize the total cost as shown in equation (3).

$$\text{Total Cost } c(S) = c(S_i) + \ldots + c(S_k) \quad (3)$$

The *k*-means algorithm identifies '*k*' number of centroids and then assigns every data point to the nearest cluster. The *'means'* in the k-means refers to averaging of the data; that is, finding the centroid. We define a target number '*k*', which refers to the number of centroids that need to be considered in the data set. A centroid is the imaginary or real location representing the center of the cluster [23].

As people of 'Kuttanad' taluk does not like to go far from their home and hence allocating a single gruel center is not a feasible solution. Also, preparation and serving of food for a huge population will be a major issue. Hence 74 gruel centres of 'Kainakary' Grama Panchayath are reduced to two solutions by trial and error approach based on the mobility of the people reaching the permanent gruel centres. Based on the survey conducted in the affected region to understand the difficulty of individuals, two solutions were arrived by considering 6 permanent gruel centres and 8 permanent gruel centres by the *k*-means clustering technique. The number of permanent gruel centres were arrived by considering various factors like distance to the permanent gruel centres from the region where the affected people are living, the number of families that can be served from each gruel centre, the maximum inventory that can be stored in the gruel centre, proximity of the gruel centre to the supply sources, etc. The location of the permanent gruel centre is also ensured that it is in the elevated location which will not be affected by the flood and mobility of essential items can be possible from the Maveli / Ration stores. By studying the problem, it was observed that grouping of gruel centres into 6 and 8 cluster regions satisfy the location constraints.

4.1.1 Grouping of gruel centres into 6 cluster regions. Gruel centres are clustered into 6 groups by k-means clustering technique. For each group, there will be a centroid representing the center of the corresponding cluster where permanent gruel centre can be setup. Figure 4 shows the Google earth image of existing gruel centre locations, six clusters and its centroids (Permanent gruel centres) marked as *'A', 'B', ..., 'F'*.

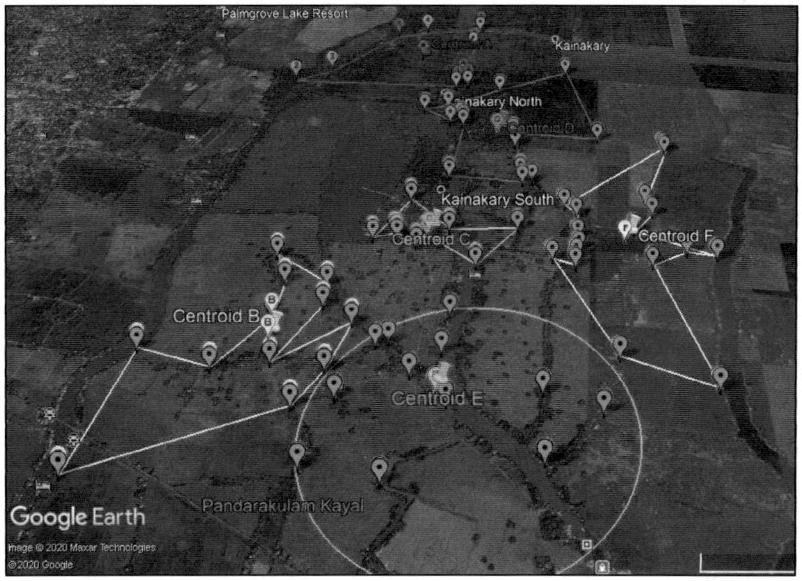

Figure 4. Grouping of gruel centres into 6 cluster regions

The existing gruel centres in the Kainakary Grama Panchayath coming under various centroids are given in Table 1.

Table 1. Permanent gruel centres ($k = 6$)

Sl. No.	The centroid of the cluster (Permanent gruel centres)	The label of gruel centres as referred to in Figure 1 and appendix 1	No. of gruel centres under each cluster
1	A	1,2,3,4,5,6,7,8,11	9
2	*B*	9,10,12,13,14,15,16,17,18,19,27,72,73,74	14
3	*C*	20,21,22,23,24,25,28,29,30,31,32,33,34,35,36,37,42	17
4	*D*	26,41,46,47,48,68,69,70,71	9
5	*E*	38,39,40,43,44,45,53,54,56,57,58,62	12
6	*F*	49,50,51,52,55,59,60,61,63,64,65,66,67	13

4.1.2 Grouping of gruel centres into 8 cluster regions The '*k*'-means clustering algorithm has been implemented by considering '*k*' = 8. Gruel centres are clustered into 8 groups. For each group, there will be a centroid representing the center of the corresponding cluster where permanent gruel centre can be setup. The location of the permanent gruel centre is also ensured that it is in the elevated location which will not be affected by the flood and also mobility of essential items can be possible from the Maveli / Ration stores. Figure 5 shows the Google earth image of existing gruel centre locations, eight clusters and its centroids (Permanent gruel centres) marked as '*A*', '*B*', ..., '*H*'.

IOP Conf. Series: Materials Science and Engineering 1132 (2021) 012013 doi:10.1088/1757-899X/1132/1/012013

Figure 5. Grouping of gruel centres into 8 cluster regions

The existing gruel centres in the Kainakary Grama Panchayath coming under various centroids are also given in Table 2.

Table 2. Permanent Gruel centres ($k = 8$)

Sl. No.	The centroid of the cluster (Permanent gruel centres)	The label of gruel centres as referred to in Figure 1 and Appendix 1	No. of gruel centres under each cluster
1	A	1,2,3,4,5,6,7,8,9,11,12,13,14,73,74	15
2	B	10,17,18,19,28,29	6
3	C	15,16,20,21,22,23,24,25,26,27,30,31,32,42	14
4	D	33,34,35,36,37,38,39	7
5	E	40,52,53,54,57,58	6
6	F	41,46,47,48,68,69,70,71,72	9
7	G	43,44,45,51,55,56,61,62	8
8	H	49,50,59,60,63,64,65,66,67	9

5. Results and Discussions

5.1. Cost of existing supply chain network

To justify the proposed approach to establishing a permanent gruel centre using a clustering approach, distance traveled by the people in the affected region from gruel centres to Maveli stores/Ration stores are computed. Distance from each of the 74 gruel centres to the nearby Ration shop or Maveli store to acquire food items determines the effectiveness of supply chain terms of getting the relief at the earliest. At present, food items are distributed from 2 Maveli stores and 9 Ration shops present in the Kainakary Grama Panchayath. A distance matrix is used to compute the total supply chain distance measure. It is found that a total of 55.88 km needs to be traveled by the people to reach the Maveli / Ration stores from the gruel centres in the present set-up. The details are shown in Appendix 1. Total cumulative distance traveled by the people to collect food items from the Maveli stores/Ration shop to the gruel centres indicates the level of difficulty in getting the relief material.

5.2. Supply chain distance measure for the proposed permanent gruel centre system

The distance matrix helps to determine the total distance to be traveled by the people from permanent gruel centres to their nearest supply sources. For both the conditions, number of permanent gruel centres as 6 and 8, the total distance is computed. The distance matrix is shown in Table 3 and Table 4. Table 3 shows the distance measure by considering the number of permanent gruel centres as 6. Table 4 shows the distance measure by considering number of permanent gruel centres as 8. The distance measure increases when number of gruel centres increased to 8.

Table 3. Distance from the centroid of clusters to the supply sources

Centroid of clusters	Supply Sources – Maveli Stores (MS) and Ration shops (RS) (Distance in km)											
	MS1	MS2	RS1	RS2	RS3	RS4	RS5	RS6	RS7	RS8	RS9	Distance to the nearest supply source
A	2.35	3.61	4.17	4.21	4.58	5.04	3.28	4.68	4.66	2.58	0.98	0.98
B	0.89	2.20	2.78	2.60	3.04	3.51	1.83	3.79	3.78	1.34	0.87	0.87
C	1.61	1.60	1.91	0.68	1.56	2.02	1.39	3.66	3.68	1.88	2.84	0.68
D	0.79	0.66	1.21	2.44	1.75	2.15	0.47	2.21	2.22	0.39	1.99	0.39
E	2.49	1.17	0.59	2.72	0.89	0.79	1.54	1.77	1.81	2.18	3.78	0.59
F	2.48	1.56	1.38	3.73	2.17	2.22	1.85	0.51	0.52	1.98	3.38	0.51
					Total distance to travel							4.03 km

Table 4. Distance from the centroid of clusters to the supply sources

Centroid of clusters	Supply Sources – Maveli Stores (MS) and Ration shops (RS) (Distance in km)											
	MS1	MS2	RS1	RS2	RS3	RS4	RS5	RS6	RS7	RS8	RS9	Distance to the nearest supply source
A	1.91	3.18	3.75	3.79	4.14	4.60	2.85	4.35	4.33	2.17	0.58	0.58
B	1.44	2.52	3.06	2.07	3.07	3.56	2.15	4.39	4.39	1.96	1.72	1.44
C	1.07	1.31	1.76	1.22	1.66	2.15	1.01	3.36	3.38	1.36	2.32	1.01
D	2.37	1.70	1.67	1.02	0.94	1.23	1.72	3.49	3.52	2.45	3.72	0.94
E	2.92	1.61	1.05	3.23	1.38	1.16	1.99	1.54	1.58	2.55	4.14	1.05
F	0.75	0.77	1.30	2.52	1.86	2.26	0.57	2.22	2.22	0.30	1.90	0.30
G	2.13	0.89	0.49	2.89	1.26	1.34	1.26	1.35	1.38	1.74	3.32	0.49
H	2.53	1.71	1.59	3.90	2.39	2.46	1.97	0.46	0.45	2.01	3.34	0.45
					Total distance to travel							6.26 km

By implementing the proposed framework and establishing permanent gruel centres, coordinating the supply of the essential items will be improved. It is also noted that the maximum distance traveled by any person from the Maveli / Ration store will be a maximum of 0.98 km in the case of establishing six permanent gruel centres. In the case of eight gruel centre, a maximum of 1.44 km distance has to be covered. The details of distance measures are shown in Table 5. It is clear that the proposed framework of setting up permanent gruel centres will reduce the total transportation distance and thereby reducing the transportation costs.

Table 5. Distance measures for the existing and proposed schemes with *k*= 6 and *k*= 8

Scheme	Distance measure	Distance value, km
Existing framework	Total distance to travel	55.88
	Max. distance from any of the temporary gruel centre to the any of the Maveli / Ration store	1.75
Proposed scheme with k = 6	Total distance to travel	4.03
	Max. distance from any of the permanent gruel centre to the any of the Maveli / Ration store	0.98
Proposed scheme with k = 8	Total distance to travel	6.26
	Max. distance from any of the permanent gruel centre to the any of the Maveli / Ration store	1.44

6. Conclusions

The location of the existing gruel centres and their large count leads to difficulty in supplying the essential food items during the time of flood. Co-ordinating the supply chain activities will be better if the demand points are reduced considerably. Also, the temporary gruel centres are not having a secure storage facility for storing groceries and other food items and is easily vulnerable to the flood. Thus, having permanent gruel centres with better facility for storing food items will suffice the need during floods.

The proposed gruel centre locations identified by the clustering method are well within the reach of the affected people. The total distance measure in a supply chain is 4.03 km and 6.26 km respectively by fixing 6 and 8 numbers of permanent gruel centres, respectively. Whereas in the exiting set-up with 74 gruel centres, the total distance from the entire gruel centres to their nearest Maveli / Ration store is 55.88 km. In the existing framework, the maximum distance from any of the temporary gruel centres to the nearest Maveli/Ration store is 1.75 km while for the proposed framework with 6 and 8 permanent gruel centres, the same has been 0.98 km and 1.44 km respectively. The clustering approach proposed in this study is justified for the efficient allocation of gruel centres considering the distance measure. The study can be extended further considering the inventory requirement at each gruel centres and the capacity allocation of the permanent gruel centres.

References

[1] Mollah A K, Sadhukhan S, Das P and Anis M Z 2018 *International Journal of Disaster Risk Reduction* **31** 1187–1198 : https://dx.doi.org/10.1016/ j.ijdrr.2017.11.018
[2] Safeer M, Anbuudayasankar S P, Balkumar K and Ganesh K 2014 *Procedia Engineering* **97** 2248–2258 : https://dx.doi.org/10.1016/j.proeng.2014.12. 469
[3] Safeer M, Anbuudayasankar S P, Saleeshya P G and Ganesh K 2016 *Int. J. Enterprise Network Management* **7** 250–259

[4] Long D 1997 *IIE Solutions* **29** 26–30
[5] Wassenhove L V 2006 *Journal of the Operational Research Society* **57** 475–489
[6] Altay N and Green W G 2006 *European Journal of Operational Research* **175** 475–493: https://dx.doi.org/10.1016/j.ejor.2005.05.016
[7] Kovács G and Spens K M 2007 *International Journal of Physical Distribution & Logistics Management* **37** 99–114: https://dx.doi.org/10. 1108/09600030710734820
[8] Charles A, Lauras M and Wassenhove L V 2010 *International Journal of Physical Distribution & Logistics Management* **40** 722–741: https://dx.doi.org/10.1108/09600031011079355
[9] Suresh M, Ganesh S and Raman R 2019 *International Journal of Agile Systems and Management* **12** 108–123
[10] Balcik B and Beamon B M 2008 *International Journal of Logistics Research and Applications* **11** 101–121: https://dx.doi.org/10.1080/ 13675560701561789
[11] Hong X, Lejeune M A and Noyan N 2015 *IIE Transactions* **47** 329–357: https://dx.doi.org/10.1080/0740817x.2014.919044
[12] Caunhye A M, Nie X and Pokharel S 2012 *Socio-Economic Planning Sciences* **46** 4–13: https://dx.doi.org/10.1016/j.seps.2011.04.004
[13] Tzeng G H, Cheng H J and Huang T D 2007 *Transportation Research Part E: Logistics and Transportation Review* **43** 673–686: https://dx.doi.org/ 10.1016/j.tre.2006.10.012
[14] Jalali R, Safari H, Momeni M and Moghadam M S 2018 *Management Science Letters* **8** 259–270: https://dx.doi.org/10.5267/j.msl.2018.4.022
[15] Forero M G and Rodriguez L G 2020 *International Journal of Industrial Engineering Computations* **11** 153–172
[16] Gupta P, Khanna A and Majumdar S 2012 *Indian Journal of Community Medicine* **37** 185–185: https://dx.doi.org/10.4103/0970-0218.99928
[17] Sreedharan V R, Kek V, Dhanya M, Anjali S and Arunprasad P 2020 *International Journal of Logistics Systems and Management* **36** 463–463
[18] Özdamar L and Demir O 2012 *Transportation Research Part E: Logistics and Transportation Review* **48** 591–602: https://dx.doi.org/10.1016/j.tre.2011. 11.003
[19] Lima F S, de Oliveira D, Gonçalves M B and Samed M M A 2014 *Journal of technology management & innovation* **9** 86–97: https://dx.doi. org/10.4067/s0718-27242014000200007
[20] Gutierrez M E and Mutuc J E S 2018 *Procedia Engineering* **212** 659–666: https://dx.doi.org/10.1016/j.proeng.2018.01.085
[21] Manopiniwes W, Nagasawa K and Irohara T 2014 *Industrial Engineering and Management Systems* **13** 398–407: https://dx.doi.org/10.7232/iems. 2014.13.4.398
[22] Macqueen J 1967 *Proceedings of the fifth Berkeley symposium on mathematical statistics and probability* **233** 281–297
[23] Osama A 2008 *International Arab Journal of Information Technology* **5** 320–325

Appendix A

Table A1 Supply chain transportation distance in the existing network

Gruel centre	Distance to the supply Sources–Maveli Stores (MS) and Ration shops (RS) from Gruel centres in km											Distance to the nearest supply source in km
	MS1	MS2	RS1	RS2	RS3	RS4	RS5	RS6	RS7	RS8	RS9	
1	2.93	4.17	4.72	4.8	5.16	5.61	3.85	5.11	5.08	3.13	1.55	1.55
2	2.69	3.65	4.14	4.91	4.75	5.14	3.43	4.07	4.03	2.65	1.54	1.54
3	2.58	3.66	4.17	4.74	4.73	5.14	3.4	4.27	4.24	2.62	1.3	1.3
4	2.92	4.2	4.76	4.67	5.15	5.62	3.87	5.26	5.23	3.17	1.58	1.58
5	2.95	4.26	4.84	4.43	5.13	5.61	3.91	5.54	5.52	3.29	1.75	1.75
6	2.57	3.88	4.46	4.14	4.77	5.24	3.53	5.16	5.14	2.91	1.37	1.37

7	1.89	3.19	3.77	3.62	4.1	4.57	2.84	4.49	4.47	2.21	0.7	0.7
8	1.75	3.06	3.63	3.48	3.96	4.43	2.71	4.39	4.37	2.08	0.63	0.63
9	1.66	2.97	3.55	3.18	3.81	4.29	2.61	4.46	4.44	2.06	0.86	0.86
10	2.21	3.47	4.03	3.15	4.14	4.63	3.09	5.14	5.13	2.69	1.66	1.66
11	1.73	3.03	3.61	3.54	3.96	4.43	2.69	4.32	4.3	2.04	0.54	0.54
12	1.37	2.66	3.23	3.3	3.61	4.07	2.32	3.94	3.93	1.66	0.25	0.25
13	1.14	2.45	3.02	3.06	3.37	3.84	2.1	3.82	3.8	1.47	0.4	0.4
14	1.01	2.31	2.89	2.95	3.24	3.7	1.97	3.72	3.71	1.35	0.49	0.49
15	0.49	1.59	2.15	1.98	2.33	2.81	1.21	3.41	3.41	1	1.48	0.49
16	0.57	1.53	2.08	1.82	2.21	2.69	1.15	3.41	3.41	1.05	1.64	0.57
17	0.76	2.02	2.59	2.31	2.8	3.28	1.64	3.72	3.72	1.26	1.14	0.76
18	0.92	2.21	2.78	2.47	2.99	3.47	1.83	3.86	3.86	1.4	1.02	0.92
19	1.45	2.46	2.98	1.89	2.95	3.44	2.09	4.37	4.37	1.97	1.87	1.45
20	1.26	0.89	1.24	1.33	1.13	1.62	0.71	2.95	2.97	1.35	2.63	0.71
21	1.38	1.1	1.42	1.12	1.19	1.67	0.92	3.15	3.17	1.52	2.72	0.92
22	1.33	1.16	1.5	1.09	1.29	1.78	0.95	3.22	3.24	1.51	2.65	0.95
23	1.22	1.22	1.61	1.12	1.45	1.93	0.97	3.29	3.3	1.45	2.52	0.97
24	1.07	1.39	1.84	1.21	1.74	2.23	1.08	3.44	3.45	1.4	2.29	1.07
25	0.92	1.4	1.88	1.37	1.86	2.34	1.06	3.42	3.43	1.28	2.12	0.92
26	0.76	0.87	1.39	1.68	1.54	2.02	0.52	2.87	2.88	0.89	2.13	0.52
27	0.48	1.47	2.03	1.89	2.19	2.67	1.09	3.33	3.33	0.96	1.6	0.48
28	1.99	2.76	3.21	1.56	3.01	3.47	2.42	4.77	4.78	2.48	2.53	1.56
29	1.98	2.73	3.18	1.5	2.96	3.42	2.39	4.75	4.76	2.47	2.56	1.5
30	1.65	2.03	2.42	0.84	2.14	2.6	1.75	4.09	4.11	2.04	2.65	0.84
31	1.73	1.99	2.34	0.66	2.01	2.46	1.73	4.06	4.07	2.1	2.8	0.66
32	1.63	1.72	2.05	0.66	1.71	2.16	1.49	3.78	3.8	1.93	2.8	0.66
33	1.95	1.73	1.93	0.48	1.43	1.84	1.59	3.74	3.76	2.17	3.2	0.48
34	2.15	2.07	2.29	0.14	1.77	2.16	1.9	4.1	4.12	2.43	3.32	0.14
35	2.4	2.31	2.5	0.12	1.93	2.28	2.15	4.32	4.34	2.69	3.55	0.12
36	3.15	2.37	2.2	1.49	1.37	1.39	2.45	3.92	3.96	3.2	4.49	1.37
37	2.47	1.57	1.41	1.43	0.6	0.82	1.69	3.19	3.22	2.45	3.85	0.6
38	2.87	1.79	1.45	1.97	0.68	0.49	2	3.05	3.09	2.78	4.26	0.49
39	2.56	1.37	0.97	2.1	0.33	0.16	1.63	2.58	2.61	2.39	3.93	0.16
40	3.16	1.93	1.44	2.58	0.96	0.47	2.22	2.71	2.75	2.97	4.53	0.47
41	1.1	0.28	0.82	2.01	1.19	1.62	0.15	2.32	2.33	0.92	2.45	0.15
42	1.45	1.06	1.34	1.14	1.08	1.56	0.92	3.1	3.12	1.56	2.8	0.92
43	1.68	0.37	0.21	2.3	0.89	1.19	0.74	1.91	1.93	1.39	2.99	0.21
44	2.09	0.77	0.19	2.5	0.81	0.94	1.14	1.79	1.82	1.78	3.38	0.19
45	2.06	0.82	0.44	2.84	1.23	1.34	1.19	1.39	1.42	1.67	3.26	0.44
46	0.97	0.58	1.08	2.53	1.7	2.07	0.5	2.04	2.04	0.54	2.13	0.5
47	0.97	0.78	1.25	2.71	1.9	2.26	0.7	2	1.99	0.48	2.02	0.48
48	1.13	0.93	1.34	2.92	2.04	2.37	0.9	1.86	1.85	0.61	2.05	0.61
49	2.94	2.03	1.8	4.18	2.55	2.53	2.33	0.04	0.08	2.44	3.78	0.04
50	2.63	1.62	1.35	3.74	2.1	2.1	1.94	0.48	0.52	2.14	3.59	0.48
51	2.49	1.33	0.96	3.37	1.68	1.67	1.69	0.91	0.95	2.04	3.57	0.91
52	2.88	1.72	1.3	3.69	1.93	1.82	2.08	0.8	0.85	2.43	3.95	0.8
53	2.69	1.46	1	3.37	1.61	1.52	1.83	1.06	1.1	2.26	3.82	1
54	3.32	2.07	1.56	3.83	1.99	1.75	2.45	1.24	1.28	2.9	4.45	1.24
55	2.49	1.33	0.96	3.37	1.68	1.67	1.69	0.91	0.95	2.04	3.57	0.91

56	2.34	1.04	0.49	2.81	1.05	1.04	1.42	1.55	1.59	1.99	3.59	0.49
57	3.3	1.99	1.42	3.45	1.6	1.27	2.37	1.76	1.81	2.94	4.54	1.27
58	2.59	1.27	0.7	2.71	0.86	0.7	1.63	1.87	1.91	2.29	3.89	0.7
59	1.76	1.05	1.15	3.24	1.97	2.18	1.24	1.21	1.21	1.25	2.68	1.05
60	1.95	1.08	1.05	3.28	1.88	2.04	1.33	1.04	1.05	1.46	2.92	1.04
61	1.98	0.92	0.75	3.07	1.58	1.72	1.23	1.15	1.17	1.53	3.07	0.75
62	2.12	0.9	0.54	2.94	1.33	1.42	1.27	1.29	1.32	1.71	3.29	0.54
63	3.25	2.53	2.38	4.72	3.17	3.18	2.78	0.62	0.58	2.73	3.87	0.58
64	2.3	1.49	1.4	3.69	2.22	2.33	1.74	0.67	0.67	1.79	3.15	0.67
65	4.2	3.23	2.89	5.29	3.52	3.36	3.55	1.23	1.24	3.69	4.98	1.23
66	1.99	1.41	1.48	3.59	2.31	2.49	1.58	1.04	1.03	1.47	2.75	1.03
67	1.9	1.57	1.75	3.7	2.56	2.78	1.67	1.31	1.29	1.39	2.5	1.29
68	0.81	0.66	1.19	2.46	1.75	2.15	0.48	2.19	2.19	0.4	2	0.4
69	0.59	1.19	1.72	2.73	2.27	2.68	0.95	2.46	2.45	0.18	1.49	0.18
70	0.95	0.84	1.3	2.74	1.95	2.31	0.74	2.02	2.02	0.44	1.97	0.44
71	0.67	0.68	1.25	2.28	1.71	2.14	0.38	2.38	2.39	0.4	1.95	0.38
72	0.26	1.46	2.03	2.55	2.44	2.89	1.13	2.94	2.93	0.5	1.18	0.26
73	1.09	2.38	2.95	3.09	3.33	3.79	2.04	3.71	3.69	1.39	0.35	0.35
74	1.39	2.63	3.19	3.44	3.62	4.07	2.31	3.79	3.77	1.6	0.01	0.01
Total distance travelled												55.88 km

Design and Analysis of a Self-balancing Bicycle Model

Nitheesh kumar G [1,*], Navneeth S [1], Suraj A [1], and Pramod Sreedharan [1]

[1] Department of mechanical engineering, Amrita Vishwa Vidyapeetham, Amritapuri, India

[*] E-mail: nitheeshkumar.g@gmail.com

Abstract. This paper discusses the design and analysis of a self-balancing bicycle model. The objective was to develop an efficient design that can be fabricated in the future. The different methods of balancing a bicycle were studied to develop an optimal design. Solidworks 17 and Ansys 18 have been used for modeling, simulation, and analysis of the structure. The use of a Control Moment Gyroscope (CMG) to balance the bicycle model was studied and the results show that the effect of precession increases with an increase in rpm and the weight of the flywheel. Thus, a bicycle model actuated with CMG is far more stable and less prone to accidental tilts and toppling than one without. The studied data can be used for future research.

1. Introduction

With the evolution of mankind, safety has become a necessity over luxury and comfort in this fast-moving world of transportation. And the implementation of safety in motorcycling always stayed a challenge. Recent studies [1, 2] show that most MTW (Motorized Two-Wheeler) accidents are caused by instability of the vehicle, increasing through the decades, and only an efficiently designed self-balancing two-wheeler can decrease this trend. Moreover, facilitating the commute and recreation of the physically disabled community [3], by efficiently balancing a two-wheeler, was found in demand. Thus, studied the different methods to balance a bicycle with the less human intervention [4].

From further research, it was concluded that CMG can efficiently serve the purpose of balancing a bicycle [4, 5]. CMG is an attitude control device generally used in spacecraft attitude control systems [6]. It consists of a spinning rotor, attached to two gimbals, producing large constant angular momentum. The vector direction of angular momentum can be changed for a two-wheeler by rotating the spinning rotor. A gyroscopic reaction torque orthogonal to both rotor spin and gimbal axes is produced as a result of precession torque applied by the spinning rotor. It amplifies torque because a small gimbal torque input produces a large control torque to the two-wheeler. This method can provide a large torque even though the energy consumption of CMG is very high (as the flywheel is spinning all the time).

For an efficient and manufacturable design, mathematical modeling and design calculations play a major role [7]. Recently, several automobile manufacturing giants have come up with self-balancing technology in their two-wheelers [8] to assist the new era of mankind.

As an initial step to the development of a self-balancing two-wheeler, studied the various parameters that effect the balance of the bike due to the influence of CMG in the system using Solidworks and portrayed the analysis related to its development using Ansys software.

IOP Conf. Series: Materials Science and Engineering 1132 (2021) 012014 doi:10.1088/1757-899X/1132/1/012014

2. System Description

The project was initiated with a 2D sketch of the prototype which then helped to develop a 3D model of the same by using Solidworks. The development of the 3D model was succeeded by the finite element analysis of the individual components using Ansys software. The design was dynamically analyzed using motion analysis in Solidworks. Further, the static analysis of the components verified the structural rigidity for manufacturability. The result of the work was obtained as graphs from the motion analysis.

The main objective was to study the self-balancing nature of a two-wheeler aided with a CMG. Thus, studied the effects of the speed of rotation of the flywheel (part of CMG) on a two-wheeler, with a finite moment of inertia and the time it takes for the setup to completely topple (horizontal to the ground).

3. Mathematical Modelling

Concepts and Formulae:

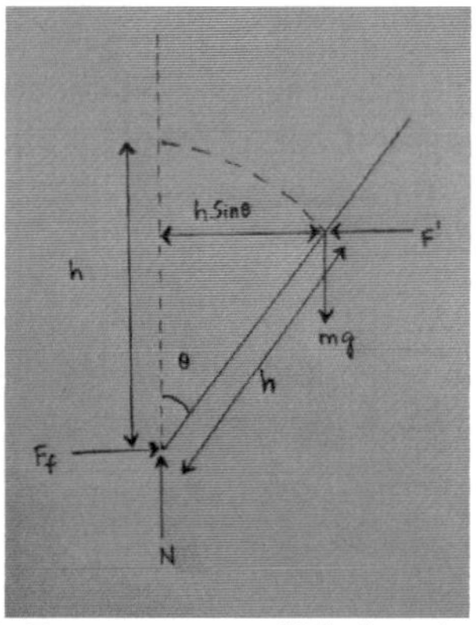

Figure 1. Forces on a self-balancing bicycle

According to 'Figure 1', the mass of the system is m kg, distance/location of the center of mass (m) from the ground is h (m). When the angle of tilt is θ, then the torque induced τ is given by

$$\tau = m\, g\, h \sin\theta \tag{1}$$

and the Reactive Gyroscopic Torque (τ') given as

$$\tau' = m\, g\, h \sin\theta \text{ (opposite direction)} \tag{2}$$

also,

$$\tau' = I\, \omega\, \omega p \tag{3}$$

where I is the moment of inertia of the disc, ω the angular speed of the disc, and ωp the precession speed of the disc.

On comparing equation (2) and equation (3),

$$m\, g\, h \sin\theta = I\, \omega\, \omega p \tag{4}$$

Therefore,

$$\omega p = m\, g\, h \sin\theta / I\, \omega \tag{5}$$

It is known that maximum frictional force can be written as

$$F'max = \mu\, N = \mu\, m\, g \tag{6}$$

Where μ is the coefficient of friction and N the normal reaction force.
Hence, at Limiting Condition

$$m\, g\, h \sin\theta = F'h \cos\theta \tag{7}$$

on comparing equation (6) and equation (7), the result is

$$m\, g\, h \sin\theta = \mu\, m\, g\, h \cos\theta \tag{8}$$

IOP Conf. Series: Materials Science and Engineering 1132 (2021) 012014 doi:10.1088/1757-899X/1132/1/012014

Therefore,

$$\mu = \tan\theta \tag{9}$$

where μ is the coefficient of friction and θ the angle of tilt produced.

The following values are taken from real-world objects available in the market for the best result in the simulation and also for better manufacturability.

Mass of DC motor = 1.06 kg
Mass of battery (12v) = 1.3 kg
Mass of wheels (10'') = 0.8 kg
Mass of Flywheel Disc = 4.165 kg
Mass of Frame = 2.504 kg
Mass of chassis = 2.3 kg
Miscellaneous mass (Nuts, bolts, etc) = 1.1 kg
Total mass = 1.06+1.3+0.8+4.165+2.504+2.3+1.1 = 13.234 kg

Centre of Mass (CM) from the ground, h = [(CMwheel * Mwheel) + (CMchassis * Mchassis) + (CMframe * Mframe) + (CMdisc * Mdisc) + (CMhub motor * Mhub motor) + (CMbattery * Mbattery) + (CMlinkage * Mlinkage)] / Total Mass (10)
Where CM = Centre of Mass

Calculation of center of mass,
Centre of mass from the ground, (h) = (12.7*0.81) +(25.2*2.3) +(30.7*2.3) +(27.2*4.16) +(18.8*1.06) +(1.3*40.7) +(1.1*18.8)] / 13.234
= 26.087 cm
= 0.26087 m

Moment of inertia of the disc (I) can be written as

$$I = mr^2/2 \tag{11}$$

Where m is the mass of the flywheel and r the radius of the flywheel. Hence
= 4.16*(0.075*0.075)/2
= 0.0117 kg/m^2

The simulation was carried out with a variety of RPMs for comparison. Used rpm value '3350', based on a motor commonly available in the market.

Speed of disc, N = 3350 rpm and so, the Angular Speed of Disc, ω = 350.81 rad/sec.

The disc mentioned here is the flywheel (part of CMG), which was initially planned to manufacture in-house. But later on, decided to go with the most easily available and commercially viable option. The vehicle is designed for the maximum tilt angle equal to 30°.
Therefore, the highest precision speed of the disc, as in equation (5), is

$$\omega p = mgh\sin\theta / I\omega$$
= 13.234*9.81*0.26*sin (30) / 0.0117*350.811
= 4.11 rad/sec

And the highest gyroscopic reaction torque, as in equation (3), is

$$\tau' = I\omega\omega p$$
= 0.0117*350.81*3.479
= 16.85 Nm

IOP Conf. Series: Materials Science and Engineering 1132 (2021) 012014 doi:10.1088/1757-899X/1132/1/012014

4. Design
4.1. Assembly drawing

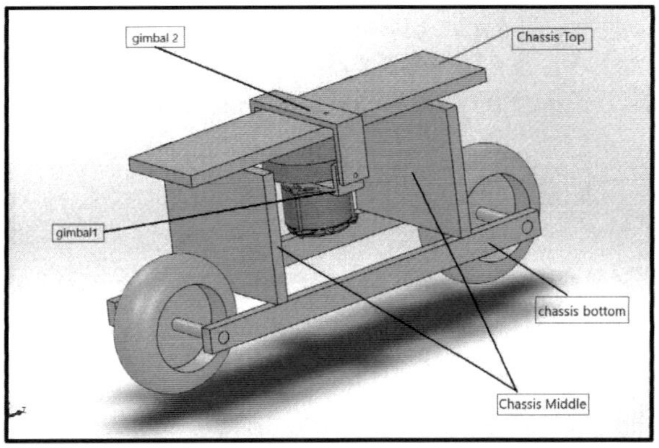

Figure 2. Assembly Drawing of the Prototype

First, a free body diagram in pen and paper was made. It involved many iterations with minimalism and sleek design as a target. The 2D image was first drawn in CAD, and then the part and assembly drawing in Solidworks ‘Figure 2’. Detailed descriptions of all Solidworks parts are given in ‘Table 1’.

Table 1. Details of Solidworks part drawings

Sl. No	Part	Part Description
1	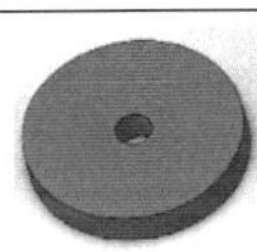	**Flywheel.** A disc-shaped solid piece of mild steel with a bore at its geometric center. This is an essential part of the CMG as its rotation helps to maintain the gyroscopic effect needed to keep the two-wheeler’s balance.
2	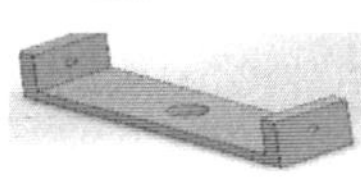	**Gimbal 1.** This part made of mild steel is necessary to give the second axis of the gyroscope the required free movement to facilitate the production of precession force.
3		**Gimbal 2.** This part attaches the whole gyroscope to the main chassis, giving the CMG its second axis. This is also made of mild steel.
4		**Shaft.** It connects the flywheel to the motor. Helps retain the rotation. This is made of mild steel.
5		**Motor.** A DC motor of maximum speed 3000 rpm. The speed and acceleration of the motor can be controlled using Arduino.
6		**Chassis Top.** A well-sized plank made of wood makes the top portion of the prototype chassis. The gyroscope is connected to this.
7		**Chassis Middle.** The plate that makes up the middle portion of the chassis. It is also made up of wood to reduce the weight and also for the ease of manufacturability.
8		**Chassis Bottom.** The plank makes up the lowest portion or base of the chassis. The middle chassis and the wheels are connected to this part.
9		**Wheels.** The wheels of desired size 10 inches. This can be purchased from a regular retailer in the market.

5. Results and discussion

5.1. Static Analysis

To test the strength of the materials under static load, the finite element method analysis of all the chassis components was done. It helped to understand the limitations of the designed chassis based on stress and buckling. The intention of this was to understand the amount of load and weight that each component can withstand individually, so as it can be manufactured efficiently. These 'Table 2' and 'Table 3' are the properties we considered for our FEM analysis models in Ansys Software.

Table 2. FEM wood properties [9]

WOOD			
Sl. No	**Property**	**Value**	**Unit**
1	Density	0.56	g cm^{-3}
2	Young's Modulus	6963	MPa
3	Poisson's Ratio	0.415	
4	Bulk Modulus	1.3653 e^{10}	Pa
5	Shear Modulus	2.4604 e^{9}	Pa
6	Tensile ultimate Strength	31	MPa

Table 3. FEM Mild steel properties

MILD STEEL			
Sl. No	**Property**	**Value**	**Unit**
1	Density	7.85	g cm^{-3}
2	Young's Modulus	2.1 e^{5}	MPa
3	Poisson's Ratio	0.303	
4	Bulk Modulus	1.7766 e^{11}	Pa
5	Shear Modulus	8.0583 e^{10}	Pa
6	Tensile ultimate Strength	31	MPa

5.1.1. PART 1: Gimbal 1. These models 'Figure 3' and 'Figure 4' represent the total deformation and equivalent stress distribution of the gimbal that is under the direct load of the motor and flywheel. These simulations are magnified to a significant factor (4.1 e^{3}) to show the deformation.

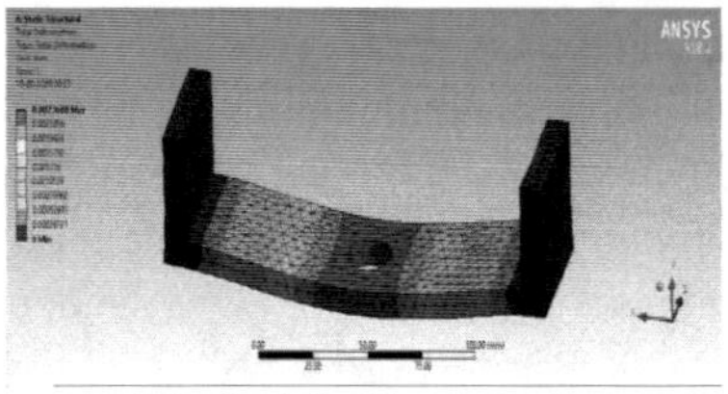

Figure 3. Gimbal 1- Total Deformation

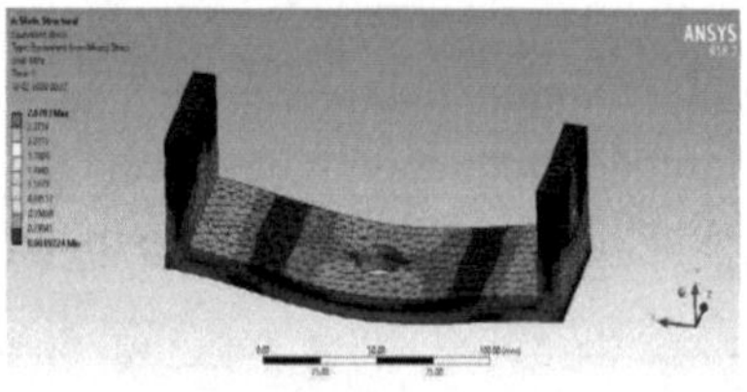

Figure 4. Gimbal 1 - Equivalent Stress

5.1.2. PART 2: Gimbal 2. This gimbal 'Figure 5' and 'Figure 6' supports the weight of the motor, flywheel, and gimbal-1. It is supported at the bottom surface by the 'chassis top' portion.

Figure 5. Gimbal 2 - Total Deformation

Figure 6. Gimbal 2 - Equivalent Stress

5.1.3. PART 3: Chassis Top. This is the top portion of the chassis, supporting both the gimbals. Wood being the material used, especially considered the forces, stress, and deformation that this may undergo. FEM results, 'Figure 7' and 'Figure 8', reconfirmed the ability of pre-determined dimensions to support the components.

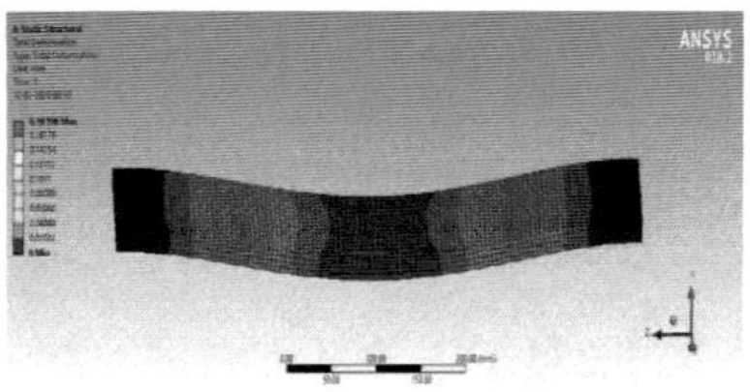

Figure 7. Chassis Top - Total Deformation

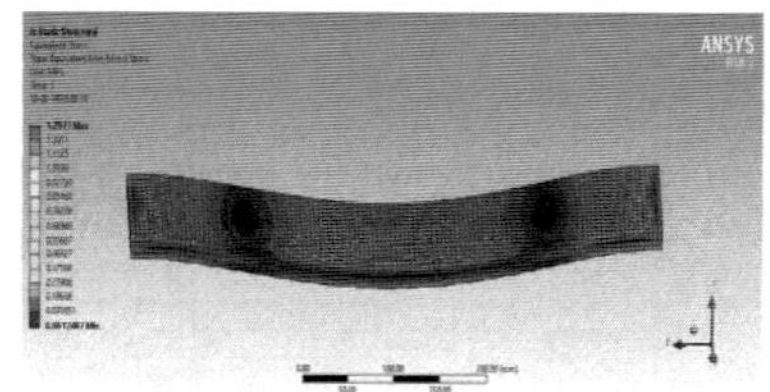

Figure 8. Chassis Top - Equivalent Stress

5.1.4. PART 4: Chassis middle. This, 'Figure 9' and 'Figure 10', the plywood component that occupies the front and back of the chassis, is not under much force as each of these experiences only half of the total force acted upon by the 'Chassis Top' of the body.

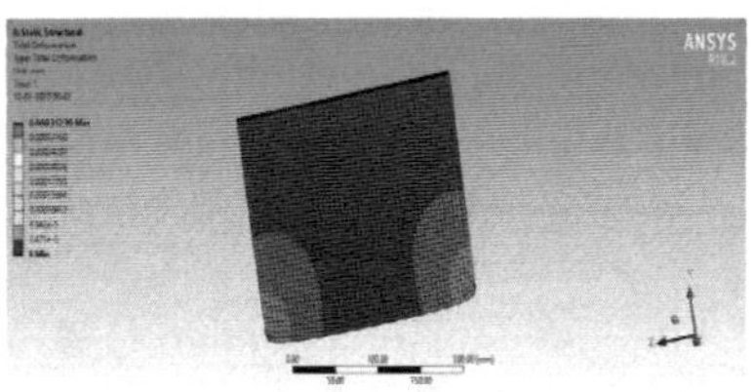

Figure 9. Chassis Middle - Total Deformation

Figure 10. Chassis middle - Equivalent Stress

5.1.5. PART 5: Chassis Bottom. The bottom chassis components 'Figure 11' and 'Figure 12' are connected to the wheels and also support the entirety of the chassis forces. So, the thickness and rigidity of this component were always under scrutiny. But here too, the FEM analysis shows that the pre-determined dimensions are well within the ability to hold the prototype together.

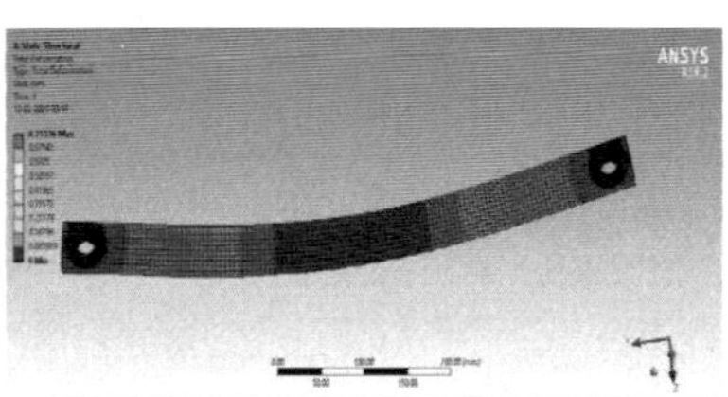

Figure 11. Chassis Bottom - Total Deformation

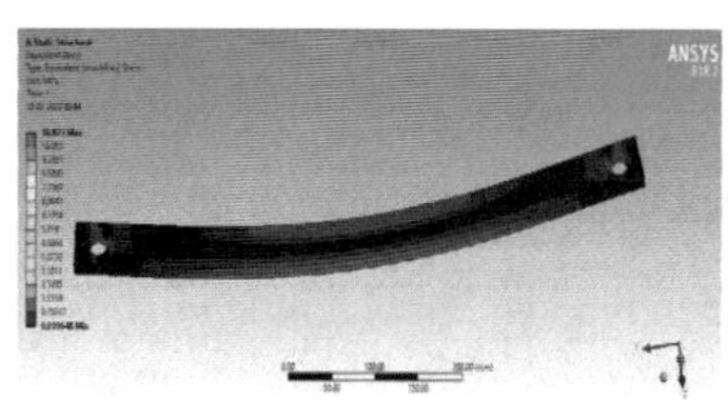

Figure 12. Chassis Bottom - Equivalent Stress

5.1.6. Analysis Result. From the analysis, the maximum amount of distortion that the part undergoes (in mm), the maximum stress formed (in MPa) under the effects of applied forces, support force, and from its weight were found 'Table 4'. Also given a considerable amount of extra force for a decent margin of safety in the analysis.

Table 4. FEM analysis result

Part	Maximum Deflection (in mm)	Maximum Stress (in MPa)
Gimbal 1	0.00236800	2.67030
Gimbal 2	0.00035759	0.94964
Chassis Top	0.18198000	1.29770
Chassis Middle	0.00031239	7.90910
Chassis Bottom	0.75536000	10.82700

IOP Conf. Series: Materials Science and Engineering 1132 (2021) 012014 doi:10.1088/1757-899X/1132/1/012014

5.2. Dynamic Analysis

The simulation of the prototype in Solidworks 'Figure 13' and 'Figure 14' gave a rough picture of how the prototype will work in real life. The motion analysis also gave the option to try out various scenarios in the prototype design such as using constant and varying rpm, with and without external force, fixed and loose gimbal axes, and many more. Each of these scenarios can be an individual project on its own. The main focus was on knowing and establishing the effect of a working gyroscope in a two-wheeler model.

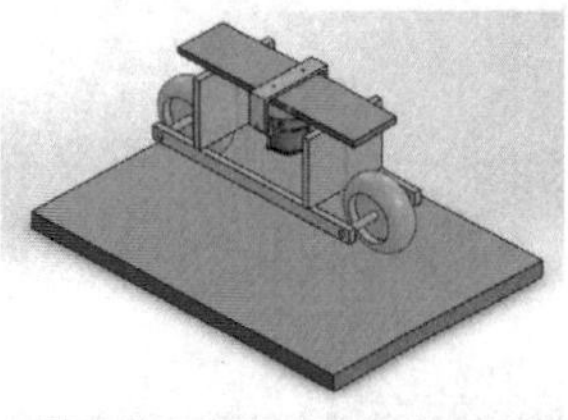

Figure 13. Model orientation at 0 seconds

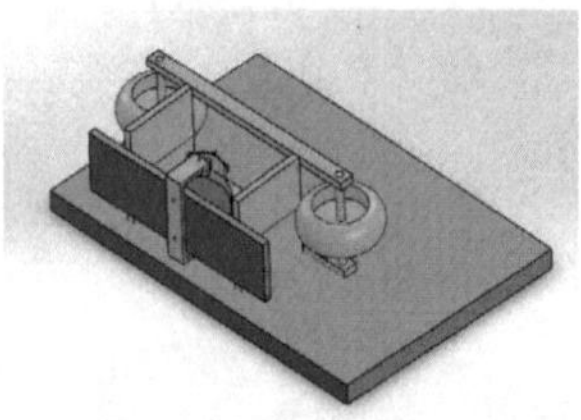

Figure 14. Model orientation at 3.2 seconds

5.2.1. Case 1: Stationary Flywheel (0 rpm). This case is to establish and analytically study the normal day two-wheelers which do not have a gyroscope with them. This Tilt vs Time graph 'Figure 15' clearly shows the time taken to fall from 90 to 0 degrees is about 1.28 seconds.

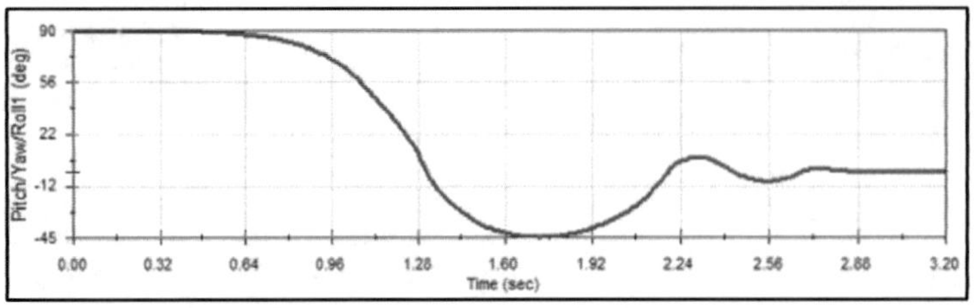

Figure 15. Tilt vs Time (0 rpm)

5.2.2. Case 2: Flywheel at 1000 rpm. When the flywheel is in a constant rotation of about 1000 rpm, the tilt of the body is slowed down due to the gyroscopic effect. The precession force acts on the horizontal axes of the gimbal. The Tilt Vs Time graph 'Figure 16' shows that the tilt starts at about 1.6 seconds and reaches 0° at 2.3 seconds (approximate value).

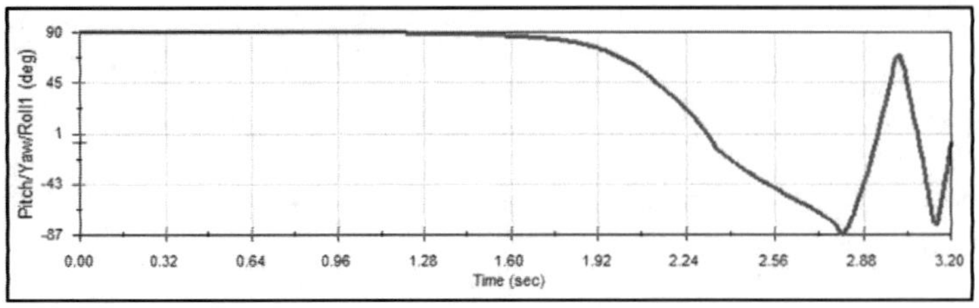

Figure 16. Tilt vs Time (1000 rpm)

5.2.3. Case 3: Flywheel at 3000 rpm. The maximum rpm of the motor, chosen for fabrication, is 3000rpm and the result of the Tilt vs Time graph 'Figure 17' was found for the same. The result shows that the greater the rpm of the flywheel, the longer it takes the body to tilt to 0°.

In the simulation, even after the 3.2-second mark, the body has tilted only about 32°.

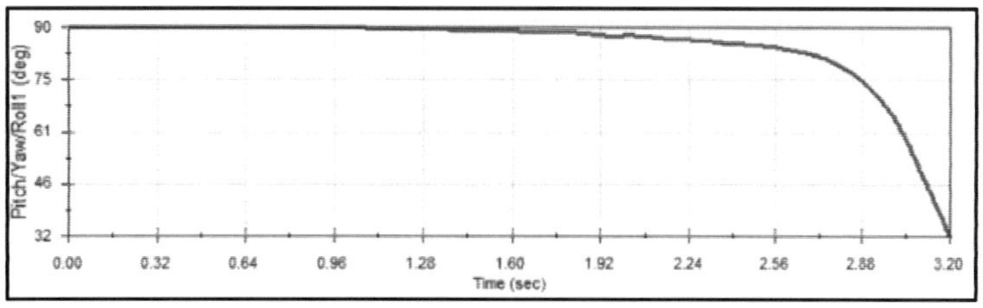

Figure 17. Tilt vs Time (3000 rpm)

5.2.4. Analysis Result. The dynamic analysis results are shown in 'Table 5'.

Table 5. Dynamic Analysis Result

Sl. No	RPM	Tilt Angle	Time taken
1	0	0-90	1.28
2	1000	0-90	2.30
3	3000	0-90	3.20

6. Conclusion

The result attained from the dynamic analysis shows that the time taken for the bicycle model to completely topple (0 to 90 degrees) increases with an increase in rpm of the flywheel and thus shows that the effect of precession increases with an increase in rpm and the weight of the flywheel. Further, the static analysis result reconfirms the appropriate material selection and the ability of each component to withstand the foresighted loads to manufacture efficiently. Thus, this study proves that the use of a gyroscope as an actuator can and is a plausible method of making a self-balancing bicycle and is far more stable and less prone to accidental tilts and toppling than one without a gyroscope. This prototype can be manufactured and the data studied can be used for further research.

References

[1] Taraknath Taraphdar, Wasil Rasool Sheikh and Sourav Kumar 2014 Motorized Two Wheeler Accidents Caused by Animals: A Study *ISEMT-2014* 103

[2] George Yannis, Constantinos Antoniou, Petros Evgenikos, Panagiotis Papantoniou, and Alan Kirk 2012 Characteristics and Causes of Power Two Wheeler Accidents in Europe *Procedia-Social and Behavioral Sciences* **48** 1535-44

[3] https://en.wikipedia.org/wiki/Physical_disability

[4] N. Tamaldin, H.I.M. Yusof, M.F.B. Abdollah, G. Omar, and M.I.F. Rosley 2017 Design self-balancing bicycle *MERD'17* 160-161

[5] Qusai Suwan, Shirine El Zaatari, and Kareem Kassem 2017 Design and Evaluation of an Autonomous Self-balancing Bicycle *16th FEASAC*

[6] https://en.wikipedia.org/wiki/Control_moment_gyroscope

[7] Hemashree Kakar 2018 Design Calculations Of A Two Wheeler Self Balanced Vehicle *IRJET* **5** 2066-75

[8] https://global.yamaha-motor.com/about/design/concept/motoroid/
https://www.bmw-motorrad.com/en/experience/stories/brand/vision-next-100.html

[9] Md. Ashraful Alam, Vikram Yadama, William F. Cofer and Karl R. Englund 2012 Analysis and evaluation of a fruit bin for apples *J Food Sci Technol-2012*

Porous superhydrophobic membranes as safe bubble absorbers for hydrocarbon industry

Kiran Raj M

Department of Mechanical and Mechatronics Engineering, University of Waterloo, Canada

kiranrajm@outlook.com

Abstract. Superhydrophobic surfaces, which repels water droplets falling on them is a hot topic in the interfacial engineering for their wide range of applications from self-cleaning to thermal management. Recently, porous superhydrophobic surfaces are introduced to the front by incorporating the element of diffusion of gases along with the extreme non-wettability of the surface. Interestingly, they exhibit superior bubble absorption capabilities in an underwater situation which is complementary to a droplet impinging on the same surface in an air medium. In the present work, we examine closely, an experimental paradigm describing the physical aspects of such an absorption event and delineate the nature of evolution of the most important parameter, the contact line. The results provide insight into the efficient development of underwater bubble absorbers for hydrocarbon industry for a safe transfer of gases from deep sea oil rigs.

1. Introduction

Superhydrophobic surfaces which are extreme repellants of water have gained much attention in the recent past due to their applications in the area of self-cleaning [1,2], anti-icing [3,4], thermal management [5,6] and microfluidics [7,8] to name a few. They are characterized by low surface energy that can repel any water droplet that interacts with it, thus the name 'superhydrophobic'. Typically, a water contact angle (commonly used method to characterize the wetting of a surface) value of more than 90^0 is termed as hydrophobic and those with less than 90^0 is called 'hydrophilic'. However, for extreme contact angles, typically more than 150^0 are called 'superhydrophobic'. Interestingly, it is not only the surface material properties that dictate the contact angle, but also the structural features on the surface which creates a particular roughness profile on the surface. A rough surface can aid in the formation of air pockets when a droplet is placed over it and thus providing a cushioning effect. This will facilitate an easy detachment of the droplet and thus it can roll of easily. In nature, it is seen in the lotus leaves, as an evolutionary adaptation of the aquatic plants to repel water droplets on it to survive during splashing and heavy rain and thus sometimes is also referred to as the 'lotus leaf effect'. Recently, several researches have aimed at biomimetically engineering such structures on surfaces like polymers to create the same effect of this repellency [9,10]. In terms of the wetting configuration, they are classified as the Cassie-Baxter state of wetting where the structure induced cushioning results in the roll-off opposed to the Wenzel state where the roughness elements are invaded by the water droplet and not showing a roll-off and is rather pinned to the surface [11]. These are quantified through the dynamic contact angle measurement where an advancing contact angle (angle measured while the contact line just start moving out of the droplet and invades fresh and unwetted surface) and receding contact angle (angle measured while the contact line just start to retract back to an already wet surface) are estimated and their difference yields the contact angle hysteresis. In brief, a surface with high contact angle hysteresis corresponds to a more wetting surface where the roll-off is minimal whereas the opposite will give a high roll-off and thus easy detachment of the droplet. In a dynamic situation like the impact, the superhydrophobic surfaces show remarkable bouncing and elastic behavior, entirely different from that of a hydrophilic surface where spreading is observed.

During underwater exploration of the hydrocarbons from deep sea, it is necessary to have specialized gas collection mechanisms for faster removal and to mitigate dangers of explosion. In this regard, porous membranes are helpful in absorbing the bubbles over a large area. For such a scenario, a porous superhydrophobic membrane can serve as an excellent candidate for the bubble absorption. It is a complementary configuration of a droplet falling on a solid surface in air medium where the role of gravity on the droplet is played by the buoyancy on a rising bubble. There are a few works focused on the concept of terminating the bubbles using superhydrophobic sponges[12,13], but a detailed study including the observable wetting parameters were not to be found in the literature. In this work, we carefully experiment such a system in its simplest terms and extract the most important parameter: the contact line and its evolution with time. The physical reasoning behind the temporal variation of the contact line is explained form the fundamental principles of force balance.

2. Materials and methods

2.1. Experimental setup

We used a Polytetrafluoroethylene (PTFE) based porous membrane of thickness b (Porex, USA) housed within a Poly(methyl methacrylate) (PMMA) cavity as the bubble absorber. A perforated PMMA sheet was used as the back support for the membrane as shown in Figure 1. The entire cavity is housed inside a water tank which is fixed using a chemical stand and can vary the angle at which it is placed with respect to the horizontal axis (α) and at a height H as shown in Figure 1. At the bottom of the tank, a custom-bubble dispenser made from surgical needle and luer-lock is placed and air is slowly pumped at a flow rate of Q into it externally from a syringe pump. This results in the dispensation of air bubbles from the dispenser with a diameter $d \approx 1.75$ mm at a velocity of $u \approx 0.3$ m/s. This droplet is then impacted on the membrane and starts to spread with a contact line length of L and an air contact angle θ which are a function of time. Quickly, the bubble is absorbed into the membrane and disappears. A tubing connects the cavity to the atmosphere through the water and thus facilitates the expulsion of air absorbed as bubbles, to the outside air at the atmospheric pressure p_{atm}. A high-speed camera (Photron FASTCAM SA6) at a resolution of 1024 x 1024 square pixels is used to capture the bubble impact and subsequent events of spreading and absorption at 5000 frame per second. Later, the images are analyzed using an in-house developed code in MATLAB.

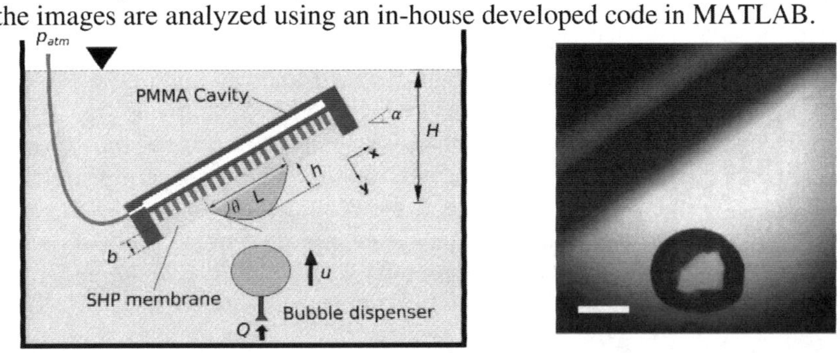

(a) (b)

Figure 1. **(a)** Experimental setup showing the bubble absorption mechanism. The cavity is immersed inside a water filled tank at an angle α and at a depth H below the reference of the free water surface. A custom-made bubble dispenser release air bubbles with diameter d, which due to buoyancy, ascents at a velocity u and impacts on the membrane and spreads with a time varying contact angle θ and contact line L. **(b)** Photograph of a bubble about to impact the membrane. Scalebar 1mm.

IOP Conf. Series: Materials Science and Engineering 1132 (2021) 012015 doi:10.1088/1757-899X/1132/1/012015

2.2. Surface characterization

As the material should be characterized for its structural and topological aspects, we have carried out the scanning electron microscopy (SEM) as shown in Figure 2. At a 500x magnification, the micro/nano features are visible on the surface, which are responsible for the structural superhydrophobicity offered by the Cassie-Baxter wetting state. Further, the roughness profile is estimated from the atomic force microscopy (AFM) as shown in Figure 3 and the average roughness value (R_a) was found to be 4.5 µm.

Figure 2. Scanning Electron Microscopic image showing the hierarchical micro/nano features that are responsible for the superhydrophobic nature of the surface.

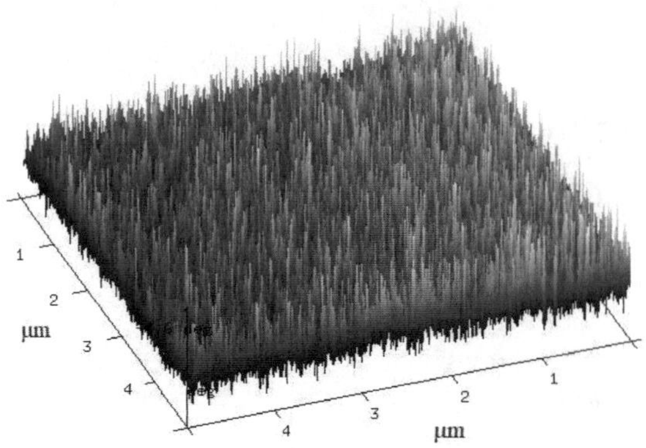

Figure 3. Atomic force microscopic profile to estimate the topology and the average roughness of the membrane surface.

3. Results and discussion

The contact line evolution with respect to time is plotted in Figure 4 with error bars representing the standard deviation from four sets of experiments. Upon contact with the surface, L is quickly increased as the bubble spreads and reach a maximum point. At the same time, the absorption takes place and the bubble is eventually absorbed into the membrane completely with the gas (air) transported through the tubes and eventually expelled to the atmosphere. As contact line is the most important parameter along with the contact angle in any wetting scenario, the time evolution of the contact line can give a clue about the parameters like absorption time and the droplet geometry. It is not so obvious to gain any physical insight from the Figure 4 and thus the same is plotted in log scale in Figure 5 from which we can extract the slope of the temporal evolution easily by fitting a linear curve.

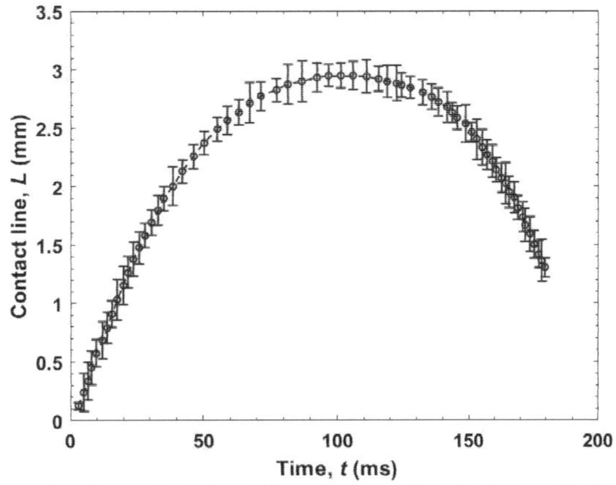

Figure 4. The contact line (L) evolution of the bubble with time.

From Figure 4, we obtain a slope of ½ in the initial regime of spreading from the fitted line. This can be explained from the balance of hydrostatic pressure: $\rho g\mathrm{H}$ and dynamic pressure: $\rho\left(\frac{dL}{dt}\right)^2$ as the gravitational effect surpasses the surface tension effect. Here, ρ is the density of water and g is the acceleration due to gravity and V is the volume of the bubble. Comparing the two pressures, $\rho g\mathrm{H} \sim \rho\left(\frac{dL}{dt}\right)^2$, we can estimate the scaling for L as: $L \sim (gV)^{\frac{1}{4}} t^{\frac{1}{2}}$. This elucidates the source the scaling relationship as observed from the experiment. However, one of the serious implications for this scaling here is in the prefactor $(gV)^{\frac{1}{4}}$ that the volume of the bubble should stay constant for the period. Although it stays constant for some period when the penetration has not started significantly, this is obeyed whereas towards reaching the maximum point, we can observe the deviation from the curve. After the peak, the sudden decrease is exponential, implying a quick removal of the bubble from the surface.

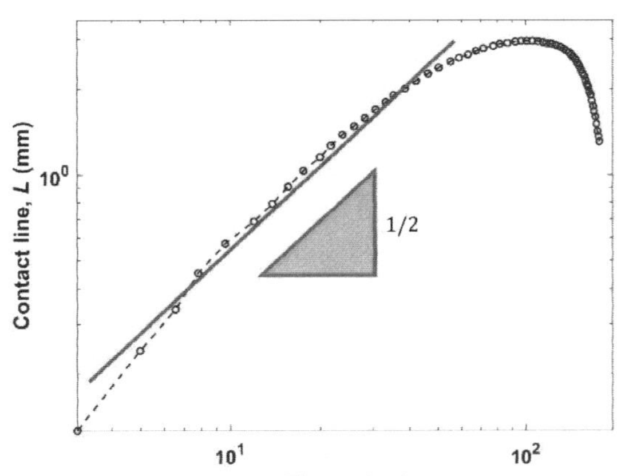

Figure 5. The log plot of Figure 4 showing the linear scale with the slope of ½ in the initial regime of spreading (marked red) signified by the physical forces in action.

The absorption time (τ) is plotted with respect to the impact angle in Figure 6. This is the time taken by the bubble to get completely absorbed into the membrane from the first instance of touching. It is found that with an increase in the impact angle, τ is increased almost linearly within the range of 0-60^0. Beyond this, the bubbles may evade the membrane and will just slide off, without any attachment points. As the bubble approaches the membrane, it squeezes the water film that is between the bubble and the membrane and thus get deformed by itself in the process. This elongates the bubble and thus have a larger area of contact while it touches. Also, there is a time delay from the first instant of touching of the droplet and a complete flattening on an inclined membrane compared to a flat membrane. This can increase the absorption times significantly as the angle increases.

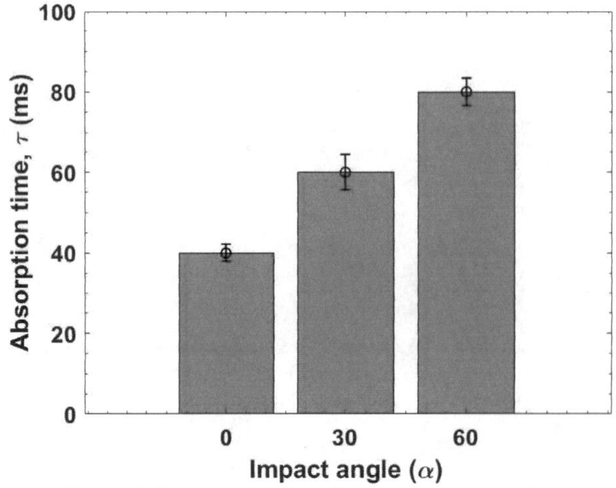

Figure 6. The variation of absorption time τ with respect to the impact angle α.

4. Conclusion
The present work investigates the physics behind the interaction of a bubble with a porous membrane. With an aim to provide fundamental insights into the physical mechanisms behind the bubble impact, spreading and the subsequent absorption into the membrane. It is a starting point for future works to study the deeper aspects of bubble absorption in underwater medium and innovate safer designs for the absorption of hydrocarbons in oil and natural gas industry, to reduce the fatalities in offshore oil rigs.

5. Acknowledgements
The author acknowledges the assistance from Prof. Yap Choon Hwai of the Department of Bioengineering, Imperial College London, through the grant NMRC/OFIRG/0060/2017 during the tenure at the National University of Singapore (NUS), Singapore.

References
[1] Bhushan B, Jung Y C and Koch K 2009 Self-cleaning efficiency of artificial superhydrophobic surfaces *Langmuir* **25** 3240–8
[2] Fürstner R, Barthlott W, Neinhuis C and Walzel P 2005 Wetting and self-cleaning properties of artificial superhydrophobic surfaces *Langmuir* **21** 956–61
[3] Farhadi S, Farzaneh M and Kulinich S A 2011 Anti-icing performance of superhydrophobic surfaces *Appl. Surf. Sci.* **257** 6264–9
[4] Cao L, Jones A K, Sikka V K, Wu J and Gao D 2009 Anti-icing superhydrophobic coatings *Langmuir* **25** 12444–8
[5] Miljkovic N and Wang E N 2013 Condensation heat transfer on superhydrophobic surfaces *MRS Bull.* **38** 397–406
[6] Chen Z, Dong L, Yang D and Lu H 2013 Superhydrophobic graphene-based materials: surface construction and functional applications *Adv. Mater.* **25** 5352–9
[7] Mumm F, van Helvoort A T J and Sikorski P 2009 Easy route to superhydrophobic copper-based wire-guided droplet microfluidic systems *ACS Nano* **3** 2647–52
[8] Tropmann A, Tanguy L, Koltay P, Zengerle R and Riegger L 2012 Completely superhydrophobic PDMS surfaces for microfluidics *Langmuir* **28** 8292–5

[9] Dey R, Bhandaru N, Mukherjee R, Chakraborty S and others 2014 Tunable hydrodynamic characteristics in microchannels with biomimetic superhydrophobic (lotus leaf replica) walls *Soft Matter* **10** 3451–62
[10] Bandyopadhyay S, Dey R, Raj M K, Bhandaru N, Mukherjee R and Chakraborty S 2017 Mixing characteristics in microchannels with biomimetic superhydrophobic (Lotus leaf replica) walls *Microfluid. Nanofluidics* **21** 152
[11] Marmur A and Bittoun E 2009 When Wenzel and Cassie are right: reconciling local and global considerations *Langmuir* **25** 1277–81
[12] Chen X, Wu Y, Su B, Wang J, Song Y and Jiang L 2012 Terminating marine methane bubbles by superhydrophobic sponges *Adv. Mater.* **24** 5884–9
[13] Ma R, Wang J, Yang Z, Liu M, Zhang J and Jiang L 2015 Bioinspired gas bubble spontaneous and directional transportation effects in an aqueous medium *Adv. Mater.* **27** 2384–9

IOP Conf. Series: Materials Science and Engineering 1132 (2021) 012016 doi:10.1088/1757-899X/1132/1/012016

Steady State Thermal and Structural Simulation of SS 316L Bipolar Plates used in 250 Watt PEMFC

Robin Raju and Vikas Rajan*

Department of Mechanical Engineering, Amrita Vishwa Vidyapeetham, Amritapuri, India

*vikasrajan@am.amrita.edu

Abstract: Design and optimisation of bipolar plates play a crucial role in determining the overall efficiency of the PEM (proton exchange membrane) fuel cell. These plates are designed in such a way that the PEM fuel cell achieves the best water and thermal management capability. The conventionally used graphite-based bipolar plates are getting replaced by metallic plates for powering space applications and portable devices due to its good mechanical and electrical properties. In this paper, we present the thermal and structural simulation of the stainless-steel type 316L bipolar plates used in 250W PEM fuel cell after the preheating process but before the supply of the fuel into the stack.

Keywords: *PEM fuel cell, SS 316 L, Bipolar plates*

1. Introduction

In a proton exchange membrane fuel cell (PEMFC), bipolar plates (BP) play a crucial role. BP has several functions in a typical cell stack like maintaining uniform conveyance of hydrogen and oxygen across the active surface area of the membrane electrode assembly, conducting the electrical output from one cell to the other, supporting thin membranes and electrodes to form the perfect fuel cell stack, etc. The flow channels for the fuel, air and coolant are either milled, stamped or wire cut (EDM) on it [1,2]. A typical BP must be designed, optimised and manufactured to obtain the best water management and thermal management within the cell. Water and thermal management are crucial in a PEMFC stack because only in the optimum hydrated state the, cell gives the best performance and during flooding and dry-out the cell gives poor output [3]. Graphite and graphite-based composites are the commonly used materials for manufacturing BP, since graphite have reasonable electrical conductivity and high thermal conductivity. The graphite-based BP lacks good machinability and mechanical strength. Hence, they cannot be reduced in size beyond a limit, thereby making them bulkier in volume [4]. In the recent trend of technology, fuel cells are used in space applications and portable devices; so, the size of the cell stack is a major issue. In order to solve this problem, researchers came up with the concept of metallic BP as a good alternative for graphite [5]. BP made of metals have good thermal conductivity, electrical conductivity, good mechanical strength, easy machinability and high-speed mass production is possible [6,7]. Due to the high machinability of metals, we can reduce the thickness of the BP & this will help us to reduce the overall size of the fuel cell stack. The most widely preferred metal for BP is stainless steel (SS). The type 316L SS is recommended by many researchers and engineers due to its good thermal, electrical and mechanical properties [6]. The BP plays a crucial role in heat management of the cell stack. The coolant flow channels on a BP are designed in a way to evenly cool the MEA of the cell. The most commonly used membrane in a PEMFC is NAFION and the best performance of the cell occurs within a temperature range of 80^{0}C to 100^{0}C [8,9]. The fuel H_2 is delivered to the anode side of the MEA and catalytic splitting of H_2 into protons and electrons takes place at anode. The newly formed protons travel through the membrane and reach the cathode side where O_2 is fed. At cathode, O_2 combines with both protons and electrons to produce water. This overall reaction produces both electrical and heat energy. A schematic representation of a PEMFC is shown in Figure 1.

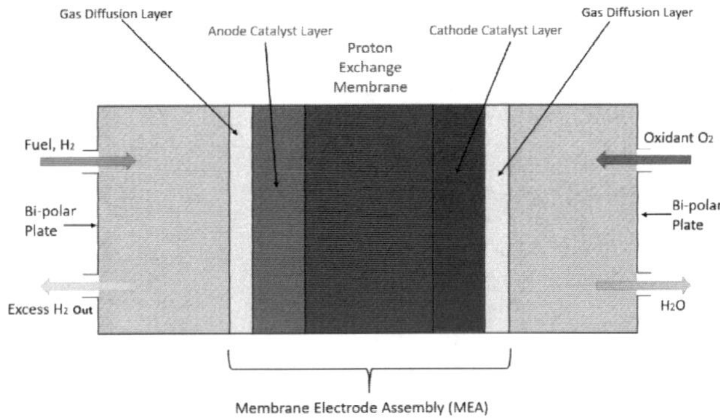

Figure 1: Parts of a PEM Fuel cell

Since the optimum temperature range of PEMFC is nearer to 80°C, we have to preheat the fuel cell stack before supplying the fuel [10–12]. This can be achieved by many ways like solar preheating, circulating hot gases or water through the flow channels and manifolds in the BP, or by the usage of heater rods [13,14]. We took the case of preheating with metallic rods for our study. This preheating gives rises to thermal stresses in BP and in our work, we are focusing to simulate the induced thermal stress and temperature distribution of a simple BP made of SS 316L using COMSOL Multiphysics 5.6.

2. Problem and Model Description

As mentioned above the BP is preheated to achieve the optimal operating temperature range. Only after attaining this temperature range, we will start supplying fuel to the cell stack; this helps to reduce fuel wastage and to attain the best efficiency. This preheating induces thermal stresses on the metallic plate and in this paper; we try to simulate the induced thermal stresses and temperature distribution of the preheated SS 316L BP after attaining steady state and before the actual supply of the fuel into the stack. To simplify the problem, we chose a simple symmetrical BP. The geometry of the plate is given in Figure 2. Our BP has an outer dimension of 24 x 20 cm^2 with an active area of 49.5 cm^2 and the thickness of the plate is 1 cm.

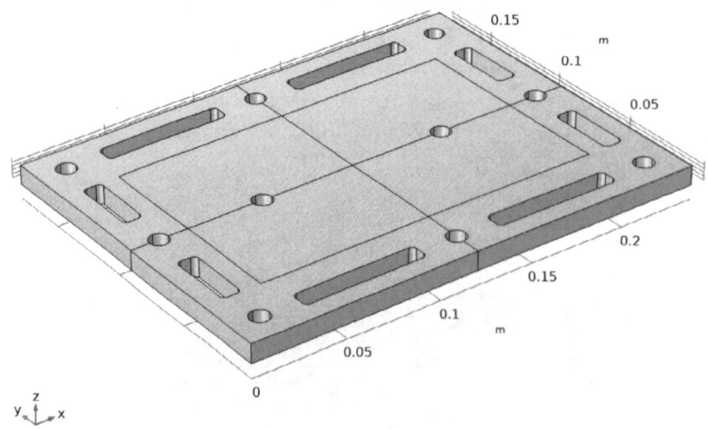

Figure 2: Bipolar plate

To reduce the computational requirements and computing time we chose to reduce the size of the plate into one in eighth of its actual size. This action is justifiable because of the symmetry of the BP we considered. This reduction in size is summarised in the Figure 3.

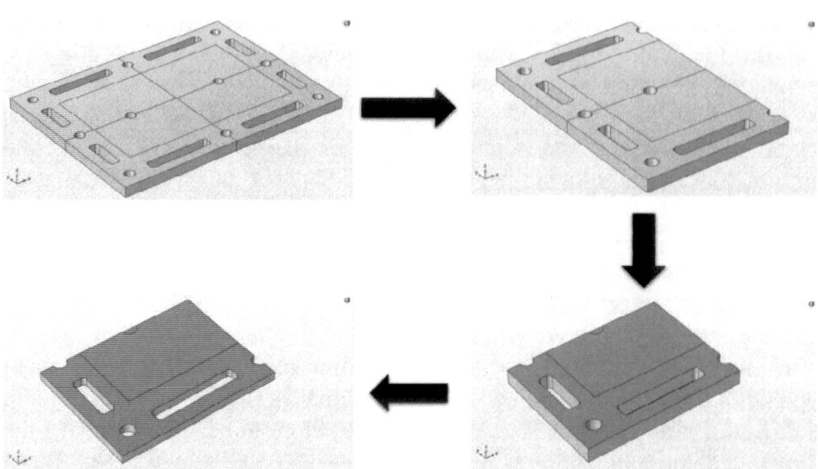

Figure 3: Scaling down to 1/8th of its original size

The reduced model is divided into three zones, active zone where the MEA comes into contact with the metal plate, heat source zone which comprises the aluminium rod, and finally the outer zone where manifold of the gas and fuel are built. In our problem, we took a fuel cell stack consisting of 25 BPs. The electrical heat input was given to the BP through the two aluminium rods placed at the mid axis of the plate. The input heat was 200 W for the entire stack assembly. The plate is made of stainless steel 316L which comprises both active and outer zone and the heat source zone is made of aluminium. The heat transfer through the gas manifold was taken to be 50 W/m^{2}K and to the surrounding as 5 W/m^{2}K. The material properties of aluminium were directly taken from the material library of COMSOL and the properties of SS 316L was manually provided.

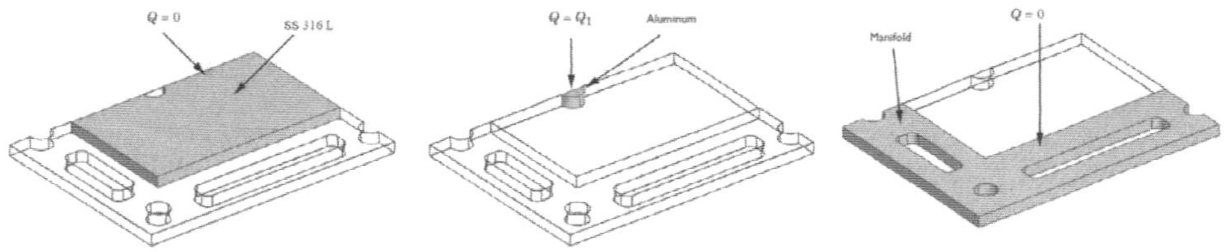

Figure 4: Different zones in the BP

Table 1: Basic properties of BP materials

Material	Density (g/cc)	Modulus of elasticity (GPa)	Coefficient of thermal expansion (1/K)	Thermal conductivity (W/m.K)	Specific heat (kJ/kg K)	Electrical Resistivity (nΩ·m)	Poisson ratio
SS 316 L	7.99	193	$16*10^{-6}$	16.2	0.5	740	0.27
Aluminium alloy	2.7	70	$23*10^{-6}$	237	0.9	26.5	0.35
Graphite (with binder)	2.2	12	$4.7*10^{-6}$	250	0.8	$18*10^{3}$	0.28

Table 2: Parameters for the stack assembly

Parameter	Value
Input heat to 25 plates	200 Watt
Heat transfer coefficient to the surrounding	5 W/m^2K
Heat transfer coefficient to the manifold	500 W/m^2K
Pressure at the Active zone (P_1)	10^6 N/m^2
Pressure at the Outer zone (P_2)	$7.99x10^6$ N/m^2

3. Thermal and Structural Analysis

Our study focusses on thermal & structural analysis of SS 316L BP. We assume that the heat input (200W) was equally distributed among all plates in the stack through the aluminium rods and we assume that the power from the heat source is uniformly distributed across the heat source zone. The active zone and outer zone do not have any heating element in them so the heat generated from those zones is zero.

$$\nabla.(-k.\ \nabla T) + Q\_{source} = 0 \qquad (1)$$

Eq(1) represents the steady-state heat equation (Fourier's law). Q_{source} and K represent heat source and thermal conductivity. Since we are studying the case of the BP when it attained steady state and prior to starting of the fuel cell, the losses that are caused by the power generation and other losses are absent. In our problem, symmetric and convective conditions are the only boundary conditions for the heat analysis. The convective and symmetric boundaries are illustrated in figure 5 & figure 6.

IOP Conf. Series: Materials Science and Engineering 1132 (2021) 012016 doi:10.1088/1757-899X/1132/1/012016

The symmetric boundary conditions imply that the flux through the boundary is zero.

$$(-k.\ \nabla\ T).n = 0 \qquad (2)$$

The convective boundary conditions set the heat flux proportional to the temperature difference between the fluid outside and the temperature at the boundary. Thus, we can express this convective boundary condition as:

$$(-k.\ \nabla\ T).n = h(T_{ext} - T_{f;uid}) \qquad (3)$$

Where, h denotes the convective heat transfer coefficient.

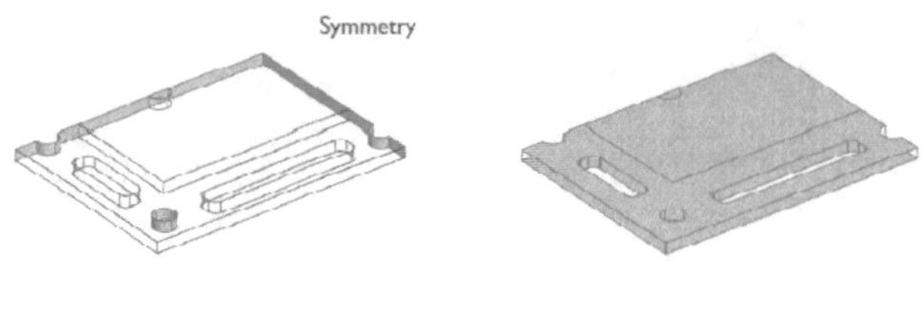

Figure 5: Symmetry boundaries

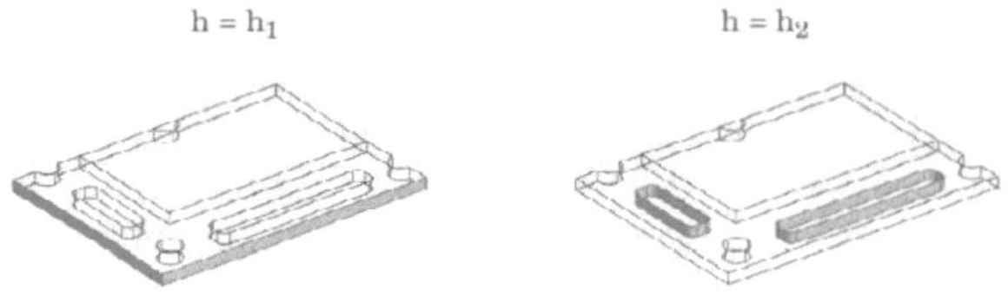

Figure 6: Convective boundaries

When the plate is heated, the plate thermally expands causing thermal induced structural stress to develop. The thermally induced loads are proportional to the temperature difference between the actual and reference temperature. The displacement perpendicular to the symmetric planes are set to zero as we apply the constraints over them as shown in Figure 7 and Figure 8

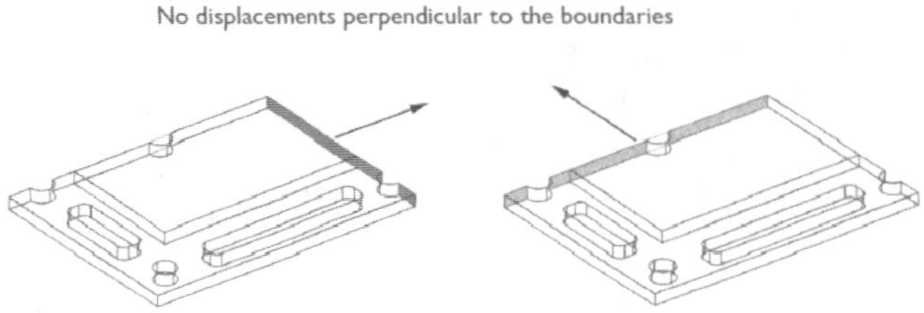

Figure 7: No displacement through symmetric boundaries

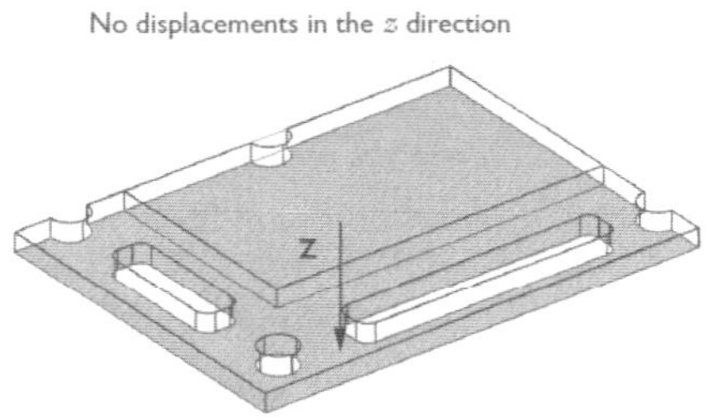

Figure 8: No displacement in downward Z-direction

In the manifold and in the active part of the BP, the load applied by the tie rods on the stack is applied as can be seen in Figure 9. The constraints and load on every other boundary are assumed to be negligible. The only questionable assumption in this sense is the possible loads and restrictions that the tie rods impose on the holes in the BP. It would be possible to pair a model for the tie rods with a detailed model of the BP in a more detailed analysis. The strain on these distinct parts would probably show different gradients. These assumptions are made to simplify the problem.

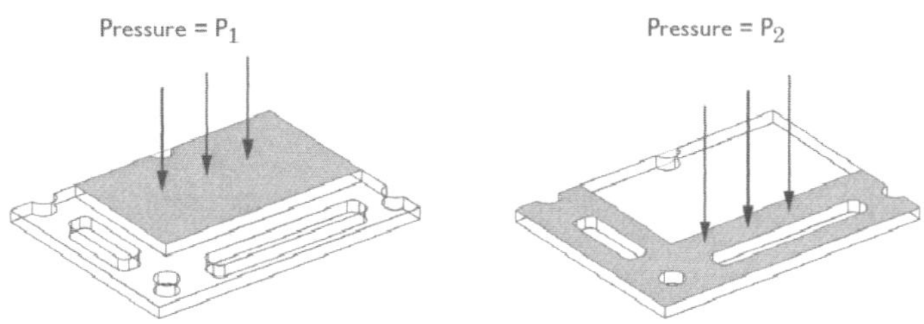

Figure 9: Pressure due to the rod connection

4. Grid Independency

The meshing was carried out using the in-built meshing tool of COMSOL Multiphysics 5.6. We used free triangular mesh on the top surface and swept it across the remaining portion. The maximum element size of 0.0024 m, maximum element growth rate 1.3, curvature factor 0.2 and resolution of narrow regions equals 1, was given. We compared the results obtained from various mesh sizes and we chose an inbuilt mesh size of extremely fine in the COMSOL mesh tool as it gave correct results. The skewness value of the mesh was in the desirable range. Hence, this meshing was enough for doing the computational analysis of our symmetric BP. The Extremely fine mesh is shown in Figure 10 and the grid independency test is illustrated in Fig 11.

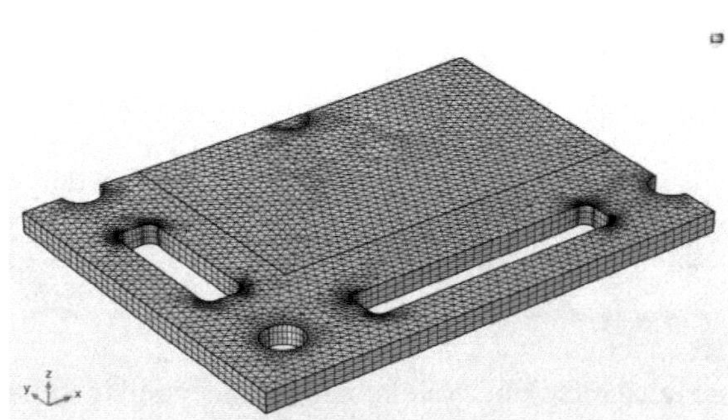

Figure 10: Mesh

IOP Conf. Series: Materials Science and Engineering 1132 (2021) 012016 doi:10.1088/1757-899X/1132/1/012016

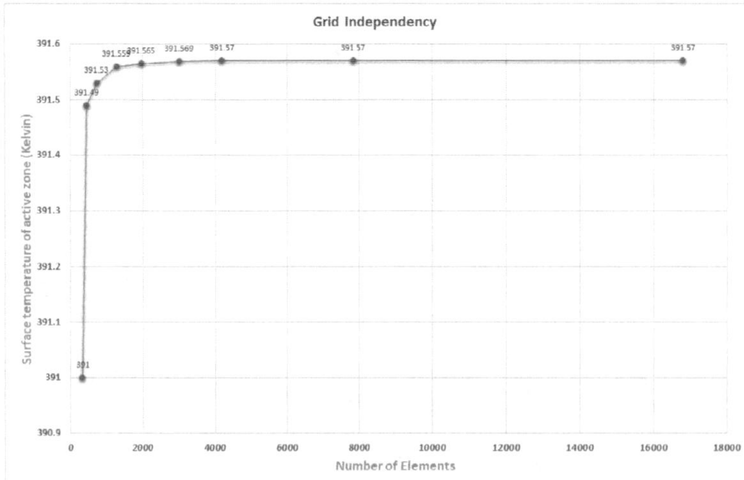

Figure 11: Grid independency

5. Results and Discussion

Using COMSOL Multiphysics 5.6, the thermal and structural simulation of a simple BP made of SS 316L was carried out. The reference temperature and pressure were set to be 293 K and 101.325 KPa. Usually, BP has manifolds for gases and coolant but in our case, we provided only gas manifolds to make the problem less complex. The geometry of the BP was symmetric. Hence, we could scale down the size of the plate to one in eighth of its actual size. Few input parameters were directly given to the system to avoid extra computing requirements and no initial values were given to the computation. The properties of SS 316 L were manually fed to the solver and all other material properties were taken from the material library. The temperature distribution across the plate was plotted and we can see a good temperature distribution across the active zone where the MEA comes into contact with the BP and the temperature drop towards the end of the active zone was only about 10 degree Celsius. This helps the MEA to function efficiently due to the efficient temperature distribution in metal plates. The button channel is larger in size than the side channel hence the temperature is lesser in the button channel due to more convective heat dissipation. The temperature distribution of the plate is provided in Figure 12.

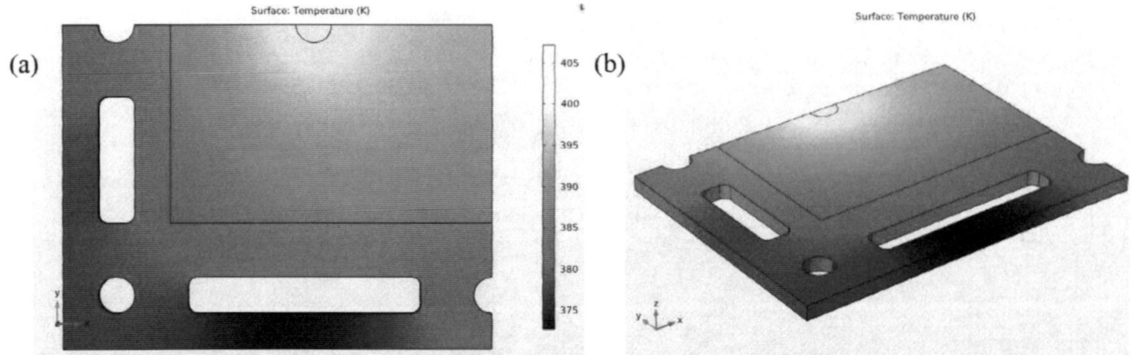

Figure 12: Temperature distribution in (a) Top surface and (b) Plate (isometric view)

The maximum temperature gradient was observed at the interface between the active zone and the outer zone. This leads to the fact that more thermal induced stress gradient is at this interface and we can also see that the magnitude of stresses nearer to the button channel is lower than the magnitude of stress at the side channel as shown in Figure 13.

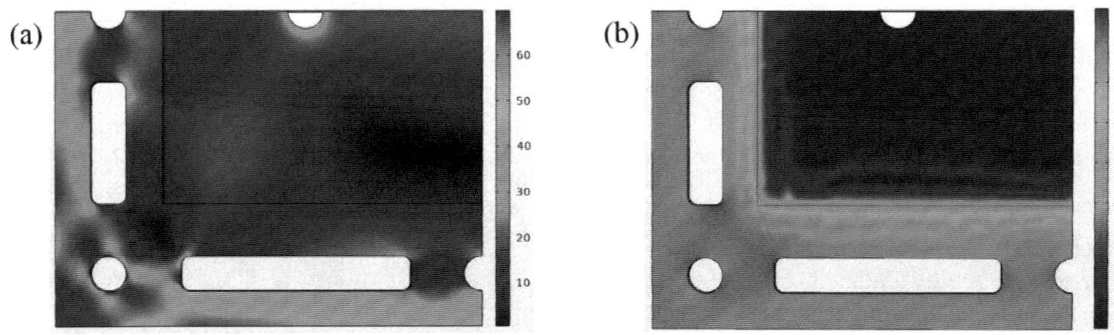

Figure 13: von Mises stress in (a) SS316L and (b) Graphite

Using von Mises stress, we can determine whether a material will yield or fail for the given conditions. Under loaded condition, if the von Mises stress of the given material is equivalent or predominant than the yield stress under simple tension, the material will yield or fail. The yield strength and other mechanical properties of the graphite is poor compared to metals like SS 316L. Hence, from the mechanical strength and stress handling point of view, the metallic BP have significant upper hand compared to conventional graphite-based plates. As shown in the Figure 13 (a), the stress nearer to the heat source zone and the outer zone are higher than that of the stress at the active zone. By turning off the thermal expansion in the solver, we can calculate the external pressure loads and it is found that the thermally induced loads are greater in magnitude than that of external pressure loads.

The results we obtained shows that the loads acting on the plates are far long away from the critical values. However, the thermal expansion can be considerably more significant. Hence, these expansions cannot be neglected. The expansion along the Z- direction is given in Figure 14. The whole plate attained an average temperature of 368.58 K and active area attained 391.57 K .The SS 316L plates have more thermal expansion than the graphite plates. The thermal expansions near the heat zone were higher than other areas. Even though these expansions are in µm range, they can significantly damage the MEA of the PEMFC.

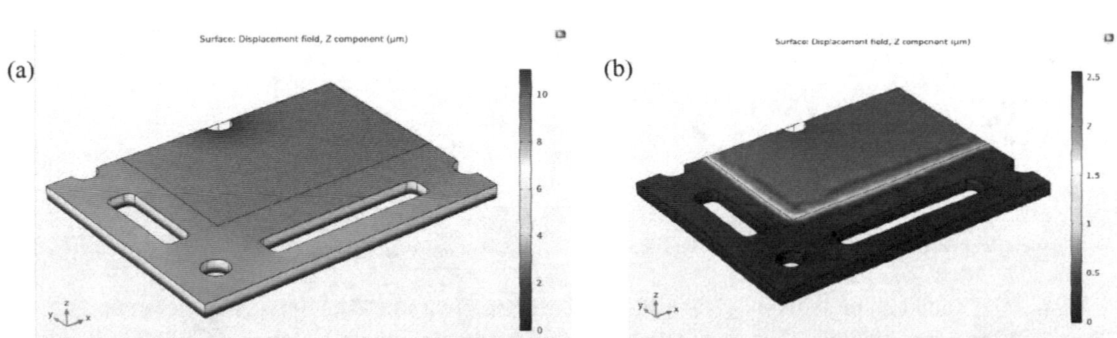

Figure 14: Expansion in the Z- direction (a) SS 316L (b) Graphite

While designing a BP we must keep these displacements to the minimum such that there is no hindrance or halt in the membrane process of separating the hydrogen and oxygen at the cell compartments. If the thermal expansion is very large it can significantly damage the MPL (microporous layer) and to the extreme, the entire MEA itself can get damaged.

6. Conclusion

Fuel cells are a promising area in the field of alternative fuels for powering portable devices, space application as well as power-plants. Among fuel cells, the most widely used cell is the PEM fuel cell because of its low temperature operations and higher efficiency. The heat and water management of the PEMFC must be optimised to get the best performance. Hence, the design and optimisation of the BP is crucial. The conventionally used graphite-based BP does not hold good mechanical strength and machinability. The lack of these properties makes the graphite-based BP bulkier in size. For portable devices and space application, we prefer compact designs with good mechanical and electrical properties. The stainless steel is the best alternative for graphite as it has got all the desirable properties required for a good BP. In our work, we provided a thermal and structural simulation of a stainless-steel type 316L used in 250 W PEMFC and the results obtained are discussed. The thermal expansion of the metallic plates is higher than that of graphite, so we must design the plates precisely such that this thermal expansion does not damage the MPL and the whole MEA of the PEM fuel cell. The SS BP is a good alternative for graphite-based BP, but the thermal expansion of metals raises issues regarding the durability and performance of the MEA. So we must focus on developing materials with good mechanical and electrical properties whose thermal expansion can be regulated to the minimum even at elevated temperatures.

References

[1] Hu Q, Zhang D, Fu H and Huang K 2014 Investigation of stamping process of metallic bipolar plates in PEM fuel cell - Numerical simulation and experiments *Int. J. Hydrogen Energy* **39** 13770–6
[2] Hung J C, Yang T C and Li K C 2011 Studies on the fabrication of metallic bipolar plates Using micro electrical discharge machining milling *J. Power Sources* **196** 2070–4
[3] Barbir F 2005 Stack Design *PEM Fuel Cells* (Elsevier) pp 147–206
[4] Cooper J S 2004 Design analysis of PEMFC bipolar plates considering stack manufacturing and environment impact *J. Power Sources* **129** 152–69
[5] Asri N F, Husaini T, Sulong A B, Majlan E H and Daud W R W 2017 Coating of stainless steel and titanium bipolar plates for anticorrosion in PEMFC: A review *Int. J. Hydrogen Energy* **42** 9135–48
[6] Wang H and Turner J A 2010 Reviewing metallic PEMFC bipolar plates *Fuel Cells* **10** 510–9
[7] Peng L, Liu D, Hu P, Lai X and Ni J 2010 Fabrication of metallic bipolar plates for proton exchange membrane fuel cell by flexible forming process-numerical simulations and experiments *J. Fuel Cell Sci. Technol.* **7** 0310091–9
[8] Kandlikar S G and Lu Z 2009 Fundamental Research Needs in Combined Water and Thermal Management Within a Proton Exchange Membrane Fuel Cell Stack Under Normal and Cold-Start Conditions *J. Fuel Cell Sci. Technol.* **6** 0440011–04400113
[9] Zawodzinski T A, Derouin C, Radzinski S, Sherman R J, Smith V T, Springer T E and Gottesfeld S 1993 Water Uptake by and Transport Through Nafion® 117 Membranes *J. Electrochem. Soc.* **140** 1041–7
[10] Zhan Z, Yuan C, Hu Z, Wang H, Sui P C, Djilali N and Pan M 2018 Experimental study on different preheating methods for the cold-start of PEMFC stacks *Energy* **162** 1029–40
[11] Lalitha Priya R, Ramachandran T and Suneesh P V. 2016 Fabrication and characterization of high power dual chamber E. coli microbial fuel cell *IOP Conf. Ser. Mater. Sci. Eng.* **149**
[12] Madhavan A, Prasad M, Girish S, Shetty K S and Nair B 2018 Caulobacter crescentus as a

novel exoelectricigen in a dual chambered microbial fuel cell (MFC) *Proc. 2017 IEEE Int. Conf. Technol. Adv. Power Energy Explor. Energy Solut. an Intell. Power Grid, TAP Energy 2017* 1–4

[13] Uzun A, Bokor B and Eryener D 2020 PEM fuel cell performance with solar air preheating *Int. J. Hydrogen Energy* **45** 34654–65

[14] Madhavan A, Nandakumar V, Shetty K S and Nair B 2016 Optimization of microbial fuel cell (MFC) operated with waste water as substrate *Int. Conf. Electr. Electron. Optim. Tech. ICEEOT 2016* 4403–6

IOP Conf. Series: Materials Science and Engineering 1132 (2021) 012017 doi:10.1088/1757-899X/1132/1/012017

Multi-objective optimization of turbulent jet impingement cooling system : A thermal and thermodynamic study

Prasanth Anand Kumar Lam

Department of Mechanical Engineering, NIT Warangal, Warangal, India - 506 004

E-mail: prasanth_anand@nitw.ac.in

K Arul Prakash

Department of Applied Mechanics, IIT Madras, Chennai, India - 600 036

E-mail: arulk@iitm.ac.in

Abstract. In this paper, an optimization methodology for turbulent air jet impingement cooling system is reported. The continuity, Reynolds-Averaged Navier-Stokes, turbulent kinetic energy (k), turbulent dissipation rate (ϵ) and energy equations are solved using the Streamline Upwind/Petrov-Galerkin (SUPG) Finite Element (FE) Method. In this study, Reynolds number (Re_j) and channel height (H/L) are selected as design parameters. The influence of design parameters on turbulent flow, heat transfer and entropy generation distribution for the system are presented and discussed. As Re_j increases and H/L decreases, the magnitude of overall Nusselt number and global total entropy generation increases. Subsequently, the Multi-Objective Genetic Algorithm (MOGA) was introduced to achieve optimal impingement system configurations with maximum heat transfer and minimum entropy generation.

1. Introduction

Heat transfer enhancement through impinging jets has gained significant interest due to various engineering applications *viz.* thermal drying, cryosurgery, aircraft wing anti-icing, electronics cooling and gas turbine blade cooling etc. [1, 2, 3]. Over the last few decades, several experimental, theoretical and computational studies were reported on turbulent impinging jets [4, 5, 6, 7]. Afroz and Sharif [8] has recently studied turbulent fluid flow and thermal transport for both round as well as annular impinging jet flow using realizable k-ϵ model. It was reported that, in comparison to the round impinging jet, annular impinging jets have produced an improved heat transfer and distribute heat more evenly over the impingement surface.

In all the above reported studies, jet impingement heat transfer over a continuous hot surface is addressed. However, in certain applications, such as data center cooling, micro-sized electronic cooling equipment and micro-electro-mechanical systems (MEMS), thermal loads are imposed as independent heating elements with intermittent thermal boundary layer [9, 10]. Forster and Weigand [11] performed an experimental and computational analysis to predict thermal transport on the heated concave surface through an array of impinging jets with reference to cooling of gas turbine blade's leading edge.

IOP Conf. Series: Materials Science and Engineering 1132 (2021) 012017 doi:10.1088/1757-899X/1132/1/012017

Bejan [12] was first to report thermodynamic analyses of numerous convection heat transfer systems. Over the last few years, numerous studies have been published on the generation of entropy in jet configurations impinging [13, 14, 15]. Esmailpour et al. [16] have recently investigated the thermodynamic performance of pulsatile jet impingement flow over flat plate. Maximum viscous and thermal entropy generation has been shown to occur at low nozzle-impingement plate hight. To the best of authors' knowledge, in spite of its significance in thermal control of electronic applications, turbulent heat transport and entropy generation minimization in impinging jet system through multi-objective design optimization has not yet been documented, which form the motivation to this research work.

The main purpose of this work is to report momentum transport, thermal energy transport and thermodynamic performance of a turbulent air jet impinging system at wide ranges of design parameters *viz.* Reynolds number and channel height. Furthermore, as objective functions, $\overline{Nu_{ov}}$ and $S_{tot,\Omega}$ are chosen. Subsequently, MOGA design optimization procedure is employed to achieve optimal solutions with maximal heat transport and minimal entropy generation.

2. Methodology

As illustrated in Fig. 1, the air jet impingement system is a channel with a series of discrete heated elements, mounted along the impinging substrate, identical to Lam and Prakash [17].

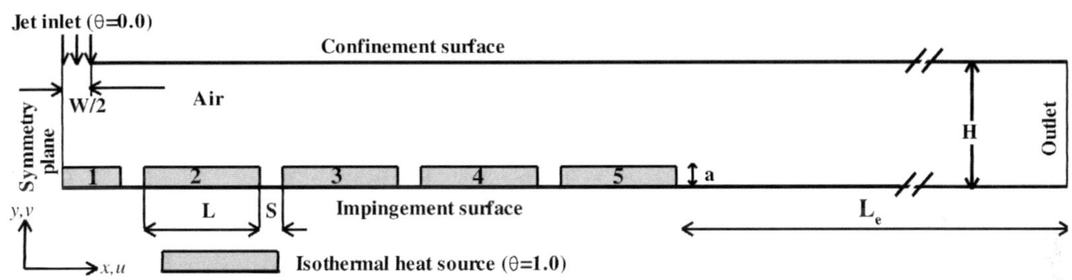

Figure 1. Physical Domain of interest

The non-dimensional equations governing turbulent forced convection are as follows [17]:

Continuity Equation:

$$\frac{\partial u_{n,i}}{\partial x_i} = 0 \tag{1}$$

Momentum Equations:

$$\frac{\partial u_{n,i}}{\partial t} + u_{n,j}\frac{\partial u_{n,i}}{\partial x_j} = -\frac{\partial p_n}{\partial x_i} + \frac{1}{Re_j}(1+\nu_{t,n})\left\{\frac{\partial}{\partial x_i}\left(\frac{\partial u_{n,i}}{\partial x_j} + \frac{\partial u_{n,j}}{\partial x_i}\right)\right\} \tag{2}$$

Here, the turbulent viscosity ($\nu_{t,n} = c_\mu \; Re_j \; \frac{k_n^2}{\epsilon_n}$) is estimated through Standard k - ϵ turbulence model. Further, near wall dynamics are estimated through standard wall functions [18].

The equations governing turbulent kinetic energy (k_n) and turbulent dissipation (ϵ_n) are as follows:

Turbulent Kinetic Energy (k_n):

$$\frac{\partial k_n}{\partial \tau} + \left[u_j \frac{\partial k_n}{\partial x_j} \right] = \frac{1}{Re_j} \left[\frac{\partial}{\partial x_j} \left(\frac{\nu_{t,n}}{\sigma_{k_n}} \frac{\partial k_n}{\partial x_j} \right) \right] + G_n - \epsilon_n \tag{3}$$

Turbulent dissipation (ϵ_n):

$$\frac{\partial \epsilon_n}{\partial \tau} + \left[u_j \frac{\partial \epsilon_n}{\partial x_j} \right] = \frac{1}{Re_j} \left[\frac{\partial}{\partial x_j} \left(\frac{\nu_{t,n}}{\sigma_{\epsilon_n}} \frac{\partial \epsilon_n}{\partial x_j} \right) \right] + \frac{\epsilon}{k_n} (C_{1\epsilon} G_n - C_{2\epsilon} \epsilon) \tag{4}$$

Energy Equation:

$$\frac{\partial \theta_n}{\partial t} + u_j \frac{\partial \theta_n}{\partial x_j} = \frac{1}{Re_j \ Pr} (1 + \alpha_{t,n}) \left(\frac{\partial^2 \theta_n}{\partial x_j \ x_j} \right) \tag{5}$$

Here, G_n, $\nu_{t,n}$ and $\alpha_{t,n}$ are stated as

$$G_n = \frac{\nu_{t,n}}{Re_j} \left[2 \left\{ \left(\frac{\partial u_{n,j}}{\partial x_j} \right)^2 \right\} + \left(\frac{\partial u_{n,i}}{\partial x_j} + \frac{\partial u_{n,j}}{\partial x_i} \right)^2 \right] \ ; \ \nu_{t,n} = C_\mu Re_j \frac{k_n^2}{\epsilon_n} \ ; \ \alpha_{t,n} = C_\mu Re_j \frac{Pr \ k_n^2}{\sigma_t \ \epsilon_n} \tag{6}$$

Also, the model coefficients are:

$$C_\mu = 0.09, \quad \sigma_k = 1.0, \quad \sigma_\epsilon = 1.3, \quad \sigma_t = 0.9, \quad C_{1\epsilon} = 1.44, \quad C_{2\epsilon} = 1.92 \tag{7}$$

The local entropy generation is defined as follows [19]:

$$\mathrm{S}_{tot} = S_\theta + S_\psi = \left[\left(\frac{\partial \theta_n}{\partial x} \right)^2 + \left(\frac{\partial \theta_n}{\partial y} \right)^2 \right] + \chi_n \left\{ 2 \left[\left(\frac{\partial u_n}{\partial x} \right)^2 + \left(\frac{\partial v_n}{\partial y} \right)^2 \right] + \left(\frac{\partial v_n}{\partial x} + \frac{\partial u_n}{\partial y} \right)^2 \right\} \tag{8}$$

In Equation.8, S_θ is contribution of entropy generation from heat transfer irreversibility, and S_ψ is from fluid friction irreversibility.

The pertinent non-dimensional parameters are defined as, Prandtl number $(Pr) = \frac{\nu}{\alpha}$, Reynolds number $(Re_j) = \frac{UW}{\nu}$, local Nusselt number $(Nu = \frac{L}{W} \frac{\partial \theta}{\partial n})$ and local Bejan number $(Be = \frac{S_\theta}{S_\theta + S_\psi})$. Also, irreversibility distribution ratio (χ_n) is assumed to be unity [19].

The boundary conditions are as follows:
a. Jet inlet:

$$u_n = 0.0; \ v_n = -1.0; \ \frac{\partial p_n}{\partial n} = 0; \ \theta = 0; \ k_n = 1.5 \ I^2; \ \epsilon_n(x) = (k_n^{1.5} C_\mu^{0.75}) / \chi \ x \tag{9}$$

Here, Turbulent intensity $(I) = 10\%$, Von-Karman constant $(\chi) = 0.42$, $\lambda = 0.09$ and $E = 9.743$.
b. Solid surfaces:

$$u_n = v_n = 0; \ \frac{\partial p_n}{\partial y} = 0; \ \theta = 1.0; \tag{10}$$

c. Outlet:

$$p_n = 0, \ \frac{\partial f}{\partial x} = 0; \quad f = (u_n, \ v_n, \ \theta_n, \ k_n, \ \epsilon_n) \tag{11}$$

d. Symmetry:

$$u_n = 0.0; \frac{\partial f}{\partial x} = 0.0; \quad f = (v_n, \theta_n, p, k_n, \epsilon_n) \tag{12}$$

The equations governing turbulent forced convection are solved through an in-house algorithm based on SUPG-FE method [20]. The detailed formulation is presnted in Arul Prakash et. al. [21]. Further, the validation of in-house algorithm and grid resolution study are presented in Lam and Prakash [17]. A detailed mesh resolution study was performed with three grids such as 456×60, 560×100 and 684×140 along x and y directions respectively. The deviation of $\overline{Nu_{Ov}}$ and $S_{tot,\Omega}$ between grids, 560×100 and 684×140 is observed to be marginal and equal 3.75% and 4.89% respectively. Hence, a grid with 560 × 100 elements is selected for FE simulations.

3. Results and discussion

In this section, momentum transport, thermal energy transport and thermodynamic performance of turbulent air jet impingement system is reported at wide ranges of Re_j and H/L. Further, Nu_{ov} and $S_{tot,\Omega}$ are choosen as objective functions and MOGA has been implemented for achieving optimal configuration with a simultaneous heat transfer enhancement and minimal entropy generation.

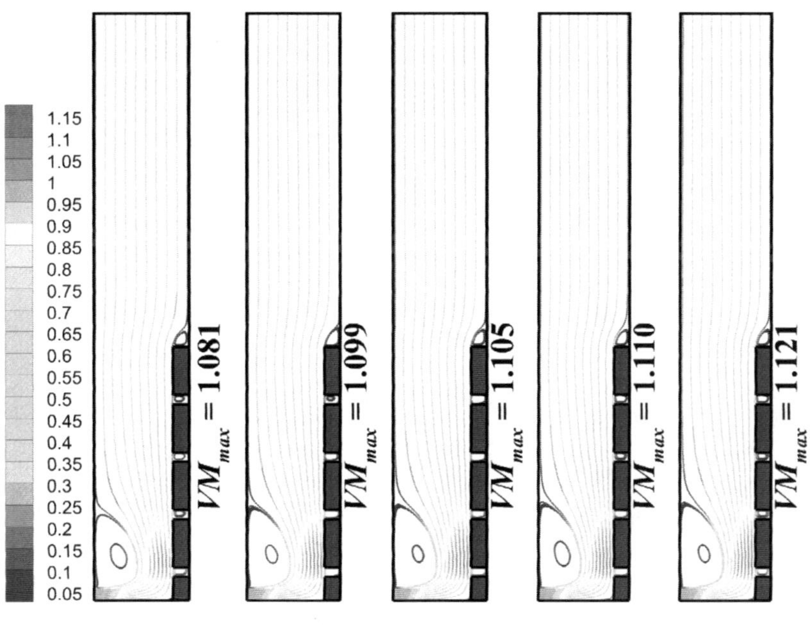

Figure 2. Effect of Re_j on Streamlines at $H/L = 0.75$ for (a) $Re_j = 5000$, (b) $Re_j = 10000$, (c) $Re_j = 15000$, (d) $Re_j = 20000$ and (e) $Re_j = 25000$

Figure 2 illustrates streamlines at $H/L = 0.75$ for different Reynolds number ($Re_j = 5000$ - 25000) values. It is noticed that, in cavities between the heated components, tiny vortices are developed. Also, a stagnation area is developed when fluid from the jet inlet impinges on the first heated element, allowing pressure to build up, resulting fluid to convect into adjacent heated elements. A primary recirculation bubble is noticed near the jet inlet and confinement surface. This is mainly due to the shear flow caused by jet on surrounding fluid. Finally, as

Re_j increases, a small rise in maximum velocity magnitude value (VM_{max}) is observed due to increased convection, as seen in Fig. 2.

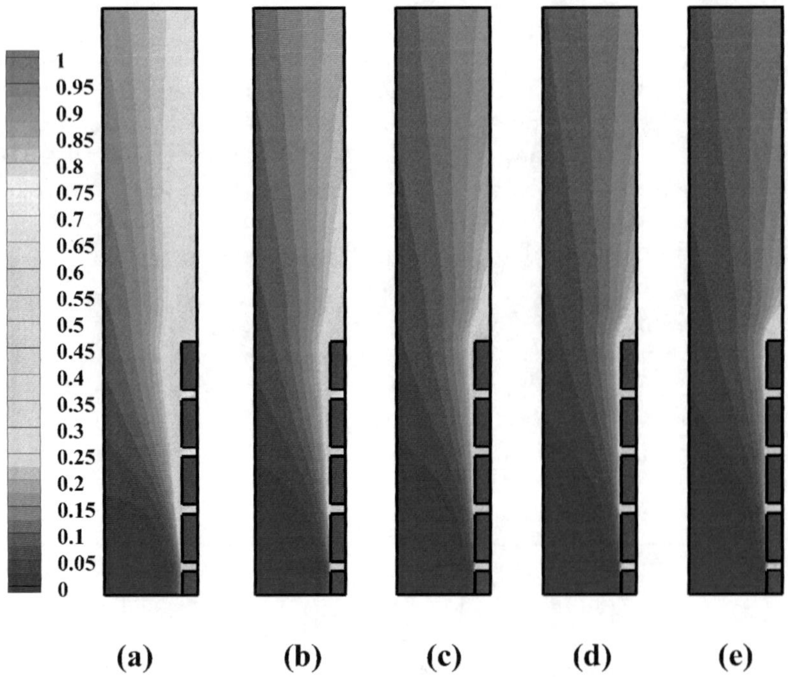

Figure 3. Effect of Re_j on Isotherms at $H/L = 0.75$ for (a) $Re_j = 5000$, (b) $Re_j = 10000$, (c) $Re_j = 15000$, (d) $Re_j = 20000$ and (e) $Re_j = 25000$

The associated isotherms at $Re_j = 5000, 10000, 15000, 20000, 25000$ in the turbulent air jet impingement system are presented in the Fig. 3. The thermal boundary layer thickness is nominal along the top surface of the first heated element (stagnation zone), and starts to grow in streamwise direction over the neighboring heated elements. Also, at both leading ($0.045 \leq X_P \leq 0.05$) and trailing ($0.045 \leq X_P \leq 0.05$) edges of all the heated elements, thermal boundary layer thickness is high as illustrated in Fig. 3. In addition, as the number of Reynolds increases, along all the individual heated components, the respective thermal gradient are observed to rise due to increased convection, which is in agreement with observed streamlines.

Turbulent Kinetic Energy (TKE) distribution for turbulent jet impingement system is demonstrated in Fig. 4. TKE is defined through root-mean-square (RMS) of velocity fluctuations and is governed by a production term (G_n), turbulent dissipation (ϵ_n), mean flow convection ($\frac{D\,k}{D\,t}$), and turbulent transport ($\Delta.T'$). From Fig. 4, TKE is high in the zones of shear-driven interaction caused by jet on surrounding fluid. This may be due to the domination of turbulence production (Gn). The magnitude of TKE is also substantial in the stagnation region. In addition, TKE approaches to zero downstream the fifth heated component due to fluid flow deceleration. This may be due to negligible velocity fluctuations and improved TKE dissipation to the ambient fluid. Finally, as expected with increase in Re_j, the maximal values of TKE (k_{max}) is observed to increase.

Figure 5 illustrates the turbulent viscosity contours and observed a maximal value at the same locations where high TKE is noticed. Also, as expected, with increase in Re_j, the maximal value of turbulent viscosity ($\nu_{t,max}$) is found to increase as demonstrated in Fig. 5.

From Fig. 6, it is found that, the magnitude of S_θ is significant over the upper surface of first heated element (stagnation zone), caused by higher temperature gradients, which is in

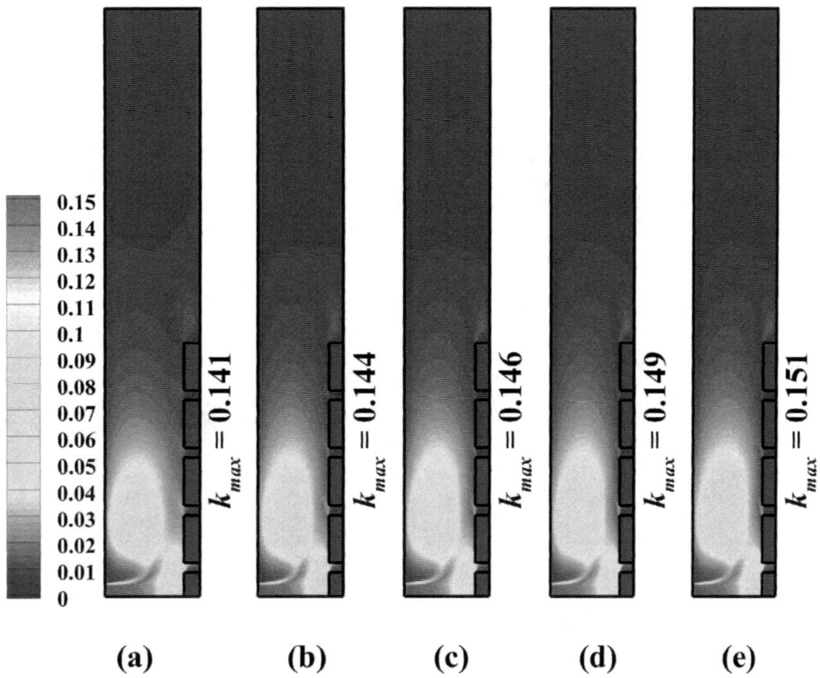

Figure 4. Effect of Re_j on turbulent kinetic energy at $H/L = 0.75$ for (a) $Re_j = 5000$, (b) $Re_j = 10000$, (c) $Re_j = 15000$, (d) $Re_j = 20000$ and (e) $Re_j = 25000$

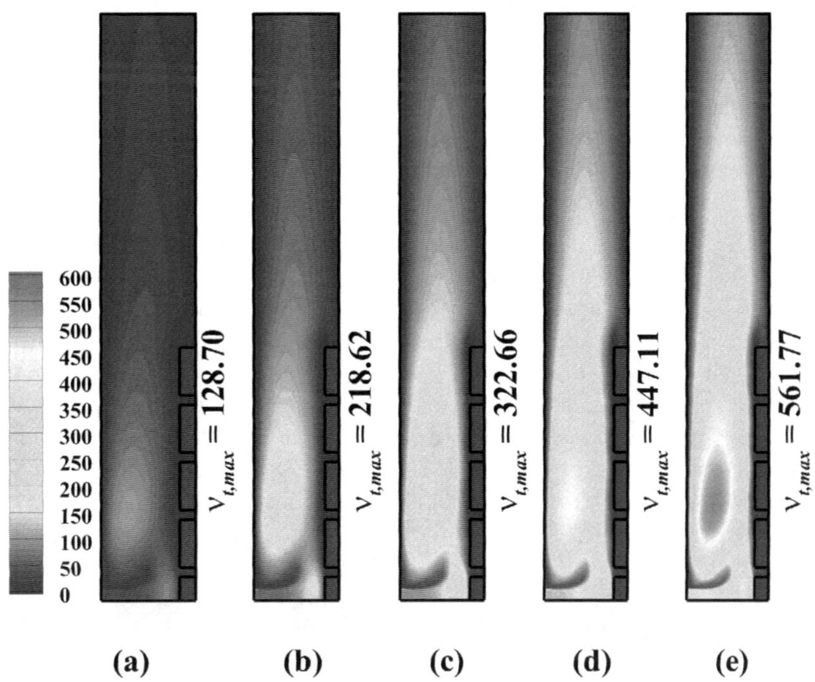

Figure 5. Effect of Re_j on turbulent viscosity at $H/L = 0.75$ for (a) $Re_j = 5000$, (b) $Re_j = 10000$, (c) $Re_j = 15000$, (d) $Re_j = 20000$ and (e) $Re_j = 25000$

accordance with observed streamlines and isotherms. It is noticed that, as Re_j increases, the zones of prominent S_θ values also decreases. Further, as Re_j increases, a significant increase in $S_{\theta,max}$ observed.

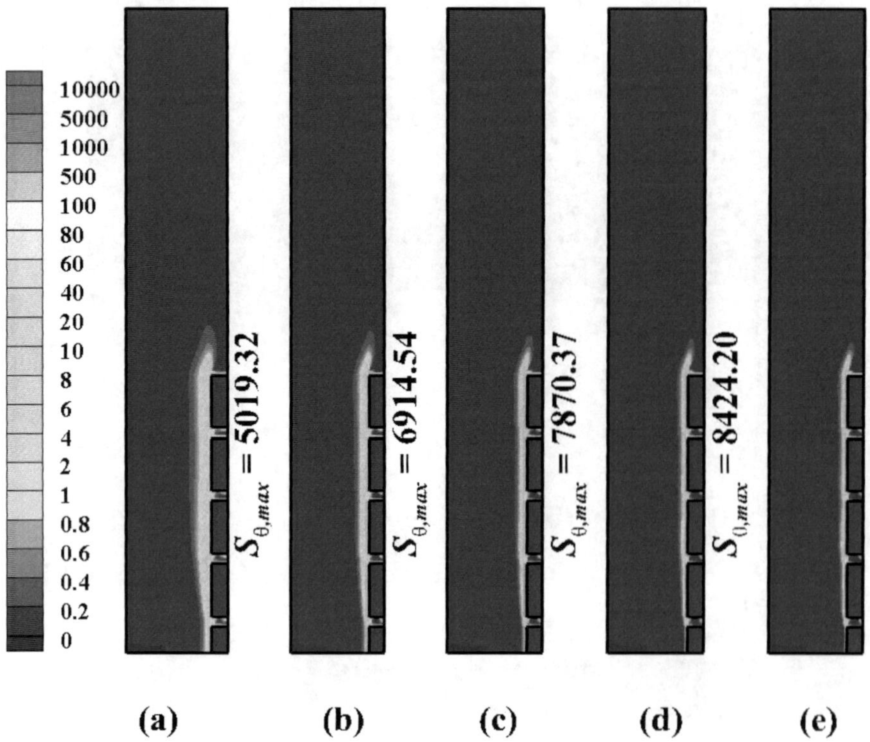

Figure 6. Effect of Re_j on $S_{\theta,\Omega}$ at $H/L = 0.75$ for (a) $Re_j = 5000$, (b) $Re_j = 10000$, (c) $Re_j = 15000$, (d) $Re_j = 20000$ and (e) $Re_j = 25000$

From Fig. 7, it is noticed that, in the zones of shear-driven flow caused by jet on ambient fluid, the magnitude of S_ψ is significant due to increased velocity gradients. Also, in stagnation zone, S_ψ is observed to be maximum due to improved viscous dissipation. Also, in the proximity of all the solid surfaces, S_ψ is significantly is fould to be higher. Finally, the magnitude of $S_{\psi,max}$ is noticed to raise with increase in Re_j. This could be attribute to increase in convection caused by an enhanced velocity magnitude and velocity gradients with increasing Reynolds number.

Figure 8 demonstrates corresponding contour maps of local total entropy generation ($S_{tot} = S_\theta + S_\psi$) at different values of Reynolds number (Re_j = 5000 - 25000). From Fig. 8, it is noticed that, at the locations of high velocity and temperature gradients, S_{tot} is significant due to contributions from S_θ and S_ψ. Furthermore, as Re_j increases, $S_{tot,max}$ is also found to increase.

Figure 9 depicts the variation of local Nusselt number (Nu) distribution on the periphery of heated elements at $H/L = 0.75$ for Re_j = 5000 - 25000. From Fig. 9, as Re_j increases, a noticiable augmentation in Nu is observed on the top sources of all heated elements. Also, as Re_j increases, Nu along side (left and right) surfaces of heated element increases. This is attributed to enhanced shear-driven convection of air trapped in the cavities between heated elements.

Figures 10 (a) - 10 (e) show the influence of Re_j and H/L on $\overline{Nu}$ for all heated elements. As Re_j increases from Re_j = 5000 to 25000, $\overline{Nu}$ for all the heated elements due to improved

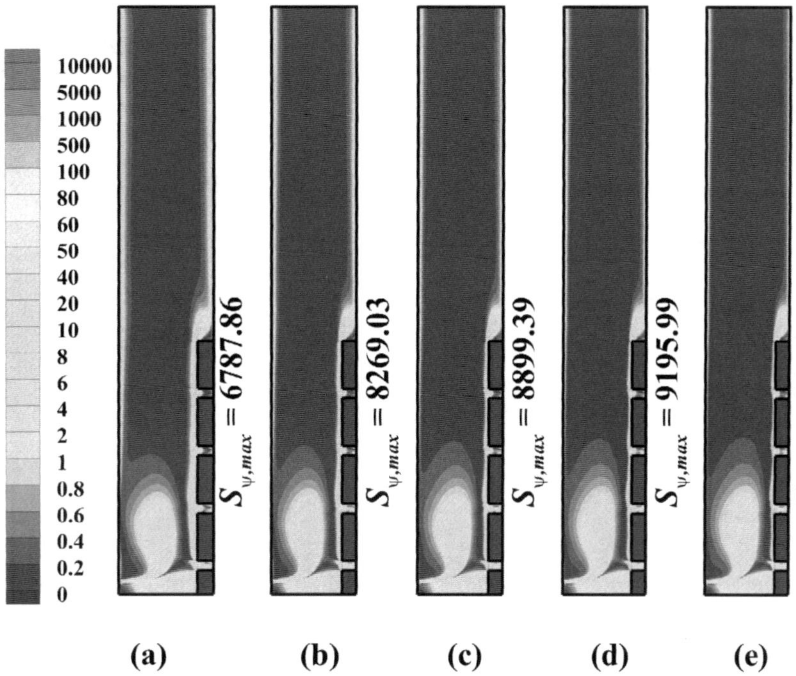

Figure 7. Effect of Re_j on $S_{\psi,\Omega}$ at $H/L = 0.75$ for (a) $Re_j = 5000$, (b) $Re_j = 10000$, (c) $Re_j = 15000$, (d) $Re_j = 20000$ and (e) $Re_j = 25000$

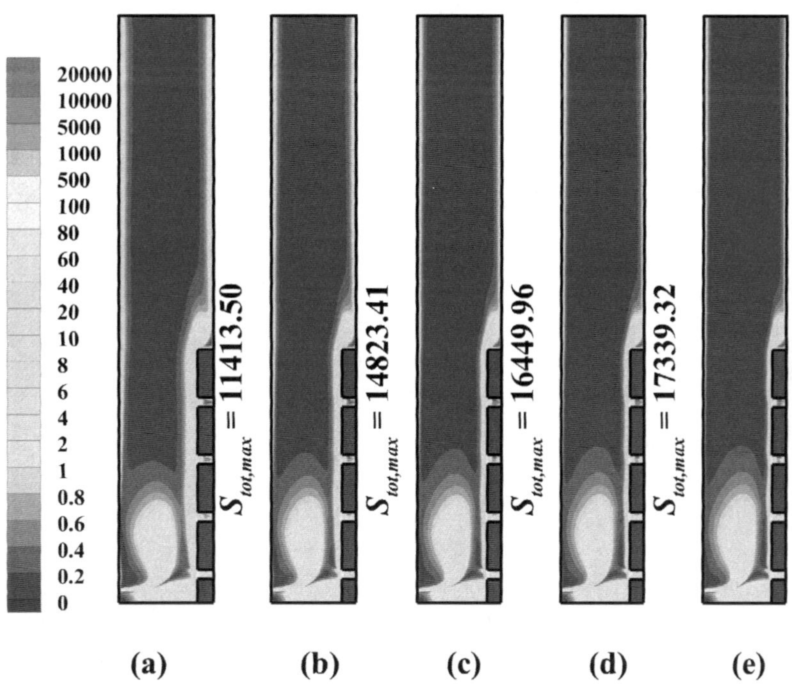

Figure 8. Effect of Re_j on $S_{tot,\Omega}$ at $H/L = 0.75$ for (a) $Re_j = 5000$, (b) $Re_j = 10000$, (c) $Re_j = 15000$, (d) $Re_j = 20000$ and (e) $Re_j = 25000$

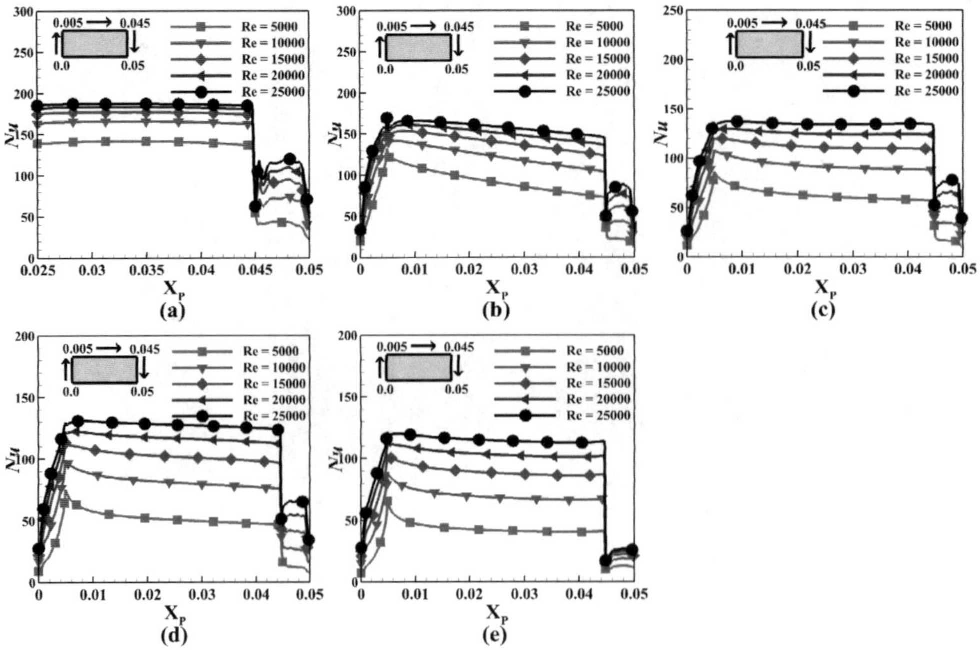

Figure 9. Effect of Re_j on local Nu distribution on the periphery of (a) First, (b) Second, (c) Third, (d) Fourth and (e) Fifth heated element

convection. Further, the effect of H/L on $\overline{Nu}$ for all heated elements is noticed to be negligible.

Figures 10 (g) - 10 (i) illustrate the effect of Re_j and H/L on global entropy generation. It is noticed that, as Re_j increases from Re_j = 5000 - 25000, the magnitude of $S_{\theta,\Omega}$ and $S_{\psi,\Omega}$ increases. This may be attributed to enhanced thermal and velocity gradients. Finally, for all the values of Re_j, it is found that, as H/L increases, the magnitude of $S_{\theta,\Omega}$, $S_{\psi,\Omega}$ and $S_{tot,\Omega}$ decreases.

4. Optimization

In order to achieve optimal designs for turbulent air jet impingement system, $\overline{Nu_{ov}}$ and $S_{tot,\Omega}$ are chosen as objective functions such that heat transfer is maximized at minimal entropy generation. For this purpose, response surfaces are derived for $\overline{Nu_{ov}}$ and $S_{tot,\Omega}$ with the design parameters (Re_j and H/L). The surrogate models defining the response surfaces for $\overline{Nu_{ov}}$ and $S_{tot,\Omega}$ are as follows:

$$\overline{Nu_{ov}} = 57.43+0.0064\,Re-39.64\,H/L+1.588\times10^{-4}\,Re\times(H/L)-1.128\times10^{-7}\,Re^2+13.22\,(H/L)^2 \tag{13}$$

$$S_{tot,\Omega} = 1991.14+0.081\,Re-4254.102\,H/L-0.0153\,Re\times(H/L)-1.094\times10^{-6}\,Re^2+2357.36\,(H/L)^2 \tag{14}$$

The R^2 values for $\overline{Nu_{ov}}$ and $S_{tot,\Omega}$ are 0.9978 and 0.9952, respectively. This suggests that, at 95% confidence limit, the derived surrogate models could be used in lieu of finite element simulations.

Furthermore, the accuracy of derived surrogate models (Equations 13 and 14) are examined through a pairity plot as shown in Fig. 11. It is noticed that, from Figs. 11 (a) and 11 (b) the

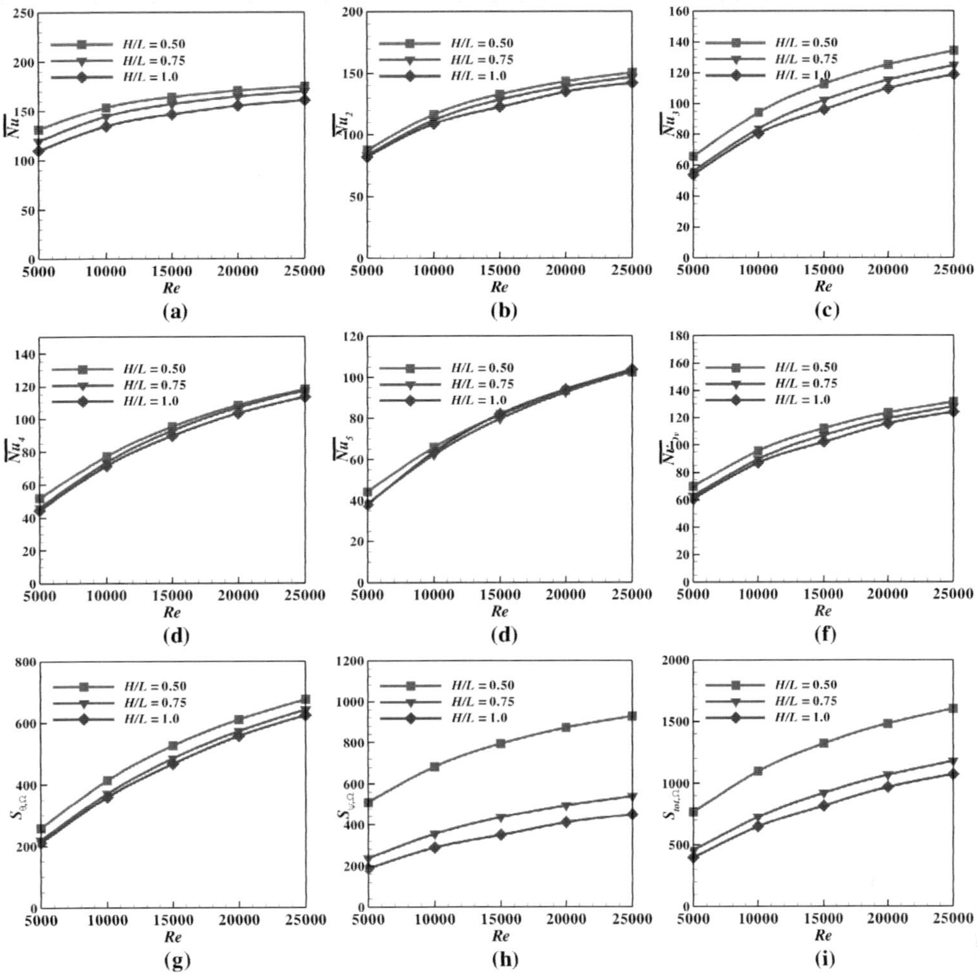

Figure 10. Effect of design parameters (Re_j, H/L) on Average Nusselt number and global entropy generation: (a) Nu_1, (b) Nu_2, (c) Nu_3, (d) Nu_4, (e) Nu_5, (f) Nu_{Ov}, (g) $S_{\theta,\Omega}$, (h) $S_{\Psi,\Omega}$, (i) $S_{tot,\Omega}$

deviations of surrogate models ($\overline{Nu_{ov}}$ and $S_{tot,\Omega}$) are within a $\pm$ 15 % deviation. Hence, the surrogate models for $\overline{Nu_{ov}}$ and $S_{tot,\Omega}$ are comparable with the results from FE computations. Therefore, to predict optimal designs through Multi-Objective Genitic Algorithms (MOGA), the derived surrogate models could be employed instead of FE Computations.

The multi-objective optimization problem is formulated as:

Maximize : $\overline{Nu_{ov}}$ = f(Re_j, H/L) ; **Minimize** : $S_{tot,\Omega}$ = f(Re_j, H/L)

Subjected to : $5000 \leq Re_j \leq 25000$ and $0.5 \leq H/L \leq 1.0$.

In order to the run the optimization problem, the population size is assumed to be 1000 with number of generations is equal to 1000 and crossover fraction = 0.7. Then, other variables such as function tolerance and Pareto-front population fraction are equal to 10^{-5} and 0.8 respectively.

Figure 12 shows the Pareto-optimal frontier in functional space featuring optimal designs with maximal heat transfer and minimal entropy generation. In the Pareto front, 80 optimal configurations are found rade-off between $\overline{Nu_{ov}}$ and $S_{tot,\Omega}$. It is noted that, a significant improvement in $\overline{Nu_{ov}}$ is always followed by a proportional rise in $S_{tot,\Omega}$ and vice versa on the Pareto optimum frontier. Further, in Fig. 12, it is seen that, the simulated designs through

IOP Conf. Series: Materials Science and Engineering 1132 (2021) 012017 doi:10.1088/1757-899X/1132/1/012017

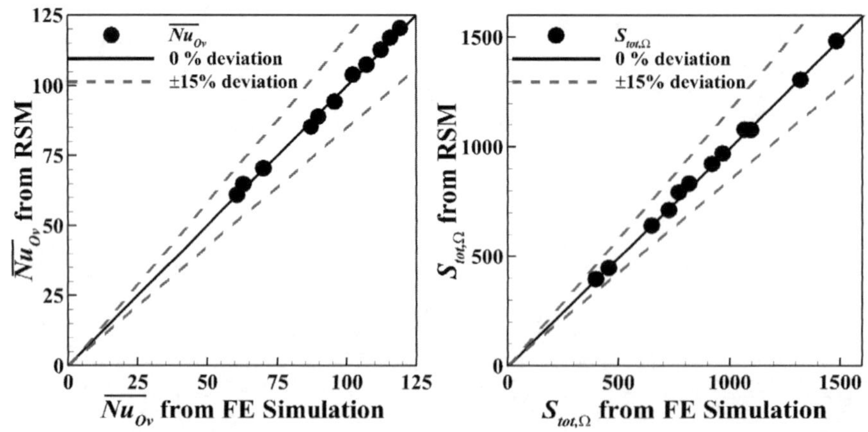

Figure 11. Comparison of results from FE simulation and Response Surfaces: (a) $\overline{Nu_{ov}}$ and (b) $S_{tot,\Omega}$

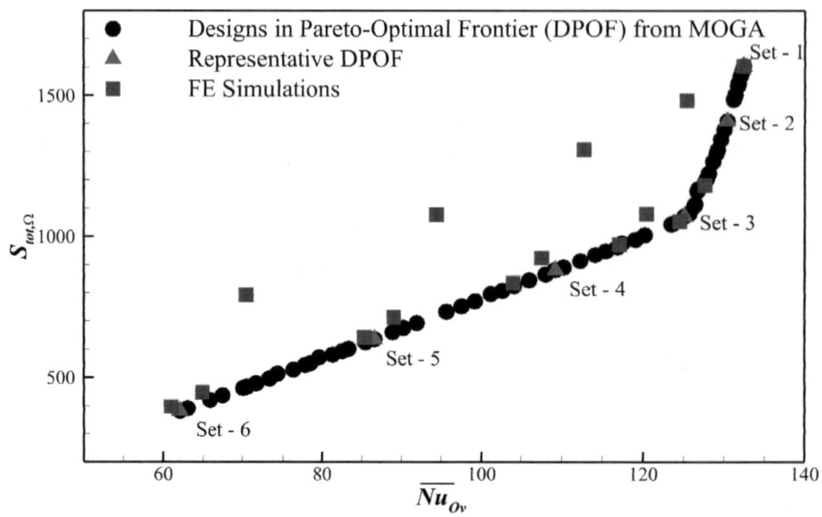

Figure 12. Pareto frontier for objective functions ($\overline{Nu_{ov}}$ and $S_{tot,\Omega}$)

in-house algorithm are scattered above the pareto-optimal frontier and no one is found below the same. It is also to be noticed that, none of the configurations in the frontier are preferable to others, as each design is a global Pareto-optimal solution, where both $\overline{Nu_{ov}}$ and $S_{tot,\Omega}$ fulfilled simultaneously.

For this reason, for six sets of representative Pareto-optimal designs (Re_j and H/L), FE simulations are performed as shown in the Table. 1. From the table, it is observed that, both $\overline{Nuov}$ and $Stot,\Omega$ from FE simulations and Pareto frontier of MOGA (as shown in Fig. 12) are comparable, suggesting excellent accuracy of RSA and MOGA.

5. Conclusions

In this work, an optimization methodology was presented to obtain optimum designs for turbulent air jet impingement system for maximal heat transfer and minimal entropy generation. An in-house algorithm based on Streamline Upwind/Petrov-Galerkin (SUPG) Finite Element (FE) Method was employed to solve continuity, Reynolds-Averaged Navier-Stokes, turbulent

Table 1. Comparison of the results ($\overline{Nu_{ov}}$, $S_{tot,\Omega}$) from MOGA and FE Simulation for six representative Designs in Pareto-optimal Frontier (DPOF)

DPOF	Re_j	H/L	$\overline{Nu_{ov}}$		$S_{tot,\Omega}$	
			MOGA	FEM	MOGA	FEM
1	25000.00	0.50	132.40	131.70	1604.23	1602.14
2	25000.00	0.59	130.39	129.17	1409.33	1406.81
3	24185.65	0.87	125.08	123.39	1071.57	1050.15
4	16275.95	0.89	109.23	107.59	880.67	876.39
5	10038.85	0.91	86.58	81.62	635.52	634.54
6	5000.00	0.92	62.08	62.13	380.70	369.97

kinetic energy (k), turbulent dissipation rate (ϵ) and energy equations were solved to predict momentum transport, thermal energy transport and thermodynamic performance at different values of Re_j and H/L. It was noticed that, as Re_j increases and H/L decreases, the magnitude of $\overline{Nu_{ov}}$ and $S_{tot,\Omega}$ increases. Subsequently, $\overline{Nu_{ov}}$ and $S_{tot,\Omega}$ were choosen as objective functions and surrogate models were derived through the Response Surface Approximation. Then, the Multi-Objective Genetic Algorithm (MOGA) was introduced to obtain optimal designs of impingement system with maximal heat transfer and minimal entropy generation. Finally, six representative pareto-optimal designs (Re_j and H/L) were chosen and FE simulations were performed for those representative designs. It was found that, the performance parameters, $\overline{Nuov}$ and $Stot,\Omega$ from FE simulations and Pareto frontier of MOGA are comparable, suggesting excellent accuracy of RSA and MOGA.

References

[1] Zuckerman N and Lior N 2006 *Advances in Heat Transfer* **39** 565–631
[2] Dewan A, Dutta R and Srinivasan B 2012 *Heat Transfer Engineering* **33** 447–460
[3] Bode F, Patrascu C and Nastase I 2020 *Thermal Science* 227–227
[4] Gardon R and Akfirat J C 1965 *International Journal of Heat and Mass Transfer* **8** 1261–1272
[5] Park T, Choi H, Yoo J and Kim S 2003 *International Journal of Heat and Mass Transfer* **46** 251 – 262
[6] Cafiero G, Discetti S and Astarita T 2014 *International Journal of Heat and Mass Transfer* **75** 173 – 183
[7] Nebuchinov A S, Lozhkin Y A, Bilsky A V and Markovich D M 2017 *Experimental Thermal and Fluid Science* **80** 139–146
[8] Afroz F and Sharif M A 2018 *Applied Thermal Engineering* **138** 154–172 ISSN 1359-4311
[9] Lam P A K and Prakash K A 2015 *Energy Conversion and Management* **89** 626 – 643
[10] Lam P A K and Prakash K A 2016 *Energy Conversion and Management* **111** 38 – 56
[11] Forster M and Weigand B 2021 *International Journal of Thermal Sciences* **164** 106862 ISSN 1290-0729
[12] Bejan A *Entropy generation through heat and fluid flow* ISBN 9780471094388
[13] Ruocco G 1997 *International Communications in Heat and Mass Transfer* **24** 201 – 210
[14] Shuja S, Yilbas B and Budair M 2007 *International Journal of Numerical Methods for Heat and Fluid Flow* **17** 677–691
[15] Lam P A K and Prakash K A 2017 *International Journal of Heat and Mass Transfer* **108, Part A** 880 – 900
[16] Esmailpour K, Bozorgmehr B, Hoseinalipour S M, Mujumdar A s and Lewis R 2015 *International Journal of Numerical Methods for Heat & Fluid Flow* **25**
[17] Lam P A K and Prakash K A 2017 Thermal and thermodynamic analyses of impingement cooling system with turbulent jet arrays *Proceedings of the 24th National and 2nd International ISHMT-ASTFE Heat and Mass Transfer Conference (IHMTC-2017)* (Begel House Inc.)
[18] Launder B and Spalding D 1974 *Computer Methods in Applied Mechanics and Engineering* **3** 269 – 289
[19] Lam P A K and Prakash K A 2014 *International Journal of Heat and Mass Transfer* **69** 390 – 407
[20] Brooks A N and Hughes T J R 1982 *Computer Methods in Applied Mechanics and Engineering* **32** 199–259
[21] Arul Prakash K, Biswas G and Rathish Kumar B V 2006 *International Journal of Heat Mass Transfer* **49** 4633–4652

IOP Conf. Series: Materials Science and Engineering 1132 (2021) 012018 doi:10.1088/1757-899X/1132/1/012018

Experimenting to optimize topology parameters of a reinforced metal matrix composite using Taguchi technique of design of experiments

Praveen maruthur

Projects manager, Acharya Analytics, Kerala

Abstract. This experimental study is carried out to optimize topological parameters like wear rate and frictional force of an aluminium alloy AA2618 which is a metal matrix of composition Al-93.7%, Si-0.18%, Cu-2.30%, Mg-1.60%, Fe-1.1%, Ni-1.0%, Ti-0.07%. The aluminium alloy is reinforced with titanium di-oxide of 5% by weight. Taguchi design of experiment is used to carry out optimization study, taguchi method is he most reliable common and successful experiment technique for process performance improvement and raise quality of product. As per taguchi technique parameters chosen are sliding distance, speed and load with each property for 3 levels, so the test consists of 27 numbers according to L27 orthogonal array. Optimal parametric condition of minimum wear and frictional force are obtained through signal to noise analysis. For verification of results confirmation experiments is also conducted.

1. Introduction

Aluminium metal composite known as AMCs have emerged into primary material in heavy engineering applications such as brake drums, engine cylinder liners, piston head, as they have good wear resistance and load bearing capacity thanks for the reinforcement of ceramic particles into AMC's because of which they have good properties [1]. Along with AMCs, PR-MMCs – particulate reinforced metal matrix composites have low density, high strength and stiffness and goof isotropic properties [2-3]. Now alloy materials like these have taken material science studies into new perspectives and because of their industrial application many engineering sectors have shown promising growth and ventured with high efficiency and minimum wear and service sections.Mechanical and material characteristics along with tribo contact condition and surface interaction are the most common tribological characters or parameters that adversely affect the discontinuously reinforced aluminium composites causing high friction and wear performance. Sannino et al, in his paper they have studied about the most common factors that are associated with above mentioned tribological parameters. Lower frictional force and coefficient of friction along with high resistance and seizure pressure were reported when studies were carried out on effect of particulate reinforcement on aluminium alloy [5-7]. In another study by Ramachandra et al [8] they concluded in the study that with addition of SiC particles erosive wear resistance and sliding wear resistance have enhanced even though corrosive resistance dropped. In another comparative study carried out in elevated temperature aluminium alloy Al 7075 reinforced with glass fibre along with base metal aluminium, the alloy showed better wear resistance than base metal [9]. Uyyuru et al [10] studied the effect of reinforcement volume fraction and size distribution of aluminium composites, Gomez de salazan et al [11] investigated the effect and influence of heat treatment over wear behaviour of aluminium composites, whereas Guo et al [12] did experimentations to study the effect of aging on wear of metal matrix composite. Huge number of researches are carried out to study and understand behaviour, performance of composite materials. This study is carried to study and valuate optimized results of tribological parameters of titanium dioxide reinforced aluminium metal matrix using taguchi's design of experiment methodology.

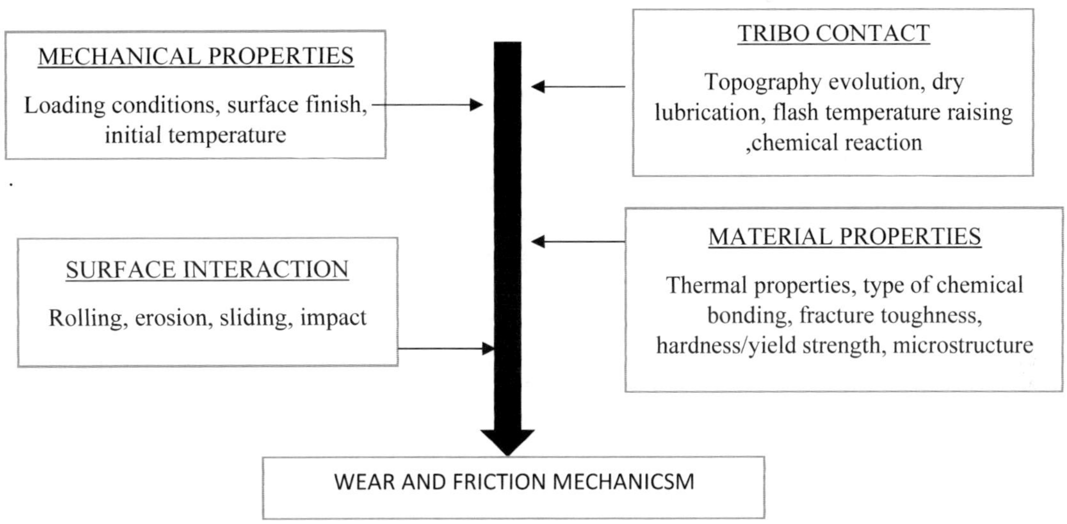

Figure 1: Tribological properties, cause and effect

2. Experimentation

2.1. Materials

AA2618 is a wrought aluminium alloy which is used as matrix material in the preparation of composite material. The AA2618 Aluminium alloy was developed for high temperature applications such as aircraft and automobile engine components which has the composition of Al-93.7%, Si-0.18%, Cu-2.30%, Mg-1.60%, Fe-1.1%, Ni-1.0%, Ti-0.07% and this alloy can withstand even a temperature of 204°c. Fe and Ni gives structural stability at high temperature along with the help of precipitation and dispersion hardening operations performed on it.

TiO_2 is used as reinforcement material, which has very common utilization in many engineering applications. Liquid vortex method was implemented to add 5% by weight of TiO_2 Aluminium alloy matrix.

2.2. Orthogonal array

Sliding speed in m/s, load in N, sliding distance in 'm' are the three parameters chosen for experiment and each parameter was analyzed three times each as shown in table 1 in order to find the nonlinear behavior

I. Sliding speed and load'
II. Sliding speed and distance
III. Load and distance

IOP Conf. Series: Materials Science and Engineering 1132 (2021) 012018 doi:10.1088/1757-899X/1132/1/012018

These are the selected two factor interactions. An L27 orthogonal array with 26 degree of freedom was selected with first column given to sliding speed, second coloumn to lead and the fifth coloumn assigned to sliding distance. The remaining coloumn were assigned to their interactions.

Table 1 Factors and details

Factors	Code	Units	Level 1	Level 2	Level 3
Sliding speed	S	m/s	1.2	2	3.1
Load	L	N	18	30.4	39
Sliding distance	D	M	700	1400	2100

2.3. Experimental procedure

pin on disc type friction and wear monitoring test rig as shown in fig 2 was used in dry sliding condition of composite to study its tribological performance. The lever mechanism applies normal force on the stationary specimen while the disc rotates, and the counter body is made of hardened steel.

Orthogonal array of taguchi design of experiment was followed to conduct wear test and it is studied. Wear rate as a function of the sliding velocity, load and sliding distance. Wear rate and frictional force are noted while conducting the experiments. To make experimental test rig clean from debris and left over after finishing experimental run discs are cleaned using acetone. Mean response values of two experimental run are tabulated into Table 2.

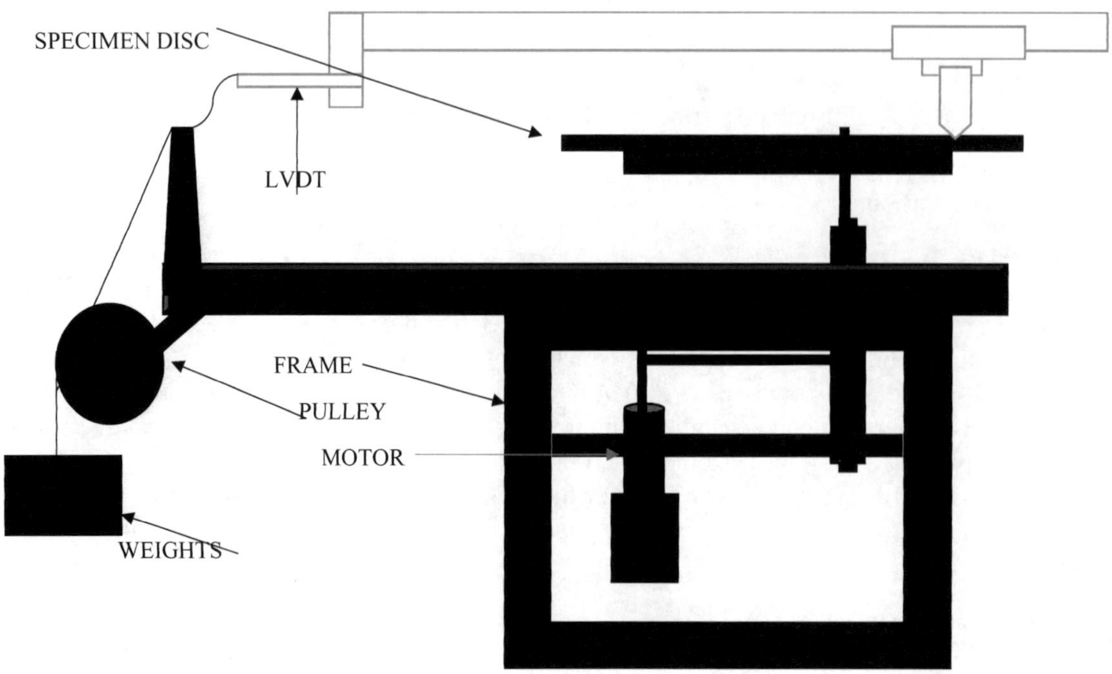

Figure 2: Test rig

Table 2: Orthogonal array

Test	sliding speed m/s	Load N	Sliding distance m	Wear rate mm^3/N-m	Force N	S/N ratio (db) wear rate	S/N ratio (db) Friction force
1	1.2	18	700	23	7.50	-26.60	-17.58
2	1.2	18	1400	56.5	7.15	-34.19	-17.22
3	1.2	18	2100	62.0	7.90	-34.98	-17.98
4	1.2	30.4	700	26.0	11.60	-27.62	-20.28
5	1.2	30.4	1400	56.0	11.45	-34.11	-20.90
6	1.2	30.4	2100	81.0	11.40	-37.37	-20.86
7	1.2	39	700	78.0	11.40	-36.84	-20.85
8	1.2	39	1400	81.0	13.58	-37.27	-22.22
9	1.2	39	2100	90.0	16.43	-38.18	-23.81
10	2	18	700	18.0	7.30	-24.57	-17.38
11	2	18	1400	55.5	6.59	-34.04	-14.26
12	2	18	2100	74.0	4.87	-36.50	-19.50
13	2	30.4	700	26.0	9.60	-27.62	-18.82
14	2	30.4	1400	61.0	8.87	-34.84	-19.66
15	2	30.4	2100	82.0	9.86	-37.38	-21.21
16	2	39	700	41.0	11.65	-31.46	-22.10
17	2	39	1400	64.0	13.30	-35.25	-22.04
18	2	39	2100	83.0	13.20	-37.48	-13.48
19	3.1	18	700	22.0	4.30	-26.23	-16.40
20	3.1	18	1400	46.0	6.50	-22.44	-16.02
21	3.1	18	2100	52.0	6.10	-34.48	-19.74
22	3.1	30.4	700	35.0	9.90	-30.12	-19.42
23	3.1	30.4	1400	44.0	9.50	-32.06	-19.54
24	3.1	30.4	2100	52.0	9.65	-33.48	-22.40
25	3.1	39	700	33.5	13.80	-29.75	-22.95
26	3.1	39	1400	65.0	13.00	-35.39	-22.86
27	3.1	39	2100	71.5	12.90	-36.20	-22.13

3. Results and discussion

3.1. Signal to noise ratio

For the comparative study under the experimentation, observations are transformed into signal to noise ratio depending on the type of parameters analysed the during the study S/N ratio are very broadly classified. Hera as the optimization parameters chosen are wear rate and frictional force, they come under "smaller is better" type quality characteristic and for them S/N ratio is formulated as

$$\eta = -10 \log_{10} \frac{1}{n} \sum_{i=1}^{n} y_i^2 \tag{1}$$

n = number of tests in trial, here n = 2

A high value of S/N ratio gives the idea that signal is higher than random effects of noise factors and always during the experimentation highest S/N ratio of parameter are chosen, the S/N ratio calculated using equation (1) is tabulated in Table 2.

IOP Conf. Series: Materials Science and Engineering 1132 (2021) 012018 doi:10.1088/1757-899X/1132/1/012018

3.2. Anova

Analysis of variance are used to analyse influence of parameters chosen sliding speed, sliding distance and applied load on performance characteristics. Table 3 and table 4 shows the results of ANOVA.

The percentage contribution factor is calculated using the largest total sum of squares values. From table 4 it is available that distance travelled (P=56.02%) is the most significant factor, while other two parameters contributes 17.69 % and 6.629 %. The table 4 and 5, both has got values of interactions between sliding speed*load, speed*distance, load*distance too.

Table 5 gives idea about variance of frictional force of composite. The test result gives information that load applied on the test material is the most significant factor which causes frictional force, whereas sliding speed and distance has negligible contribution.

Table 3: ANOVA values of sliding wear

Factors	DOF	Seq ss	Adj ss	Adj MS	F	P	% contribution
Sliding speed, S (m/s)	2	983.011	983.011	491.05	5.58	0.020	6.629
Load, L(N)	2	2300.425	2300.425	1152.2	14.40	0.002	17.69
Distance, D (m)	2	6868.935	6868.935	3433.4	44.94	0.000	56.02
S*L	4	544.571	544.571	135.40	1.82	0.118	2.95
S*D	4	420.224	420.224	104.11	1.72	0.215	1.90
L*D	4	193.357	193.357	47.59	1.31	0.543	1.00
Error	8	597.386	597.386	73.89	0.55		7.78
total	26	11915.194	11915.194				

Table 4: ANOVA values of frictional forces

Factors	DOF	Seq ss	Adj ss	V	F	P	% contribution
Sliding speed, S (m/s)	2	10.416	10.416	10.416	2.67	0.073	2.694
Load, L(N)	2	208.640	208.640	208.640	72.87	0.000	84.472
Distance, D (m)	2	1.826	1.826	1.826	0.58	0.435	0.105
S*L	4	1.784	1.784	1.784	0.23	0.749	0.180
S*D	4	3.914	3.914	3.914	0.71	0.424	1.334
L*D	4	2.487	2.487	2.487	0.51	0.565	0.746
Error	8	10.356	10.356	10.356			7.155
total	26	242.701	242.701	242.701			

3.3. Concurrent optimization of properties

This experiment is carried out to simultaneously optimize the parameters, wear rate and frictional force. The lowest wear rate is recorded at tenth run, when parameters were at second level of sliding speed and first level of load and distance.

Frictional force was least at nineteenth run, when parameters are at third level of sliding speed and first level of load and distance.

Harrington's desirability function method has been used for multipurpose optimization and represented as

$$h_i = \exp [- \exp (-y_i)] \qquad (2)$$

single value desirability as

$$H = \{ \prod_{i=1}^{n} h_I \}^{1/n} \qquad (3)$$

Results are tabulated into Table 5 for comparison.

Table 5: concurrent optimization

Optimal levels			Sliding wear rate	Individual desirability H1	Frictional force	Individual desirability H2	Combined desirability H
S	L	D					
2	18	700	18	0.999	7.3	0.999	0.999
3.1	18	700	22	1	4.3	0.995	0.997

3.4. Confirmation experimentation

Experiments of confirmation is carried to verify that optimal solutions obtained would yield to quality improvement under study.

Table 6 shows comparative values of predicted wear rate and functional force with the actual error is difference between actual and predicted values of S/N ratio.

Table 6: verification

Sliding wear rate			Frictional force		
	Predicted optimum	Actual optimum		Predicted optimum	Actual optimum
Parameters level	S:3.1, L: 18, D: 700	S:3.1, L: 18, D: 700	Parameters level	S:3.1, L: 18, D: 700	S:3.1, L: 18, D: 700
Wear rate	22	21	Wear rate	4.3	4.4
S/N ratio	-26.2364	-25.8584	S/N ratio	-13.4852	-14.7564
Predicted error	0.286103		Predicted error	0.15235	
Confidential limit, 2σ	0.446		Confidential limit, 2σ	0.327	

4. Conclusion

1. Taguchi technique can be successfully used to analyse and predict sliding wear and frictional force as parameters of metal matrix composite
2. The study of variance of sliding wear rate gave evidence to conclude that distance travelled by the specimen is the important factor to be analysed above load and speed.
3. In case of study of variance of frictional force gave conclusive evidence that load is the important factor to be analysed above speed and distance.

References

1. Debdas Roy, Bikramjit basu, Amitava basu mallick. Tribological properties of Ti-aluminide reinforced Al-based In-situ- metal matrix composite. *Intermetallics* 13(2005) 733-740.

2. Mingzhao tan, Qibin Xin, Zhenghua Li, B Zong. Influence of SiC Al_2O_3 particulate reinforcements and heat treatments on mechanical properties and damage evolution of Al-2618 metal matrix composites. *Journal of material science* 36 (2001) 2045-2053
3. Nitsham A E. New application for aluminium based metal matrix composites. 1997. Vol. 54. *Light metal usage*. USA
4. A P Sannino, H J Rack. Dry sliding wear of discontinuously reinforced aluminium composite: review and discussion. *Wear* 189 (1995) 1-19
5. R N Rao, S Das, P V Krishna. Experimental investigation on the influence of SiC particulate reinforcement in Al alloy composites
6. Prasanna kumar M, Sadashivappa K, Prabhukumar G P, Basavarajappa S. Dry sliding wear behaviour of Garnet reinforced metal matrix composite. *Material science,* Vol 12, No.3.2006. PP. 209-213
7. G Ranganath, S C Sharma, M Krishna. Dry sliding wear of Garnet reinforced Zink/ Al MMCs. *Wear.* 251 (2001) 1408-1413
8. Ramachandran M, Radhakrishna. Sliding wear, slurry erosive wear and corrosive wear of Al/SiC composites. *Material science.* Vol 24. No.2/1.2006.
9. R K Uyyuru, M K Surappa, S Bruesthaug. Effect of reinforcement volume fraction and size distribution on the tribological behaviour of Al- composite. *Wear.* 260(2006) 1248-1255
10. J Guo, X Yuan. The aging behaviour of SiC /Gr/6013 Al composites in T-4-T-6 treatments.
11. J H Hatch (ed), Aluminium: properties and physical metallurgy, ASM, Metals park, OH, (1988),371.
12. James C, Williams. Progress in structural materials for aerospace systems. *Acta materials.* 51 (2003) 5775-5799.

Development and fabrication of fused deposition modelling 3D printer

Samson Wilson, Rahul Thomas, Nimmi Mary, Eric Tom Bosco, Ajith Gopinath*

Department of Mechanical and Automobile Engineering, Christ Deemed to be university, Bengaluru

*Email: ajith.gopinath@christuniversity.in

Abstract. The usage of 3d printing has been a widely applicable in various branches such as aerospace, medical, food industry and recently even in construction of houses. As we can see 3d printing has become a revolutionary invention and is becoming very common in manufacturing of various machining parts. So as we know there are various forms of 3d printing technologies and each of them have their own advantages and disadvantages. One of the main disadvantage is the quality of finishing on the 3d printed parts. Apart from this the cost of 3d printing has also been one of the major disadvantage. The purpose of this projects is to design and fabricate a 3d printer which would increase the quality of the parts manufactured and also make it cost efficient. This quality product can be achieved by decreasing the layer thickness. The other ways to increase quality of the product is fabricate the 3d printer in such a way that the temperature of the nozzle is within the given range all through the process. This project is expected to result in the fabrication of the 3d printer considering all the above features.

Keywords: 3D printing, Aerospace, Manufacturing, Design, Fabrication

1. Introduction

Manufacturing of difficult parts has become a major issue in the industries. The accuracy and precision of the parts made by the use of subtractive processes isn't up to the level which company desires. Hence to eradicate these difficulties additive manufacturing is introduced which helps in conversion of 3D design models into actual physical objects. The author explains how the Fused Deposition Modelling (FDM) cartesian co-ordinate based 3D printer should be designed and developed. PLA (Poly Lactic Acid) is used in this project for the 3D printer be- cause of it's often use. The 3D Modelling is done using CATIA V5 R20 software is converted into STL (Stereo lithography) file. CURA software is used to convert the STL file into G-codes which id then sent to PRONTERFACE software. The PRONTERFACE software guides the instructions to the Cartesian coordinate based 3D printer and 3D Modelled design are printed to physical part as a result [1].

3D printing is a method of non-traditional machining and a technique becoming popular nowadays. Additive manufacturing is a way in which object by laying successive layers of material till the entire object is made. There are many different kinds of 3D printers and they make use of different kind of printing methods, technologies and also different kinds of materials. This paper explains in detail about

5 different kind of 3D printers. They are: Stereolithography (SLA), Fused deposition modeling (FDM), Selective Laser Sintering (SLS), Laminated object manufacturing (LOM) and Digital Light Processing (DLP). The research paper gives a basic introduction to the concept of various 3D printers with their advantages, limitation, and application and compares each of them to different standards such as dimensional accuracy, surface finish, material used and post processing requirements [2]. Additive manufacturing (AM) has been nowadays widely used in many industries like aerospace, automotive, biomedical, and energy industries. Presently, with the quick growth of this technology, there are a huge number of 3D printing approaches in the market. Each and every method has its own advantages, disadvantages and applications. Materials used for printing too are divided into many groups such as ceramics, metals, polymers, and cements. In this paper they discus on how to build a 3D printer based Fused Deposition Modelling technique. They divide all the process involve in building of FDM based 3D printer into many steps and study on it. They also used Arduino Mega as a microcontroller in this project [3]. 3D Printing or Additive manufacturing is an innovative method of manufacturing parts right from CAD design model by using layer by layer material approach. This manufacturing technique can produce fully condensed metallic parts in a very short time with high precision. The main features of additive manufacturing are part complexity, part consolidation, freedom of part design and light weighting are gathering specific interests in metal additive manufacturing for aerospace, oil and gas, marine and automobile applications. Directed Energy Deposition and Powder Bed Fusion Technologies are mainly used in mainly Metallic 3D printing. Powder bed fusion is a technique in which each powder bed layer is carefully fused by using energy source like laser. This is the most auspicious additive manufacturing technology that can be used for manufacturing small, complex and low volume metallic parts. This review grants outline of 3D Printing technologies, applications, advantages, challenges, disadvantages, materials and economics of 3D metal printing technology[4].

The research paper on 3D printing has become a distinguished topic in today's technological conversation. In the paper, author explains how a 3D printer works and discuss on additive manufacturing or 3D printing. The paper explains of the advantages of 3D printing compared to conventional methods of manufacturing. The author explains in detail about different application of 3D printer and how it is affecting our life presently. It also clarifies how 3D printing has become an essential part in various industries and the way in which solved many existing problems. Lastly the imminent potential of 3D printing or additive manufacturing has been drawn by the author [5]. Fused Deposition Modelling (FDM) is an Additive Manufacturing Method for printing 3D designed objects layer by layer successively. The key objective of this research is to develop a low-cost 3D Printer using readily accessible materials and traditional methods for fabrication which can be used to print the objects confined in a 300 x 250 x 300 (in mm) printing space. According to the market survey done by the authors, they arrived at an inference that 3D printers are available at a price about Rs.50,000 to Rs.60,000 in Indian market. Originally, they designed a 3D Printer entirely in 3D Modelling Software SOLIDWORKS and studied each part and chose a readily available material suitably to develop a cost effective 3D printer. The researchers have explained step by step on making a cost effective FDM 3D printer[6]. This research paper is on various types of 3D printing and several materials used in 3D printing. It also explains in detail about 3D printing and its significance in today's world. Selective Laser Sintering, Fused Deposition Modelling, Stereolithography and Laminated Object Manufacturing are some of the several processes used in 3D printing explained in the research paper. It gives a detail description on the concept and principle of each 3D printing process. It also elucidates on different types of materials used in 3D printing process. They are: Acrylonitrile Butadiene Styrene (ABS), High Impact Polystyrene (HIPS) and Poly Lactic Acid (PLA). It clarifies on the properties, advantages, limitation and application of the several materials used in 3D printing [7].

The 3D printing technology is widely used in the current manufacturing industries. The 3D printer although efficient has its disadvantages. It takes 4-12 hours to complete a work piece and almost a day for printing a complex or a big shape. The paper talks about the methods proposed to improve the overall speed and ac- curacy of the printer [8]. 3D printing usually works in a way that whatever the feeding material, it will get printed. If the material needs to be changed, the whole filament needs to be changed. The paper proposes to create new materials and designs like composites using fused fiber deposition method by using a drop down platform which will allow the printing of new materials [9].

The 3D printing technology has revolutionized the engineering industry and many other industries like medical and construction industries. The paper depicts the different processes involved in the 3D printing and the different types of options available in the market based on price, workspace, accuracy and speed. The advantages and drawbacks are also mentioned [10]. The manufacture of small components, big components, printed biological tissues etc. The 3D printer plays a huge role in the current manufacturing industry. Yet there are some imperfections like structural defects and pore formation associated with it. The paper talks about the methods to find such irregularities and to re- solve the problems with some modifications [11]. The paper elaborates about the concept of 3D printing and the advantages of 3D printing over conventional methods in the engineering field and other industries and how the end product is superior to conventional methods like machining, casting. It also mentions the ways adopted to print in low cost methods [12].

2. Methodology for the Study

The given flowchart displays the methodology used in the construction of FDM 3D printer. The initial step is to select one of the additive manufacturing processes among many described in the literature survey. Then a suitable mechanism is selected for X, Y and Z axis movements, considering various factors such as simplicity of design, cost of fabrication, accuracy, synchronization etc. Once the mechanism is carefully chosen the following step is integration of electronics and software. Finally, the machine is fabricated. The final step is, synchronization of electrical, mechanical and software elements of the machine.

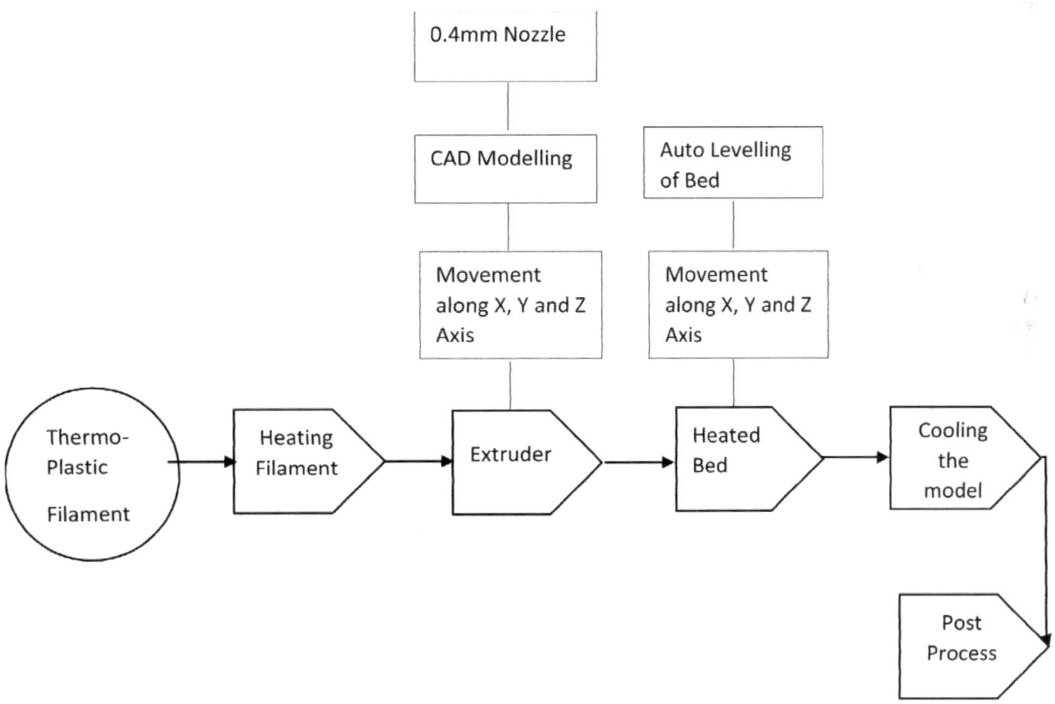

Figure 1. Flowchart for the functioning of the 3D printer

3. Apparatus

3.1. Process Selection

The next step is to decide which coordinates are supposed to move in which direction (X Axis, Y Axis, Z Axis). This selection would depend on the same factors earlier considered such as cost, capacity etc. After the mechanism is selected the next process is to choose the appropriate electronics, materials for the fabrication and finally then synchronizing all the components together so that they work in tandem. The FDM starts with a CAD Model where a required model to be manufactured is designed using a

IOP Conf. Series: Materials Science and Engineering 1132 (2021) 012019 doi:10.1088/1757-899X/1132/1/012019

CAD designing software such as SolidWorks or CATIA. This model with the required dimensions is converted into a STL file format. The designed model is then sent to slicing software's such as CURA, Astroprint, Craftprint etc. where the model is divided into small layers up to which the nozzle can print. These layers decide the surface finish of the manufactured model. Lower the layer thickness better the surface finish it will have. Now after all these processes the model is ready to be manufactured, the manufacturing starts with a thermoplastic material in a loop is fed to the nozzle through a PTFE Tube. THE hot end heats this thermoplastic material to a glass change temperature and then exits out through the nozzle. The Thermos plastic material used in our 3d printer is PLA type filament.

3.2. Mechanism Selection

There are different ways in which the mechanism of a 3d printer can be built, some of them are SCARA, Cartesian, Polar, Delta and various other mechanism. We have decided to go with a Cartesian type of mechanism as it is easier to fabricate and also cost effective. In this type of mechanism the bed moves in Y-Axis, the hot end moves in the X Axis and the 2 Lead Screw moves in the Z Axis up and down. All these Axis's must move with great precision in order to get good accuracy and repeatability The Z Axis moves at very low rpm as it has to complete fabrication of each layer by layer. The bed is of the type Mk2 heat bed where the bed heats up to a certain temperature so that the base layer of the model must stick to the bed so that the foundation of the model is rigid. This heat bed is bolted to a wooden piece and is also provided with a spring in order to make in more flexible. Both Z axis lead screws must move in tandem with each other otherwise the precision will be affected. The construction of a Cartesian mechanism type of 3d printer in simpler than the rest of the mechanism making it one of the most preferable type of mechanisms for an FDM type of 3d printer.

The components used for the project are considered with the following factors:-

1. Reducing the fabrication cost of the 3D Printer
2. Making the 3D printer economical for household applications
3. The product manufactured provides good dimensional accuracy
4. To achieve repeatability

The components are classified based on the following criteria.

1. Electronics
2. Mechanical
3. Softwares

3.3. Electronics

3.3.1. Controllers. The main part of the electronics list is the controller which is brain of the system, this decides which and when the Axis's should move depending upon the desired model. 44 This basically controls all electronic components used in the 3d printer. The 3d printing technology uses Arduino microcontroller as their basic choice of microcontrollers. In this project we have used Arduino Mega as our choice of controller as it is easier to program the codes and also provide good response to the input given.

3.3.2. Stepper Motor. The stepper motor plays a vital role in the working of thee 3d printer as it help in the movement of the axis to required precision. The type of 3d printer we have used is a NEMA 17 type of stepper motor. The main significance of a NEMA 17 stepper motor is that it provides more torque than other variants of the NEMA stepper motor. The motor is selected depending upon the specifications we need for the working of the 3d printer. For this project we have used 5 NEMA-17 stepper motor's (1 for Y-Axis, 1 for X-Axis, 2 for Z-Axis and 1 for the extruder).Some of the specifications of the NEMA 17 are : - Rated Voltage = 12V DC - Current = 1.2A - Step angle = 1.8 degree - Holding Torque = 3.2 kg-cm

3.3.3. Endstops. The end stops we have used for this project is a mechanical type of end stop. The reason we have used a mechanical type of end stops is because it is cheaper than other type of end stops such as Optical end stops. The purpose of using an end stop is to define the maximum distance the axis are supposed to move. Once it touches the mechanical end stop it stops the movement of the motor. There are other types of end stops such as read switches where the contact is cut off once a magnet is present in front of it. The end stop is easy for connection as it requires only 2 wires that is to be connected at the normally open and the Common points.

3.3.4. Heated Bed. One of the major disadvantage of using FDM type of 3d printer is warping. The use of a heated bed reduces the formation of warping and improves quality of the model. The type of heated bed we have used is an Mk2 heat bed, the heated bed comes with a thermistor which helps to distribute the heat evenly throughout the heat bed. The temperature of the heat bed is around 110 degree-Celsius which helps to create a rigid base for the model. The dimensions of the heat bed is 214x214mm which will provide enough work area to create medium sized models. The heat bed can be used for the working of various materials such as ABS, PLA. One of the major advantage of the heat bed is that it provides heat throughout the model all through the process thereby not allowing warping to take place.

3.3.5. Hot end. A hot end is the part in the 3d printer which heats the filament (PLA) and pushes it to the nozzle. The temperature of the hot end is around 230 degree-Celsius and also contains a 100k thermistor which can be used for high temperature performance. It comes with a cooling fan which helps to maintain the temperature throughout the process. The nozzle 47 present in the hot end has a diameter of 0.4 mm which also contributes to the accuracy and surface finish of the product. This hot end is capable of working with materials such as PLA, ABS and PETG. There is a heat sink present behind the hot end for the excess heat to escape to the environment.

3.3.6. Stepper Motor Driver. Stepper motor drives are small components which act as a medium of communication between the Arduino controller and the stepper motor. The signal for the stepper motors are given by the motor drive. There are 4 motor drives (1 for X-Axis, 1 for Y-Axis, 1 for Z-Axis and finally 1 for the extruder) these motor drives are connected to the Ramps. A heat sink is stuck to the top of the stepper motor drives so that the heat emitted from the drives can escape to the environment and thereby not damage the circuit board.

3.3.7. Power Supply. The power required for the working of the 3d printer is provided by using an external power supply which provides a voltage of 12V and current of 30A. The Normal AC Power supply of 220 V cannot be used to power the 3D printer as it will damage the circuit board, hence an external power supply is used.

3.4. Mechanical

3.4.1. Timing belt. A Timing Belt is a belt containing teeths which mesh with the teeths of the pulley. The main purpose of a timing belt is to provide a linear motion which converted by using a rotational movement of the pulley. Each tooth of the timing belt should match with the tooth of the pulley otherwise this may cause slipping of the belt leading inaccuracies. The belt we have used has a Pitch of 2mm, width of 6mm and a length of 2m which is sufficient for the movement of the Y-axis and X –axis.

3.4.2. Pulley. A Pulley is a disc like structure which helps to convert the rotary motion into a linear motion. They are used usually to carry heavy loads making them easier to lift. The pulley consists of teeth's which mesh with the the teeths of the timing belt thereby avoiding slippage. The bore diameter of the pulley is 5mm having 20 teeths.

3.4.3. Lead screw. Lead screw are long rods which have threaded pitch all through the rod. The Z-axis movement for the 3D printer works with the help of this lead screws. They provide uniform motion on both sides of the Z-axis. The lead screw we have used has a diameter of 8mm,length of 300mm and a pitch of 2mm. The material of the lead screw is stainless steel.

3.4.4. Smooth shaft. Smooth shafts are used for the smooth movement of the X and Y axis, they provide

IOP Conf. Series: Materials Science and Engineering 1132 (2021) 012019 doi:10.1088/1757-899X/1132/1/012019

the linear motion. The smooth shaft we have used has a diameter of 8mm, length of 400mm and the material is made up of mild steel.

3.4.5. Linear bearings/Radial bearings. Linear bearings are bearings which are meant to slide along the shaft. The bearings we have used have a Bore diameter of 8mm and an outter diameter of 15mm.these bearings consists of lubrication inside providing lesser friction with the surface of the rod. The radial bearing unlike linear bearing reduces rotational friction and also help in radial and axial loads. The bearings we have used have a diameter of 8mm and the dimensions are 8mm*22mm*7mm.

3.4.6. PLA Filament. There are many filament materials which could be used but we have used PLA filament. The reason for using PLA filament is that it has many significance than other materials. Some of these significance are they are easier to print as they can flow through the nozzle easily. Also the printing temperature of PLA is much lower making it convenient to print. The surface finish offered by PLA filament is better compared to other materials such as ABS. Other materials find it difficult to do the after processing while it is easier in the case of PLA filament. They are also considered to be environmentally friendly making them Biodegradable and also less toxic than other materials.

3.5. CAD Tools

The tools used for designing the part which is to be manufacture are called as CAD tools (Computer Aided Designing). There are various designing software's available which will help in the designing of the parts, some of them are SolidWorks, CATIA, and CREO etc. The sole purpose of these tools are to design a part which is to be manufactured to a given dimension. The Computer Aided Design is used to increase the designer's efficiency to develop a model to their required configuration. These designing software's also help in achieving interchangeability of parts in a machine. These software's help to provide dimensional tolerance and also check for any faults in the design without actually having to build a prototype model. Hence they help in saving time and also reduces the wastage of material. These software's not only provide a deep insight about the model to be manufactured but also help in making any changes.

3.6. CAM Tools

The computer Aided manufacturing tool is used to convert a CAD file to a format with which the machine will be able to understand and proceed for manufacturing. The software used for this project is a slicing software, this software will divide the designed model into very fine layers. Slic3r is the software used in this project. Slic3r mentions the layer thickness while it is 3D printed material is manufactured, the surface finish will depend on the layer thickness. The nominal thickness given while manufacturing is 0.1mm. The CAD model is converted into an STL file and contains G codes to make it machine friendly.

3.7. Prototype & testing

The testing of the 3d printer on basis of the objective can be done by checking its dimensional accuracy. The model which has been printed is compared with its CAD model for its dimension. The model created is measured using a scale and the difference between the actual dimensions will show us the efficiency of the 3d printer to print with accuracy. The other testing would be for the surface finish of the model which has been created. The tested model would be compared to a Model which had been created on a High 54 Prized 3D Printer. Achieving a surface finish similar to that of the high prized 3D Printer will decide whether the objective is achieved or not. One other Testing method will be to check for any Warping. Warping is a problem associated with temperature difference causing the edges to bend. Achieving highly finished edges without any bending will give the result of the test.

4. Results and discussion

For the fabrication of Fused Deposition Modeling (FDM) 3D Printer the budget is calculated and estimated to be Rs 12198. The whole project budget only consists of material cost. Aluminum Frame is obtained from scrap material to reduce cost of production. Freeware software's like Arduino, Slic3r and

IOP Conf. Series: Materials Science and Engineering 1132 (2021) 012019 doi:10.1088/1757-899X/1132/1/012019

Pronterface were used to avoid expensive cost. Creality 3D Ender is a 3D Printer which has a Build Volume of 220 x 220 x 250 mm cost Rs 21,000 and is one of the cheap 3D printer currently available in the market. We have built the FDM Printer which works with maximum efficiency and minimum budget. We have used Arduino Mega 2560 and Ramps 1.4 as the microcontroller. Our project cost atmost Rs. 12200 which is even Rs 7000 less than the product present on the market. This result helps us in achieving our 1st prioritized objective of fabricating a low cost 3D printer. The part which was printed in the 3D printer was compared to a high cost 3D printer and checked for its surface quality. We were able to see an accuracy similar to the one of the high cost 3D printer hence satisfying our objective of achieving good surface finish and dimensional accuracy. It was also observed that there was no cases of warping leading to uneven edges while printing the parts.

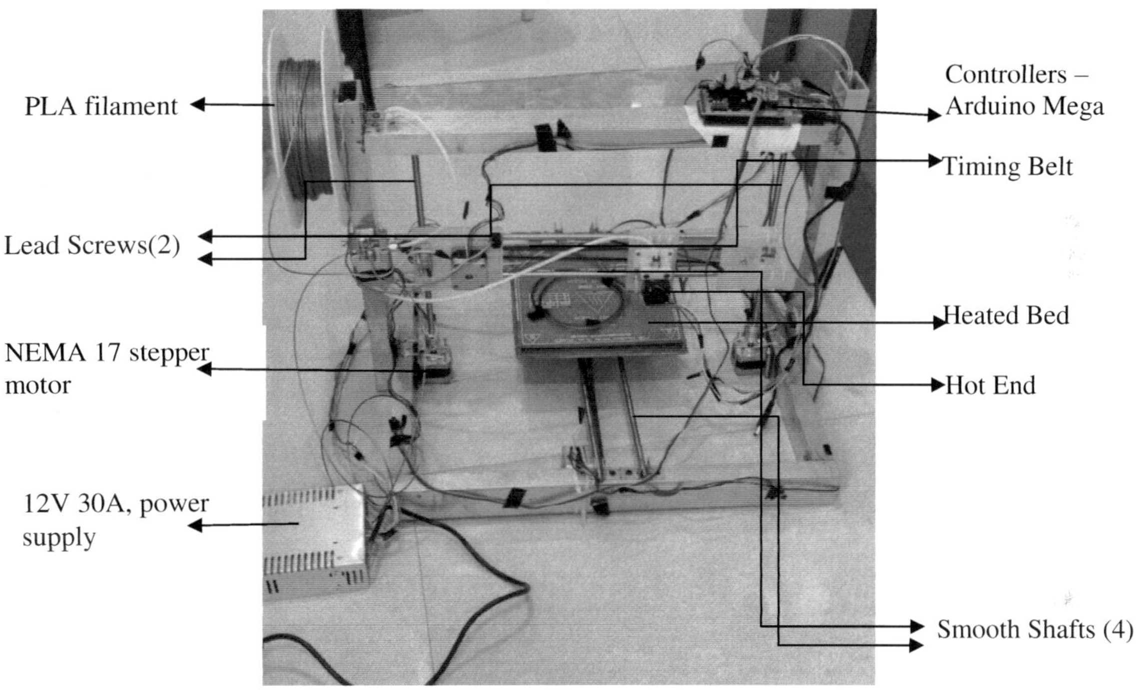

Figure 2. The fabricated FDM 3D printer

The fabricated fused deposition modeling 3D printer is shown in the figure 2. All the components are labelled for the easy identification. The total cost incurred for the project has been mentioned in detail in the table1. Table 1 shows the cost effectiveness of the 3D printer manufactured. If taken up in a large scale the cost can be further reduced and 3D printers will then be a regular household equipment.

Table 1. Budget of the project.

No.	Item	Price	Quantity	Amount
1.	Shaft Coupler	215	2	430
2.	Timing Belt	379	2 meters	379
3.	Stepper Motor Nema 17	650	5	3285
4.	Smooth Rod	650	4	2600
5.	Screw Thread	599	2	1198
6.	Endstop	100	3	300
7.	MK8 Extruder	549	1	549

8.	Pulley	249	2	498
9.	HeatBed MK2 with thermistor	1050	1	1050
10.	J Head Hotend V6	699	1	699
11.	Arduino Mega	600	1	600
12.	Ramps 1.4	400	1	400
13.	Motor Drive	30	6	180
14.	Linear Bearing	25	8	200
15.	Radial Bearing	30	6	180
16.	Hardware parts	300	-	300
17.	Aluminium Frame	NIL	2x53cm , 2x45cm , 3x43cm	NIL
18.	PLA Filament	400	250 gram	400
19.	Power Supply	900	1	900
TOTAL COST				12198

Table 2 is the specification chart of the 3D printer fabricated. The build volume, method and nozzle diameter and many other specifications are mentioned in the table2.

Table 2. Specification of the 3D Printer

Specification	
Build Volume	200mm x 200mm x 150mm
Method	Fused Deposition Modelling
No. of Extruders	1
Power Supply	DC 12V 30A
Connectivity	USB
Filament Diameter	1.75mm
Nozzle Diameter	0.4mm
Filament Material	PLA
Print File Type	STL, G-Code
Machine Size	450mm x 530mm x 430mm
Machine Weight	5kg

5. Conclusion

A Cost Effective 3d printer was developed and fabricated. 3mm thickness aluminum frames were assembled together using multiple bolts and nuts for the fabrication of the 3d printer. The softwares used for the development of the 3d printer are Arduino software, CATIA for the creation of the 3d MODEL, SLIC3R software for slicing the CAD models in various layers, Pronterface for calibration and execution of the 3D printer to print the models. The Build Volume of the 3D printer is 200mm*200mm*150mm. The number of extruders used in this 3D printer is one. The machine size is around 450mm*530mm*430mm and weighs around 5 Kg. As the 3D Printer is light weight, it is portable and easy to handle. The 3d model was printed and compared to 3D objects printed on a High Cost 3D printer. The cost Effective 3d printer can be used in small scale industries and can even be used in household for various educational purposes. The nozzle of diameter 0.4mm used in the fabrication of this 3D printer gives it more dimensional accuracy. The printing material used in the 3D printer is PLA as it provides better strength and operates at higher temperature. The cost effectiveness makes the 3D printer affordable and desirable to more customers. In the era of additive manufacturing advancements a cost effective, affordable and efficient 3D printer can pave way for new opportunities and even the ordinary people could reap the benefits of the same.

References

[1] D.Dev Singh and Gopi Rahul 2018 *Int J of Mechanical and Production Engineering Research and Development*. **8**, 263-270

[2] Shiwpursad Jasveer and Xue Jianbin 2018, *Int J of Scientific and Research Publications*, **8**, 2250-3153

[3] Ngoc-Hien Tran, Van-Cuong Nguyen and Van-Nghia Nguyen 2017, *Int J of Engineering and Advanced Technology*. **6**

[4] Thomas Duda, L. and Venkat Raghavan 2016, *Int Federation of Automatic Control* **49**, 103-110

[5] Siddharth Bhandari and B Regina 2014, *Int J of Computer Science and Information Technology Research*, **2**, 378-380

[6] Vinod G. Surange and Punit V. Gharat 2016, *Int Research J of Engineering and Technology* **03**, 2395- 0072

[7] Vinod G. Gokhare, Dr. D. N. Raut and Dr. D. K. Shinde 2017, *Int J of Engineering Research and Technology* **6**

[8] Kalpesh Dapurkar, Rahul Singh, Sitaram Gawas, Rahul Ugale and Dipak Shelar, 2019, *Int J of Engineering Research and Technology* **06**

[9] P. Dudek 2013, *Archives of Metallurgy and Materials*, **58**

[10] Taha Hasan Masood Siddique, Iqra Samiy, Malik Zohaib Nisarz, Mashal Naeemx, Abid Karim, Muhammad Usman 2019, *Second Int C on Latest trends in Electrical Engineering and Computing Technologies*

[11] Alexandru Pirjan and Dana-Mihaela Petrosanu, *Romanian Economic Business Review*, **7**, 360-370

[12] Evgeniy G. Gordeev, Alexey S. Galushko and Valentine P. Ananikov 2018, *Russian Science Foundation*

Product design and development of a reading aid for the visually impaired

Ajith Gopinath[1*], Shibin Babu[1], Allen Bob Claudius[1], Shaik Mohammed Ismail[1], Sandeep.O.S[2]

[1]Department of Mechanical and Automobile Engineering, Christ Deemed to be university, Bengaluru

[2]Department of Mechanical Engineering, Adi Shankara Institute of Engineering and Technology, Kalady, Ernakulam, Kerala

[*]Email: ajith.gopinath@christuniversity.in

Abstract. One of the main hurdles for the blind are to read, easily and there by learn and graduate. Various researches are happening to reduce the effort required for the blind to read in terms of aids which can speak out the written data. The current research targets to design and fabricate a reading aid for the visually impaired. This device will help them to identify the text written in any book or page by scanning the text and speaking it out to them with the help of a mobile application called KNFB reader which helps to read anything to blind or dyslexic people. This project utilizes the fact that majority of the people use smartphones, which can be used for the advantage of the blind people. Based on a survey conducted among the blind people with the help of an NGO Anuprayas all the criteria for the design of reading aid, in the current project were selected. Four sample designs were prepared after taking the criteria for reading aid into consideration. Forced Decision matrix (FD) analysis and Decision alternative ratio System (DARE) analysis, which are tools to rank the properties based on the priority, were conducted to identify the most influential parameter for the design. The parameters were ranked based on these analyses. The ergonomics, cost and durability were the most important properties desired in the product. This product can be deployed for the effective utilization by public at railway stations, bus stations, libraries etc. This will enable the blind to be self-independent and have the information available for them whenever they need it.

Keywords: Blindness, Reading aid, visually impaired, Forced Decision matrix method, Decision alternative ratio System

1. Introduction

According to WHO 2017 statistics, about 285 million people are visually impaired worldwide. Only 27 percentage of blind and partially sighted people of working age are in employment and 39 percentage of blind and partially sighted people of working age say they have some or great difficulty in making ends meet according to the Royal National Institute Of Blind People. There are many more aids for visually impaired to help track their movements (Electronic travel aids), know about their surroundings (Position locator devices), read texts (Optical character reader) etc [1]. The need for a design that is flexible to compensate for changes in human condition is essential. Principles like affordability and emotional durability is important when it comes to designing an aid for the visually impaired [2]. The path of technological development is difficult to predict but they should be focused in making the tools more cost efficient, portable and light [3].

Consumer involvement during designing is an area that needs further research [4]. Further switching between sketching and CAD modeling had a huge impact during concept design and development of the tool [5]. There is a need to understand product design typicality in order to know about the market potentials of any tool [6]. Additive manufacturing gives an element of freedom when comes to designing and manufacturing over traditional methods [7]. Research done to help visually impaired goes in vain due to the aid being too expensive for the masses to use[8][9].The purpose of our study is to provide a solution to social problem faced by visually impaired.

2. Methodology

2.1. Criteria Identification

A survey was conducted among the visually challenged population to identify some of the most desirable properties for them in a reading aid. The aim of the project and the vision was explained to them and their inputs on the design were collected and a chart was made which contained all the identified parameters of the design.

2.2. Sample Design Creation

Several designs were made on CAD which can accommodate a book and phone in such a way that the phone can scan a whole page on the book.

2.3. Prioritization of Requirements

The parameters that were identified are prioritized using a scale in ascending order using the help of FD analysis and DARE analysis.

2.4. Selection of Design

The designs that were made are rated according to the prioritized parameters and the highest rated design is selected for optimization and modification.

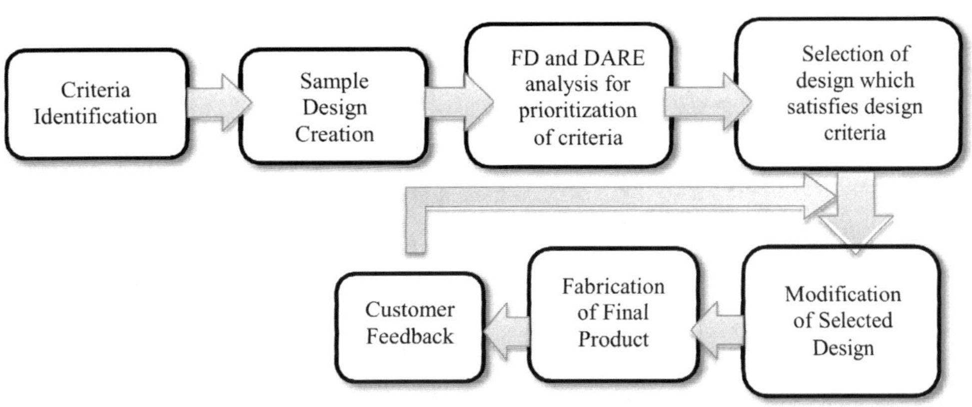

Figure 1. Flow Chart of the experimental procedure

2.5. Modification and Optimization

The selected design is modified to make it more customer friendly and Design Study is conducted using the software to make the design lighter and stronger and also to make sure less material is used to manufacture it.

2.6. Fabrication of Model

The final optimized design is manufactures using 3Dprinting.

2.7. Customer Feedback

The fabricated model is put to test by the customers and the feedback is taken to consider modifications. In the end, after the complete process, the final model is obtained which meets the expectations of the customer and is also an optimized model.

2.8. Customer Survey

Customer surveys help in development of a product to an extreme extent as these products are made for these customers. From customer surveys one can understand the needs or requirements of the consumer. All the details are stored in Table-1 and these details are further used to analysis.

Table 1: List of required feature from the point of view of the customer

No.	Requirement	Description
1	Setup	The amount of effort required for the initial assembly of the scanning tool should be minimal.
2	Cost	The amount at which the scanning aid will beavailable on the market should be affordable.
3	Weight	The tool should be light in weight
4	Size/Compactness	The scanning aid should be portable
5	Surface Quality	The tool should have a good design aesthetically and should have a good finish
6	Durability	The tool should be strong enough to withstand ill usage and shocks due to drops.

IOP Conf. Series: Materials Science and Engineering 1132 (2021) 012020 doi:10.1088/1757-899X/1132/1/012020

7	Compatibility with Phones	The scanning aid should be able to accommodate different aspect ratios of different phones
8	Ergonomics	The tool should be comfortable to handle and should have good grip

3. Results and Discussion

3.1. Sample Design Creation

This design in model A as shown in Figure 2(a) consists of no moving parts and can be assembled like a puzzle.

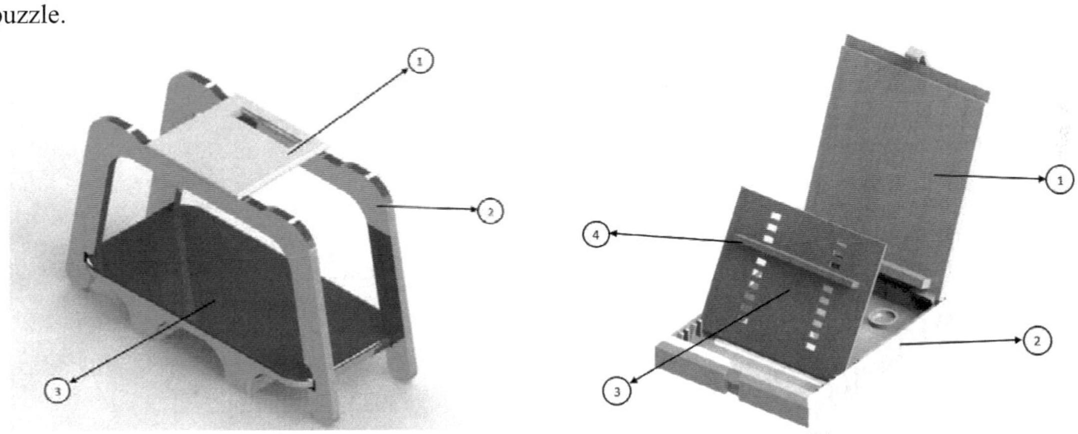

(a) (b)

Figure2. (a) Model-A (b)Model-B

The book is placed on part-3 and the phone on part-1. Part-2 provides structural strength and also helps maintain a minimal distance from the book for the phone to focus and scan the book. This design satisfies durability criteria as well as cost effectiveness. However, since the movement is limited, thicker books can't be focused on by the phone making it more suitable for single page scanning which gives this design a downside on ergonomics and flexibility. The model B shown in figure 2(b) is an upgraded version of Model-A. It includes moving parts and is foldable making it more portable. Part-3 and part-4 constitute an adjustable phone holder. Part-4 can be removed and inserted in several slots to change the size of the holder making it possible to accommodate bigger phones. The book to be scanned is placed on part-1. Part-1 folds into part-2 and can be carried around like a box. This design is cost effective and portable but less ergonomic than Model-A and also less durable as movable parts tend to have lesser life. The distance between the phone and the reading material still can't be adjusted.

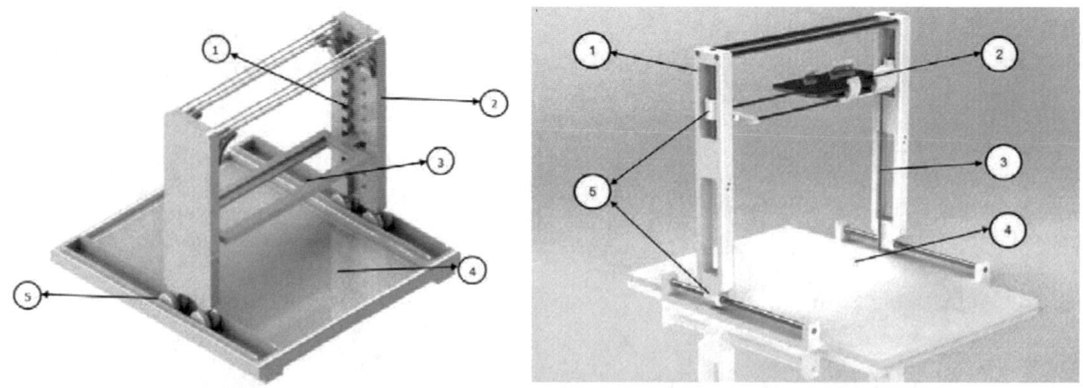

(a) (b)
Figure3. (a) Model-C (b) Model-D

This model C as shown in figure 3(a) contains more parts and also moving parts than the other three models. The phone can be places on part-3 and the height can be adjusted from the book to help focus thicker books. The book is placed on part-4 and the phone can be moved along x and y axes using rollers (part-5) and the phone holder can be locked in place using part-1. The design C is flexible and slightly less durable than models A and B. This model C is not portable and not very ergonomic as a visually impaired person will not be able to judge if the phone is focusing on the book or not. In the model D as shown in figure 3(b) there are several design changes that have been made keeping Model-C as the initial design. To increase the ergonomics, guide-rods (part-3) have been incorporated on the phone holder (part- 2). This makes sure that the phone always maintains a fixed distance from the book to ensure clear focus. Slots have been made in side supports (part-1) to help the blind user to hold and move part-1 along the x- axis with the help of the linear guide bearings (part-5). The y-axis movement is made possible using linear bearings(part-5).The weight of the model is also less as optimization is done to remove unwanted material without compensating on safety. This model is flexible, cost effective, ergonomic and durable. There are only few things the model compensates on, and they are, size, portability and setup. The design of part-2 enables it to hold phones of almost any size.

3.2. Prioritization of the Customer Requirements

The Prioritization of customer requirements is done to determine relative importance of each feature. It is done in two stages:

3.2.1. FD Analysis. is used to determine the importance of the features. In this method, items arecompared in pairs as in a league tournament. [8] Of the pair, the item having the higher value to thecustomer is given a rating of 1 and the other 0. When all the pairs have been compared, the total scoresgiven to each of the items are summed in the total column. Priority is given to each item according to itstotal points, as shown in the column titled Ranking. Table 5 shows the FD analysis of the Scanning aid.

Inference of the aforementioned FD analysis suggests that ergonomics of the design has number one priority when compared with other qualities. Even though most of the customers prefer to put the quality of weight above ergonomics, the latter earned the highest rank as it is highly preferred over mostof the other qualities.

Table 2: Forced Decision Matrix

Function Description		**Comparative Study**								**TOTAL**	**RANK**
		A	B	C	D	E	F	G	H		
A	Setup		0	0	0	0	0	1	0	1	5
B	Cost	1		1	1	1	0	1	0	5	2
C	Weight	1	0		0	0	1	1	1	4	3
D	Size/Compactness	1	0	1		0	1	1	0	4	3
E	Surface quality	1	0	0	0		0	1	0	2	4
F	Durability	1	1	1	0	1		1	0	5	2
G	Compatibility with phones	0	0	0	0	0	0		0	0	6
H	Ergonomics	1	1	0	1	1	1	1		6	1

Inference of the aforementioned FD analysis suggests that ergonomics of the design has number one priority when compared with other qualities. Even though most of the customers prefer to put the quality of weight above ergonomics, the latter earned the highest rank as it is highly preferred over most of the other qualities

3.2.2. DARE Analysis. is an abbreviation for Decision alternative ratio System. DARE is the method toquantify the degree of importance. Once the priorities have been determined by the FD method, it isrelatively easy to compare the importance of two functions which are adjacent to each other in priority. Beginning with the item highest in priority ranking, it is determined how much more important that item isto the item that is directly below in priority ranking. The item lower in priority is given the factor 1, & thehigher priority item, a number higher than 1 depending on how much more important it is to the functionbelow it in ranking. From the DARE analysis, as shown in table 3, one can understand the importance of aquality over the quality positioned below it from the FD analysis conducted above.

Table 3: DARE Analysis

Function Description	**Comparative Scores**							**Score**	**Cumulative Weightage**	**Weightage %**
Ergonomics	1.2							1.2	12.71	23.94
Cost	1	1.7						1.7	10.63	19.94
Durability		1	1.7					1.7	6.24	11.73
Weight			1	1.2				1.2	3.67	6.90
Size				1	1.5			1.5	3.06	5.75
Surface Quality					1	1.2		1.2	2.04	3.84
Setup						1	1.7	1.7	1.7	3.20
Compatibility with phones							1	1	1	1.88
Total									53.1	100

Total number of people surveyed was regarded as 100% and their individual answers were used to determine how important a quality is over the next one. The weightage of the qualities can now be used to decide which model can be selected for further modification and optimization.

3.3. Selection of Design

Model-D as shown in figure 4, satisfied most of the customer requirements. As per FD and DARE analysis Ergonomics and cost were main criteria that people looked for. The final design was not only relatively cheaper but also more ergonomic. Major advantage of this design is the phone holder which takes care of position of the camera and the guide rods make sure the required amount of distance is maintained between the surface to be scanned and the camera.

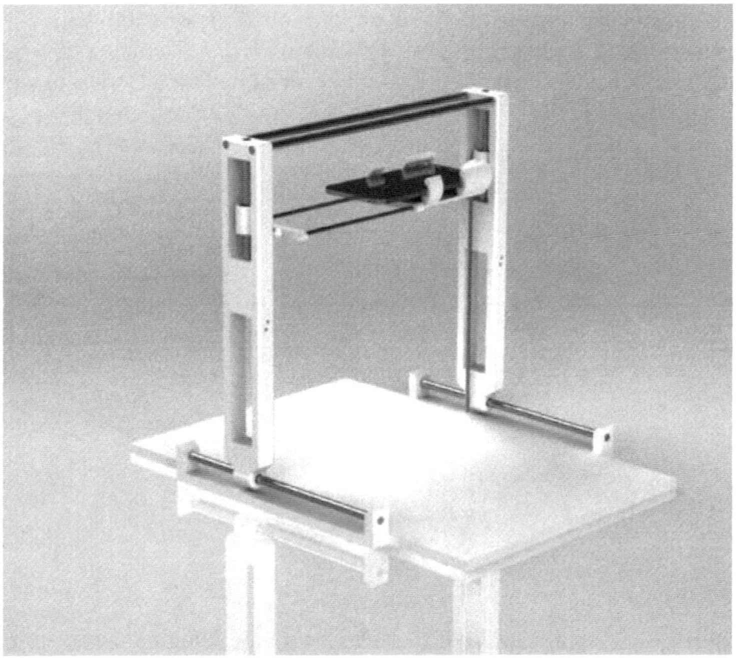

Figure 4. Model-D

3.4. Modification and Optimization of Design

Changes were made to the initial design of the side supports by removing material in the load bearing elements and making them of a T-type cross section. A T-cross section beam exhibits qualities of good stiffness and low weight. Fillets were added to sharp corners to enhance the durability and cycle life. The changes in the design after design optimization are shown in the figure 5.

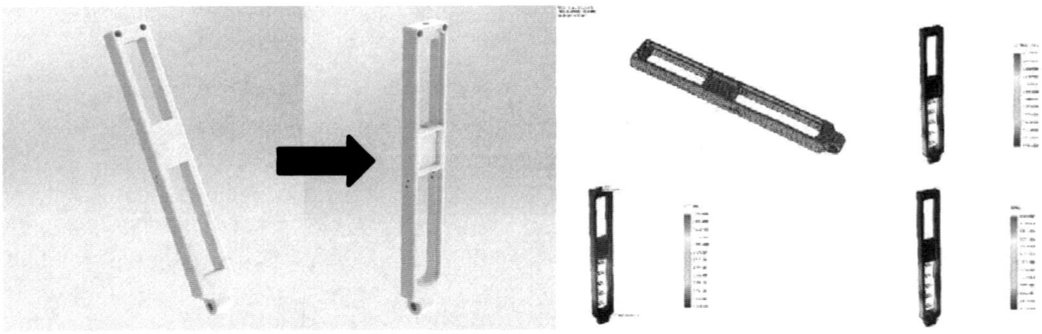

(a) (b)

Figure 5.(a) Optimization of the model. **(b)**Optimization of the part1.

The modified design performed well in all the loading conditions with very small resultant displacement and stress values much below yield point. The applied load ranges between 30N to 40N per item. The modified design was found to be structurally resilient and able to survive the load applied.

Figure 6(a) and (b) shows the analysis of the parts which was done for optimization of the design. Since all the parts performed well under the application of load and the design optimization was achieved for all the components of the hearing aid. The part 2 and 3 design optimization and the analysis is shown in the figure 6.

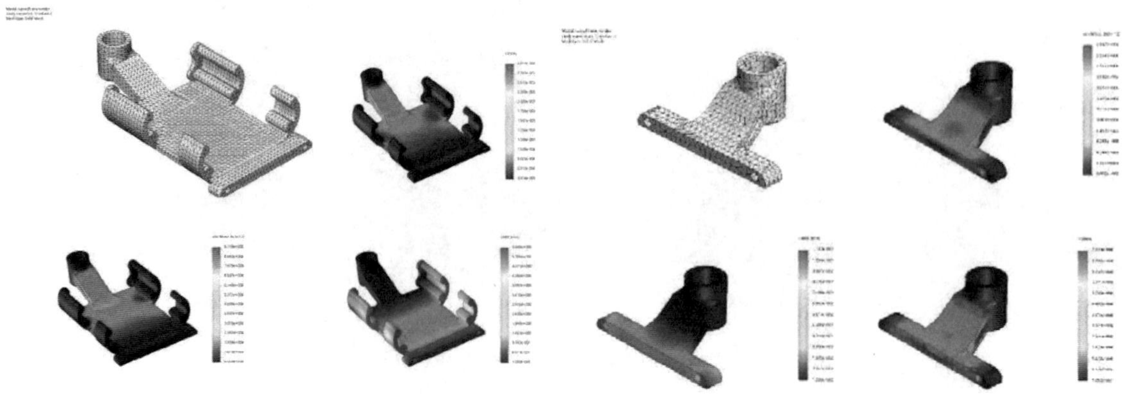
(a) (b)

Figure 6.(a) Optimization of thepart 2. (b)Optimization of the part3.

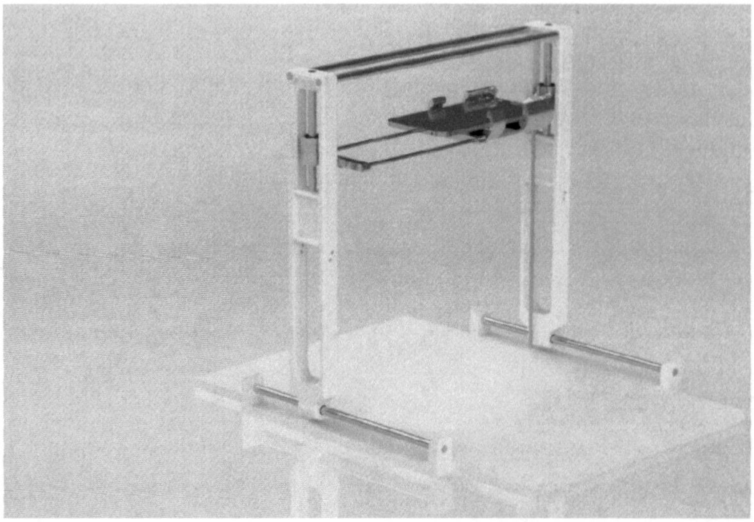

Figure 7. Final Optimized Design.

Figure 7 depicts the selected design after undergoing part level modification and analysis, an optimized model which is durable, can withstand major loading conditions, light in weight and has good stiffness properties was achieved without compensating on the ergonomics. This design satisfies all the criteria adapted for the design development process and was found to be most ergonomic.

3.5. Fabrication of the Product

In order to accommodate the side supports over the small print bed of the 3D Printer it had to be divided into two parts as shown in figure 9, which were later printed in PLA and joined together using stainless steel bolts. The Parts of the Phone holder were also 3D Printed in PLA. The parts were then saved and converted to STL file format for the subsequent fabrication of the product.

The components were then 3D printed with the PLA material and was assembled together to obtain a well performing device. The mobile phone was successfully held in the slot and with the KNFB reader the text was easily readable. The device worked efficiently and is expected to be of great service to the blind people in reading books and any texts they feel to read.

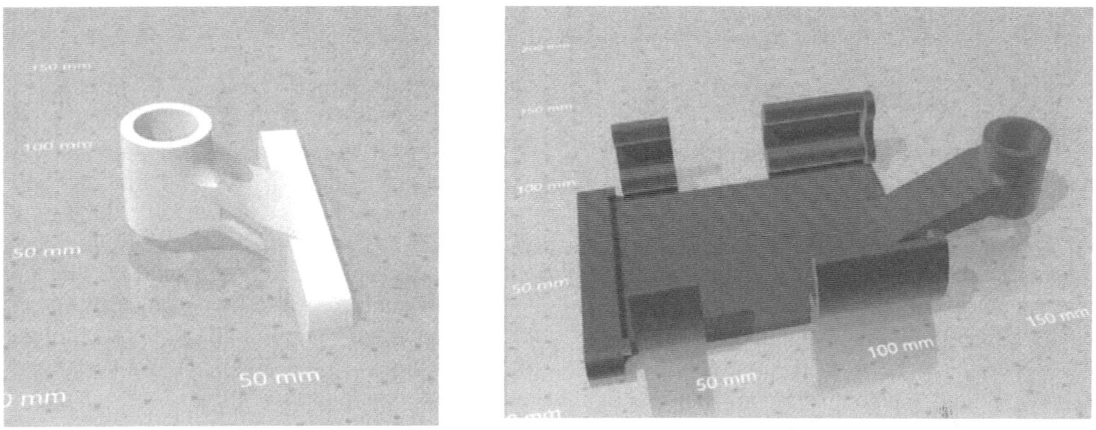

(a) (b)

Figure 8(a) Final Optimized Part Model for Part 3 (b) Final Optimized Part model for Part 2

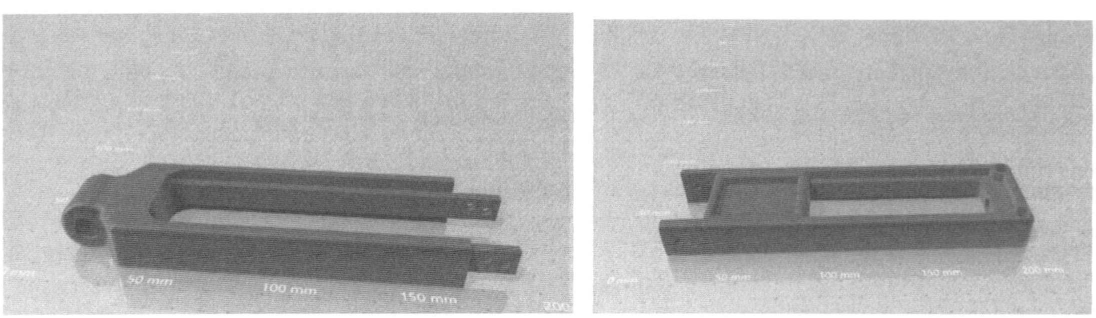

(a) (b)

Figure 9. (a) Final Optimized Part Model for Part 1 (view 1) (b) Final Optimized Part Model for Part 1(view2)

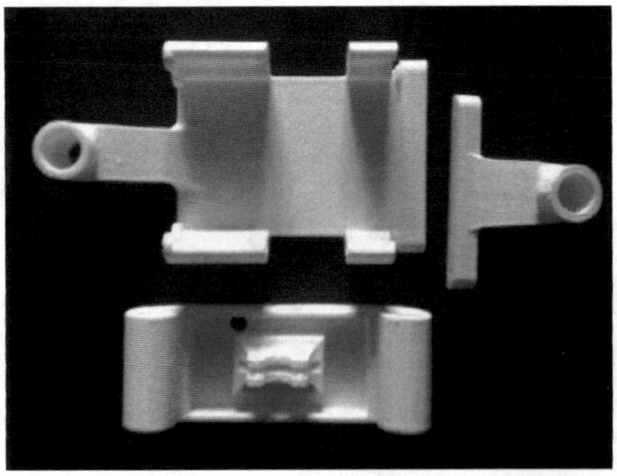
(a)

(b)

Figure 10. (a)(b)Fabricated parts of the product.

For the base a 3 mm plywood (500 mm×650mm) was used as its sturdy and reduces the cost. Linear bearings and guide ways were used to provide linear motion for the side supports (X-Direction) and the phone holder (Y-Direction). The Hearing aid for visually impaired was successfully designed and optimized. This optimized design was fabricated by fused deposition modeling and was successfully assembled. The product was found to be fully functional and was tested for its functionality. The KNFB reader app installed in the mobile phone captured the text and converted it into speech for the visually challenged person. The Minimum requirement for the Smart phone camera is 5 Mega pixel resolutions. The focal distance for most Smartphone is 22-28 cm which has been considered while designing the apparatus. The scanning methods available in the market are complex and most of the designs are similar to the Model-B of the paper; which has rotational DOF which the visually impaired may find difficult to operate. The current design is neither complex nor has unwanted DOF. Furthermore, the FD and DARE analysis helped the design to be catered to the people. The user-friendly design of the product and the cost effectiveness are some of the important factors which were considered while designing the apparatus. The entire setup was manufactured under 5,000 rupees which for a prototype is comparatively cheaper. The KNFB READER app is a freeware and is available in the play store. The hearing aid will be a valuable addition to the public places such as library, railway stations and bus stations where the blind people could make use of the this product to get proper information read out to them without any external help.

4. Conclusion

The product, hearing aid for visually impaired, was successfully designed, optimized and fabricated. This product was designed keeping in mind various difficulties the blind people have to face in order to get information from written literature. This problem was taken up as one of the primary concerns for the design and development of this product. In order to understand various design considerations, a survey with the blind was conducted by which the important aspects for the aid were identified. Further by Forced Decision (FD) matrix method and Decision alternative ratio System (DARE) analysis the most important/relevant property required in the product was identified. Sample designs were made by taking these parameters into consideration. From the available sample designs an optimized design was selected, which satisfies all the condition/criterion of FD and DARE analysis. All the CAD models were compared and the model in which the highest ranked properties are getting satisfied was selected as the optimized design. The selected design prototype was fabricated with fused deposition modeling and was successfully tested for the functionality. The design is ergonomically designed to be handled easily. It has a mobile platform which can be moved up and down to adjust the scanning range of the mobile application. It is not electrically powered and has to be handled by the person himself. The final design was fabricated using additive manufacturing and with the help of smartphone camera and KNFB reader mobile application text from books were scanned and converted to speech. This Product will serve as a helping hand to the blind in public places such as libraries and bus stations and is a first step towards helping the blind people become independent.

Acknowledgement
The authors of this paper are greatly indebted to our co-author Late Mr. Shibin Babu who was an integral component of this research group. His contributions to this paper were monumental and without which this research would not have succeeded. The authors also would like to thank NGO Anuprayas who provided us with this idea and the necessary feedback to make this project happen.

References
[1] Elmannai W and Elleithy K 2017 Sensor-Based Assistive Devices for Visually-Impaired People: Current Status, Challenges, and Future Directions *Sensors (Basel, Switzerland)* **17,** 565.
[2] Nickpour F and Sullivan C O 2016 *Designing around people* (Cham:Springer), 45-54.
[3] Olabanji O M and Mpofu K 2014 *Procedia CIRP* **23**, 264-269.
[4] Teng Shih Y, Sher W and Taylor M 2017, *Design Studies* **53**, 47-77.
[5] Mayer S and Landwehr J R 2018, *Design Studies* **54,**146-161
[6] Bousbia-Salah M, Bettayeb M and Larbi A 2011, *J. of Intelligent & robotic systems* **64,** 387-400
[7] Mullen R J, Duhig J, Russell A, Scarazzini L, Lievano F and Wolf M S 2018, *Patient education and counseling* **101,** 1351–1367.
[8] Wai Michael Siu K 2013, *Facilities* **31,** 314-327
[9] Yurtay N, Bicil Y, Celebi S, Cit G and Dural D 2011, *The Turkish Online J Of Educational Technology* **10**.

IOP Conf. Series: Materials Science and Engineering 1132 (2021) 012021 doi:10.1088/1757-899X/1132/1/012021

Heat Transfer Effect of Louvered Baffle in Jet Impingement Heat Transfer with Cross-flow

Sabu Kurian*, Tide P S and Biju N [1]

Department of Mechanical Engineering, School of Engineering, Cochin University of Science and Technology, Cochin, Kerala-682022, India

*email: sabukurian74@gmail.com

Abstract. The objective of this research is to numerically investigate heat transfer characteristic of a louvered baffle assisted multi-jet impingement of air on a heated plate subjected to constant heat flux and cross flow. Two h/d configurations were considered for the present study. An array of jets with 3 x 3 configurations discharging from round orifices of diameter d=5 mm and with jet-to-target plate distance ranging from 2d to 3 d were studied. SST k-ω turbulence model was used for numerical simulation to examine the effect of blow ratio, louver angle and baffle clearance on heat transfer. Blow ratios of 0.25, 0.5, 075 and 1.0 and louver angle 30°, 45° and 60° were considered for CFD simulations. In all the cases, heat transfer increases with blow ratio and blow ratio with 1 has the maximum heat transfer. It is also noticed that heat transfer rate increases with decrease in louver angle h/d ratio.

1. Introduction

Jet impingement has been widely used in many fields such as turbine cooling, rocket launcher cooling and electronics cooling due to its intensive heat transfer characteristics. The air-jet impinging on a normal small surface will enhance heat transfer than any other conventional cooling techniques. Impinging jets have the three different zones as shown in Figure 1. a) core jet, b) free jet, c) stagnation point. Core jet has the same velocity that of the nozzle exit velocity and velocity decrease is observed due to the entrainment of outside air into the jet.

In multiple air jet impinging, the effect of nozzle geometry was experimented by Sinan Caliskan et al. (1). Heat transfer effects of rectangular and elliptical jets impinging over a target plate at Reynolds numbers varying from 2000 to 10000 were investigated by them. They found that, due to the augmented entrainment of surrounding air, performance of elliptical jets is better than rectangular jets. Nakod et al. (2) investigated the heat transfer characteristics of circular jet impinging over a finned flat plate and vortex generator. Cube type fins and equilateral triangle type vortex generators were used in this experiment. It is revealed from this study that vortex generators are better than cube shape finned surface. The effect of nozzle inlet geometry on heat transfer was experimentally and numerically investigated by M F Koseglu and S. Baskaya (3) and found that elliptical and rectangular jets are superior to other geometries. Mixing characteristics and entrainment phenomenon of interrupted jets were studied by Harekrishna Yadav et al.(4). They found that frequency and magnitude of pulsation jet determine the length of fully developed zone which characterizes the mixing and entrainment. Nusselt number distributions of a staggered five jet nozzle array were investigated by Jung-Yang san and Jenq-Jye Chen (5). They investigated about the influence jet to jet spacing and h/d ratio on heat transfer. Junsik Lee et al.(6) experimentally studied about the cross-flow effect with varying jet to jet spacing and h/d ratio on heat transfer. They noticed the deterioration of heat transfer with the increase of cross-flow.

Hollworth and Durbin investigated the application of jet impingement in electronics cooling [7]. Jet impingement on the vehicle windscreen for moisture removal was investigated by Roy et al. [8]. Jet impingement for the cooling of a grinding process was investigated by Babic et al. [9] and the methods of controlling the vortex flow of a free jet were reported by Hussain and Zaman [10], Ho and Huang [11]. Liu and Sullivan [12] have showed that, the heat transfer characteristics of a jet will be enhanced when the jet is excited acoustically at certain frequencies. Hwang et al. [13] investigated about the methods to regulate the vortex roll-up in the jet flow. Heat generation and dissipation in a grinding process were investigated numerically by Lavine and Jen [14], [15], Jen and Lavine [16], [17] and Liao et al. [18] and proposed a new model. Smoke wire technique was employed by Fleischer et al. [19] to visualize the initiation and development of vortices in an impinging jet flow.

Weigand and Spring [20] investigated the heat transfer behavior of a multi jet impingement and analyzed the results with single jet. Suitability of different CFD models for multiple impinging jet system was also analyzed by them. Influence of the initial cross-flow on both heat transfer was studied by Florschuetz et al. [21, 22] and Florschuetz and Su [23, 24]. They found that the influence of the initial cross-flow depends primarily on h/d ratio and the blow ratio. They also observed the overall decrease in heat transfer due to cross-flow. Roa et al. [25] and zu et al. [26] reported that SST k-ω model is best suitable for modeling multiple impinging jets due to the good agreement between numerical and experimental data at fair costs.

Heat transfer can be further augmented by providing ribs, baffles, vortex generators etc in the flow domain. Lot of experiments have been noticed with ribs and vortex generators. Insertion of baffles may be a contributing factor, especially when cross flow is present. It has been noticed that baffles create extra turbulence in the flow [27]. Turbulence is directly proportional to heat transfer in most cases.

From the elaborate literature review, it can be concluded that, the fundamental parameters affecting multi jet impingement heat transfer is known. But role of baffles and louvers were not properly investigated in the past.

This study focuses on numerical and experimental investigation on the heat transfer properties of 3 x 3 jet array impingement over a flat plate with cross-flow and louvered baffle. Investigation was carried out at different cross-flow velocities, target plate distance (h/d) and louver angles. A clearance of 1 mm is provided between the baffle and target plate which is constant throughout this research. Thermocouples are deployed at different locations of the target plate to measure the temperature. Constant velocity air jets are used throughout the experiment. Cross-flow velocities have been varied at four levels. Mean while louver angle and target plate distance varied and 3 and 2 levels respectively. Numerical techniques were used to simulate the flow patterns under different conditions by using suitable turbulence models and the predictions were compared with experimental measurements. Commercial CFD package ANSYS Fluent is used to conduct the numerical study.

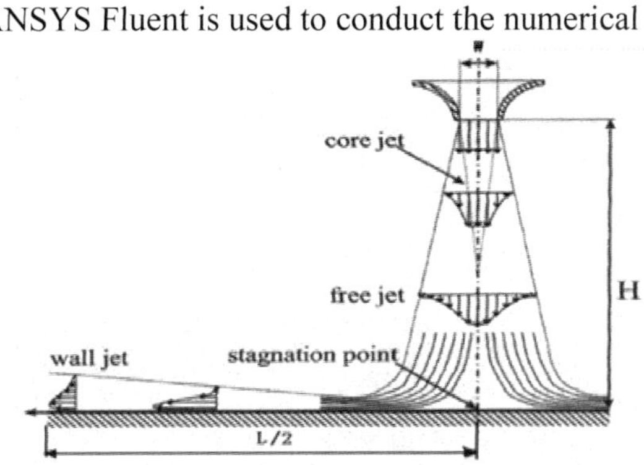

Figure 1 Stages of impingement

2. Experimental setup and procedures

Important parts of the present test facility are test section fitted with data acquisition system, and system to develop high speed jet flow, nozzle, cross-flow arrangement and an impingement surface. The schematic diagram of the test facility is shown in Figure 2. Blower (3 hp) draws air from atmosphere and supplies to the air pre-heater, where it is pre-heated to a fixed temperature of 305 K One portion of the air moves to the plenum chamber to produce air jets, while another portion passes to the cross-flow duct. Separate orifice-meters and manometers are provided in the plenum chamber and cross-flow duct pipe line. The target plate is an aluminium flat impingement surface instrumented with thermocouple and foil type 50 W foil type heater, is mounted on a z traverse which can be moved in vertical direction. Movement of the target plate in the vertical direction is used to fix the h/d ratio. Dimension of the target plate is 120 x 120 mm. Temperature of the target plate is measured with 18 K-type thermocouples deployed at the different locations. DAQ is connected to a personal computer and all thermocouples, to record the real time temperature. All temperature and flow measuring devices are NABL calibrated. Nine nozzle of five mm diameter is used. The length of the nozzle is 10mm. The cross section of the cross-flow region is 120 x 10 mm. Baffle with louvers are used with louver angle 45^0. Estimated uncertainty of jet velocity is ±5.385 %.

Blow ratio is defined as the ratio of cross-flow velocity to jet velocity. Experiments were conducted with four blow ratios ranging from 0.25 to 1.0. Four louvered baffles having a length of 60 mm are attached alternatively on either side of the confinement. Louver angle for the experiment is fixed as 45°.

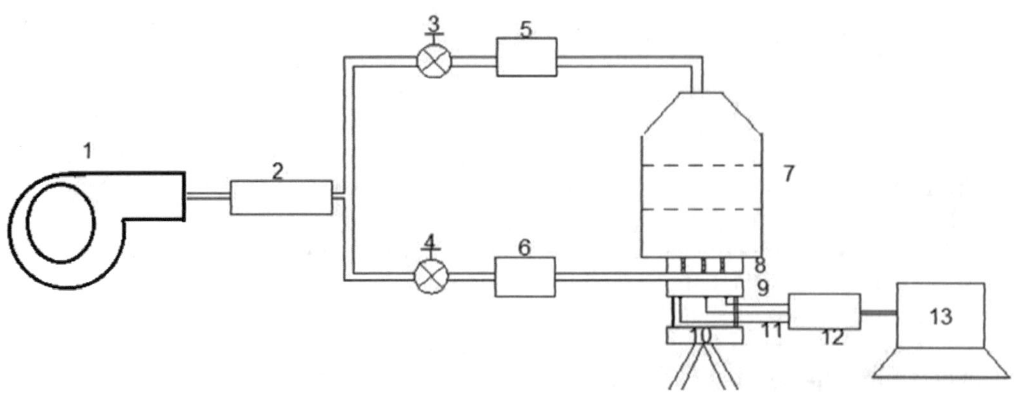

Figure 2 Schematic diagram of experiment setup [Parts list: 1. 3 hp blower, 2. Air Pre-heater, 3. & Valves, 5, 6. Orifice-meters, 7. Plenum chamber, 8. Nozzle plate, 9. Target plate, 10. Z-traverse, 11 Thermocouples, 12. DAQ, 13. Computer]

3. Numerical work

3.1. Geometry and boundary conditions

The domain of the CFD model is shown in figure 4. The domain is a three dimensional space consist of an inline 3 x 3 nozzles array, a target plate of 120 x 120 mm, made of aluminium and louvered baffles. Of the four sides, two sides are open for the entry and exit of cross-flow, while other two sides are confined. Air inlet into the domain is through 9 nozzles with diameter, d=5 mm and through the cross-flow inlet section. Domain discretisation is done through structured hexahedral mesh. Grid study was conducted by varying the cell size from 0.97 million to 1.87 million cells which shows that grid having 1.3 million cells was the best suited for present study. Viscous, steady, incompressible, and three dimensional are the initial assumptions of the study. Also gravitational effects, radiation and viscous dissipation are neglected. Air properties such as specific heat, density, and thermal conductivity are considered constant.

A heat flux of 3472 W/m^2 is supplied to the target plate. All the confinement walls of the domain are adiabatic in nature. 'Velocity inlet' boundary conditions were given at the nozzle inlet and cross-flow inlet. Pressure outlet is the exit boundary condition at the exit of domain. Air velocity at the inlet of nozzle is fixed as 38 m/s and cross-flow velocity varies with blow ratio. Temperature of air at the inlet is kept at 305 K. For the numerical analysis, h/d considered is 2 and 3and louver angles are 30^0, 45^0, 60^0.

Commercial CFD package, ANSYS Fluent is used to carry out the numerical simulation. Second order discretization scheme was used for the pressure, continuity, and momentum. For pressure velocity coupling, standard coupled algorithm is adopted. Turbulence model selected is *SST* k-ω as it provides a good balance between accuracy and computational time, steady state impingement is assumed throughout this simulation. To show temperature and Nusselt number in numerical results, a diagonal line is considered over the heated plate as shown in Figure 3.

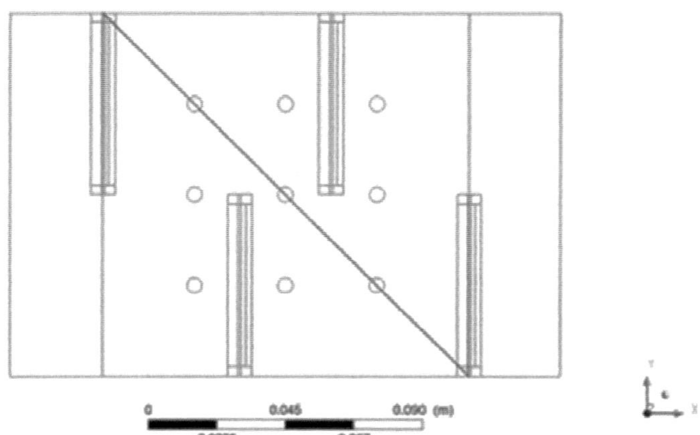

Figure 3 Wireframe model with location of louvered baffles and jets and diagonal line of study

3.2. *Grid study*

Grid independence study is performed to reduce the influence of the number of grids on the computational results. It is always a good practice to follows this for every geometry which is tedious. Grid independence is achieved at 1314937 elements. Table 1 showing average temperature variation at different grid elements, temperature at .55mm and .5mm is almost same. So current study deals with edge element of 0.55 were chosen as a safe limit for the created model.

Table 1 Grid independence study

Edge sizing	0.65mm	0.6mm	0.55mm	0.5mm
Average Temperature (K)	328.24	327.656	323.609	323.65
No of cells	972154	1120796	1319521	1877774

3.3. *Validation*

Numerical results are validated against the corresponding experimental study, by considering the temperature at specific location of target plate. Temperature obtained from the experimental and numerical investigation is shown in Figure 4. Louver angle of 45^0 and blow ratio 0.25 is the case considered for the validation. It can be concluded that experimental results are in close agreement with numerical result and experimental results and the maximum deviation is limited to ±4.38 %.

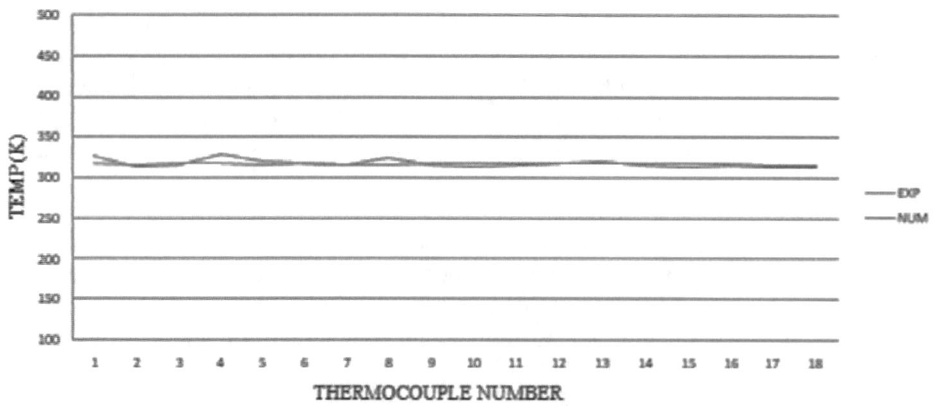

Figure 4 Validation for temperatures at different locations of the plate

4. Results and discussion

4.1. Experimental results

Average surface temperature for the blow ratio 1, 0.75, 0.5 and 0.25 determined from experiment at h/d of 2 and louver angle 45° are shown in the Table 2. It is evident that, as the blow ratio increases, the average temperature decreases and hence the heat transfer increases. The high velocity air coming from the nozzle mixes with cross-flow air and passes through the baffles which create turbulence above the target plate which results in more heat transfer.

Table 2 Experimentally determined average surface temperature

	BR=1	BR=0.75	BR=0.5	BR=0.25
Average surface temperature (K)	316.167	318.472	320.732	324.532

Figure 5 shows the variation of surface temperature along the reference diagonal line with respect to the changes in the blow ratio. In all the cases, as the blow ratio increases, surface temperature at the salient points also increases. General trend in the changes of surface temperature is almost same and hence, we could infer that blow ratio is the important factor which determines the surface heat transfer.

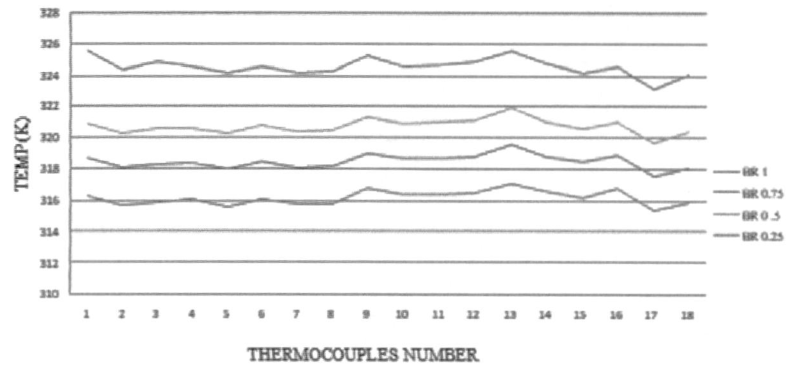

Figure 5 Temperatures at different locations of the plate at various blow ratios

Average surface heat transfer coefficient (h) in W/m^2K and surface Nusselt number (Nu) is estimated using the following relations $q = h\,A\,(T_s - T_\infty)$ and Nu=(hd)/k, where q = Heat flux,3472.2 W/m^2, A =

Area of plate (120 x120mm), Ts = Average surface temperature from 18 thermocouples, (K), T_∞= Ambient temperature, 305K., d = Jet diameter,5mm and k = Thermal conductivity of fluid, W/m K.

Experiments were performed for the validation purposes only and the rest of the experiments were conducted numerically.

4.2. Numerical results

Total number of cases considered for the present study is 24, keeping blow ratio, louver angle and h/d ratio as the variable. Blow ratio considered at 4 levels such as 0.25, 0.5, 0.75 and 1.00, louver angle at 3 levels such as 30°, 45° and 60° and h/d at 2 levels such as 2 and 3.

From Table 3, it is evident that, heat transfer is directly proportional to blow ratio. An appreciable Increase in Nusselt number can be noticed between the blow ratio 0.5 and 0.75 while the increase in Nusselt number between the blow ratio 0.75 and 1.00 is minimal. This is due to the increased turbulence along with the increase in cross-flow velocity. An increase in Nusselt number can be noticed with decrease in louver angle. This happens due to the increase in cross-flow deflection. Increasing h/d ratio from 2 to 3 brings decrease in Nusselt number. As the h/d ratio increases, the jet travel distance increases and downward shift in the impingement location causes the delay in impingement. Increase in h/d ratio decreases the jet momentum is also a factor which cause decreased heat transfer rate.

Table 3 Numerically determined average surface temperature and Nusselt number

h/d ratio	Louver Angle	Blow Ratio	Average Temperature (K)	Average Nusselt Number
2	30°	0.25	319.203	50.51022
		0.5	313.479	84.60864
		0.75	306.632	439.5813
		1	306.596	449.4967
	45°	0.25	320.229	47.10728
		0.5	314.682	74.09592
		0.75	307.164	331.5142
		1	307.092	342.9239
	60°	0.25	321.687	42.99135
		0.5	316.356	63.17336
		0.75	308.163	226.8089
		1	308.097	231.6425
3	30°	0.25	321.164	44.38237
		0.5	314.251	77.54802
		0.75	306.912	375.2075
		1	306.826	392.8788
	45°	0.25	322.015	42.1626
		0.5	315.337	69.40086
		0.75	307.238	320.5526
		1	307.222	322.8608
	60°	0.25	322.959	39.94636
		0.5	316.126	64.4793
		0.75	307.631	272.6707
		1	307.624	273.3981

4.3. Streamline

For better understanding the nature of flow, velocity streamlines of cross-flow and air jets are shown separately in Figures 6 and 7 corresponding to blow ratio 0.25, 0.5, 0.75, 1.00. Figure 6 exhibits the variations in cross-flow velocity due to the effect of louvered baffles. Stream lines originating from the jets are kept hidden. Presence of baffles largely affects the flow dynamics in the domain. In all the cases, it is evident that, as the flow passes across the baffle, there is acceleration in the velocity.

Figure 7 shows the velocity stream lines of the jet air movement, keeping the cross-flow steam lines hidden. Stream lines originating from jets seems to be quite complex. Flow emanating from jets gets deflected toward the downstream direction, especially in the region where jets get exposed directly to cross-flow. But jets protected by the baffles show better impingement on the target plate. Velocity of the flow gets accelerated towards the downstream in all the cases. Figure 8 shows the flow dynamics of combined jet flow and cross-flow.

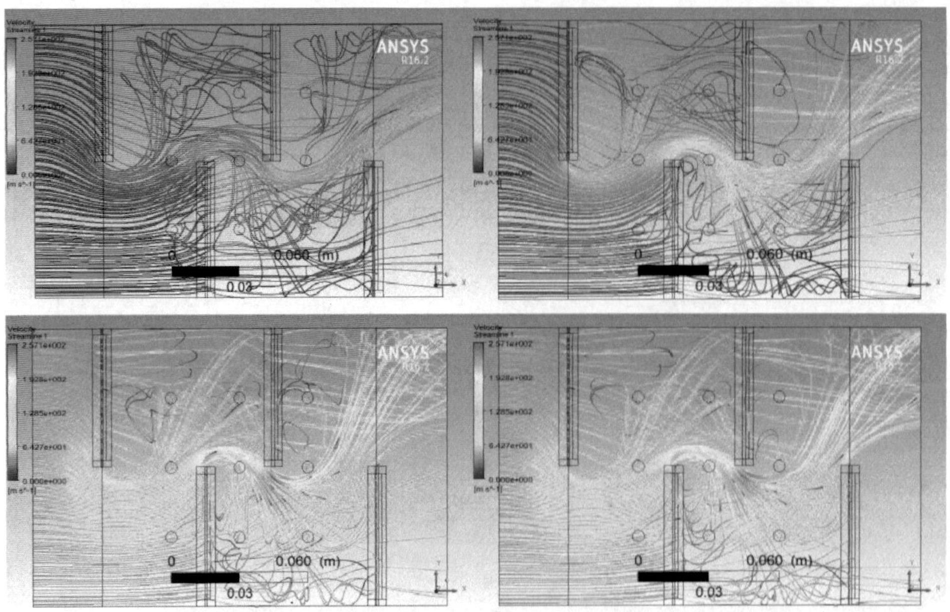

Figure 6 Velocity stream lines of cross flow alone for different blow ratios

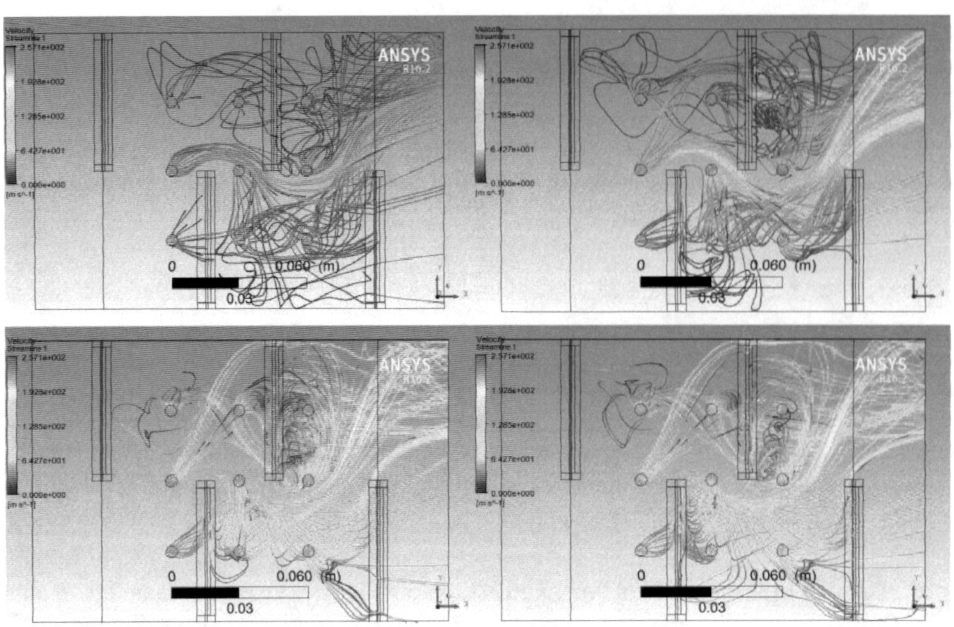

Figure 7 Velocity stream lines of jet flow alone for different blow ratios

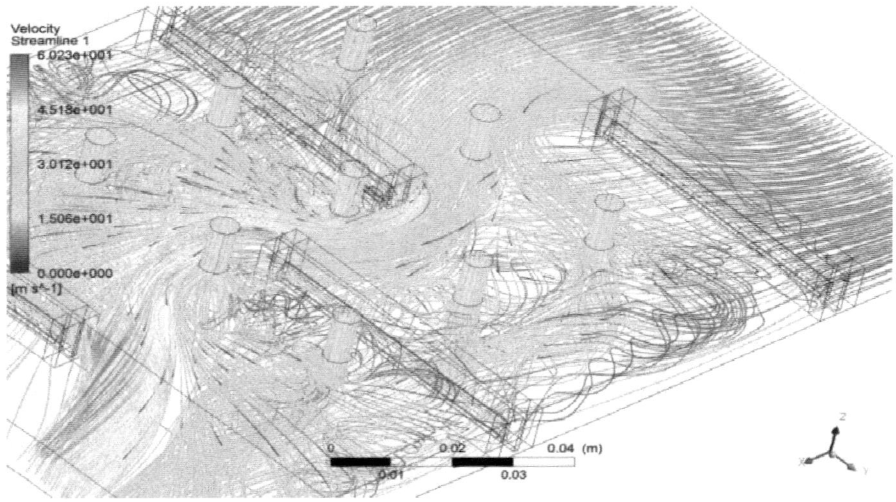

Figure 8 Velocity stream lines with cross-flow and jet flow

4.4. Temperature contours

Figure 9 and 10 shows the surface temperature contours for blow ratio 0.25, 0.5, 0.75 and 1.00 at h/d ratio 2 and 3 and constant louver angle 30°. For h/d ratio 2 and blow ratio 0.25, temperature drop at the point of jet impingement is clearly visible. As the blow ratio increases, temperature drop beneath the jet location slowly disappears showing the wash away of impinging jets by the cross-flow. It could be inferred that, heat transfer due to jet impingement deteriorates as cross-flow velocity increases. Cross-flow becomes the dominating factor for increased heat transfer, as blow ratio increases. For h/d =3 and blow ratio 0.25, temperature drop at the point of jet impingement is less visible compared to previous case. Momentum loss due to the increased jet travel distance is the reason for that. This comparison reveals that, jet impingement cooling with l cross-flow is best suited for low h/d applications.

Presence of baffle creates two distinct region of temperature difference in target plate. In the upstream side, baffle creates blockage in the cross-flow passage causes decrease in flow velocity and increase in temperature. Flow through baffle clearance and louver hole creates increased velocity towards the downstream of baffle. Increased jet velocity scores away the boundary lay and causes increased heat transfer and reduced temperature. This phenomenon is same in both h/d ratios considered in this research.

4.5. Temperature variations along reference line

Figures 11, 12 and 13 shows the local surface temperature on the target plate at louver angle 30°, 45° and 60° along diagonal line considered as reference for this study. Axis of three diagonal jets passes through reference line. For blow ratio of 0.25, jet impingement at the target plate is evident, represented by the sharp dip in temperature. Decrease in temperature at the center jet impingement location is less sharp because of larger mass flux concentration in that area causes sweep away of jets in that location. As blow ratio increases, temperature drop at the impingement location decreases due to the larger interruption by the cross-flow on jets. For blow ratio 0.75 and 1.00, the temperature profile is almost flat and same which is due to the extensive sweep away of incoming jet flow. For higher louver angle and lower blow ratios, temperature drop at the impingement locations become shallow as shown in Figures 12 and 13. But secondary impingement from the louvers is covered by larger area. For larger blow ratio, temperature profile is almost same at all louver angles.

Figure 14 represents the graphical representation of overall performance of the jet impingement with louvered baffles. Among the considered cases, jet impingement with h/d 2, louver angle 30° and blow ratio 1has the lowest average surface temperature (306.596 K). Meanwhile, jet impingement with h/d 32, louver angle 60° and blow ratio 0.25 has the highest average surface temperature (322.959 K).

IOP Conf. Series: Materials Science and Engineering 1132 (2021) 012021 doi:10.1088/1757-899X/1132/1/012021

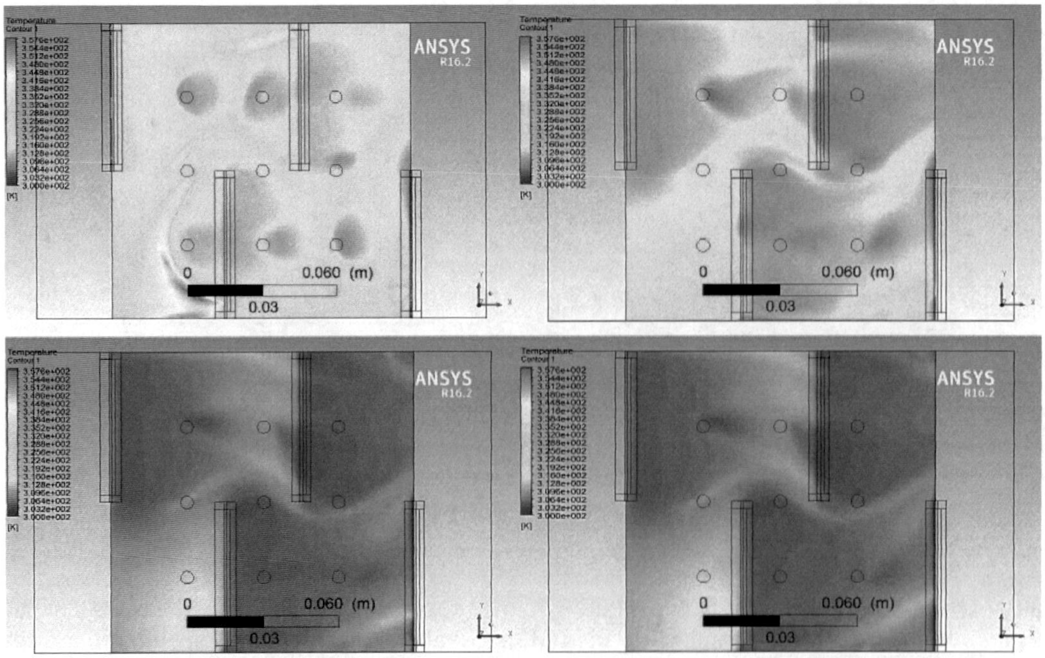

Figure 9 Temperature contours for different blow ratios at h/d=2 and louver angle 30°

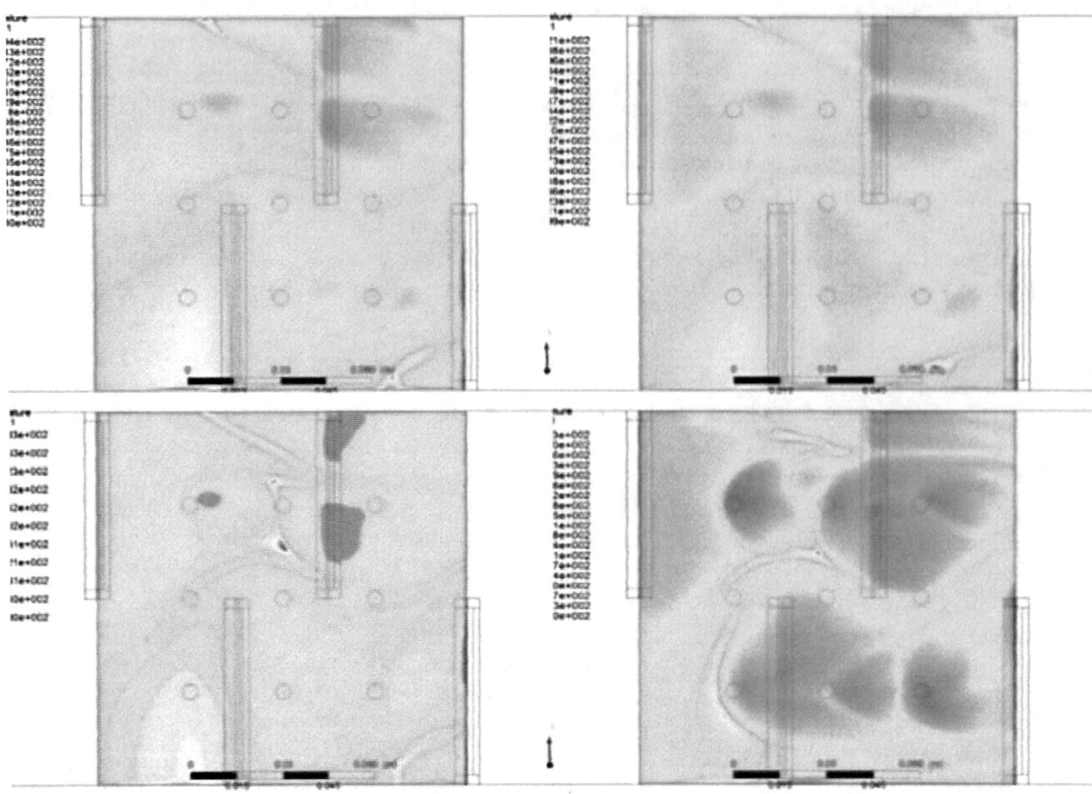

Figure 10 Temperature contours for different blow ratios at h/d=3 and louver angle 30°

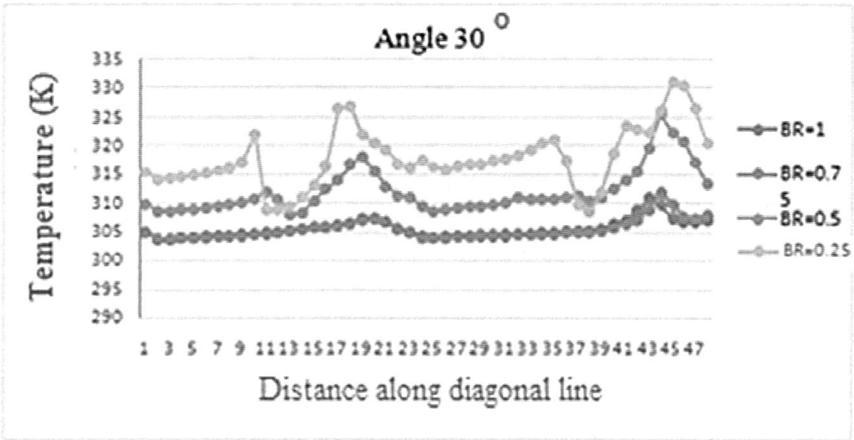

Figure 11 Temperature along diagonal line for h/d = 2 and louver angle 30°

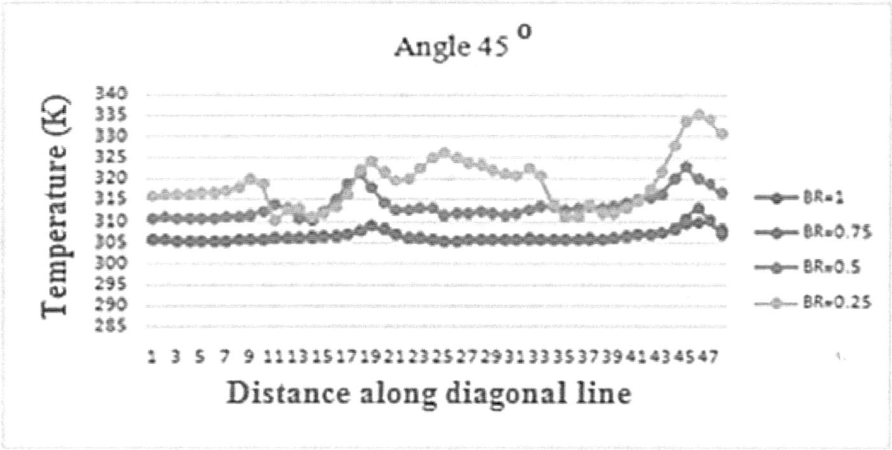

Figure 12 Temperature along diagonal line for h/d = 2 and louver angle 45°

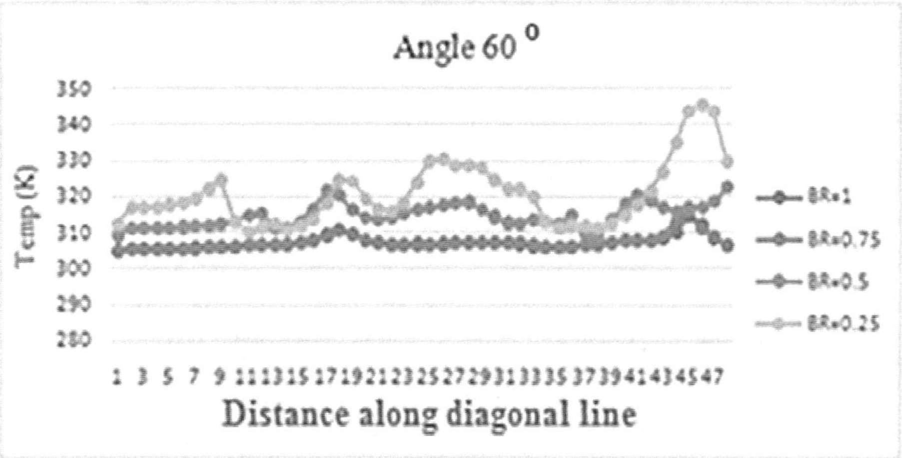

Figure 13 Temperature along diagonal line for h/d = 2 and louver angle 60°

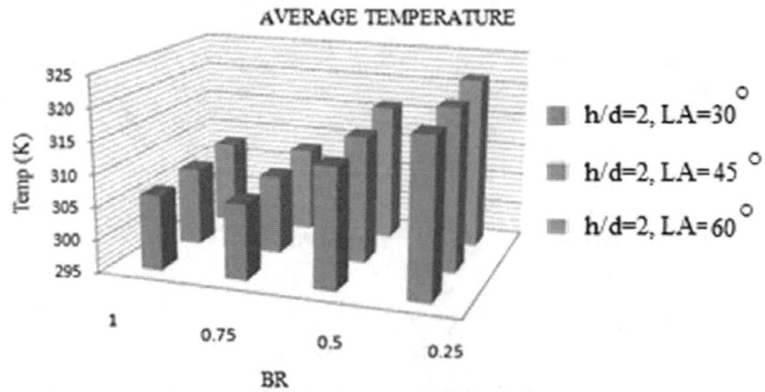

Figure 14 Comparison of average temperature at different blow ratios and louver angles at h/d=2

5. Conclusion

Numerical and experimental investigations of jet array impingement on a flat plate with louvered baffle and cross-flow subject to constant heat has been carried out. Effect of blow ratio, h/d ratio and louver angle on heat transfer were the factors considered in the investigation. From the detailed study, it has been found that, numerical investigation is in good agreement with the results obtained from the experiment. Maximum deviation of average surface temperature between experimental and numerical results limited to 3.55%. Important observations derived from this study are:

- As the blow ratio increases, heat transfer from the target surface increases. This is due to the increased turbulence as the flow passes through the baffles. As the blow ratio increases, cross flow become the dominating factor in heat transfer. This is due to the sweep away of impinging jets at higher blow ratio.
- Nusselt number increases with decrease in louver angle. This happens due to the increase in cross-flow deflection. Decrease in louver angle causes creates a secondary impingement zone immediately after the baffle with will increases the heat transfer.

Increasing h/d ratio from 2 to 3 brings decrease in Nusselt number. As the h/d ratio increases, the jet travel distance increases and downward shift in the impingement location causes the delay in impingement. Increase in h/d ratio decreases the jet momentum is also a factor which cause decreased heat transfer rate.

References

[1] Sinan Caliskan, SenolBaskayaand Tamer Calisir, "Experimental and numerical investigation of geometry effects on multiple impinging air jets," *International journal of heat and mass transfer* (2014)75 pp. 685-703.

[2] P M Nakod, S V Prabhuand R P Vedula, "Heat transfer augmentation between impinging circular air jet and flat plate using finned surfaces and vortex generators," *Experimental thermal and fluid science 32* (2008)1168-1187.

[3] M F Koseoglu and S Baskaya, "The role of jet inlet geometry in impinging jet heat transfer, modeling and experiments,"*International journal of thermal science49*(2010)1417-1426

[4] Harekrishna yadav, Amit Agrawal and Atul Srivastava, "Mixing and entrainment characteristics of a pulse jet," *International journal of heat and fluid flow000*(2016)1-13

[5] Jung-Yang San andJenq-JyeChe, "Effects of jet to jet spacing and jet height on heat characteristics of an impinging jet array," *International journal of heat and mass transfer71*(2014)8-17

[6] Junsik Lee, ZhongRen, Phil Ligrani, Dae Hee Lee, Michael D.Foxand Hee-Koo Moon, "Cross flow effects on impinging array het transfer with varying jet-to-target plate distance and hole spacing," *International journal of heat and mass transfer75*(2014)534-544.

[7] Hollworth, B. R and Durbin M., "Impingement cooling of electronics," *ASME Journal of Heat Transfer* (1992) 114, pp. 607- 613

[8] Roy S, Nasr K, Patel P and AbdulNour B, "An experimental and numerical study of heat transfer of an inclined surface subject to an impinging air flow," *International Journal of Heat and Mass Transfer*,(2002) 45, pp. 1615 - 1629.

[9] Babic, D. M., Murray, D. Band Torrance, A. A,"Mist jet cooling of grinding processes," *International Journal of Machine Tools and Manufacture*,(2005) 45, pp. 1171 – 1177

[10] Hussain, A. K. M. Fand Zaman, K. B. M. Q., "Vortex pairing in a circular jet under controlled excitation. part 2. coherent structure dynamics," Journal of Fluid Mechanics(1980), 101, pp. 493 - 544.

[11] Ho, C. Mand Huang, L. S., "Sub harmonics and vortex merging in mixing layers," *Journal of Fluid Mechanics*,(1982) 119, pp. 443 – 473

[12] Liu, Tand Sullivan, J. P., "Heat transfer and flow structures in an excited circular impinging jet," *International Journal of Heat and Mass Transfer*,(1996) 39, pp. 3695 – 3706

[13] Hwang, S. D, Lee, C. Hand Cho, H. H., "Heat transfer and flow structures in axisymmetric impinging jet controlled by vortex pairing," *International Journalof Heat and Fluid Flow*,(2001)22, pp. 293 – 300

[14] Lavine, A. Sand Jen, T. C., 1991, Thermal aspects of grinding: heat transfer to work piece, wheel and fluid," *ASME Journal of Heat Transfer*,(1991)113, pp. 296 -303

[15] Lavine A. S., "An exact solution for surface temperature in down grinding, "*International Journal of Heat and Mass Transfer*, (2000)43, pp. 4447 – 4456

[16] Jen, T. Cand Lavine, A. S., "A variable heat °ux model of heat transfer in grinding: model development," *ASME Journal of Heat Transfer*,(1995)117, pp. 473- 478

[17] Jen, T. C and Lavine, A. S., "A variable heat °ux model of heat transfer in grinding with boiling," *ASME Journal of Heat Transfer*, (1996) 118, pp. 463 – 470

[18] Liao, Y. S and L. S. Y. Y. T. H., "A thermal model of the wet grinding process," *Journal of Materials Processing Technology,* (2000)101, pp. 137 – 145

[19] Fleischer, A. S., Kramer, Kand Goldstein, R. J., Dynamics of the vortex structure of a jet impinging on a convex surface," *Experimental Thermal and Fluid Science*,(2001)24, pp. 169 – 175

[20] Bernhard Wigand, Sebastian Spring," Multiple Jet Impingement — A Review", Heat Transfer Research, 2011, Vol. 42, No. 2, pp. 101-142

[21] Florschuetz, L. W., Metzger, D. E., Su, C. C., Isoda, Y., and Tseng, H. H. Jet Array Impingement Flow Distributions and Heat Transfer Characteristics. Effects of Initial Cross-flow and a Non-uniform Array Geometry, NASA-CR-3630, 1982

[22] Florschuetz, L. W., Metzger, D. E., and Su, C. C. Heat transfer characteristics for jet array impingement with initial cross-flow, ASME J. Heat Transfer, 1984, Vol. 106, pp. 34–41.

[23] Florschuetz, L. W. and Su, C. C. Heat Transfer Characteristics within an Array of Impinging Jets. Effects of Cross-flow Temperature Relative to Jet Temperature, NASA-CR-3936, 1985

[24] Florschuetz, L. W. and Su, C. C. Effects of cross-flow temperature on heat transfer within an array of impinging jets, ASME J. Heat Transfer, 1987, Vol. 109, pp. 74–82

[25] Rao G. A., Kitron-Belinkov M., and Levy Y. Numerical analysis of a multiple jet impingement system. In: Proc. ASME Turbo Expo 2009: Power for Land, Sea, Air, June 8–12, Orlando, Florida, USA, GT2009-59719, 2009

[26] Zu Y. Q., Yan, Y. Y., and Maltson, J. D. CFD prediction for multi-jet impingement heat transfer. In: Proc. ASME Turbo Expo 2009: Power for Land, Sea Air, June 8–12, Orlando, Florida, USA, GT2009-59488, 2009

[27] Berner, Durst, F., and McEligot, D. M., 1984, "Flow Around Baffles," ASME journal of heat transfer, Vol. 106, pp. 743-749

Performance analysis of single slope solar still integrated with refrigeration systems

V Aneesh*, **S Vishnu, Arun P Das and M S Ajith**

Department of Mechanical Engineering, Adi Shankara Institute of Engineering and Technology, Mattoor, Kalady, Ernakulam, Kerala, India-683574

*email: aneesh.me@adishankara.ac.in

Abstract. The over misuse of existing new water supplies has become a major issue in numerous pieces of the world. Hence, recycling of water is necessary to meet the requirements of the growing population. Solar water purification is one of the viable technologies. This work is focused on the productivity enhancement of a single slope solar still by using two types of refrigeration system such as thermoelectric refrigeration system and vapour compression refrigeration system. In the modification of the existing solar still setup, the extra cooling was achieved by adding thermoelectric modules in the external condenser which is used for increasing the condensation area and thus the rate of evaporation. In the second set up, a vapour compression refrigeration system is coupled with the solar still in such a way that the evaporator coil of the refrigeration system is placed inside the external condenser for providing extra cooling effect and the condenser of the refrigeration system is placed in the basin of solar still for enhancing evaporation rate of the basin water. The results show that these modifications enhanced the performance and productivity of the solar still. The enhancement in the productivity of the passive solar still using thermoelectric refrigeration system was found to be 45.77% more and for solar still with external condenser integrated with vapor compression refrigeration system was found to be 167.7% more as compared with the solar still alone.

1. Introduction

Seas and oceans add up to about 97% of the water on Earth which is saline, remaining 2% of the water is frozen in ice sheets as glaciers, a simple 1% of the water is accessible for the life of plants and other creatures. The over misuse of existing new water supplies is turning into an issue in numerous pieces of the world. Because of increasing population and human activities the water is getting depleted and polluted. Desalination of saline water would be one of the viable solutions for the water scarcity [1]. Desalination of water using solar energy is being done naturally throughout the year in the Earth's atmosphere and it is directly used for the drinking purpose [2]

Solar still is used for solar distillation, which is a process of distilling pure water from saline water by using solar energy. Solar still can be classified as passive and active. Conventional (Simple) solar still are passive solar still and solar stills with additional equipment like preheater, additional condenser, for enhancing the performance are active solar still. Malik et al. [3] was published a review paper on various factors affecting for the improvement of productivity in passive solar distillation. Later, Tiwari [4] had worked on active solar distillation. The evaporation rate within the solar still depends upon the temperature distinction between glass cover and the glass cover. In order to enhance the yield of the distilled water, it is desirable to have the basin temperature as high as possible and the glass cover temperature as low as possible [5]. This was the motivation for the present research work on to enhance the productivity of the solar still by increasing the temperature difference between the condensation and evaporation of the solar still.

The main objective of this work is to maximize the rate of evaporation of the saline water inside the basin and the rate of condensation of the vapour at the inner surface of the glass cover [6]. H.E.S. Fath recommended that for enhancing the productivity of solar still, the basin temperature, evaporation and condensation surface area should be maximised and glass temperature, heat losses through side walls should be minimised. Active solar still has any one of the following: solar collectors [7-14] or water heater or waste heat recovery for preheating the basin of solar still, the use of internal and external condensers for enhancing the rate of condensation or applying vacuum (partial) inside the solar still to enhance the evaporation processes, and cooling the glass cover for decreasing the glass cover temperature hence the temperature gradient between the glass and the basin water increases eventually increasing the rate of evaporation.

Murugavel et al. [13] proposed different methods of improving the productivity experimentally on single basin passive type solar still. Inclination and the direction of inclination of the cover depend on latitude of the location. Results showed that a solar still with glass cover plate of 3 mm thickness gives 16.5% more production than the glass cover of 6 mm thickness. Experiments with deep basin reveal that the productivity of the still decreases with an increase in depth of water during daylight. Monowe et al. [17] put forward a new design of a portable thermal–electrical solar still with an external reflecting booster and an outside condenser. H.N. Singh et al. [20] suggested the monthly performance of passive and active solar stills for different Indian climatic conditions. From different water depth, the annual yield is at its maximum when the condensing glass cover inclination is equal to the latitude of the place. Badran [19] optimized thickness of the solar still glass cover. The efficiency of the conventional passive solar still is very low and hence, improvement in its design is the need of the hour. But only few works has been reported on increasing the condensation rate by using refrigeration systems like vapour refrigeration system [5] and thermoelectric refrigeration [22].

In this work, main focus on increasing temperature gradient between the condenser and evaporator part and that leads to the productivity of solar still was increased by increasing the condensation rate by means of thermoelectric refrigeration and increasing the condensation area by attaching the external condenser having thermoelectric module. Incorporation of external condenser resulted in the condensation of evaporates partially on the glass cover and external condenser.

2. Working Principle

2.1. Conventional Solar Stills (CSS)

Simple solar still [Figure 1.] consist of a basin, glass cover, base liner etc. The incident solar radiation is transmitted through the glass cover of the solar still after partial reflection and absorption. The transmitted radiation is further reflected and absorbed by the basin water and finally reaches the blackened surface of the basin where it gets mostly absorbed. After absorption of solar radiation by the blackened surface, most of the thermal energy is absorbed by the water mass, and the rest, which is very small, is lost to the atmosphere through bottom insulation. Thus, the water gets heated and move upward due to convection and buoyancy. And eventually it reached the inner surface of the glass cover. The heat loss from the evaporate through radiation, convection and evaporation, occurs between the water and glass cover. The evaporated water gets condensed on the inner surface of the glass cover after releasing the latent heat of condensation. This condensed water trickles into the distillate output under the effect of gravity [22].

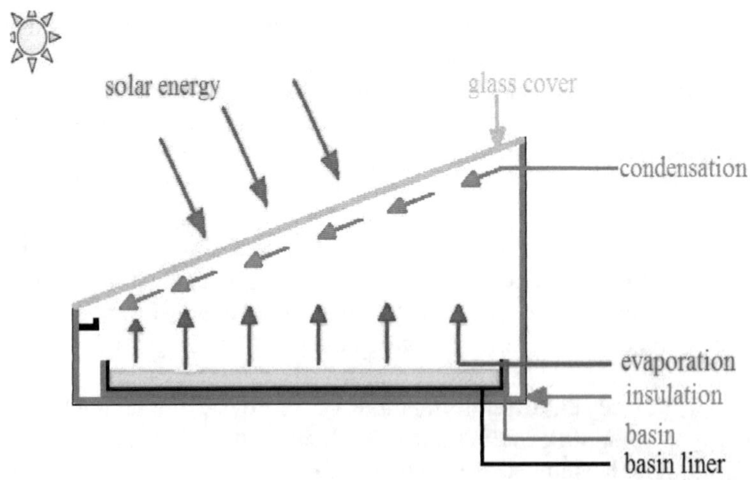

Figure 1. Schematic representation of working of a single slope solar still

2.2. Thermoelectric refrigeration
Early 19th century scientists, Thomas Johann Seebeck and Jean Charles Athanase Peltier, first discovered the phenomena that are the basis for today's thermoelectric industry. Seebeck found that if a temperature gradient exists across the junction of two dissimilar conductors, electric current would flow. Peltier, on other hand, learned that passing current through two dissimilar electrical conductors, caused heat to be either emitted or absorbed at the junction of the materials. Usually thermoelectric module consists of P-type and N-type semiconductors. Figure 2. shows the schematic diagram of thermoelectric module.

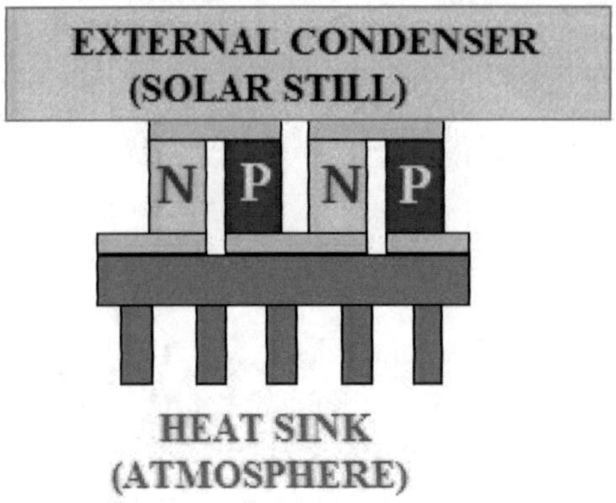

Figure 2. Schematic diagram of thermoelectric module

The main advantage of thermoelectric refrigeration is that, it has no moving parts; hence it has long life and noise free. Most of the thermoelectric modules are compact and simple in construction. Thermoelectric refrigeration has low cooling capacity. In thermoelectric refrigeration, no refrigerant is used and hence there is no leakage. Thermoelectric module required less DC voltage is, so it can be connected with PV panels directly.

2.2.1. Vapour Compression Refrigeration System. Refrigeration and air conditioning is used for cooling products or a building environment. The refrigeration or air conditioning system (R) transfers heat from a cooler low-energy reservoir to a warmer high-energy reservoir at the expense of energy. The vapour compression cycle is used for refrigeration in preference to gas cycles; making use of the latent heat enables a far larger quantity of heat to be extracted for a given refrigerant mass flow rate.

IOP Conf. Series: Materials Science and Engineering 1132 (2021) 012022 doi:10.1088/1757-899X/1132/1/012022

This makes the equipment as compact as possible. A simple vapour compression refrigeration system consists of a compressor, condenser, an expansion valve and an evaporator as shown in Figure 3. The low temperature, low pressure vapour at state 2 is compressed by a compressor to high temperature and pressure vapour at state 3. This vapour is condensed in the condenser at constant pressure to state 4 and then passed through the expansion valve where, the vapour is throttled down to a low pressure liquid and passed on to an evaporator, where it absorbs heat from the surroundings from the circulating fluid (being refrigerated) and vapourizes back into low pressure vapour at state 2. This cycle then repeats.

2.3. Single Slope Solar Still with External Condenser having Thermoelectric Modules

The saline water from the solar still evaporates by absorbing solar insolation and the resulted vapour is partially transported to the inner glass surface due to the convection current formed in the space between the water surface and glass cover. The other part of the vapour moves to the external condenser chamber due to the pressure difference between the small evaporator volume and the larger condenser chamber volume and due to the comparatively low temperature created in the condenser chamber with the use of thermoelectric module. Figure 4. shows the schematic design of Single slope solar still with external condenser having thermoelectric modules. The diffused vapour at the condenser chamber gets condensed, propped up by the thermoelectric module and the distillate was collected from both the solar still side and the condenser chamber side. Figure 5. shows configuration of the thermoelectric modules with side walls of external condenser. Fins and fans were used to increase the heat disposal rate at the hot side thermoelectric modules. Figure 6. shows actual thermoelectric module

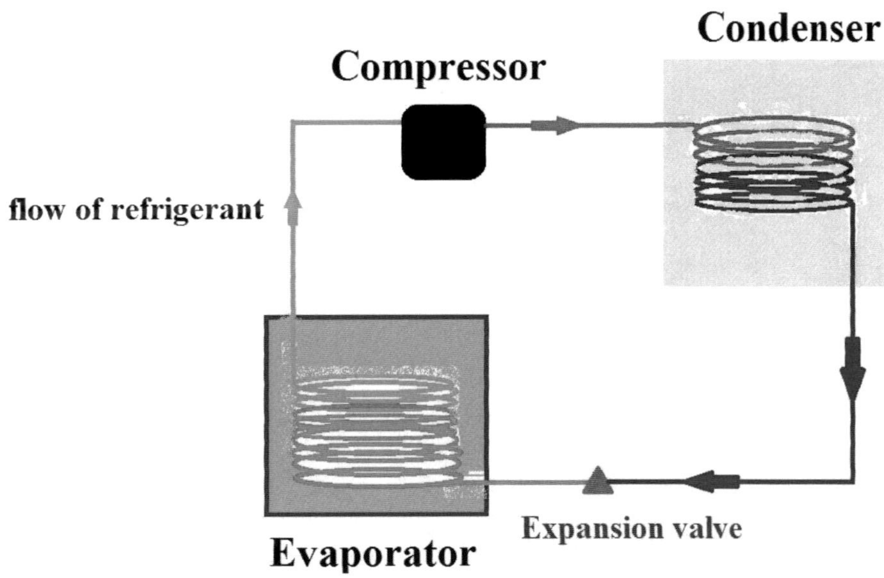

Figure 3. Schematic representation of vapour compression cycle

2.4. Single Slope Solar Still with External Condenser Integrated with Vapour Compression Refrigeration System

The saline water from the solar still evaporates by absorbing solar insolation and the heat liberated by the condenser of refrigeration system, which is immersed in the basin of the solar still. The resultant vapour is partially transported to the inner glass surface due to the convection current formed in the space between the water surface and glass cover. The other part of the vapour moves to the external condenser chamber due to the pressure difference between the small evaporator volume and the larger condenser chamber volume and due to the comparatively low temperature created in the condenser chamber with the use of cooling effect of vapour compression refrigeration system (i.e., evaporator of refrigeration system placed inside the external condenser of the solar still). The schematic design of single slope solar still with external condenser integrated with vapour compression refrigeration system is shown in Figure 7. The diffused vapour to the condenser chamber gets condensed with the help of cooling effect of vapour compression refrigeration system and the distillate was collected from both the solar still side and the condenser chamber side. Thus, coupling the solar still with the vapour compression refrigeration cycle will have following significant advantages:

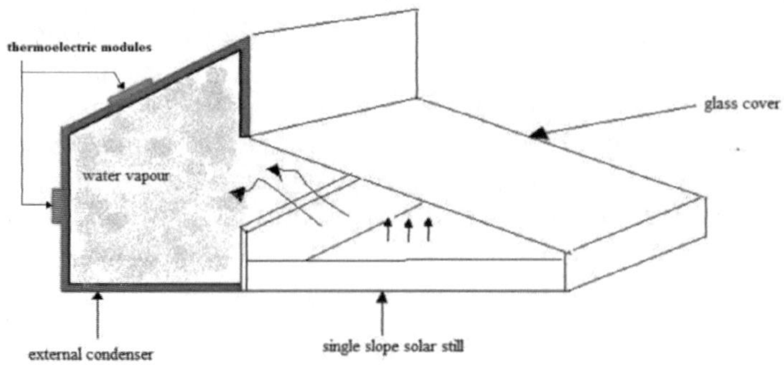

Figure 4. Schematic design of Single slope solar still with external condenser having thermoelectric modules.

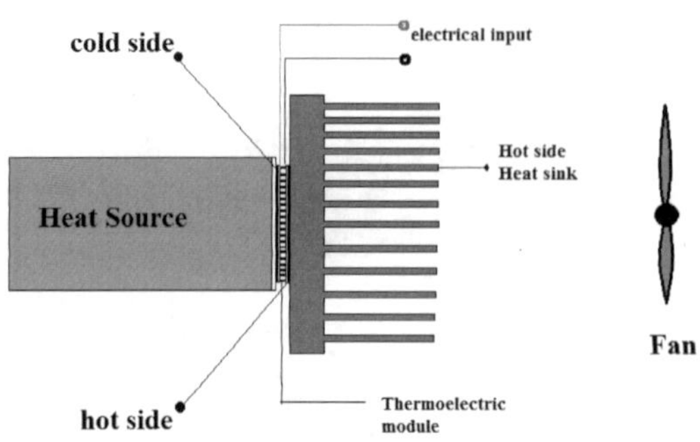

Figure 5. Thermoelectric module assembly

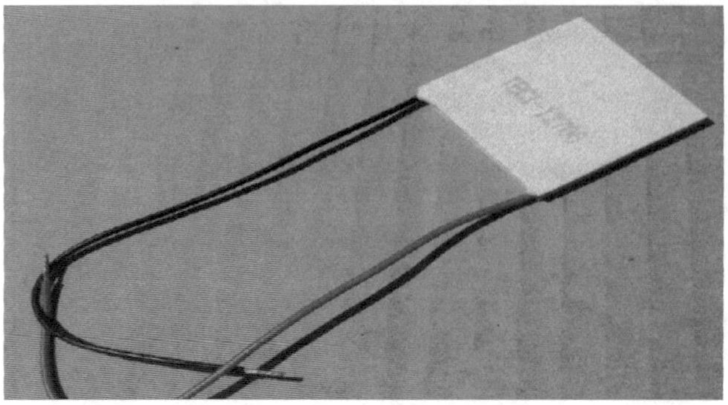

Figure 6. Thermoelectric Peltier sensor Module

The evaporator of the refrigeration unit is used to extract the latent heat of the evaporation from the water vapour in the solar still.The heat rejected in the condenser of the refrigeration unit which is dipped inside the basin of the solar still is utilized for heating the feed saline water. Both these factors lead to an increase in the rate of the condensation and hence the fresh water produced.

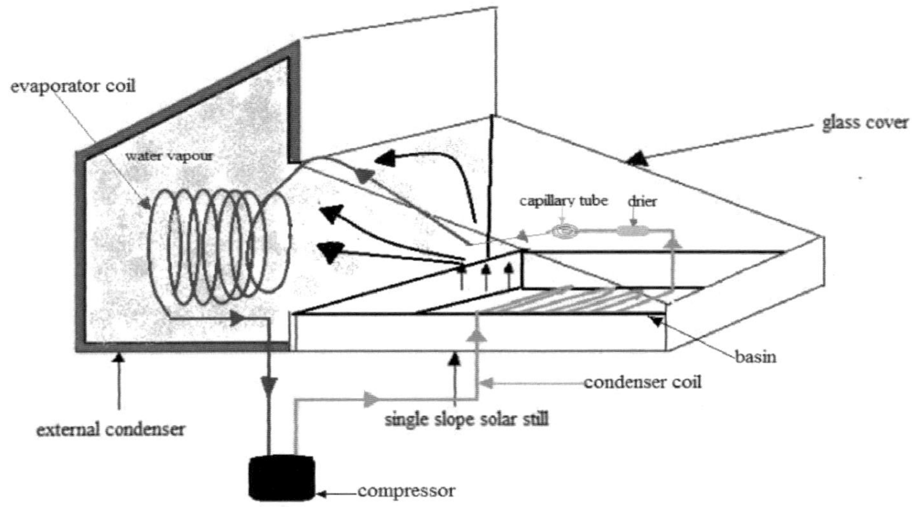

Figure 7. Schematic design of Single slope solar still with external condenser integrated with vapour compression refrigeration system

Table 1. Thermoelectric Module Specifications

Maximum cooling capacity	80 W
Input Current	8.5 A
Input Voltage	15 V
Thickness	2 mm
Dimensions	40 × 40 ×44 mm

2.5. Design of Solar Stills

2.5.1. Passive Solar Still. Passive single slope solar still with basin area of 0.6 m X 0.4 m was fabricated using GI sheet. Basin was painted with black plaint as basin liner for increasing the absorption rate of the basin. A 3 mm thick glass cover with 12^0 inclinations was fixed at top of the vertical walls of the solar still. Side walls of the still were fabricated using plywood and 3 mm thick thermocol was used to minimise the heat loss from the solar still. The vacuum sealing was done with silicon. J-type thermocouples were installed having a range of 0°C to 350°C. The details of solar still is given below

Table 2. Solar still Specifications

Length of still absorber basin	0.6 m
Breadth of still absorber basin	0.4 m
Inclination of glass cover	120 (latitude of Calicut)
Glass cover thickness	3 mm
Depth of water in the basin	2.5 cm

3. Results and Discussion

Experiments were conducted at the solar energy centre, NIT Calicut during April, 2014 on the days when the average solar radiation was almost equal (similar pattern of global radiation variation) excluding the cloudy days. The experiments were conducted on solar still alone, solar still with external condenser having thermoelectric modules and solar still with external condenser integrated with vapour compression refrigeration system. For all the experiments, a basin water depth of 2.5 cm was maintained. The average data of 5 days from 9.00 am to 5.00 pm was taken. The distilled water was collected from the solar still side and the condenser chamber side separately. The thermoelectric module and cooling fans was powered using DC power supplier and compressor of the vapour compression refrigeration system was powered by AC supply. A comparative study of the cumulative yield over the day in different experiment setups at constant water depths is discussed.

3.1 Hourly variation of yield with respect to time

The variation of yield with respect to time with solar still alone, still coupled with thermoelectric module and still coupled to the refrigeration system is shown in Figure 10. The hourly collection of yield started from 9 am and reached the maximum at 2.00 pm and then decreased for all three setups. When the experiment was done with solar still alone, the maximum yield was found to be 90 ml., for the solar still coupled with thermoelectric module, the maximum yield was 125 ml. and for solar still coupled to the refrigeration system, the maximum yield was found to be 225 ml and is the highest among the three different modes of operation.

At 9.00 am, the solar still integrated with vapour compression refrigeration system gave a yield of 15 ml from the condenser side, which was due to the additional heat liberated from condenser of refrigeration system immersed in the basin. It is clear from the results that the hourly yield of the distilled water was always maximum for solar still integrated with vapour compression refrigeration system as compared to the other modes of operation.

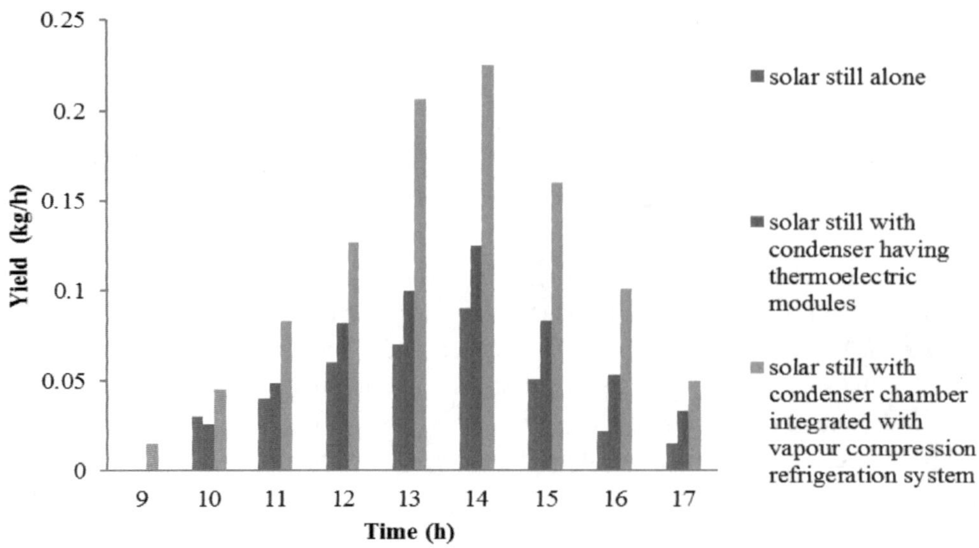

Figure 10. The hourly production rate of passive solar still and modified solar still.

3.2. Comparison of hourly yield collected from solar side and condenser side

Figure 11. shows the comparison between hourly yield of distilled water from solar still side and the condenser side for solar still with external condenser having thermoelectric module. The hourly yield of distilled water from both solar still side and condenser side increased as solar insolation increases and reaches a maximum at the noon and reduces at evening. In early hours, the yield from solar still was found to be more as compared to the condenser side. This may be due to the fact that in early hours the space between the basin and glass cover was less saturated, so the evaporation and condensation rate of water is proportional and hence the mass of water drawn to the external condenser was less. But towards the noon, the evaporation rate was much more than the condensation rate of solar still. So the space between the basin and glass cover got saturated, eventually the vapour pressure increased and the water vapour moved to the external condenser chamber due to the comparatively low temperature created in the condenser chamber with the use of thermoelectric module. In the afternoon, the yield rate in the solar still side increased slightly than the condenser side. This was because the low temperature of the glass cover (condensation surface of solar still) increased the condensation rate.

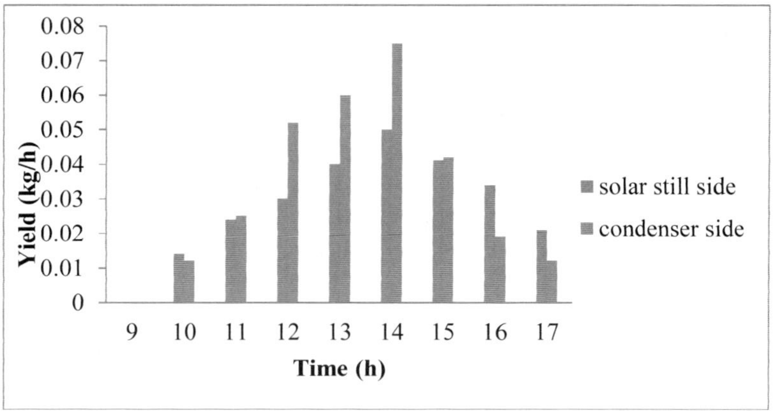

Figure 11. Comparison of hourly yield collected from solar side and condenser side of solar still with external condenser having thermoelectric modules.

In the second case; solar still with external condenser integrated with vapour compression refrigeration system, the hourly yield of distilled water from the condenser side was always much more than the solar still side as shown in Figure 12. This might be due the extra cooling effect of the refrigeration system. In this case, the evaporator coil which was placed inside the external condenser was in direct contact with evaporates of the solar still, so quick cooling of the condenser is possible. But in the first case, cold side of the thermoelectric modules was in contact with the surface of the external condenser. The surface of the condenser had to be cooled first and then the space inside the condenser had to be cooled which would obviously take time for cooling. The production rate in this setup is much more due to the fact that extra heat energy is used to heat the basin water of the solar still. In this case also the water yield from both solar still side and condenser side increased as solar insolation increased and reached a maximum at the noon and reduced at evening.

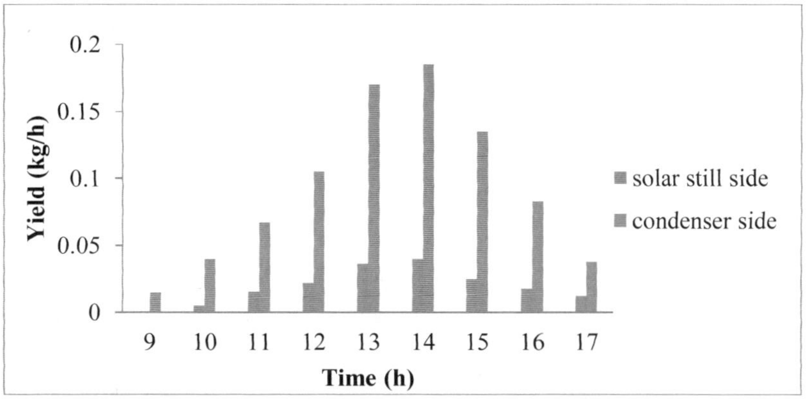

Figure 12. Comparison of hourly yield collected from solar side and condenser side of solar still with external condenser integrated with vapour compression refrigeration system

3.3. Variation of productivity for different modes

Figure 13 shows the variation in daily productivity for all the three different modes. The daily productivity from the solar still alone is 378 ml, solar still with external condenser having thermoelectric module is 551 ml. and the maximum productivity obtained with solar still with external condenser integrated with vapour compression refrigeration system is 1012 ml. From the experiments, there was an enhancement in productivity of 45.77% when the modification of solar still done with thermoelectric refrigeration. For solar still with external condenser integrated with vapour compression refrigeration system, the productivity enhancement was 167.7% than the solar still alone. So, this setup gave the best productivity among the two-refrigeration system at the tested conditions. Figure 14. shows the variation of yield of the solar still with respect to the time and solar radiation. The hourly yield for all three setup is directly proportional to the incident solar radiation. The production of hourly yield was highest when the solar radiation is extreme and it decreased with the intensity of solar radiation. But yield rate in the afternoon session was much more for all the three setups than the morning session; this is because at morning the basin water at atmospheric temperature and the incident solar energy is used for the sensible heating of the basin water. In afternoon the basin water was already hot so the incident water is used for the evaporation process.

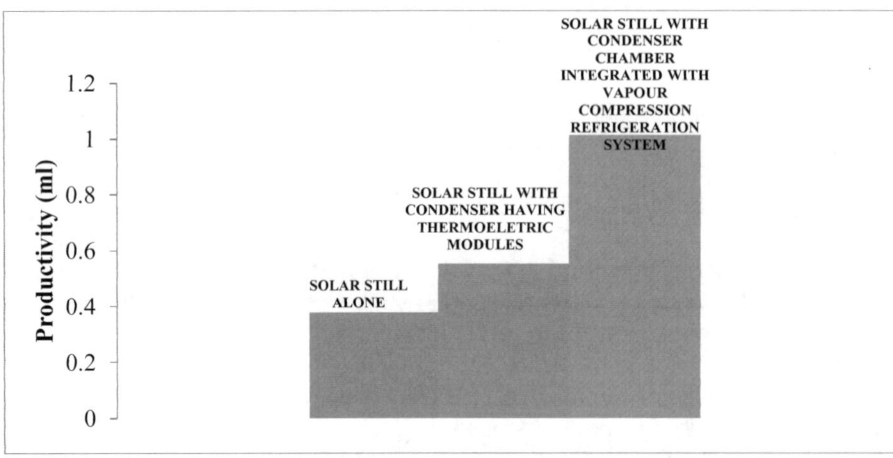

Figure 13. Variation in productivity for different modes

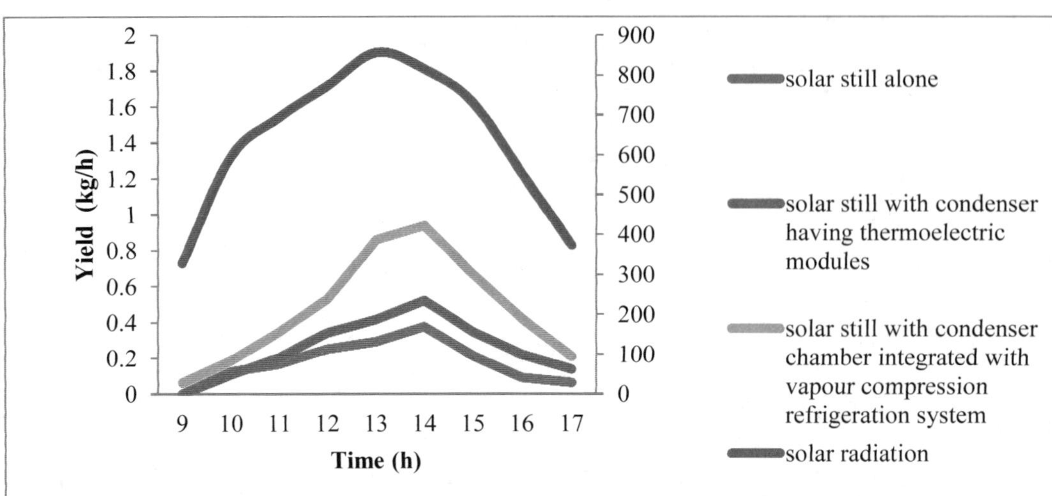

Figure 14. Variation of yield with respect to time and solar insolation

4. Conclusions

The current solar energy distillation plants are relatively inexpensive, especially useful where there is a need for small plant exists. However, there is still much room for innovation and improvement. It is well known that solar distillation exhibits a considerable economic advantage over other salt water processes, because of cost free energy and reduced operating costs.

The operation of a solar distillation system coupled with refrigeration systems have been investigated experimentally. Comparison of the output for different setups was studied. Experiments are conducted at constant water depths of 2.5 cm. It is observed that solar still is highly efficient when it was coupled with vapour refrigeration system as compared the thermoelectric refrigeration system for the climatic condition of Calicut. Solar still with external condenser integrated with vapour compression refrigeration system gives the maximum daily yield of 1012ml. The daily productivity in the solar still alone was 378 ml where as in the case of solar still with external condenser having thermoelectric module was 551 ml.

Enhancement in productivity of passive solar still by using thermoelectric refrigeration system was 45.77%. For solar still with external condenser integrated with vapour compression refrigeration system, the productivity enhancement was 167.7% compared with the solar still alone. So this setup gives the best productivity among the two refrigeration system. The hourly yield for all three setup is directly proportional to the incident solar radiation. The production of hourly yield is highest when the solar radiation is extreme and it decreases with the intensity of solar radiation.

Acknowledgements
The authors are grateful to MED, National Institute of Technology Calicut for financial support during the project.

References

[1] G.N. Kathpalia, R. Kapoor, Water Policy and Action Plan for India 2020: An Alternative [Report prepared for The Planning Commission, Government of India], 2002
[2] Ravishankar Sathyamurthy, S.A. El-Agouz b, Vijayakumar Dharmaraj: Experimental analysis of a portable solar still with evaporation and condensation chambers, Desalination **367** (2015) 180–185
[3] M.A.S. Malik, G.N. Tiwari, A. Kumar and M.S.Sodha. 1982. Solar Distillation. Pergamon Press, UK,
[4] G.N. Tiwari, R. Kamal, K.P. Maheshwari and R.L. Sawhney, eds., 1992. In: Recent Advances in Solar Distillation. Wiley Eastern, New Delhi, Chap. 2.
[5] Sabah A. Abdul-Wahab, Yousuf Y. Al-Hatmi, 2012. Study of the Performance of the Inverted Solar Still Integrated with a Refrigeration Cycle. Procedia Engineering **33**, pp.424 – 434,
[6] H.E.S. Fath, 2000. Encyclopaedia of Desalination and water Resources (DESWARE), ELOSS Publishers, section 10.2.5.3.
[7] A.K. Singh and G.N. Tiwari, 1993. Thermal evaluation of regenerative active solar distillation under thermosyphon mode. Energy Conversion and Management, vol. **34,** Issue 8, pp. 697–706.
[8] G.N.Tiwari, Madhuri and H.P. Garg, 1985. Effect of water flow over the glass cover of a single basin solar still with an intermittent flow of waste hot water in the basin. Energy Conversion and Management, vol. **25**, pp. 315-322.
[9] A.J.N. Khalifa, A.S. Al-Jubouri, M.K. Abed, 1999. An experimental study on modified simple solar stills. Energy Conversion & Management, vol. **40**, pp. 1835-1847.
[10] M. Abu-Qudais, B.A.K. Abu-Hijleh and Othman N. Othman, 1996. Experimental study and numerical simulation of a solar still using an external condenser. Energy, vol. **21**, Issue 10, Pages 851–855.
[11] H. Al-Hussaini and I.K. Smith, 1995. Enhancing of solar still productivity using vacuum technology. Energy Conversion and Management, vol. **36**, Issue 11, pp. 1047–1051.

[12] S.A. Lawrence, S.P. Gupta and G.N. 1990. Tiwari, Effect of heat capacity on the performance of solar still with water flow over the glass cover. Energy Conversion and Management, vol. **30**, Issue 3, pp. 277–285.
[13] K. Kalidasa Murugavel, Kn.K.S.K. Chockalingam, K.Srithar, 2008. Progresses in improving the effectiveness of the single basin passive solar still. Desalination, vol. **220**, pp. 677–686,
[14] O.O. Badran, 2007. Experimental study of the enhancement parameters on a single slope solar still productivity. Desalination, vol. **209**, pp. 136-143.
[15] T. Arunkumar, R. Velraj, D.C. Denkenberger, Ravishankar Sathyamurthy, K. Vinoth Kumar, Amimul Ahsan, Productivity enhancements of compound parabolic concentrator tubular solar stills. Renewable Energy **88** (2016) 391-400.
[16] Aneesh Somwanshi, Anil Kumar Tiwari, Performance enhancement of a single basin solar still with flow of water from an air cooler on the cover, Desalination **352** (2014) 92–102.
[17] P. Monowe, M. Masale, N. Nijegorodov, V. Vasilenko, 2011. A portable single-basin solar still with an external reflecting booster and an outside condenser. Desalination , vol. **280**, pp. 332–338.
[18] Javad Abolfazli Esfahani, Nader Rahbar, Mehdi Lavvaf, 2011. Utilization of thermoelectric cooling in a portable active solar still An experimental study on winter days. Desalination, vol. **269**, pp. 198–205.
[19] O.O. Badran, H.A. Al-Tahaineh, 2005. The effect of coupling a flat-plate collector on the solar still productivity. Desalination, vol. **83**, pp. 137–142.
[20] H.N. Singh, G.N.Tiwari,2004. Monthly performance of passive and active solar stills for different Indian climatic conditions, Desalination, vol. **168**, pp. 145-150.
[21] Ahmed Z. Al-Garni, 2012. Productivity Enhancement of Solar Still Using Water Heater and Cooling Fan. J. Sol. Energy Eng. vol. **134**(3), 031006.
[22] A. EI-Bahi, D.Inan, 1997. A Solar still with minimum inclination, coupled to an outside condenser. Desalination, vol.**123**, PP.79-83.

IOP Conf. Series: Materials Science and Engineering 1132 (2021) 012023 doi:10.1088/1757-899X/1132/1/012023

Aerodynamic Characteristics of Baseball and Tennis Ball using CFD Analysis

S Keerthekesh Nadar, T Amrit, S Sanjay Srinivas and K Balaji

Department of Mechanical Engineering, Amrita School of Engineering, Coimbatore, Amrita Vishwa Vidyapeetham, India.

k_balaji@cb.amrita.edu

Abstract- The objective of the project is to determine the aerodynamic characteristics of baseballs and tennis balls using Computational Fluid Dynamics (CFD) methods. Most realistic initial and boundary conditions are used to simulate each type of ball and to identify key design aspects that can be applied and modified to enhance the performance of the ball and hence improve the game. The basic conservation equations in fluid dynamics are applied to the domain of study to plot the results. For the inlet, different values of velocity based on the motion of the ball in the game were given. A moderate value of surface roughness was given to model the effects of change in the surface between the two balls and the delay in boundary layer separation subsequently affecting the distance travelled by the ball was seen. By observing the values of lift and drag coefficients we can validate computational results with experimental results obtained from reference journals, thus proving that computational simulation has a place in predicting the trajectory and behaviour of moving balls in real-time. The integration of spin in the simulation also yielded results that showcased the Magnus effect, this visualization is generally not observable on experimental methods since it involves complex processes to simulate spin, hence the trajectory of balls under different shots and pitches has been observed. We further developed an idea that could also be used to determine the trajectory of a dual-axis spinning ball. This would help in building the performance of athletes in the game.

1. Introduction

Baseball and Tennis are the most popular games played across the world. The 'ball' is common to both the games and it influence and control the nature of the game under different events. So, to understand the behaviour of the ball and to make the game even more entertaining, the study of aerodynamic characteristics of these different balls are necessary. The study of different balls is quite complex due to the presence of seams, stitches, and other geometrical parameters that directly influences the fluid domain characteristics during different maneuver. So, the player usually tries to predict this nature and changes his game according to the situation. Lift and drag coefficients (Cd and Cl) are two important factors that help us to quantify this behaviour of the ball under different given conditions. The boundary layer formation and pattern directly influence the aerodynamic forces experienced by a body [2]. The orientation of the ball concerning free stream velocity or the tangential direction of the ball movement is very crucial in deciding the trajectory of a ball.

The authors carried out an experimental analysis of baseballs in a wind tunnel to study the drag force at different free stream velocities at four different seam positions [4]. A support system with a sensor to measure force was developed to measure drag force. Another experimental study on similar grounds showed variation in the flow visualization technique. A novel infrared flow visualization technique was used [6].

Another research is a broader and covers experiments on football, cricket ball, baseball, golf ball, and a tennis ball [10]. Since it was purely experimental, minute details such as boundary layer separation are not accurately found. There are two types of pitching techniques in baseball; one is by spinning the ball while pitching and another with almost zero spins which is also known as a knuckleball. In another work the experimental results of the coefficient of drag to influence the design of his project to attain optimum performance is studied [7].

The differences in trajectory caused by spin on the tennis ball by pitching it against a smooth ball and observing the results is seen in [7]. The experiments are carried out in a wind tunnel in the range of 80,000 < Reynolds number < 250,000. To create spin, the ball is suspended by a 3.5 mm diameter string and rotated close to 3000rpm. A similar procedure is followed in the numerical analysis conducted by [8], where he studies the boundary layer development and wake structures at different L/D ratios of a cylinder. Whereas [13] study the variations in the coefficient of drag (Cd) over a cycle as well as the time-averaged value of Cd of different L/D ratios. Further study gives an idea about fluid flow over a body at a low Reynolds number and its corresponding aerodynamic performance [12].

The study that [1] did begins by analysing the flight trajectory equations and balancing the gravitational force with the centripetal force of the spinning ball and the aerodynamic forces of drag and lift. Since all the results were those of experimental studies, it allowed verification of the computational results with data from the experiments.

2. Methodology

For this study, we have used the numerical approach for solving the physical governing equations. In the case of ball aerodynamics, we mainly solve for two governing equations which are continuity equation for incompressible flow (as the flow over a ball is always incompressible) and the Navier's Stokes equation. These two equations help us in predicting the flow field passing over a ball during its flight. But solving those analytically is almost impossible due to the complex design of balls. So, only numerical or experimental methods can be equipped to solve these kinds of problems. Model to study the variation in aerodynamic efficiency by providing incremental values of Reynold's number and iterating is studied in [11]. On the other hand, [14] use a method of study which extracts results based on the variation of the position of the body in the study. Due to the lack of time, resources, and facility, we have decided to perform a few computational fluid dynamics simulations at different conditions and tried to understand the dynamics behind it to correlate with experimental results and establish the usage of CFD for predictive purposes based on the accuracy of results.

2.1. Geometry and Pre-processing

The model has been approximated by neglecting the stitches due to involvement of complex meshing and processing requirements. Therefore, we used roughness height equal to the stitches thickness. The seam of the tennis ball is neglected during analysis as previous experimental research papers have proven that the aerodynamic effects of seam are negligible on a tennis ball moving at a speed of 25m/s. For meshing, analysis, and extraction of results, Ansys Workbench and Fluent software are used. A fluid domain is created that encloses the ball. This fluid domain will represent the region of free stream airflow over our ball. So, this is our domain of study. Before further proceeding, the ball needs to be removed out of the domain and captured as space which the fluent software further assumes as the walls of the ball. To do that we will be using a Boolean operation that subtracts the ball from the fluid domain. So, all the surface features are traced in the fluid domain. Then it becomes just one integrated structure with all the surface features engraved in the fluid domain. Few of these techniques and methods have been developed from [9].

Table 1. Domain & mesh specifications

	Baseball	**Tennis ball**
Ball diameter	72mm	65mm
Computational domain	Cylindrical	Cuboidal
Domain dimensions	Radius = 250 mm & Height = 1250 mm	1000 mm x 800 mm x 800 mm
Mesh size	3 mm	3 mm
Mesh shape	Tetrahedron	Tetrahedron

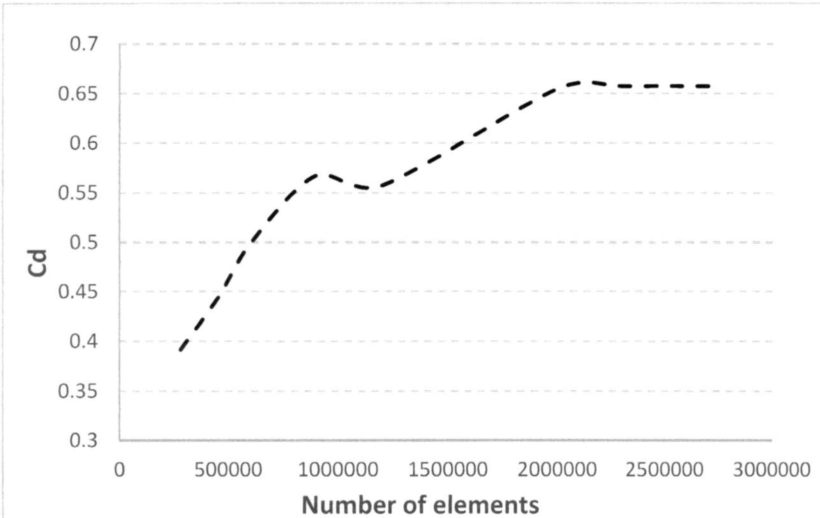

Figure 1. Grid independence test plot

This plot shown in Figure 1. describes the change in Cd values as the number of elements is gradually increased. Results had been extracted for 9 different mesh sizes in the range of 3 – 100 mm. It was observed that as the number of elements approached 3 mm the slope of the plot remained constant with increase in the number of elements henceforth exhibiting grid independence.

2.2. *Setup*

After establishing and checking for the mesh quality, initial conditions are defined and solver models are set up to establish the method of solving the physical governing equations. A pressure-based solver is used as the flow is incompressible, and a transient state solver was employed to see how the flow field evolves concerning time. Further a standard k-epsilon model with a standard wall treatment function is chosen so as to account for turbulence in the flow because this model is generally good for external flow over complex geometry, has a good convergence rate, and has relatively fewer memory requirements. Other models generally do not converge at faster rates. In our case low-pressure gradients are observed, so k-epsilon is better for these situations. And the scope of our study lies in regions beyond the walls as well, in these cases balls pitching characteristics under different rotating conditions were studied and it is observed how streamlines are like when a ball moves in a time frame. So, keeping all these conditions in mind we chose to go with the k-epsilon turbulent model for our simulation as also indicated by [9] in his research work. Standard k-epsilon models are efficient for high Reynolds number flows and can be beneficial in the abstraction of data in flows that have rotational components in them. Finally, when compared to realizable and RNG model, a standard k-epsilon model is more stable. At high Reynolds number, due to seam orientation, the boundary layer separation reaches past the critical regions and results in turbulent flow after the flow separation point. This model also provides better results for observing free-shear layers and wake regions. Using this model will help in predicting the motion of an object by studying its wake region.

IOP Conf. Series: Materials Science and Engineering 1132 (2021) 012023 doi:10.1088/1757-899X/1132/1/012023

2.3. Boundary Conditions

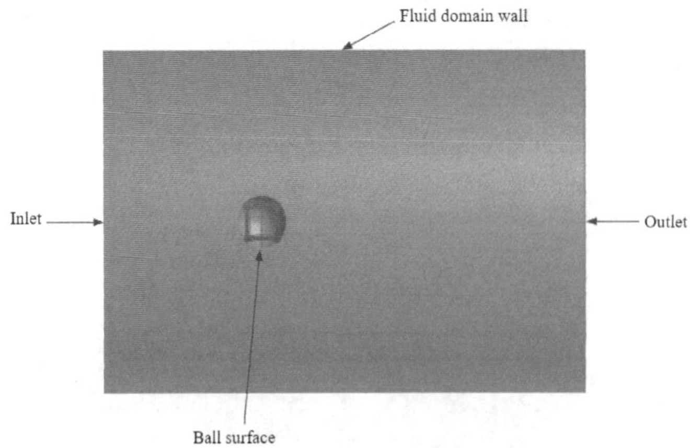

Figure 2. Boundaries

2.3.1. Inlet: This is the region where the fluid (here, air) enters the domain. So, we have given the velocity inlet boundary condition in this region. The velocity defined was normal to the boundary which is along the negative direction of the Z-axis. The turbulence at this boundary was specified as having 1%intensity and with a 0.5 m of length scale. One thing we have to keep in mind is that here, the fluid has motion instead of the ball because the effect caused by either of the aspects is the same.
2.3.2. Ball surface: This is the region where the ball interacts with the flowing fluid. So, for Navier's Stokes equation at this region, we will define the no-slip condition at the wall. For simulating knuckleball conditions, we will be having a stationary wall with the no-slip condition but for curve balls and other spinning ball conditions, we will have to define the no-slip condition with moving walls.
2.3.3. Outlet: In this region pressure boundary condition is chosen over velocity boundary condition keeping in mind that the velocity characteristics of the fluid stream after passing over the ball cannot be predicted. We know that the outlet pressure will be at 1 atm considering the outlet to be in an ambient environment, and all other turbulence parameters are further defined by inlet conditions.
2.3.4. Fluid domain wall: This is the outermost region of fluid flow. Here, we will define a zero-shear boundary condition with a stationary wall. We define zero shears because the effect of this wall shouldn't be affecting the fluid flow and it has been checked that this wall lies at the inviscid region of flow, which means the viscosity effects are almost negligible at this region.

Table 2. Simulation specifications of baseball and tennis ball

	Baseball	**Tennis Ball**
Solver type	Transient Pressure based	Transient Pressure based
Viscous model	Standard k-epsilon	Standard k-epsilon
Inlet air velocity	8 – 39 m/s (along X-axis)	25 m/s (along X-axis)
Outlet pressure	1 atm = 101325 Pa	1 atm = 101325 Pa
Outer wall	Specified shear = 0 Pa	Specified shear = 0 Pa
Ball surface	Moving wall: 15 - 70 rev/s (Spin)	Moving wall: 50 rev/s (Spin)
Roughness coefficient	0.3 - 0.8	0.1
Roughness height	0.762 mm	2 mm

IOP Conf. Series: Materials Science and Engineering 1132 (2021) 012023 doi:10.1088/1757-899X/1132/1/012023

2.4. Solution

After setting up all the boundary conditions now we have to set up all the solver settings which will help us to solve our problem according to the input conditions. Initialization plays an important role in solving the governing equations, so here we have used hybrid initialization for more realistic initialization. It simplifies the process of computing values, and we have used coupled model and second-order upwind schemes for solving momentum, turbulent kinetic energy, turbulent dissipation, and energy equations. These solver settings were cross verified with research paper [9] before employing them.

2.4.1. Case at which the baseball spins in counter clockwise direction about X-axis or this type of pitch is also known as topspin

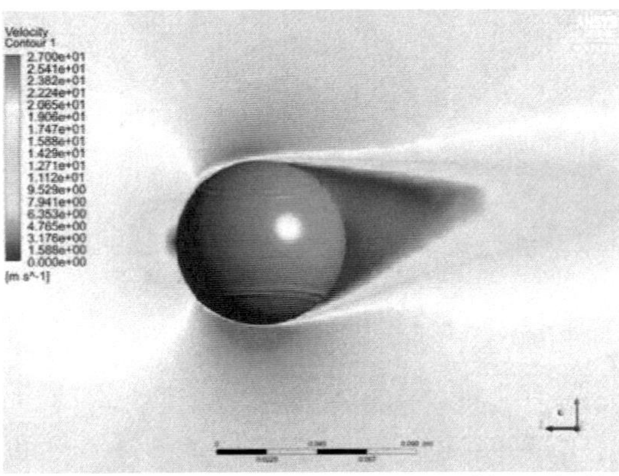

Figure 3. Velocity contour of base ball

Figure 4. Boundary layer near stitches of base ball

In this case the baseball has been numerically simulated for the topspin condition. The boundary layer separation at the ball surface can be visualized in Figure 4. And the wake region and the velocity distribution can be identified from the Figure 3. For, the topspin condition the Figure 4 also shows how the seam and stitches induces a non-uniform boundary layer and which leads to prior flow separation.

2.4.2. Case at which the baseball spins in clockwise direction about X-axis or this type of pitch is also known as backspin

Figure 5. 3D streamline representation

IOP Conf. Series: Materials Science and Engineering 1132 (2021) 012023 doi:10.1088/1757-899X/1132/1/012023

The high-pressure region at the bottom of the ball and low-pressure region at the top of the ball leads to lift. And due to this we can clearly see the lift in the baseball or let's say the positive magnus effect in Figure 5.

2.4.3. Numerical simulation over baseball for different pitches.

The simulation has been performed to replicate a few pitching styles; the extracted results give us better visualization for these types of pitches. The following are few types of common pitches using in the baseball game.

Figure 6a. Front view of a screwball

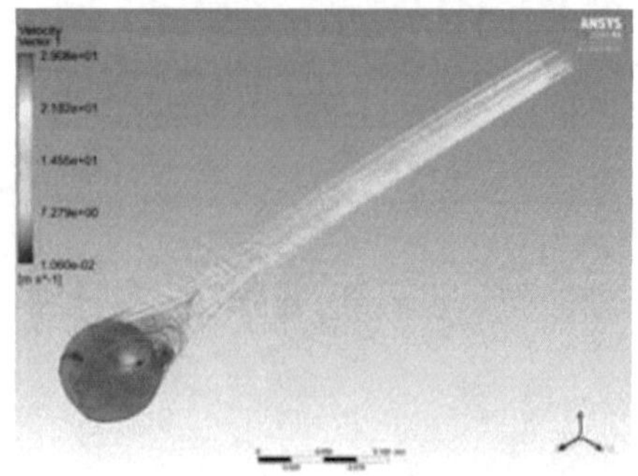

Figure 6b. Isometric view of a screwball

In this case the baseball rotates about two axes, here it rotates about X and Y axes. The Figure 6a and 6b shows the deflection in the baseball path which has been caused due to spinning of the ball.

Figure 7a. Front view a slider

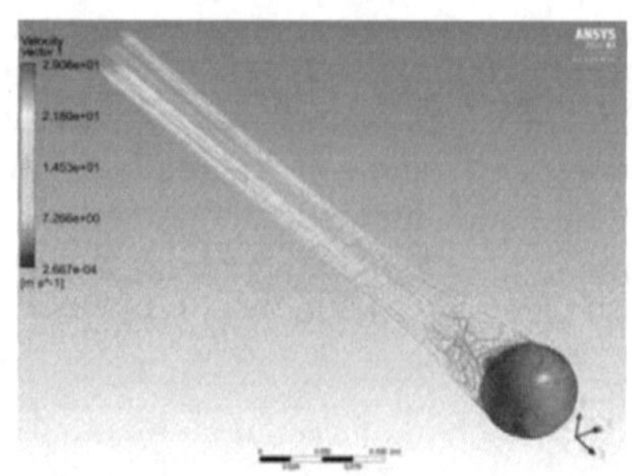

Figure 7b. Isometric view of a slider

In this case the baseball rotates about the Y axis and slides towards the right direction. The Figure 7a and 7b shows the deflection in the baseball path which has been caused due to spinning of the ball in the Y axis.

Figure 8a. Front view of a Sinker:

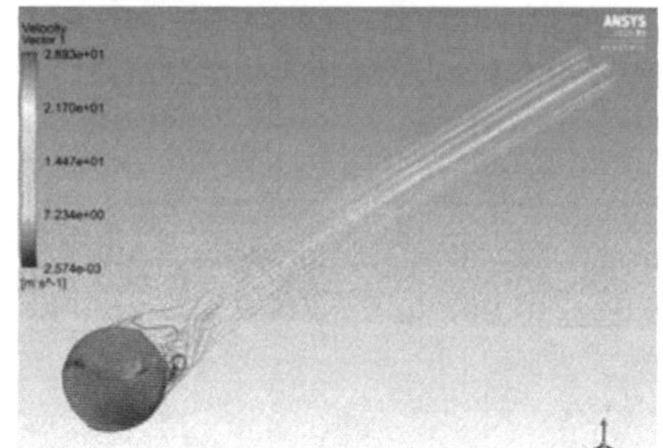

Figure 8b. Isometric view of a sinker

In this case the baseball topspin or say spins about the Y axis and starts to sink downwards. The Figure 8a and 8b shows the deflection in the baseball path which has been caused due to spinning of the ball in X axis.

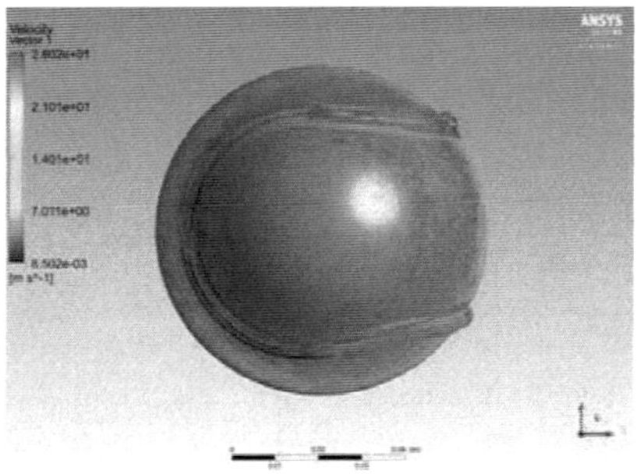

Figure 9a. Front view of a Gyroball

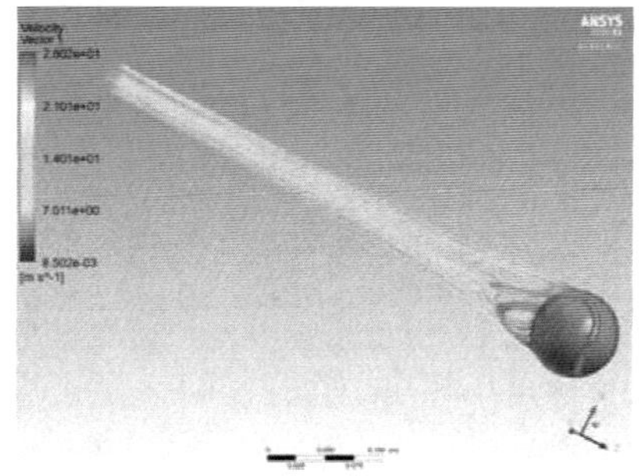

Figure 9b. Isometric view of a Gyroball

In this case the baseball rotates about the Z axis, we can see in Figure 9a and 9b that there is no change in the flight path of the baseball as it travels straight towards the batter. This pitch is unique than other spins as there is almost no lift in the baseball and no magnus effect seen in this scenario.

2.4.4. Tennis ball. Net drag along X axis:
Tennis Ball = - 0.402 N (Spinning case) & Tennis Ball = - 0.72 N (Non-spinning case)
This result conclusively shows that spinning balls have lesser drag when compared to non-spinning balls. The negative sign in these values indicate the direction of the force being applied due to spin effects.

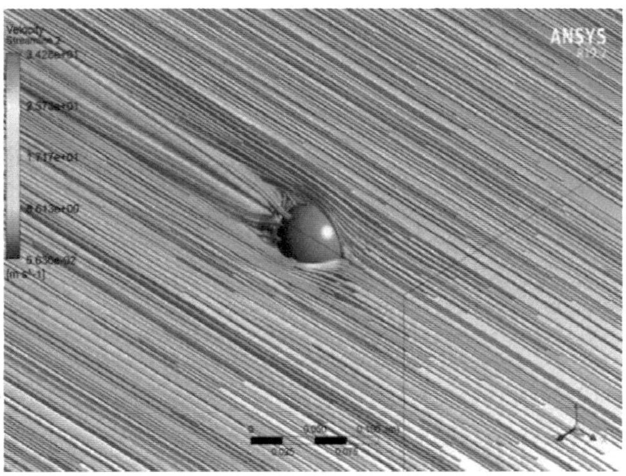

Figure 10. Streamlines in the form of tubes

Figure 10 shows streamlines from the surface of the inlet to determine velocity magnitude and direction at different points across the ball in both X-Y & X-Z plane. Deviation from a straight path explains the presence of Magnus force. This is the velocity contour in the X-Z plane. As we can see the velocity is greater at the bottom than at the top which eventually creates a pressure difference.

IOP Conf. Series: Materials Science and Engineering 1132 (2021) 012023 doi:10.1088/1757-899X/1132/1/012023

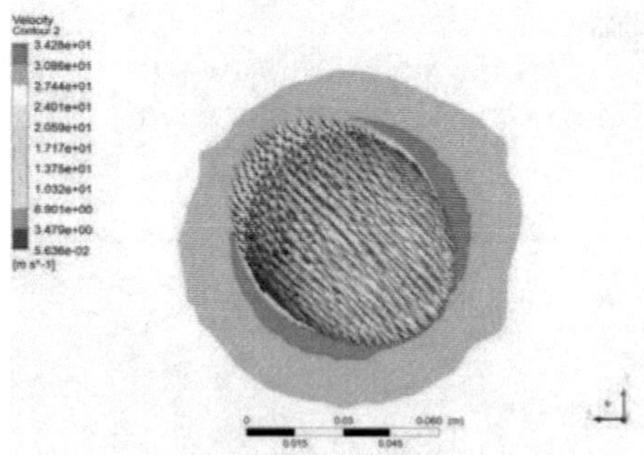

Figure11a. Front view

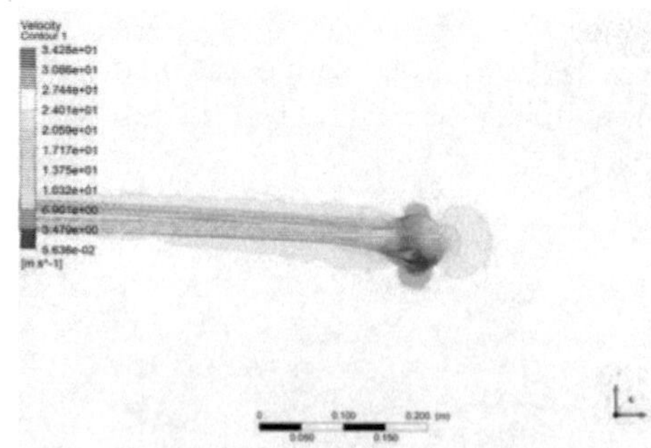

Figure 11b. Side view

Figure 11a depicts how the spin occurs concerning two axes and as it can be observed the velocity is maximum when the vector is pointed towards the flow direction (- X-axis) and minimum when pointed against it. In Figure 11b the wake region seen behind the ball has a green shade indicating the velocity of air to be lower than that of free stream velocity and stagnation region, when seen closely, would approach a color close to dark blue indicating zero flow velocity.

3. Results

3.1. For base ball, few of the simulation results were validated by comparing these results with that of the experimental results which have already been performed by other researchers. The experimental result pictures were taken from the research paper [10].

3.1.1. At flow speed of 21 m/s and the baseball rotates at a counterclockwise direction at a speed of 15 rev/sec

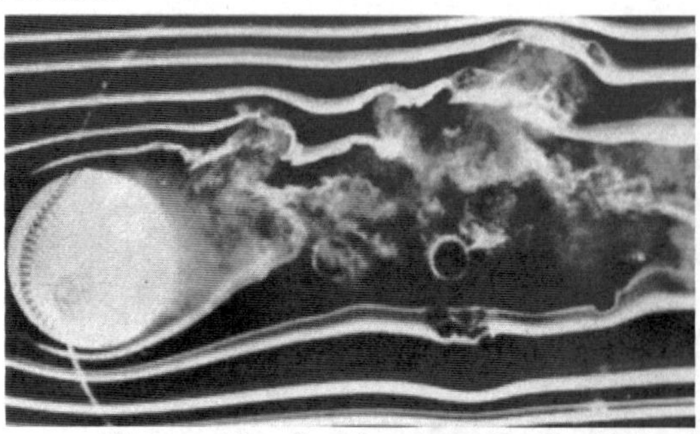

Figure 12a. Smoke photography [10]

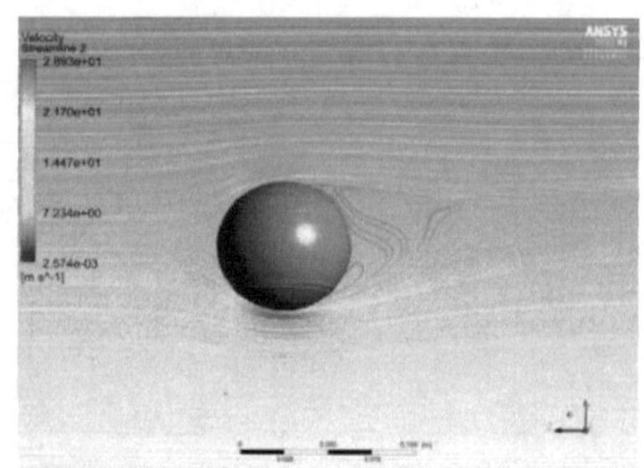

Figure 12b. Numerical simulation result

3.1.2. At flow speed of 21 m/sec and the baseball stays stationary with no spin.

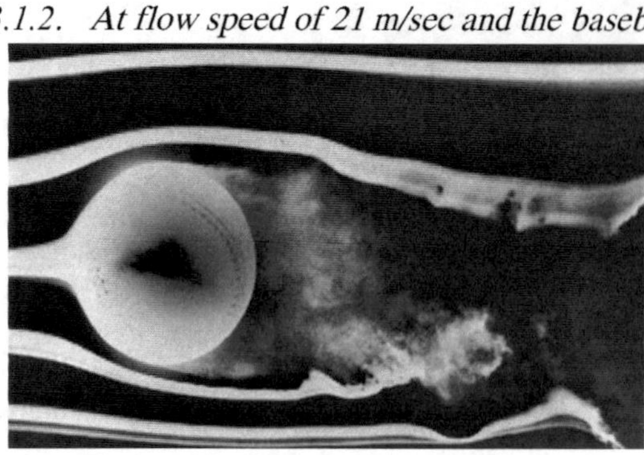

Figure 13a. Smoke photography [10]

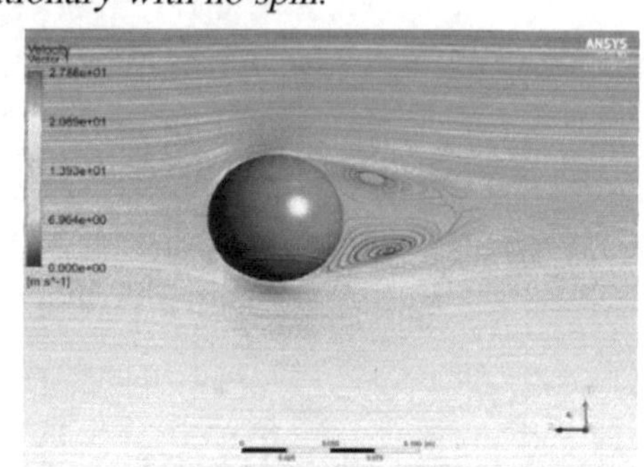

Figure 13b. Numerical simulation result

Table 3. Baseball results comparison

	Cd	Cl
Simulation results	0.317	0.113
Experimental results	0.3 - 0.4	0.1 - 0.2

The experimental results were compared with simulation results for a particular range of spinning speed and free stream velocity. The simulation results were cross verified with [10] experimental results.

3.2. Tennis ball

To verify the results obtained a few reference papers have been chosen based on their experimental validation procedure. In the paper [3], it was estimated that the maximum lift coefficient acting on the ball is approximately 0.54 moving at speed of 25m/s at the maximum spin rate of 3000 rpm. In the paper [5], Tests for the spinning conditions (250–2750 rpm) were conducted at wind speeds of 25 and 50 m/s, and data for the new tennis ball (with sufficient roughness) was observed to have drag coefficients in the range 0.6–0.7. These conclusively prove that the results obtained through simulations are genuine values and that these values can further be used to plot the 3D trajectory of a tennis ball.

Table 4. Tennis ball results comparison

			Chadwick		Haake SJ	
	Cd	Cl	Cd*	Cl*	Cd*	Cl*
Along X axis	0.657	-	-	-	0.6 - 0.7	-
Along Y axis	-	-0.338	-	0.54	-	-
Along Z axis	-	-0.337	-	-	-	-

4. Future prospects

A spinning tennis ball launched in the 1st serve, with a dual axis spin, experiences the following forces: Frontal drag force (along X axis), Axial lift force (spinning along Y axis), Lateral lift force (spinning along Z axis), Gravity (along Y axis)

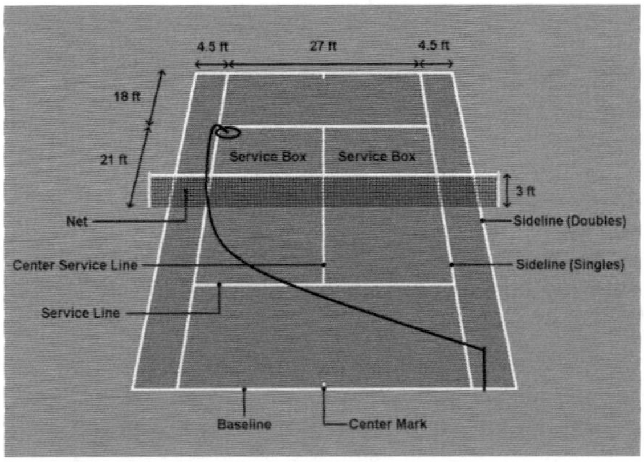

Figure 14. Trajectory representation

IOP Conf. Series: Materials Science and Engineering 1132 (2021) 012023 doi:10.1088/1757-899X/1132/1/012023

The Figure 14 shown is a pictorial representation of the trajectory a tennis ball simulated under dual spin condition. The combined equations of motion of ball and fluid dynamics can be iterated using the Runge-Kutta method. This method of plotting the trajectory by balancing forces along each axis using MATLAB simulations can be used for all types of sports balls. Finally, to induce dual axis spin as a part of experimental validation a gyro setup can be used.

Position and Time:

$$x = \frac{1}{g}\int_{\alpha}^{\tau} \frac{v^2 \cos\tau}{\cos\tau + M^*} d\tau \tag{1}$$

$$y = y_0 - \frac{1}{g}\int_{\alpha}^{\tau} \frac{v^2 \sin\tau}{\cos\tau + M^*} d\tau \tag{2}$$

$$z = -\frac{1}{2} a_z t^2 \tag{3}$$

$$t = -\frac{1}{g}\int_{\alpha}^{\tau} \frac{v}{\cos\tau + M^*} d\tau \tag{4}$$

Force along Y:

$$mv^2/R = mg\cos\tau + M \tag{5}$$

Force along X:

$$m\frac{dv}{dt} = -D - mg\sin\tau \tag{6}$$

Force along Z:

$$mv^2 / R = L;\; a_z = \frac{L}{m} \tag{7}$$

$$\frac{dv}{d\tau} = \frac{\sin\tau + D^*}{\cos\tau + M^*} v \tag{8}$$

$$D^* = C_D(\pi d^2/8mg)\rho v^2 \tag{9}$$

$$M^* = C_L(\pi d^2/8mg)\rho v^2 \tag{10}$$

Where,

$D^* = D/mg$; $M^* = M/mg$;
M = Axial lift force (Magnus force)
L = Lateral lift force (Magnus force);
D = Frontal drag force

IOP Conf. Series: Materials Science and Engineering 1132 (2021) 012023 doi:10.1088/1757-899X/1132/1/012023

5. Conclusion

Numerical simulations on both baseball and tennis balls have been carried out for predicting their aerodynamic characteristics such as lift and drag coefficient. Streamlines and contours of pressure and velocity are extracted for the domain nearest to the ball and have been visualized for required conditions. Finally, these results were compared and validated with the known experimental data which have already been published by other researchers. Future prospects of the study have also been discussed at the end of the study.

References

[1] Antonin S, 1988, *The aerodynamics of tennis balls—The topspin lob,* Am. J. Phys. **56**,138: 10.1119/1.1569

[2] Cengel Y, Simbala JM, *Fluid Mechanics - Fundamental and Applications 4th Edition*, McGraw Hill Education

[3] Chadwick, Stephen George, 2003, *The aerodynamic properties of tennis balls.* PhD thesis, University of Sheffield.

[4] Firoz A, Huy H, Harun C, Aleksandar S,2011, *Aerodynamics of baseball*

[5] Haake SJ, Goodwill SR,Carre MJ, 2007, *A new measure of roughness for defining the aerodynamic performance of sports balls*

[6] James A. Scobie, Carl M. Sangana, Gary D. Locka, 2014, *Flow visualisation experiments on sports balls,* The 2014 conference of the International Sports Engineering Association Procedia Engineering **72** 738 – 743

[7] Kumar, T.R.S., Venugopal, S. Ramakrishnananda, B., Vijay, S., 2020, *Aerodynamic performance estimation of camber morphing airfoils for small unmanned aerial vehicle,* Journal of Aerospace Technology and Management Volume **12**, Issue 1, Article number e1420

[8] Pillai, S. M., Arun Kumar, K., & Ajith Kumar, R. (2019). *Flow around a circular cylinder: Effect of twin splitter plates.* International Journal of Mechanical and Production Engineering Research and Development, ***9*(2)**, 838-47.

[9] Pouya J, P Kreun, Hady M, William WL,2014, *Computational Aerodynamics of Baseball, Soccer Ball and Volleyball*, American Journal of Sports Science **2(5)**:115-121

[10] Rabindra M, 1985, *Sports Ball Aerodynamics* Vol.**17**:151-189

[11] Rajendran, A.K.a, Shobhavathy, M.T.b, Kumar, R.A.a, 2015, *CFD analysis to investigate the effect of vortex generators on a transonic axial flow compressor stage,* ASME 2015 Gas Turbine India Conference, India; Code 119665

[12] Rajesh, S.K.T. Sivakumar, V., Ramakrishnananda, B., Arjhun, A.K., Suriyapandiyan, 2017, *Numerical investigation of two element camber morphing airfoil in low Reynolds number flows,* Journal of Engineering Science and Technology, Volume **12**, Issue 7, Pages 1939-1955

[13] Rajesh, R., & Rakesh, S. G. (2020). *Effect on the drag coefficient of various spiked cylinders during buzz phenomenon subjected to hypersonic flows.* Journal of the Brazilian Society of Mechanical Sciences and Engineering, ***42*(6)**. doi:10.1007/s40430-020-02384-5

[14] Sudhamshu, A.R., Pandey, M.C., Sunil, N., Satish, N.S., Mugundhan, V., Velamati, R.K., 2016, *Numerical study of effect of pitch angle on performance characteristics of a HAWT,* Engineering Science and Technology, an International Journal, Volume **19**, Issue 1, Pages 632-641

Mitigation of shockwave induced boundary layer separation through a secondary recirculation duct in a hypersonic flow

Kiran Joy Irimpan[1] and Anand Raj Hariharan[1*]

[1]Dept of Mechanical Engineering, Adi Shankara Institute of Engineering and Technology Mattoor, Kalady, Ernakulam, India

* anand.me@adishankara.ac.in

Abstract. A secondary recirculation duct which leaks fluid from a high pressure reattachment zone to a low pressure upstream location of a shock wave boundary layer interation zone is discussed in the below article. The mitigation can be attributed to both suction and blowing. Computational analyses of the above problem at a freestream of Mach 8.6 with various jet injection locations were studied for its effectiveness via a commercial software ANSYS Fluent19 and the results are very promising. The jet injection close to the separation point is found to be most effective in delaying separation and reducing the separation bubble.

Keywords: Hypersonic, CFD, Ansys Fluent, SWBLI, Shock Wave

1. Introduction

With humans aiming to travel faster and faster, hypersonic cruises are a reality in the near future. Such cruises would need an air breathing engine like a SCRAM jet engine for efficiency and sustainability. In SCRAM jet engines, the flow compression prior to ignition is brought about by a series of shock waves, hence are susceptible to Shock Wave Induced Boundary Layer Interaction (SWBLI). As hypersonic flows have thick boundary layers, adverse pressure gradient caused by SWBLI can result in large boundary layer separation [1]. Shock wave induced boundary layer separation can result in total pressure losses, flow unsteadiness, choking, high wall temperatures, flow non-uniformities etc. [2-5]. As SWBLI cannot be avoided most research is aimed at reducing its adverse effects like boundary layer separation

As boundary layer separation is attributed to low momentum BL and adverse pressure gradient, its mitigation techniques should involve either energizing the boundary layer, eliminating the low momentum boundary layer or reducing the adverse pressure gradient [6]. Energizing the boundary layer is done through active mass injection into the boundary layer, while elimination is done through suction or passive bleeding [5-10]. Attempts to isolate BL separation via stream wise slots and cavities were also found to be effective BL control techniques [11-12]. The cavity modifies and weakens the incident shock to a spread out lamda shock, resulting in reduced BL thickness, suppression of BL separation and improved pressure recovery across the shock wave. BL control was also achieved through the introduction of 2D bumps which weakens the incident shock wave through a simultaneous compression and acceleration effect [13-14]. Microramps and vortex generators with dimensions less than the thickness of boundary layer height was also found to energize the BL and sometimes trip the BL to turbulence resulting in delayed separation [15-16]. Another concept with a secondary recirculation jet providing both BL bleed and suction simultaneously was proposed by Li Yan *et al.* [17]. Li Yan observed that the suction slots placed downstream of the separation were efficient in BL control and that the suction and blower slot dimensions had less impact on efficiency. Further studies on duct design optimization were carried out by Zhao-bo Du *et al.* [18] and a reduction of separation bubble area of up to 81.43% was reported for a Mach number of 3. The studies on the secondary recirculation jets are still in its infancy and more studies are required, especially experiments.

An attempt to study the effect of secondary recirculation jet in a large separation bubble at a Mach number of 8.6 is attempted through a 2-D ANSYS Fluent laminar solver. A deflector with an angle of 15^0 was used as incident shock generator. High Mach number (8.6) and strong incident shock combination creates a large separation bubble of length: 15.5 cm, due to SWBLI in the present case. A simple secondary recirculation duct with varying lengths were designed with tangential suction and blowing and was studied for its efficacy in BL control. The area ratio between the suction and blower slots was maintained at 2 and this ratio ensured a chocked flow and a sonic fluid injection. The suction slot was always fixed at the peak pressure spot near the reattachment point and the blower slot position was varied.

2. Methodology

2.1 Computational Domain

In the present work, all the numerical simulations have been carried out by using commercial software ANSYS FLUENT 19 R2. The computational domain consists of a simple wedge on the top (to generate an oblique shock) and a slender sharp plate at the bottom as shown in Fig. 1.

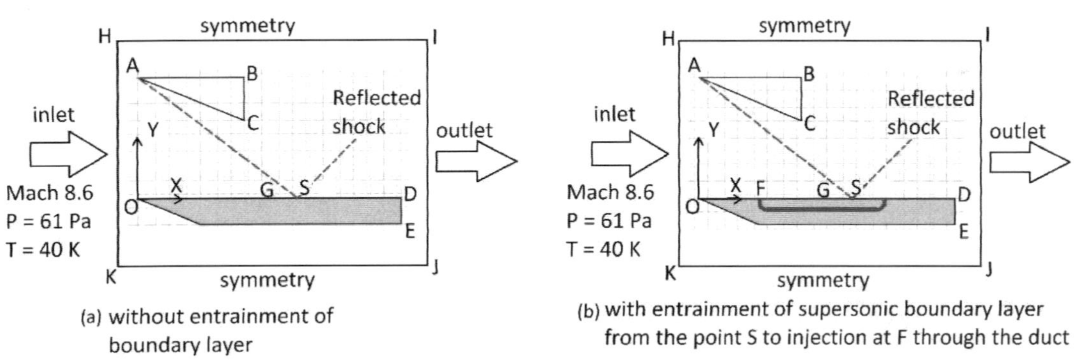

Figure.1. Computational domains used for the simulations. *AG* is the oblique shock generated by the wedge *ABC*. The wedge is oriented such that *AB* is horizontal and < *BAC* is 15°. *S* is the point of attachment of boundary layer. The domain is 32 cm long (KJ) and 22 cm wide (IJ).

The computational domain used for the simulation is 2-D such X and Y axes are pointing towards the stream wise and vertical directions respectively. The origin of the coordinate system is placed at the leading edge of the bottom plate (Marked as O in Fig. 1). The bottom plate is 25 cm long (*OD*) and the wedge is placed at 4.12 cm from the bottom plate (*OD*) 19]. The leading edge of the wedge and the bottom plate was kept in the same plane. To capture the oblique shock from the leading edge of the plate, the inlet of the domain was placed 2 cm ahead. The point of attachment (S) was identified through a simulation without secondary recirculating duct. Slots were provided at the reattachment point and the high pressure fluid was entrained to various upstream locations through a constant are duct with a convergence at the point of fluid injection (F). The suction and injection was done tangentially through curved countered ducts so as to ensure minimal disturbance and pressure loss. In the present numerical simulations, a parametric analysis was carried out by varying the distance of fluid injection from the leading edge (OF). OF was maintained at 30 mm, 60 mm and 90 mm in the three cases reported. The study was carried at a Mach number 8.6 with a static pressure and temperature of 61 Pa and 40 K, respectively.

2.2 Governing equations and boundary conditions.

The flow-field is modelled as 2-D, steady, compressible, and laminar with double precision. At inlet to the domain the freestream conditions are imposed by using *pressure farfield* boundary condition. At outlet all the flow variables are determined from the interior by extrapolation as flow is supersonic. The boundaries on the top and bottom of the computational domain is taken as *symmetry* as flow velocity is predominantly horizontal. The working fluid is air, with variable density (determined using the ideal gas equation of state) and temperature dependent specific heat (8th order piecewise polynomial). The dynamic viscosity and thermal conductivity are modelled using Sutherland's law and kinetic theory (L-J parameters -3.711 ∘A and 78.6 K, for air), respectively. All the wall boundaries are modelled as *isothermal* (T = 300 K) with *no-slip* surfaces. Time marching is carried out by Euler implicit scheme. The convective term is discretized using second order upwinding scheme, whereas central difference scheme is used for the viscous term. During the simulations the inviscid fluxes were calculated using Advective Upstream Splitting Method (AUSM). The resulting system of linear equations are solved iteratively by Gauss Seidel method. The convergence of the iterations is enhanced by algebraic multi-grid method.

2.3 Grid independence and code validation

All the numerical predictions are obtained by steady state calculations until a reasonable level of convergence is attained (residue falls below 10^{-6} based on the initialization using freestream conditions). Thereafter the convergence is assessed using mass, momentum, and energy balances. All the results reported here have a net imbalance below 0.01% of mass flow rate, momentum, and energy.

The grid sensitivity analysis was carried out by grid independence study. The computational domain is discretized using a structured quadrilateral mesh. The details of the mesh used in grid independence study is presented in the Table 1. The mesh used were progressively resolved in the boundary layer and by adaptation based on the velocity gradients in the shock wave. In the Table 1, the grid independent study is shown only for the case without entrainment. Based on this study, a fine mesh was used for the various cases of entrainment as it was found to be more capable of capturing the gradients present in the flow.

Table 1. Details of the mesh used for the grid independence study. Δx_{min} and Δy_{min} are minimum cell distance in X and Y directions, respectively and δ^* is the displacement thickness of the Boundary layer [19].

Level of resolution achieved	Number of cells	$\Delta y_{min}/\delta^*$	$\Delta x_{min}/\delta^*$	$\Delta y_{min}/\delta^*$ at SWBLI
Coarse	584,500	0.005	0.06	0.03
Intermediate	1,564,000	0.002	0.002	0.003
Fine	2,797,870	0.0001	0.00015	0.0001

The validation of the numerical simulation is carried out by comparing the static pressure distribution over the flat plate (without entrainment) with that of experimental data reported by Maitri *et al.,* [19]. The results obtained (Fig. 2) clearly shows that the trends of static pressure distribution have been captured by all the mesh schemes used. Even though course mesh as well as fine mesh scheme seems to agree well with the experimental predictions, the static pressure values are over predicted for the former.

As the over prediction of the peak pressure was more than 50 % for the coarse mesh, the fine mesh was used for other cases.

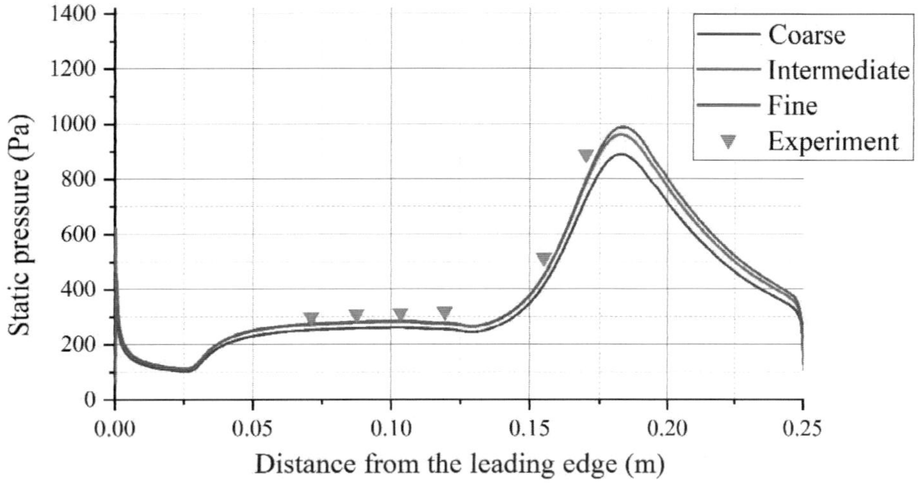

Figure 2. Comparison of numerical predictions with the experiment [19] for the static pressure distribution along the plate surface OD.

3. Results and discussion

3.1 Contours of Mach number

The distribution of Mach number across the flow-field with and without entrainment of flow is shown in the Fig. 3. The Mach contour is presented as the separated BL is a shear layer with velocity gradient. As the separated BL has low momentum, it's visible as dark blue shades in Fig. 3. The points *a, b* and *c* in Fig. 3 are the locations of separation, top of the separation bubble and reattachment of BL. The points a and c and are identified based on the distribution of skin-friction drag coefficient whereas point *b* is identified based on the change in direction of the shear layer towards the bottom wall. Mach contours in Fig. 3 also validates chocking in the duct and the sonic jet at the blower slot. The plots also verify that with tangential sonic blowing, there is minimal interaction with the outer flow.

3.2 Point of boundary layer separation

The determination of separation point of the boundary layer is based on the skin friction drag coefficient distribution along the length of the flat plate. The point of separation of BL on the flat plate is presented in Fig 4. From the figure, its observed that BL separation is delayed with the introduction of secondary recirculation duct and that the delay is substantial when the blower slots/ injectors are placed farther upstream on the flat plate. These observations can be attributed to the delay in boundary layer separation brought about by BL energization. As the distance of the blower slots from the leading edge (OF) increases, the low momentum BL has to traverse larger distance against the adverse pressure gradient and hence results in earlier separation.

3.3 Point of attachment and Length of Boundary layer separation

The point of attachment of boundary layer from the leading edge of the bottom plate (O) was found to be almost similar for all the cases (the difference is less than 2 %). The reason for almost similar location for the boundary layer attachment is due to the fact that the separated boundary layer is turned back to the bottom plate by the shock wave, generated by a common wedge. The minor differences in the point of attachment between the cases, could be due to the modification of the incident shock wave by the separation shock. The length of BL separation is calculated as the difference between the point of separation and point of attachment. The length of separation of BL with and without entrainment is presented in Fig. 5.

3.4 Area of separated bubble

For the calculation of the area, the separation bubble is approximated as a triangle with vertices at point of separation (a), point of deflection of shear layer by shock wave (b) and point of attachment (c) [19]. This approximation is shown in the Fig. 6. Percentage of separation bubble area reduction is presented in Fig.7. Results clearly shows that early energization of incoming boundary layer significantly reduces the size of the separation bubble. Separation bubble area reduction of up to 75% was observed when the blower slots were fixed at 30 mm from the leading edge.

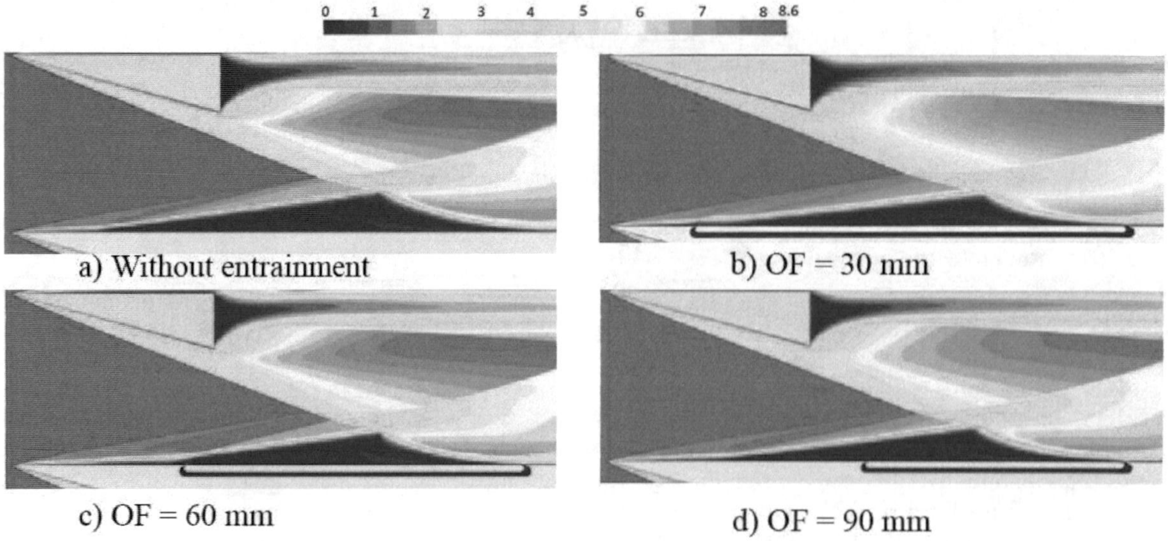

Figure 3. Contours of Mach number for the supersonic flow over flat plate with and without secondary recirculation jets.

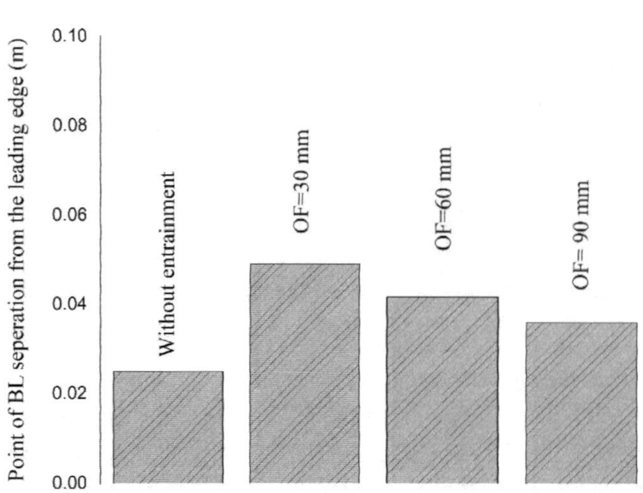

Figure 4. The point of separation of boundary layer on the bottom plate from the leading edge.

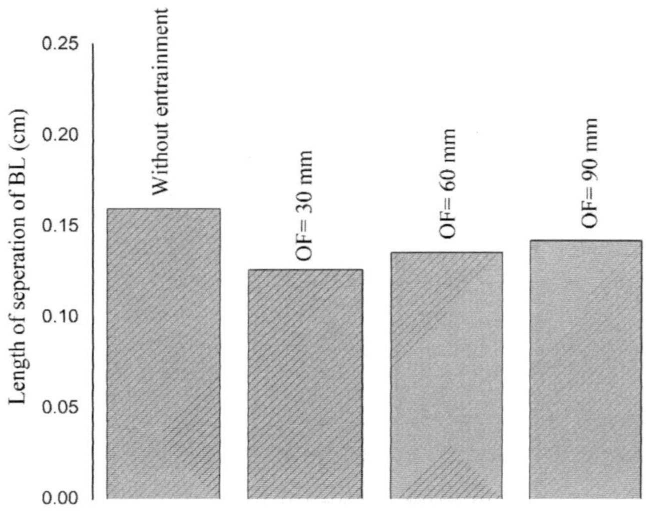

Figure 5. The length of separation of boundary layer on the bottom plate.

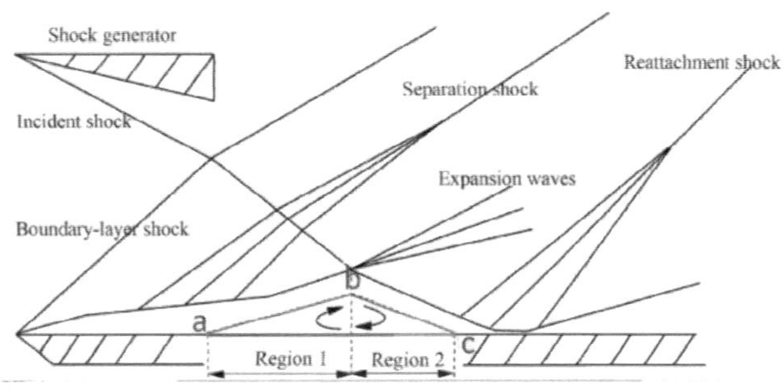

Figure 6. The approximation of a separation bubble as a triangle with vertices at the point of separation of boundary layer (a), point of deflection of shear layer by shock wave (b) and point of attachment (c) of boundary layer.

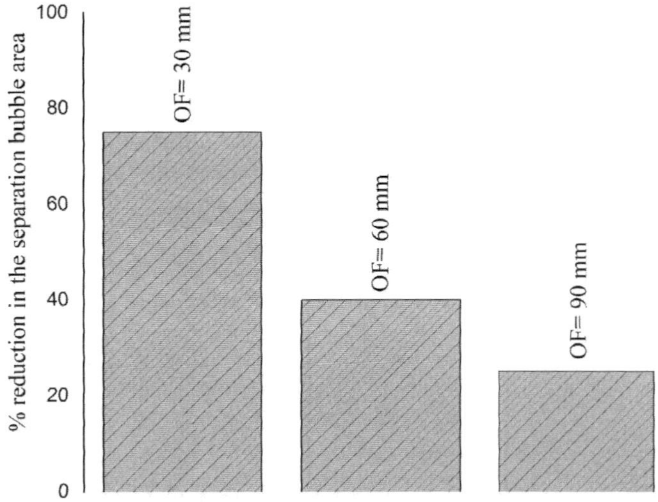

Figure 7. The percentage reduction in the separation bubble area with the introduction of secondary recirculation jets for various blower slot locations.

4. Conclusions

The present work focuses on a new passive hypersonic BL control technique using both suction and blowing. The technique involves a secondary recirculation duct connecting the high pressure zone (near reattachment) to the upstream low pressure zone. It has been found that the point of separation of boundary layer is delayed and the size of the separation bubble is significantly reduced by positioning the blower slots of the secondary recirculating ducts closer to the leading edge. Separation bubble area reduction of up to 75% was observed in a certain case when the blower slots were closer to the leading edge. The present concept can be applied to complex hypersonic intake geometries to suppress SWBLI and reduce losses.

References

[1] Anderson, J. D. (1989). Hypersonic and high temperature gas dynamics. New York: McGraw-Hill.
[2] Babinsky, H., and Harvey, J. K., Shock Wave—Boundary Layer Interactions, Cambridge University Press, Cambridge, UK, 2011;5–86.
[3] Kim, H. D., Matsuo, K., and Setoguchi, T., Investigation on Onset of Shock-Induced Separation, Shock Waves, 1996; 6(5),275–286.
[4] Murthy, S. N. B., and Curran, E. T., Scramjet Propulsion, Progress in Astronautics and Aeronautics, AIAA, Inc., Reston, VA, 2001;447–504.
[5] Lee, C. B., and Wang, S., Study of the Shock Motion in a Hypersonic Shock System/Turbulent Boundary Layer Interaction, Experiments in Fluids, 1995; 19(3): 143–149.
[6] Babinsky, H., and Ogawa, H. SBLI Control for Wings and Inlets, Shock Waves, 2008; 18(2): 89–96.
[7] Krogmann, P., Stanewsky, E., and Thiede, P., Effects of Suction on Shock/Boundary-Layer Interaction and Shock-Induced Separation, J. Aircr., 1985; 22(1): 37–42.
[9] Harloff, G. J., and Smith, G. E., Supersonic-Inlet Boundary-Layer Bleed Flow, AIAA J. 1996; 34(4): 778–785.
[10] A. Ruban,V. Menezes, S.Balasubramanian and K. Srinivasan, Shock–Boundary Layer-Interaction Control Through Recirculation in a Hypersonic Inlet, Journal of Fluids, 2020,142 (114502).
[11] Smith, A. N., Babinsky, H., Fulker, J. L., and Ashill, P. R., Shock Wave/Boundary-Layer Interaction Control Using Streamwise Slots in Transonic Flows, Journal of Aircraft, 2004; 41(3):540–546.
[12] A. Ruban,V. Menezes and S.Balasubramanian, Boundary-Layer Control for Effective Hypersonic Intake Journal of propulsion and power, 2018;34(6).;1611-1614
[13] Zhang, Y., Tan, H. J., Tian, F. C., and Zhuang, Y., Control of Incident Shock/Boundary-Layer Interaction by a Two-Dimensional Bump, AIAA Journal, 2014; 52(4): 767–776.
[14] A. Chitharenjan and V. Menezes, Hypersonic Incident Shock-Boundary Layer Interaction Control by Wavy Patches, Journal of Aerospace Engineering, 2021; 34 (1).
[15] McCormick, D. C., Shock/Boundary-Layer Interaction Control with Vortex Generators and Passive Cavity, AIAA Journal, 1993; 31 (1): 91–96.
[16] Babinsky, H., Li, Y., and Ford, C. W. P., Microramp Control of Supersonic Oblique Shock-Wave/Boundary-Layer Interactions, AIAA Journal, 2009; 47 (3):668–675.
[17] Li Yan, Han Wu, Wei Huang, Shi-bin Li, Jun Liu, Shock wave/turbulence boundary layer interaction control with the secondary recirculation jet in a supersonic flow, Acta Astronautica, 2020; 173:131-138.
[18] Zhao-bo Du, Chi-bing Shen, Yang Shen, Wei Huang, Li Yan, Design exploration on the shock wave/turbulence boundary layer control induced by the secondary recirculation jet, Acta Astronautica,2021;181:468-481
[19] M. Kshetrimayum, K. J. Irimpan and V. Menezes, Skin friction estimation on a surface under shock-boundary layer interaction, Sādhanā 2020; 45 (195).

IOP Conf. Series: Materials Science and Engineering 1132 (2021) 012025 doi:10.1088/1757-899X/1132/1/012025

Productivity Improvement – A Case Study of Hindustan Polymer Products

Akhil S K[1*] and Narendran A K[1]

[1]Dept of Mechanical engineering, Adi Shankara Institute of Engineering and Technology Mattoor, Kalady, Ernakulam, India

*akhilsankark050394@gmail.com

Abstract. In India, a large fraction of industries are labor intensive in nature which requires a large amount of labor for work and the productivity depends on the performance of workers that can vary according to many facts. By proper implementation of various techniques and methods, productivity can be improved and this article suggests a methodology that can be implemented in any labor intensive industry to improve productivity, carefully examining various aspects and making the necessary adjustments. The methodology begins with the analysis of the work environment and work study in a labor intensive industry and the results are systematically analyzed using appropriate tools such as 5S, Pareto Analysis etc. to provide better alternatives and solutions. A significant percentage of time is saved after the implementation of solutions and standardization of the operations in the company and these recommendations may be extended to other forms of industry by making suitable changes.

1. Introduction

Labor intensive industries are very common in developing countries like India. The productivity and output of these industries are strongly related to the performance of the human resources. An industry may be fully automated, semi-automated or fully labour intensive and the later demands maximum skill from the labours. Major concern related to these types of industries is that human resources will not be as efficient as machines in automated industries. Because unlike machines, humans get tired and experiences fatigue and there may be unnecessary motions and activities associated with the work, reducing the efficiency of the work done. Improving the performance of human resources is a prime concern in improving productivity, whichever the industry considered.

In the past, considerable amount of research has been performed in the area of improving productivity and efficiency in industries. Generally small scale industries are labor intensive industries and many authors investigated different methods to improve productivity and efficiency in these industries. Venugopal discussed the needs of TPM implementation in Indian SMEs and its effects on productivity, quality of product, culture of the organization, maintenance activity etc. He concluded that the process can be improved based on method study, work procedure and proper utilization of machine and material and will improve the current process by reducing the number of workstations, transportations, combining the operations and reducing the workers fatigue [1]. Azid et al. (2020) proposed a process improvement model that is enhanced by integrating motion time study and quality tools. The model suggested that by integrating two approaches to investigating motion time study and quality tools, the main causes of the bottleneck problem and the solution to production shortages can be identified and analyzed at the same time. A case study carried out showed that production improved significantly by ten percent thanks to the reduction of times and processes to complete assembly work [2]. Ricondo Iriondo et al. (2016) examined how the concepts of takt time and flow can act as levers for the introduction of lean production into the SME machine tool manufacturing sector.. The implementation of a system based on takt time in a pilot company confirmed that production efficiency, delivery time reliability and productivity have been increased [3]. Gunasekaran et al. (2000) analyzed problems of productivity and quality in a small company and the study indicated that the manufacturing techniques typically used by large enterprises can be implemented in a small scale industry also for productivity and quality improvements. The study also suggested that non- value adding activities can be considerably reduced by making minor improvements in SMEs [4].

Braglia et al. (2016) researched further about Single minute exchange of die (SMED). The proposed proposal saw a significant reduction of wasted time (in certain cases up to 87% of the original set-up time) with limited costs and very short return periods [5]. Braglia et al. (2017) also presents a new methodological approach to effectively integrate the 5 – Why Analysis - a root cause analysis tool to identify causes of problems - in a more general context to reduce installation time in general SMED programs [6]. Canan Aglan et al. (2019) suggested a four-step design method, from the parts line side presentation through in-plant vehicle routing and a continuous improvement scheme of the proposed project was developed [7]. Aliabouni et al. (2007) conducted a study about factors affecting employee productivity in a UAE construction industry and the results of the investigation of significance revealed the 14 factors that influence productivity, out of which the seven most important factors as work schedules, process control, salaries, supervision, group dynamics, climate and equipment [8]. A study by Karthick et al. (2016) presented the application of lean tools in minimizing the lead time in gear manufacturing process. The study showed that implementation of Kaizen, VSM in the industry reduced the time required for different operations and hence increasing the production rate [9].

Majority of researches done in the past contributes to improving productivity by applying various scientific tools and methods, but were implemented and studied based on a particular type of industry. A general methodology which can be applied commonly to any industry is not proposed in these works. The present work deals with proposing a methodology for the enhancement of productivity in a labour intensive industry. The approach involves systematic implementation of a series of scientific techniques to investigate the issues that hinders productivity and proposing appropriate recommendations based on industry to improve productivity. A case study is conducted at a small scale industry which is labor intensive to validate the effectiveness of the methodology.

The industry selected for the study is Hindustan Polymer Products, Kottayam, who produces 'roto gravure and flexo printed multilayer plastic pouches, sheets, covers and rolls'. The average production in the firm is about 145MT per annum with a workforce of 25 people. From the pilot study, it was evident that, the firm was facing serious productivity issues. Lack of performance standards for measuring productivity, non-streamlined methods and lack of a proper organization of resources in the firm contributed to the low productivity. In this analysis the firm is examined to investigate how change can be implemented to enhance productivity in a labor intensive industry. The productivity improvement can be achieved only when there is a real change in the mind-set of people at work in the way they look at the global business, the technology and the organizations. This change can be made only by proper implementation and utilization of technology and human resource development practices. The changes must become a company-wide cultural change and not just a process change for the productivity gains to last.

The main objective of this paper is to propose ta methodology to improve the productivity in a labor-intensive industry and investigate the results to understand the applicability of the methodology in similar industries. The framework will be used to see, how modifications can be made to improve the output and the objective is to suggest methods for improving the productivity by the direct involvement of employees. Scientific tools like SMED, 5S, Kaizen etc. is used to identify the major issues in the firm and appropriate alternatives or improvements are recommended based on the feasibility and effectiveness by considering the views of the human resources in the firm.

2. Methodology

The performance of labour intensive industries is highly dependent on the efficiency of the human resources. All the industries require human resources, but the performance of human resources mostly affects semi-automated and fully labour intensive industries and the methodology that is proposed here can be implemented in these industries.

Low productivity and efficiency of workers has been a major problem in Hindustan Polymer Products and this constrained them from meeting the demands and on time product deliveries. From the pilot study it is found that they lacked a worker performance measurement standard and the operations are not streamlined. The methodology proposed as shown in Figure 1 aims to achieve an improvement in productivity and efficiency of workers with minimum costs.

A key factor that is associated with the performance of workers is the availability of a well-defined performance measurements standard based on scientific techniques. The first step in improving productivity is to analyse whether there is a well-defined system for measuring the performance of the workers. The standards used to measure performance of the workers can be based on production cycle time or quantity of products produced in a shift and should be selected based on the type of industry.

For Hindustan Polymer products there are no standards for measuring performance of workers. Though work study, the standards is fixed for future references. Work study involves detailed analysis of all the activities of the human resources and helps to identify the drawbacks and scope for improvement. Information is also collected through onsite observations and brain storming with workers and management. A sophisticated questionnaire survey for worker is also used to investigate problems which were not identified from onsite observations.

To identify the root causes of the problem, pareto analysis is used and a cause and effect diagram is prepared. The Pareto analysis helps to categorise the problems into major causes and minor causes which affect productivity.

In this paper, implementation of SMED, 5S concepts and Kaizen tools is applied to improve the efficiency of workers which in turn contributes to the productivity. Implementation of SMED reduces

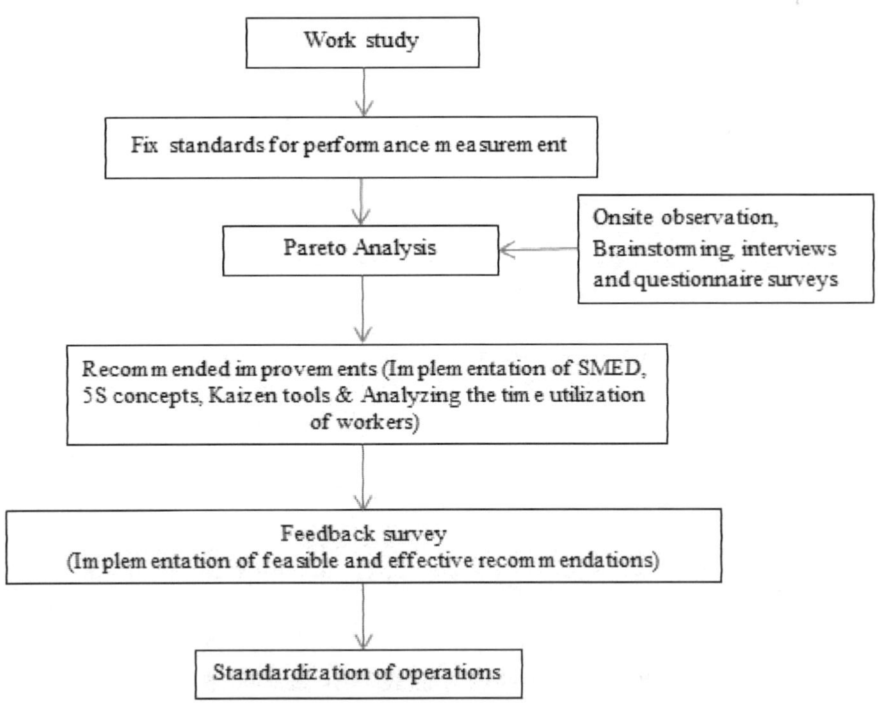

Figure 1. Methodology

the changeover time or setup times of machines. Using SMED operations related to machines is highly streamlined by eliminating unnecessary activities. 5S is an important technique which can enhance the smoothness of workflow in the firm and helps to organize a work space for efficiency and effectiveness by identifying and storing the items used, maintaining the area and items, and sustaining the new order. Kaizen tools like Kanban, 5W+1H is also implemented to improve the productivity. By implementing these concepts, performance of the workers and productivity is improved without any additional cost. Also allocation of workers and machine time utilization are studied to analyse possibility of work rotation and combination of workers in machines to maximize the time utilization.

The most notable limitation of this study is that it is based on a single small industry investigation. The most interesting results could be obtained if more similar industries were considered. Factors such as working conditions, norms, culture, and social values, as well as the level of development of the city in question, have an individual impact on the productivity and efficiency of workers in the enterprises. Another important limitation of the study was the time restriction. The study was conducted in three months (January 2019 to March 2019). As a result, the seasonal factors that could affect the production of the company were not taken into account. Future research may include data from different time periods (seasons) to analyze the variations in productivity and efficiency.

3. Results and discussion

3.1 Work study

Work study is conducted for the major product manufacturing activities. Each operation is broken down to the simplest elements to study the methods and time requirements (Figure 2 and Figure 3). After analyzing the work study results, the time standards for performance measurement for workers in each machine are fixed. Flexo printed covers and Roto gravure printed covers are the main revenue earning products in the firm and hence work study is mainly conducted for operations related to producing these products. In these products 1kg variants, which contributes to the major demand, are considered for work study.

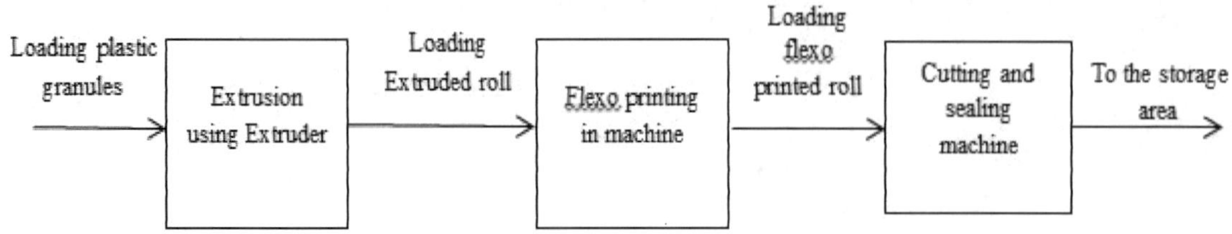

Figure 2. Flexo printing process

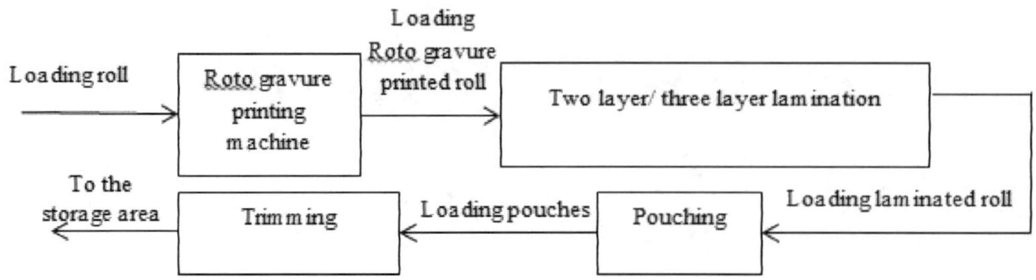

Figure 3. Roto gravure printing process

Cycle times corresponding to cycles with and without machine setup activities are found from the work study and In the cases were setup activities are considered, Roto gravure printing machine showed significant increase in cycle time (Table 1) [10].

Table 1. Time study details of 1kg products

No	Machine type	CWOT in min	CWT in min
1	Extruder	45.8	109.60
2	Flexo printer	65.1	99.90
3	Cutting and sealing machine	83.2	93.80
4	Roto gravure printing machine	74.1	216.70
5	Two layer lamination machine	115.4	162.60
6	Three layer lamination machine	220.3	283.50
7	Trimming machine	59.7	65.20
8	Pouching machine	63.6	66.90

^{CWOT} Cycle time without setup activities; ^{CWT} Cycle time with setup activities.

3.2 Implementation of SMED

From the work study it is evident that setup activities contribute a significant percentage of cycle time for each product manufacturing line. The setup activities are done by the human resources in the firm which has the potential for improvement. The SMED methodology divides the activities into internal and external activities. The prime objective of this approach is to convert internal activities into external so that, these activities can be completed without affecting the machine. This improves the time requirement for setup activities in machines and helps to eliminate unnecessary activities by the operators. The first step to implement the SMED was to divide all the activities of the company into internal and external. The second step was to analyze the internal activities to see if any of them could be transformed into external activities without affecting the total workload. The final step was to convert these internal activities into external activities. Many operations like transporting rolls into and from the storage, Acquisition of tools and materials, Arrangements of inventory materials etc. are converted to external activities which helped to reduce the total cycle time for producing the products. These activities were previously internal activities, that is, workers only performed these activities when the machines were off. Proper allocation of workforce facilitated activities while the machine was in operation. This was possible because employees had downtime while the machines were running. The results of implementation of SMED is shown in Table 2.

Table 2. Setup time saved by implementation of SMED

No	Machine type	TSTBS in min	TSTAS in min	TSTS in %
1	Flexo printing process	109.2	79	27.66%
2	Two layer laminated roto gravure printing process	198.6	106	46.63%
3	Threelayer laminated roto gravure printing process	214.6	117	45.48%

^{TSTBS} Total Setup time before implementation of SMED; ^{TSTAS} Total Setup time after implementation of SMED; ^{TSTS} Total setup time saved.

3.3 Implementation of 5S Concept and Kaizen tools

By having a systematically organized facility, a company increases the likelihood that production will occur exactly as it should. 5S concept is employed in a manufacturing firm for the purpose organizing the work environment to improve the workflow and efficiency. The work environment in the firm is not properly organized, which is evident from the work study. From the information collected by various sources like onsite observation, the major problems identified in the firm are:

1. Tools and materials were not properly arranged and hence a considerable time is wasted for finding the appropriate tools.
2. Storage area of inventory materials are at a considerable distance from the production machines, which contributes to long transportation time.
3. The pathways through which materials travel is not clear and smooth.
4. The inventory flow is not streamlined in the firm
5. A considerable time is taken for insertion and removal of inventory to and from the machines.
6. The tools, raw material, die etc. are not labelled and inventory materials are not tagged
7. No boards for showing the status production near the machines.
8. Difficult to distinguish between the WIP inventory and finished goods
9. There is inventory build-up near some machines and unavailability of inventory in some other machines at various times.

5S concept is then implemented in the firm. The tools and materials required for each of the machines are arranged near the machines, so that operator can easily acquire them. The storage area is rearranged to enable easy storage and retrieval of inventory and to reduce the distance from the working machines. Results of implementing Kanban and 5W+1H are reflected in the final standard time measurements. A significant time difference is seen in the cases of transportation time (Figure 4). All the inventories are tagged with relevant details using cards attached to them and all the tools, die etc. are properly labeled to eliminate the uncertainty in operators mind. Inventories are rearranged in storage area separating WIP materials and finished good using multi coloured tags.

3.4 Time utilization of man and machine

From Table 3, it is evident that operators in the Lamination machines and trimming machines are idle for a large part of the cycle time and they can be assigned to cutting machines and pouching machines. From the study it is evident that cutting and sealing machines and pouching machines are part of bottlenecks to the maximum output. There are two machines each for cutting and sealing and pouching processes (according to the product size requirement), but only a single operator is controlling both of them. This reduces the output of the firm due to the lack of continuous outputs from both machines. The workers from the Lamination machines and trimming machines can be assigned to this during their idle time, if proper trainings are given. With proper planning, the use of combination of workers in machines can improve the output of these machines up to 150%.

3.5 Questionnaire survey results

A questionnaire survey was conducted among the workers and management of the Hindustan Polymer Products. A total of 25 people responded to the survey. Since the sample size was less than 30, the survey

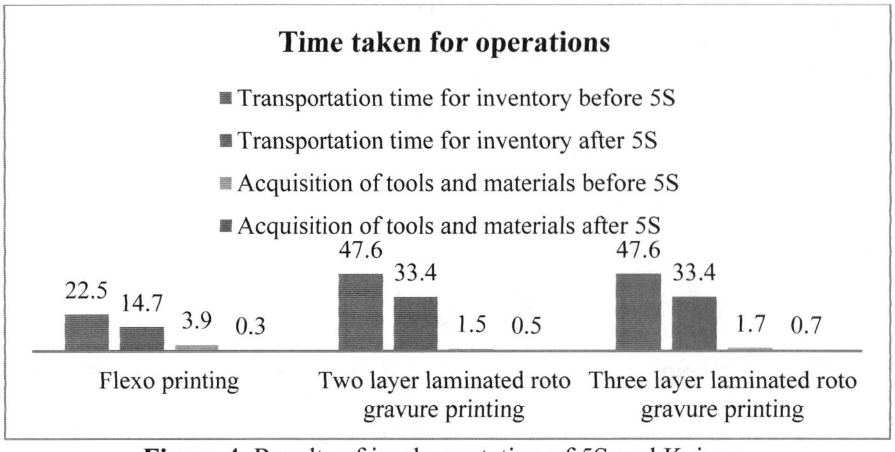

Figure 4. Results of implementation of 5S and Kaizen

Table 3. Time utilization in present situation for 1kg products (with setup)

No	Machine type	N	MWT in min	MIT in min	MMWT in min	MMIT in min	CT in min
1	Extruder	2	69.6	40	40	69.6	109.60
2	Flexo printer	2	43.9	56	56	43.9	99.90
3	Cutting and sealing machine	1	93.8	0	73	20.8	93.80
4	Roto gravure printing machine	3	151.7	65	65	151.7	216.70
5	Two layer lamination machine	2	64.6	98	98	64.6	162.60
6	Three layer lamination machine	2	87.5	196	196	87.5	283.50
7	Trimming machine	1	21.2	44	44	21.2	65.20
8	Pouching machine	1	66.9	0	49	17.9	66.90

^{N} No of operators; ^{MWT} Man working time in min.; ^{MIT} Man idle time in min.; ^{MMWT} Machine working time in min. ^{MMIT} Machine idle time in min.; ^{CT} cycle time in min.

is only considered as quick response survey, to identify major problems which were not identified through other methods and no statistical analysis was carried out. Major problems identified from the questionnaire survey are:

- Lack of backup generators for electrical power failure
- Lack of performance monitoring system
- Inefficient use of IT system in organization
- Organization doesn't take effort to provide training
- Lack of goals and standards
- Inefficient transportation facility for inventory
- No regular supervisions conducted
- Unsatisfactory arrangement of plant layout
- Improper arrangement of tools and equipment
- Lack of well-defined storage areas

Most number of negative feedbacks were obtained for Backup generators for electrical power failure, Performance monitoring system, IT system in organization, Organization take effort to provide training, Setting of goals and targets, Transportation facility for inventory and Regular supervisions. All the people who attended the survey responded poor or average for these items.

3.6 Pareto analysis and cause and effect diagram

The information collected from the work study and interview with is used for Pareto analysis. The output of Pareto Analysis is shown in Figure 5, Figure 6, Figure 7 and Figure 8.

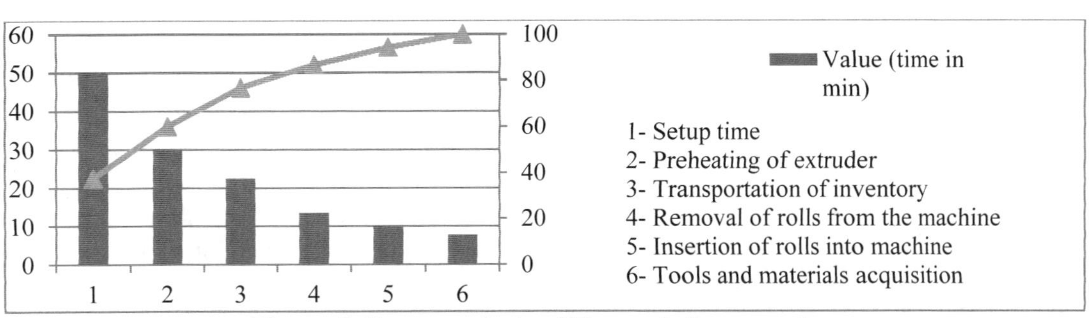

Figure 5. Pareto analysis of Flexo printing operations

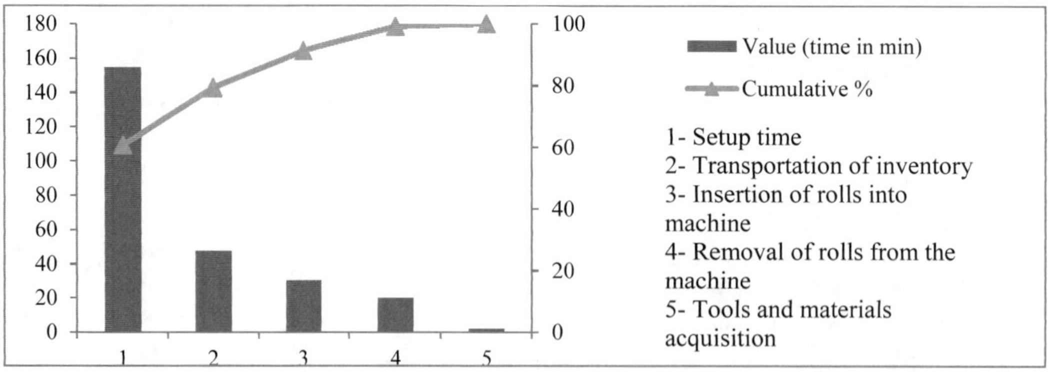

Figure 6. Pareto analysis of Two layer laminated roto gravure printing operations

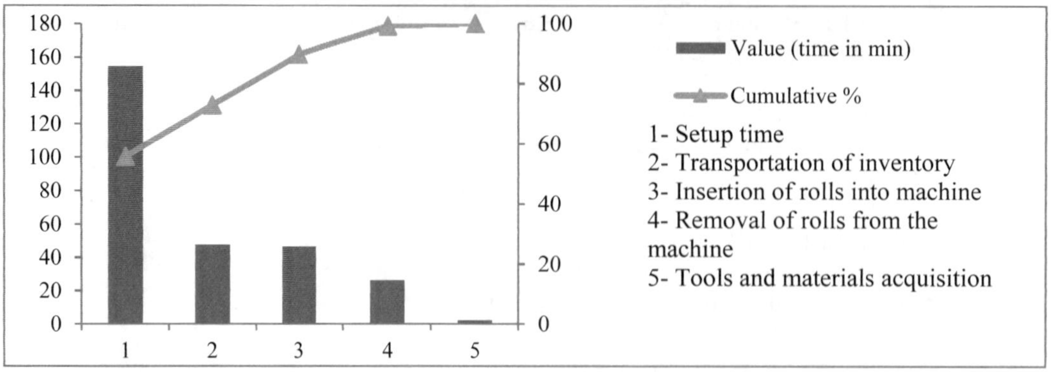

Figure 7. Pareto analysis of Three layer laminated roto gravure printing operations

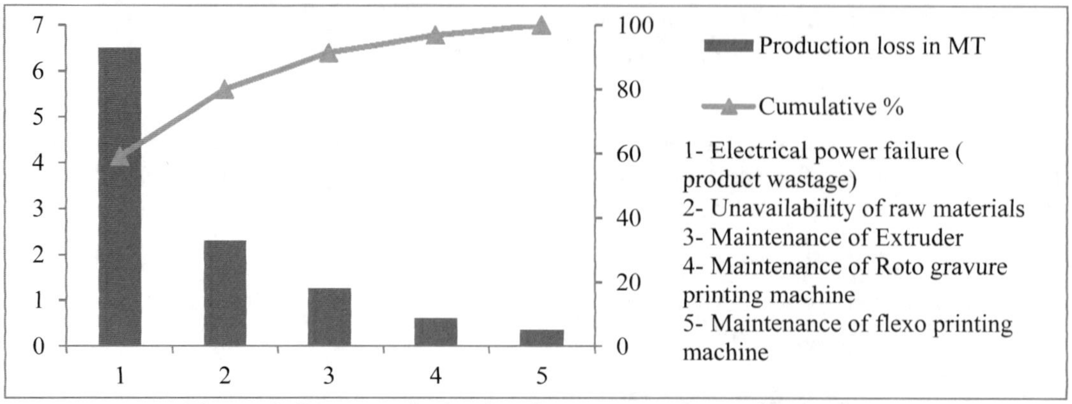

Figure 8. Pareto analysis of various reasons for production loss

From the Pareto Analysis (Figure 5, Figure 6 and Figure 7), it is evident that setup time contributes to the majority of cycle time when the machines are not operating and the same is applicable for flexo printing operations and two- and three-layered roto gravure printing operations. In the case of flexo printing, preheating time required for Extruder is also a major constraint in reducing cycle time. Transportation time for inventory materials in the firm also calls for important attentions. Electrical power failure is also another major cause of production loss (Figure 8). The cause-and-effect diagram prepared from the quantitative aspects of Pareto analysis and qualitative aspects of investigation of work environment is shown in Figure 9.

IOP Conf. Series: Materials Science and Engineering 1132 (2021) 012025 doi:10.1088/1757-899X/1132/1/012025

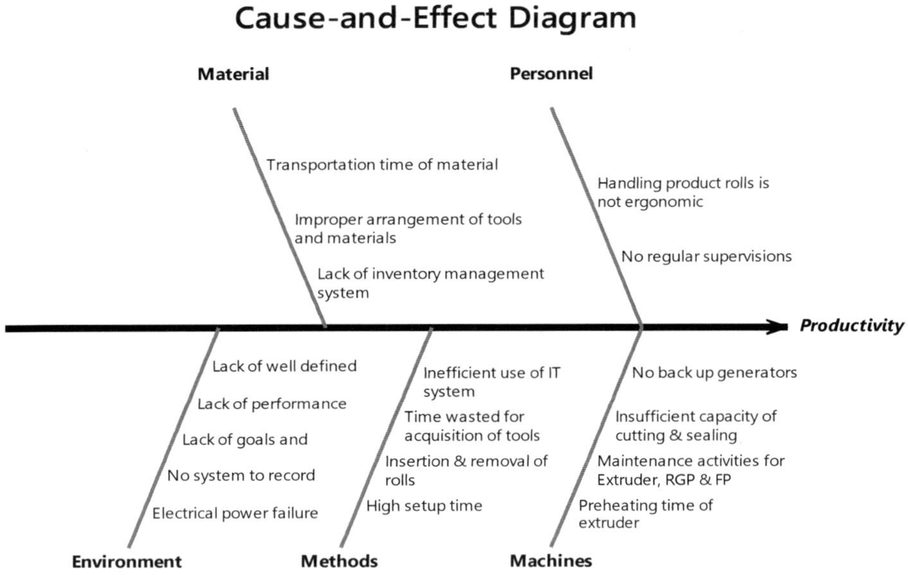

Figure 9. Cause and Effect diagram

3.7 Recommendations

The major problems that affected productivity and efficiency of workers in the enterprise was found and recommendations are given and can be implemented in the real time. The feedbacks of recommendations are also collected and assessed the feasibility and effectiveness of each of the recommendations based on workers and management opinion. All the personnel were asked to rate the recommendations from 1 to 5 (1 – low and 5 – high), for both feasibility and effectiveness. Then the sum of feasibility and effectiveness is considered for ranking the recommendations (Table 4). The recommendation with highest rank is most feasible and effective and can be implemented in short time.

Table 4. Recommendations feedback survey results

Sl. No.	Recommendations	Average score		Total
		Feasible	Effective	
1	Transport rolls from and to storage area during machine is working	4.65	4.2	8.85
2	Use engine lifter to remove and insert rolls into the machine	4.3	4.5	8.8
3	Start the preheating of extruder as soon as plant is opened in the morning	4.2	4.35	8.55
4	Use small carts to carry the rolls during production (50kg roll)	3.8	4.1	7.9
5	Re-allocate idle workers into required machines	3.2	4.5	7.7
6	Promote rotation of work and give appropriate trainings to the workers	3	4.2	7.2
7	Give electrical power backup to preheater in extruder and blowers in printing machines	2.85	4.3	7.15
8	Use combination of workers in machines	3	4	7
9	Use the shelves near the machines to store immediate inventory	3.1	3.75	6.85
10	Arrange the tools and materials required for production near each machines	3.15	3.6	6.75

IOP Conf. Series: Materials Science and Engineering 1132 (2021) 012025 doi:10.1088/1757-899X/1132/1/012025

Transportation rolls from and to storage area during machine is working and Usage engine lifter to remove and insert rolls into the machine showed maximum total score in the survey. These are the recommendations that can be readily implemented.

3.8 Standardization of operations

After considering all the recommendations and solutions from different methods, all the operations are streamlined and time for completing these operations are standardized (Table 5). Implementation of solution from work study, 5S and Kaizen, allocation of workers and brainstorming significantly reduced the cycle times for operations, which in turn improves the productivity of the firm (Table 6).

Table 5. Standard time details of 1kg product production machines

No	Machine type	STWOT (min)	STWT (min)	TSWOS (%)	TSWS (%)
1	Extruder	39.2	82.3	14.41	24.91
2	Flexo printer	59.3	78.5	8.91	21.42
3	Cutting and sealing machine	74.2	79.4	10.82	15.35
4	Roto gravure printing machine	67.4	118.5	9.04	45.32
5	Two layer lamination machine	100.4	121.1	13.00	25.52
6	Three layer lamination machine	202.3	235.7	8.17	16.86
7	Trimming machine	47.3	51.4	20.77	21.17
8	Pouching machine	50.4	53.7	20.75	19.73

STWOT Standard time without setup activities; STWT Standard time with setup activities; TSWOS Time saved without considering setup activities in percentage; TSWS Time saved considering setup activities in percentage.

Table 6. Standard time details of 1kg product production processes

No	Process	STWOT (min)	STWT (min)	TSWOS (%)	TSWS (%)
1	Flexo printing	172.7	240.2	11.03%	20.80%
2	Two layer laminated roto gravure printing	265.5	344.7	15.12%	32.60%
3	Three layer laminated roto gravure printing	367.4	459.3	12.04%	27.36%

STWOT Standard time without setup activities; STWT Standard time with setup activities; TSWOS Time saved considering setup activities in percentage; TSWS Time saved without considering setup activities in percentage.

4. Conclusions

The productivity of a labor-intensive industry is dependent upon the performance of human resources in the industry. By the application of different tools like SMED, 5S and Kaizen, a significant time difference is observed for various operations and the most predominant difference is for Roto gravure printing machine operations, which may be attributed to the long setup time. The major issues in the Hindustan Polymer Products are addressed and given proper recommendations are implemented in real time. The recommendations for enhancement of productivity are ranked on the basis of a feedback survey. Further, most relevant recommendations are economically justified. Electrical power failure is found to be the main external cause that affects production. A significant increase in profit is seen after standardization of the operations in the company. The framework used in this study can be implemented for similar kind of industries to enhance their productivity and efficiency. The study also encourages using advanced approaches in a competitive world, which offers significant potential for improving the productive efficiencies of small and medium business firms.

IOP Conf. Series: Materials Science and Engineering 1132 (2021) 012025 doi:10.1088/1757-899X/1132/1/012025

References

[1] Venugopal N S B 2015 Productivity improvement in small scale industries *International Journal of Mechanical and Production Engineering* **11(3)** 113-119

[2] Azid I A, Ani M N C, Hamid S A A and Kamaruddin S 2020 Solving production bottleneck through time study analysis and quality tools integration *International Journal of Industrial Engineering* ***27*(1)**

[3] Ricondo Iriondo I, Serrano Lasa I and De Castro Vila R 2016 Takt time as a lever to introduce lean production in mixed engineer-to-order/make-to-order machine tool manufacturing companies *International Journal of Industrial Engineering*, ***23*(2)**

[4] Gunasekaran A, Forker L and Kobu B 2000 Improving operations performance in a small company: a case study *International Journal of Operations & Production Management* **20(3)** 316-336

[5] Braglia M, Frosolini M, and Gallo M 2016. Enhancing smed: changeover out of machine evaluation technique to implement the duplication strategy *Production Planning & Control* **27(4)** 328-342

[6] Braglia M, Frosolini M, and Gallo M, 2017 Smed enhanced with 5-whys analysis to improve set-up reduction programs: the swan approach *The International Journal of Advanced Manufacturing Technology* **90(5-8)** 1845-1855

[7] Aglan C and Durmusoglu, M B 2019 A complete design methodology for lean in-plant logistics to assembly line using ad principles *International Journal of Industrial Engineering* **26(5)**

[8] Ailabouni N, Gidado K and Painting N 2007 Factors affecting employee productivity in the UAE construction industry *In Proceeding Conference for Postgraduate Researchers of the Built and Natural Environment (PRoBE)* 33-46

[9] Karthick N, Hari Prasad V, Gopi S and Balaji K 2016 Application of lean tools (VSM, SMED, KAIZEN) in minimizing the lead time in gear manufacturing process *International Journal Of Scientific & Engineering Research* **7(5)** 53-71

[10] Akhil Sankar K, G Venugopal and Ajith P M 2019 Enhancing productivity by standardization of operations in a small scale industry *International Journal of Applied Engineering Research* **14** ***(14)*** 22-27

[11] Ahmed T, Ali S M Allama M M and Parvez M S 2010 Total productive maintenance (TPM) approach to improve production efficiency and development of loss structure in a pharmaceutical industry *Global Journal of Management and Business Research* **10 (2)** 186-190

[12] Dogra M, Sharma V S, Sachdeva A and Dureja J S 2011 TPM – A key strategy for Productivity Improvement in Process Industry *Journal of Engineering Science and Technology* **6(1)** 1 -16

IOP Conf. Series: Materials Science and Engineering 1132 (2021) 012026 doi:10.1088/1757-899X/1132/1/012026

Fuzzy Probist System Reliability of Weaving Machine in Textile Industry

Jini Varghese P[1*], Dr. G Michael Rosario[2]

[1]Department of Basic Science and Humanities Department, Adi Shankara Institute of Engineering and Technology, Mattoor, Kalady, Ernakulam, Kerala, India-683574

[2]Research Centre of Mathematics, Jayaraj Annapackiam College for women (Autonomous), Periyakulam-625601, Tamilnadu, India

[*]email: jinijees@gmail.com

Abstract: In this paper, the binary state assumptions have been replaced by a fuzzy state assumption. Probist Reliability of a weaving machine is calculated with the help of fault tree diagram using the pendant triangular fuzzy number and pendant trapezoidal fuzzy numbers. The formula thus created is used to evaluate the failure rate of the machines, which would determine the reliability of the machines and maintain the quality of performance.

Keywords: Probist Reliability, Pendant fuzzy number, triangular Pendant fuzzy number and trapezoidal pendant fuzzy number, Probist series system, Probist parallel System

1. Introduction

The concept of Reliability is old as man himself. From ancient days, man had started his work on failure of the system such as unreliability of the machine and reliability evaluation of a system. Characteristics of system reliability depends on time failure, failure factors, availability of the products and probability measures.

Reliability, in its simplest form, failure does not happen between specific time intervals. The performance of a machine depends on some factors that affect the machine:

1. Capability and capacity of the machine.
2. Quality of the raw materials put into the machine.
3. Technical difficulties occurred can alter the performance and output of the machine.

Zedeh started the brief introduction about fuzzy sets [1] in 1965 to represent the data handled non-statistical certainties. Types of fuzzy numbers are defined and the vague idea about the fuzzy numbers. In [2], Yager explained the clear idea about to solving mathematical relationships. Representation of trapezoidal fuzzy numbers and the multiplication operation introduced, Rezvani.S[3]. In [4], Nagoorgani.A, a triangular fuzzy number based operation for solving linear programing problem. A new fuzzy number named as trident fuzzy numbers are introduced to fuzzy system and clarify its arithmetic operations by Praveen Prakash.A and M. Geetha Lakshmi [5]. Another fuzzy number is introduced by B. Rama and G. Michael Rosario [6], named as quadrant fuzzy number and its arithmetic operations

IOP Conf. Series: Materials Science and Engineering 1132 (2021) 012026 doi:10.1088/1757-899X/1132/1/012026

explained using types of fuzzy numbers. In [7],[8], Sub-Trident fuzzy numbers are explained and ranking is introduced. Pentagonal fuzzy numbers are explained in [9] by Avinash. J. Kamble. In [10] explained the operations fault tree analysis of a system characterized by fuzzy numbers especially by triangular intuitionistic and deeply explained applications. In [11] factors affecting the failure rate explained and discussed the different failure possibility. In [12] G. Michael Rosario and A. Dhana Lakshmi are discussed the applications of intuitionistic fuzzy equations on reliability evaluation. In [13,14] presented the reliability evaluation of weaving machine in textile industry using trapezoidal intuitionistic fuzzy numbers. In [15], Kai-Yuan Cai explained clear-cut idea about probist fuzzy numbers and series and parallel system with the help of probist reliability.

2. Probist Reliability

In the framework of probist reliability theory a probist system is based on probability assumption and the binary-state assumption:
In probability notion, in the context of probability measures the system failure behavior is characterized. In binary-state notion, there are two crisp states for a system, fully functioning system or fully failed system. The system is of one of the two states at the time. A completely functioning system means that the crisp value is one and a non-functioning system means that the crisp value is zero. However, in the fuzzy case the value is between the binary digits. Therefore, other than the two crisp states in the fuzzy state the system can work between infinite states between the two binary digits.

3. Definitions

3.1. Probist Series System

Let$R = R_1 R_2 R_3 \ldots\ldots R_n = \prod_{i=1}^{n} R_i$, be a probist system with the n components are arranged in series mode.
Suppose $R_1, R_2, R_3 \ldots\ldots R_n$ are triangular pendant fuzzy number. Write it as $R_i = (a_{i1p}, a_{i2p}, a_{i3p})$; $i = 1,2, \ldots\ldots n$

$(R_i)_\alpha = \left[a_{i2p} - \alpha^5(a_{i2p} - a_{i1p}), a_{i2p} + \alpha^5(a_{i3p} - a_{i2p})\right], \alpha \epsilon [0,1]$

$$(R)_\alpha = \prod_{i=1}^{n} (R_i)_\alpha$$

$$= \left[\prod_{i=1}^{n}(a_{i2p} - \alpha^5(a_{i2p} - a_{i1p})), \prod_{i=1}^{n}(a_{i2p} + \alpha^5(a_{i3p} - a_{i2p}))\right], \alpha \epsilon [0,1]$$

Also, we denote, $R_i = (m_i - \alpha_i, m_i, m_i - \beta_i)$

$$R = \prod_{i=1}^{n} R_i$$

$$= [\prod_{i=1}^{n}(m_i - \alpha_i), \prod_{i=1}^{n} m_i, \prod_{i=1}^{n}(m_i - \beta_i)] \quad (1)$$

Symbolically, we write trapezoidal pendant fuzzy number as, $R_1, R_2 \ldots\ldots R_n$
$R_i = (a_{i1p}, a_{i2p}, a_{i3p}, a_{i4p}), i = 1,2 \ldots\ldots n$

3.2. Probist Parallel System

Let$R = 1 - \prod_{i=1}^{n}(1 - R_i)$, be a probist system with the n components are arranged in parallel mode.
Suppose $R_1, R_2, R_3 \ldots\ldots R_n$ are triangular pendant fuzzy number. Write it as $R_i = (a_{i1p}, a_{i2p}, a_{i3p})$; $i = 1,2, \ldots\ldots n$

$(R_i)_\alpha = \left[a_{i2p} - \alpha^5(a_{i2p} - a_{i1p}), a_{i2p} + \alpha^5(a_{i3p} - a_{i2p})\right], \alpha \epsilon [0,1]$

$$(R)_\alpha = 1 - \prod_{i=1}^{n} 1 - (R_i)_\alpha$$

$$= \left[1 - \prod_{i=1}^{n}(a_{i2p} - \alpha^5(a_{i2p} - a_{i1p})), 1 - \prod_{i=1}^{n}(a_{i2p} + \alpha^5(a_{i3p} - a_{i2p}))\right], \alpha\epsilon[0,1]$$

Also, we denote, $R_i = (m_i - \alpha_i, m_i, m_i - \beta_i)$

$$R = 1 - \prod_{i=1}^{n}(1 - R_i)$$

$$= 1 - [\prod_{i=1}^{n} 1 - (m_i - \alpha_i), 1 - \prod_{i=1}^{n} 1 - m_i, 1 - \prod_{i=1}^{n} 1 - (m_i - \beta_i)]$$

$$= [1 - \prod_{i=1}^{n} 1 - (m_i - \alpha_i), 1 - \prod_{i=1}^{n} 1 - m_i, 1 - \prod_{i=1}^{n} 1 - (m_i - \beta_i)] \quad (2)$$

Now suppose $R_1, R_2 \ldots.. R_n$ are trapezoidal pendant fuzzy numbers symbolically write it as $R_i = (a_{i1p}, a_{i2p}, a_{i3p}, a_{i4p}), i = 1,2 \ldots \ldots n$

3.3. Triangular Pendant Fuzzy Number (TrPFN)

A Triangular Pendant Fuzzy Number is signified by $\tilde{A} = (a_{1p}, a_{2p}, a_{3p})$ where a_{1p}, a_{2p}, a_{3p} are real numbers and its membership function is defined as

$$\mu_{\tilde{A}}(x) = \begin{bmatrix} 1^{\frac{1}{5}} & for & x < a_{1p} \\ \left[\frac{a_{2p}-x}{a_{2p}-a_{1p}}\right]^{\frac{1}{5}} & for & a_{1p} \le x \le a_{2p} \\ 0 & for & x = a_{2p} \\ \left[\frac{x-a_{2p}}{a_{3p}-a_{2p}}\right]^{\frac{1}{5}} & for & a_{2p} \le x \le a_{3p} \\ 1^{\frac{1}{5}} & for & x > a_{3p} \end{bmatrix} \quad (3)$$

3.3.1 α-cut of a Triangular Pendant Fuzzy Number

α-cut-Triangular Pendant Fuzzy Number is given by

$$A_{p\alpha} = [a_{2p} - \alpha^5(a_{2p} - a_{1p}), a_{2p} + \alpha^5(a_{3p}\text{-}a_{2p})], \alpha \in (0,1]. \quad (4)$$

$$i = 1,2, \ldots..n$$

3.4. Trapezoidal Pendant Fuzzy Number (TrpPFN)

A Trapezoidal Pendant Fuzzy Number represented by $\tilde{A} = (a_{1p}, a_{2p}, a_{3p}, a_{4p})$ where $a_{1p}, a_{2p}, a_{3p}, a_{4p}$ are real numbers and the membership function, with a structures help,

$$\mu_{\tilde{A}}(x) = \begin{bmatrix} 1^{\frac{1}{5}} & for & x < a_{1p} \\ \left[\frac{a_{2p}-x}{a_{2p}-a_{1p}}\right]^{\frac{1}{5}} & for & a_{1p} \le x \le a_{2p} \\ 0 & for & a_{2p} \le x \le a_{3p} \\ \left[\frac{x-a_{3p}}{a_{4p}-a_{3p}}\right]^{\frac{1}{5}} & for & a_{3p} \le x \le a_{4p} \\ 1^{\frac{1}{5}} & for & x > a_{4p} \end{bmatrix} \quad (5)$$

3.4.1 α-cut-Trapezoidal Pendant Fuzzy Number

α-cut-Trapezoidal Pendant Fuzzy Number is given by

IOP Conf. Series: Materials Science and Engineering 1132 (2021) 012026 doi:10.1088/1757-899X/1132/1/012026

$$A_{p\alpha} = [a_{2p} - \alpha^5(a_{2p} - a_{1p}), a_{3p} + \alpha^5(a_{4p}\text{-}a_{3p})], \alpha \in (0,1]. \quad (6)$$

4. Failure of weaving machine

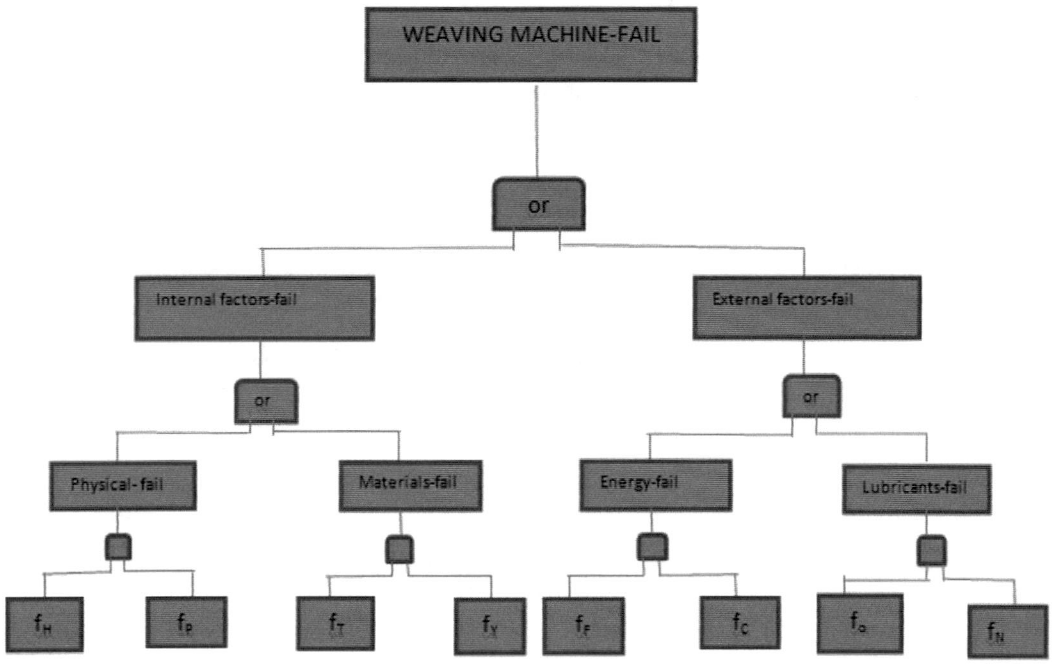

Figure. 1 Fault tree diagram of the failure rate of the weaving machine with the corresponding factors

Figure shows the failure rate of the following factors are denoted as follows

$\widetilde{f_{WM}}$ Weaving machine
$\widetilde{f_{IF}}$ Internal factors
$\widetilde{f_{EF}}$ External factors
$\widetilde{f_{PS}}$ Physical stabilities
$\widetilde{f_{RM}}$ Raw materials
$\widetilde{f_{WE}}$ Working energy
$\widetilde{f_{FO}}$ Flow of lubricants oil

$\widetilde{f_H}$ Controlled humidity
$\widetilde{f_P}$ balanced pressure
$\widetilde{f_T}$ Quality of the thread
$\widetilde{f_Y}$ Continuity of filament yarn
$\widetilde{f_F}$ Availability of the fuel
$\widetilde{f_C}$ Balanced current flow
$\widetilde{f_O}$ Shortage of lubricants oil
$\widetilde{f_N}$ Nozzles blocked

There are eight sub factors $\widetilde{f_H}, \widetilde{f_P}, \widetilde{f_T}, \widetilde{f_Y}, \widetilde{f_F}, \widetilde{f_C}, \widetilde{f_O}, \widetilde{f_N}$ .Here, $R_1, R_2, R_3, R_4, R_5, R_6, R_7, R_8$ are the corresponding reliabilities of $\widetilde{f_H}, \widetilde{f_P}, \widetilde{f_T}, \widetilde{f_Y}, \widetilde{f_F}, \widetilde{f_C}, \widetilde{f_O}, \widetilde{f_N}$ respectively.
Take a probist parallel machine, of order 8, that is,

$$R = 1 - \prod_{i=1}^{8}(1 - R_i) \quad \textbf{(7)}$$

Suppose $R_1, R_2, R_3, R_4, R_5, R_6, R_7, R_8$ are triangular pendant fuzzy numbers, then so are $1 - R_1, 1 - R_2, 1 - R_3, 1 - R_4, 1 - R_5, 1 - R_6, 1 - R_7, 1 - R_8$.
Denote $R_i = (a_{i1p}, a_{i2p}, a_{i3p})$; $i = 1,2,3,4,5,6,7,8$

And the $\alpha - cut$ is denoted as $(R)_\alpha = [1,1] - \prod_{i=1}^{8}\{[1,1] - (R_i)_\alpha\}$
$= [1,1]-\prod_{i=1}^{8}[\, a_{i2p} - \alpha^5(a_{i2p} - a_{i1p}), a_{i2p} + \alpha^5(a_{i3p} - a_{i2p})], \alpha \in (0,1]$.
Since, α-cut of a Triangular Pendant Fuzzy Number, given by
$A_{p\alpha} = [a_{2p} - \alpha^5(a_{2p} - a_{1p}), a_{2p} + \alpha^5(a_{3p}\text{-}a_{2p})], \alpha \in (0,1]$.
Here, Triangular Pendant Fuzzy number can be written as
$R_i = (m_i - \alpha_i, m_i, m_i - \beta_i)$;$i = 1,2,3,4,5,6,7,8$ (8)
Suppose, for example height of a person is m, where $\alpha, \beta$the left-right are spreads of m.
In general, fuzzy number can be represented$(m - \alpha, m, m + \beta) = (a_1, a_2, a_3)$, these types of numbers are called triangular pendant fuzzy numbers alternatively represented as(m, α, β).

(8) in (7)⇒ $R = 1 - \prod_{i=1}^{8}(1 - R_i)$
$= 1 - \prod_{i=1}^{8}(1 - (m_i - \alpha_i, m_i, m_i - \beta_i))$
$= 1 - \prod_{i=1}^{8}(1 - (m_i - \alpha_i), 1 - m_i, 1 - (m_i - \beta_i))$
$= 1 - \prod_{i=1}^{8}(1 - (m_i - \alpha_i), 1 - m_i, 1 - (m_i - \beta_i))$
$=1 - \{\prod_{i=1}^{8}[(1 - (m_i - \alpha_i)], \prod_{i=1}^{8} 1 - m_i, \prod_{i=1}^{8}[1 - (m_i - \beta_i)]\}$
$=\{1 - \prod_{i=1}^{8}[(1 - (m_i - \alpha_i)], 1 - \prod_{i=1}^{8} 1 - m_i, 1 - \prod_{i=1}^{8}[1 - (m_i - \beta_i)]\}$ (9)

Suppose,$R_1, R_2, R_3, R_4, R_5, R_6, R_7, R_8$ are trapezoidal pendant fuzzy numbers, then
$1 - R_1, 1 - R_2, 1 - R_3, 1 - R_4, 1 - R_5, 1 - R_6, 1 - R_7, 1 - R_8$ are trapezoidal pendant fuzzy numbers.
So, denote$R_i = (a_{i1p}, a_{i2p}, a_{i3p}, a_{i4p})$;$i = 1,2,3,4,5,6,7,8$; then
$1 - R_i = (1 - a_{i1p}, 1 - a_{i2p}, 1 - a_{i3p}, 1 - a_{i4p})$; $i = 1,2,3,4,5,6,7,8$
Therefore, $\prod_{i=1}^{8}(1 - R_i) = [\prod_{i=1}^{8}(1 - a_{i1p}), \prod_{i=1}^{8}(1 - a_{i2p}), \prod_{i=1}^{8}(1 - a_{i3p}), \prod_{i=1}^{8}(1 - a_{i4p})]$

So, $\boldsymbol{R = 1 - [\prod_{i=1}^{8}(1 - a_{i1p}), \prod_{i=1}^{8}(1 - a_{i2p}), \prod_{i=1}^{8}(1 - a_{i3p}), \prod_{i=1}^{8}(1 - a_{i4p})]}$ (10)

5. Numerical Evaluation

Let $R_1 = (0.1,0,2,0.3)$, $R_2 = (0.1,0,3,0.4)$, $R_3 = (0.2,0,3,0.4)$, $R_4 = (0.3,0,4,0.5)$,
$R_5 = (0.4,0,5,0.6)$, $R_6 = (0.4,0,5,0.7)$, $R_7 = (0.5,0,6,0.7)$, $R_8 = (0.5,0,6,0.8)$

$$\prod_{i=1}^{8} 1 - R_i = \left(\prod_{i=1}^{8}(1 - a_{i1}), \prod_{i=1}^{8}(1 - a_{i2}), \prod_{i=1}^{8}(1 - a_{i3}), \prod_{i=1}^{8}(1 - a_{i4})\right)$$
$$= (0.9 \times 0.9 \times 0.8 \times 0.7 \times 0.6 \times 0.6 \times 0.5 \times 0.5,$$
$$0.8 \times 0.7 \times 0.7 \times 0.6 \times 0.5 \times 0.5 \times 0.4 \times 0.4,$$
$$0.7 \times 0.6 \times 0.6 \times 0.5 \times 0.4 \times 0.3 \times 0.3 \times 0.2,$$
$$0.6 \times 0.5 \times 0.5 \times 0.4 \times 0.3 \times 0.2 \times 0.2 \times 0.1)$$

That is, $1 - R = (0.040824, 0.009408, 0.0009072, 0.000072$
Therefore, $R = (0.959176, 0.990592, 0.9990928, 0.999928)$

6. Conclusions

In this paper, different fuzzy numbers are used to evaluate fuzzy reliability. Fuzzy numbers may vary from time to time and new fuzzy numbers are introduced frequently. This is a comparison study. In this study, the reliability evaluation of a machine is one fuzzy numbers is better than the other. This machine cannot be concluded as the most reliable in the prescribed fuzzy numbers, however, we can be conclude that the given fuzzy number is more reliable than the other fuzzy numbers. In this paper two new fuzzy numbers are introduced they are triangular pendant fuzzy numbers and trapezoidal pendant fuzzy numbers. In the previous paper while solving the fuzzy reliability of a machine, this value was found, trapezoidal fuzzy numbers is better than triangular fuzzy numbers with the help of numerical evaluation. Similarly, if we evaluate trapezoidal pendant fuzzy numbers is better than triangular pendant fuzzy numbers.

References

[1] Zadeh L.A, *Fuzzy sets, Information and control*,8(1965),338-353.

[2] Yager. R, *On Solving Fuzzy Mathematical relationships, Information control*, 41, 29-55(1979)

[3] Rezvani. S, *Multiplication Operation on Trapezoidal Fuzzy numbers, Journal of Physical Sciences*, vol no-15, 17-26(2011)

[4] Nagoorgani A, *A new operation on triangular fuzzy number for solving fuzzy linear programming problem, Applied Mathematical Sciences*, 6, 525-532(2012)

[5] A. Praveen Prakash and M. Geetha Lakshmi, *A Trident Fuzzy Number and its Arithmetic Operations, International Journal of Innovative Research in Computer and Communication Engineering*, vol no-4(10) (2016)

[6] B. Rama and G. Michael Rosario, *Quadrant Fuzzy Number and its Arithmetic Operations, Journal of Computer and Mathematical Sciences*, vol no-10(7), 1466-1475 July2019

[7] A. Praveen Prakash and M. Geetha Lakshmi, *Sub-Trident Ranking Using Fuzzy Numbers, International Journal of Mathematics and its Applications*, vol no-4(2-A) (2016),143-150

[8] A. Praveen Prakash and M. Geetha Lakshmi, *A Comparative Study- Optimal Path using Trident and Sub-Trident Forms through Fuzzy Aggregation, Ranking and Distance Methods, International Journal of Pure and Applied Mathematics*, vol no-109(10) (2016), 139-150.

[9] Avinash J. Kamble, *Some Notes on Pentagonal Fuzzy Numbers, International Journal of Mathematical Archive*, Vol no-13(2), 2017, 113-121

[10] Wang, J.Q., Nie, R.R., Zhang, H.Y., and Chen, X. H.: *New operators on triangular intuitionistic fuzzy numbers and their applications in system fault analysis. Inf. Sci*. 251, 79–95 (2013)

[11] Majeed, *Cotton and Textiles- the challenges ahead*, 2009-02-12

[12] G. Michael Rosario and A. Dhanalakhmi "*Intuitionistic Fuzzy Equations and its Application on Reliability Evaluation", 'International Journal of Advanced Technology Engineering and Science', Volume* 03, *Number*03, *March* (2015)

[13] P. Jini Varghese and G. Michael Rosario, *Fuzzy reliability evaluation of weaving machine in textile industry, International Journal of Mathematical Archieve*, vol 9.no1 (2018), ISSN 2229-5046

[14] P. Jini Varghese and G. Michael Rosario, *some arithmetic operations in trapezoidal Fuzzy numbers and intuitionistic trapezoidal Fuzzy numbers, Journal of Applied Science and Computations*, vol 5-issue 122018, 40-49.

[15] Kai-Yuan Cai, *Introduction to Fuzzy Reliability, The Kluwer International Series in Engineering and Computer Science* '363,1965.

[16] Kai-Yuan Cai, *Introduction to Fuzzy Reliability, Springer Nature*,1996.

Process parameter optimization of wire EDM on weldment of Monel 400 and AISI 316 grade steel

P Hema[1] and K Aparna[2]

[1]Department of Mechanical Engineering, Sri Venkateswara University College of Engineering, Tirupati-517502, AP, India

[2]Department of Production Engineering, Sri Venkateswara University College of Engineering, Tirupati-517502, AP, India.

Abstract: The purpose of this paper is to explore the optimal set of input parameters such as pulse on time, pulse off time and wire tension in Wire Electrical Discharge Machining (WEDM) on weldment of dissimilar metals of Monel 400 and AISI 316 grade steel using brass wire. This experiment describes the difference in values in material removal rate (MRR), surface roughness and kerf width with change in three process parameters. A rectangle shaped cut is made using ZNC WEDM. Each combination of parameters is considered according to the L9 Orthogonal array. Taguchi method based grey relational analysis is used and analyzed results indicated that pulse on time has a significant effect on three output parameters. 9th experiment, i.e., pulse on time at 150μs, pulse off time at 40μs, wire tension of 14 kg-f are found to be optimal process parameters.

Keywords: Wire electrical discharge machining (WEDM), Taguchi technique, material removal rate (MRR).

1. Introduction

A non-conventional method of machining that uses thermal electrical energy for machining is Wire Electrical Discharge Machining (WEDM) as it machines with good dimensional accuracy and surface finish, it is used in machining complex shapes. It is often used in the machining of hard materials that, due to vibration issues, or otherwise difficult to machine using traditional machining processes.In Wire EDM, the effect of process parameters on material removal rate (MRR), surface roughness, kerf width are to be investigated experimentally. In order to evaluate the parameters using different approaches, different methods are used. Here, as a standard for the design of experiments, Taguchi orthogonal array was used and Grey relational analysis was carried out. In this study, the cutting factors considered are pulse on time, pulse off time and wire tension. Using Pulsed Current Gas Tungsten Arc welding using three different filler metals such as ER309L, ERNiCu-7 and ERNiCrFe-3, K.Devendranath Ramkumar,N.Arivazhagan(2014) investigated the pulsed current for joining two different Monel400 and AISI 304 metals using Pulsed Current Gas Tungsten Arc welding and obtained that parent metal Monel 400 had better tensile properties compared to other areas. Angolkar Pooja, J Saikrishna (2017) have investigated the weldability of steel grade Monel 400 and AISI 316 and mechanical properties such as tensile strength, hardness are measured. Residual stresses are measured and maximum stress tensors at the welded joint are defined. J. Jeeamalar, Dr. S. Ramabalan (2017) found that both optimization and prediction can be used by taguchi experimental design tool. Rajyalakshmi.G(2013) analyzed the optimization of process parameters of Wire Electrical Discharge

Machining performed on Inconel 825 alloy. The major performance Characteristics selected to evaluate the process are material removal rate(MRR), surface roughness (SR), and spark gap (SG) using grey relational analysis by using multi objective optimization. Applying grey relational analysis to a problem with multiple-performance characteristics is proven to be capable of objectively reflecting the relative importance of each performance characteristics. Most of the researchers have focussed on parameter optimization of wire EDM on single metal so, experimental investigation was done on a welded zone of dissimilar metals. In the present study weldment of Monel 400 and AISI 316 stainless steel is used as work piece material welded by arc welding joined by butt joint and 0.25mm^2 brass wire is used for machining. Experiments were designed and conducted based on Taguchi's L_9 orthogonal array and Grey relational analysis is performed to obtain multi objective optimization of required parameters such as Material removal rate (MRR), surface roughness, and kerf width by varying the process parameters like pulse on time, pulse off time and wire tension.

2. Experimental setup and methodology

The metals and the filler metal employed in this study were Monel 400, AISI 316 and E7018 electrode respectively. Metals were sliced to the dimensions of 120 mm x 40 mm x 8 mm. These samples were clamped firmly in the fixture designed with a copper back plate so as to avoid distortions and bending during welding. The process parameters were established from the open literatures as well as from the trial and error studies. Followed by welding, the weldments were cut into different coupons using wire cut EDM (Electrical Discharge Machining).

2.1 Machine setup before machining

The experiments were carried on the RATNAPARKHI WEDM machine .The wire was made vertical on verticality block and the work piece is mounted and clamped to the work table. A reference point was set for setting work coordinate system. The programming was done with reference to the WCS. The reference point was defined by the ground edges of work piece. The program was made for cutting operation and a profile of 26.5mm x16.5mm obtained dividing the total dimensions of work piece into equal rectangular blocks. While, performing various experiments precautionary measures were taken to reduce error.

3. Experiment procedure on wire EDM

Work piece after welding is clamped in the WEDM. The work piece is divided into L9 orthogonal array with each rectangular block of dimension 13.5X26.5X8mm. Then each rectangular block is cut by changing different input parameters according to L9 orthogonal array where 9 experiments were performed by varying the input parameters. Three input parameters were considered at three levels pulse on time at 130µs, 140µs, 150µs, pulse off time at 20µs, 30µs, 40µs and wire tension at 13kg-f, 14kg-f, 15kg-f are arranged according to L_9 orthogonal array. Machining time for cutting each rectangular array is measured. After cutting all the rectangular blocks the surface roughness of each side of rectangular block is measured using taly surf method and the average value is taken. Width of each rectangular block is measured using digital micrometer and difference between obtained dimension and calculated dimension were measured and noted as kerf width. Through rectangular holes are machined using WEDM set-up are shown in the below figure (1) by varying process parameters like pulse on time, pulse off time, wire tension with a dielectric fluid, then the corresponding response readings is recorded in the designed experiment tables as shown in below table (1) and the rectangular blocks cut were labelled in the figure (1) below. The Metal Removal Rate (MRR) is calculated based on ratio of difference in work piece weight to the product of density and machining time of material. Similarly surface roughness is measured by taly surf instrument and kerf

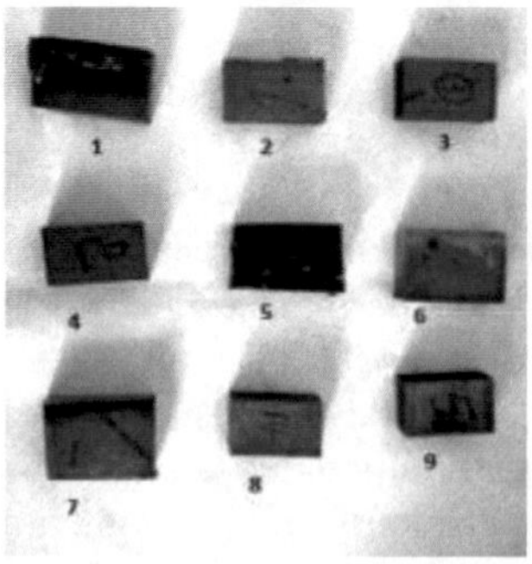

Figure1. Rectangular blocks

3.1 Calculation of process parameters

$$MRR=\frac{(Wjb-Wja)}{tþ} mm^3/min \quad (1)$$

Whereas,
W jb =Initial weight of work piece before machining in gm
W ja =Final weight of work piece after machining in gm
t =Machining time in minutes.
Þ =Density of material in gm/mm^3

MRR is calculated by weighing the work piece before machining and after machining and the difference is calculated by dividing with product of machining time and density.

The surface roughness parameter used to evaluate surface roughness is the Roughness average (Ra). This parameter is also known as the arithmetic mean roughness value, the arithmetic average or centerline average. The average roughness is the area between the roughness profile and its Centre line is measured using Taly surf instrument. After machining the blocks surface roughness of each rectangular block is measured.

Kerf width is measured in millimetres (mm). It is the measure of the amount of the material that is wasted during machining and determines the dimensional accuracy of the finishing part. Kerf width is measured using digital vernier calliper by taking the difference in width assumed and width obtained.

Table 1. Experimental Results

Experiment no	Pulse on time µs	Pulse off time µs	Wire tension (kg-f)	MRR(mm^3/sec)	Surface roughness(µm) R_a	Kerf width mm
1	130	20	13	16.123	1.90	0.699
2	130	30	14	16.20	1.93	0.645
3	130	40	15	15.53	1.964	0.6134
4	140	20	14	18.76	2.43	0.523
5	140	30	15	17.94	2.52	0.549
6	140	40	13	17.52	2.61	0.432
7	150	20	15	20.4	2.81	0.399
8	150	30	13	19.23	2.94	0.383
9	150	40	14	18.84	3.01	0.225

4. Results and discussions

MRR values which are tabulated are analyzed in MINITAB19 software and SN ratios and graphs were obtained for finding the most effective parameter during machining.

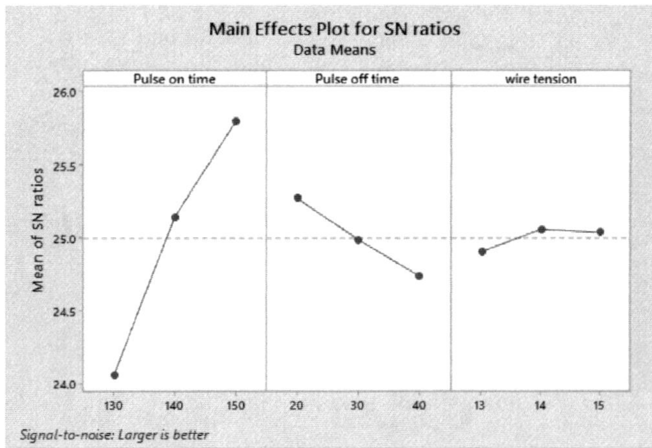

Figure 2. SN ratio graph for MRR

The S/N ratio plot for MRR in the above figure (2) has shown the significance of each parameter. It is clear from S/N ratio plots that these plots were generated based on "Larger is Better"130μs is the effective parameter for pulse on time and pulse off time at 40μs is the effective parameter and wire tension at 13kg-f is the effective parameter as the mean of SN ratios at these parameters have lesser noise.

Surface roughness values which are tabulated are analyzed in MINITAB19 software and SN ratios and graphs were obtained for finding the most effective parameter during machining. These plots were generated based on "Smaller is better".

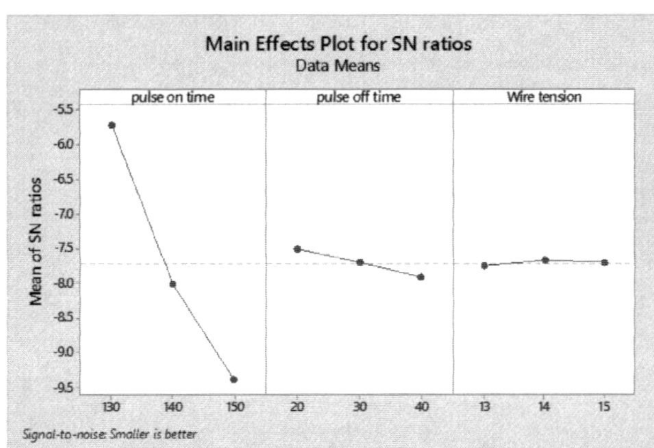

Figure 3. SN ratio graph for surface roughness

The SN ratio graph for surface roughness in above figure (3) has shown that pulse on time has less noise at 130μs pulse off time is more effective at 20μs and wire tension is more effective at 14 kg-f.

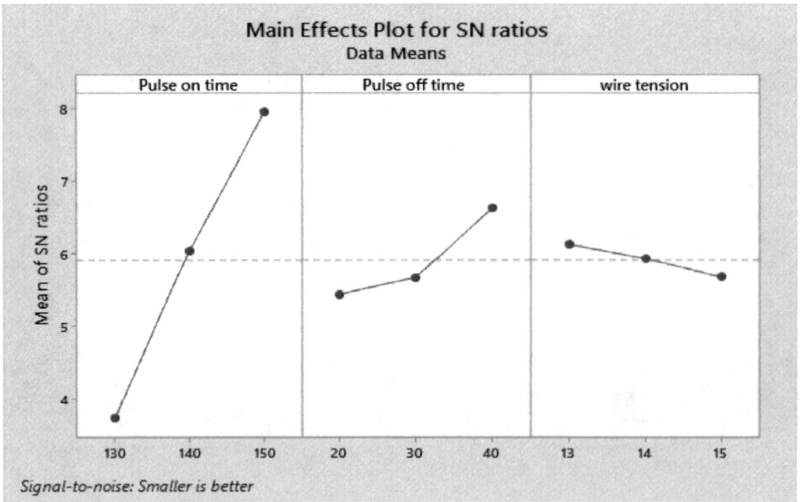

Figure 4 SN ratio graph for kerf width

The S/N ratios plot for kerf width shown in the above figure (4) has shown the significance of each parameter. These plots were generated based on "Smaller is better". Pulse on time at 130μs is the most effective parameter and pulse off time at 20μs and wire tension at 15kg-f is the most effective parameter. The S/N ratio plots indicated and predicted the optimal parameters for a single variable.

Empirical modelling of the three output parameters were done by taking 3 process parameters as input variables Regression equations are formulated for the three output parameters and R-sq values are found for three output parameters.

MRR = -7.56 + 0.1770 Pulse on time - 0.0566 Pulse off time + 0.166 wire tension. (2)

R-sq values for MRR is 97.45%

Surface roughness = -4.637 + 0.05068 Pulse on time + 0.00608 Pulse off time-0.0125 wire tension. (3)

R-sq values for surface roughness is 99.03%

Kerf width = 2.779 - 0.01584 Pulse on time - 0.00584 Pulse off time+ 0.0079 wire tension. (4)

R-sq values for kerf width is 94.99%.

Grey relational analysis is performed for multi objective optimization of parameters from the grey relational grade rankings are given to the parameters.SN ratios, normalized SN ratios, grey relational coefficients, grey relational grades are found in order. G.Rajyalakshmi (2013).

IOP Conf. Series: Materials Science and Engineering 1132 (2021) 012027 doi:10.1088/1757-899X/1132/1/012027

Table 2. Grey relational grade and rank tabulation

Exp no	Pulse on time(μs)	Pulse off time(μs)	Wire tension (kg-f)	MRR(mm^3/min)	Surface Roughness (μm)	Kerf width (mm)	Grey relational grade	Rank
1	130	20	13	16.123	1.90	0.699	0.5653	3
2	130	30	14	16.20	1.934	0.645	0.5566	4
3	130	40	15	15.53	1.965	0.6134	0.5317	6
4	140	20	14	18.76	2.43	0.523	0.5173	7
5	140	30	15	17.94	2.52	0.549	0.4640	9
6	140	40	13	17.52	2.61	0.432	0.4768	8
7	150	20	15	20.4	2.89	0.399	0.6745	2
8	150	30	13	1923	2.94	0.383	0.5411	5
9	150	40	14	18.84	3.01	0.225	0.6476	1

The above tabular column (2) shows the tabulation of grey relational grade and ranking of the experiments based on the desired output, i.e. high material removal rate, low surface roughness and lower kerf width.

5. Conclusions

An indigenous set-up of wire Electrical Discharge Machine is utilized to bring-out machining of through holes on weldment of Monel 400 and AISI 316 grade steel work pieces. These are the below conclusions from present work. Through holes are machined on work piece of thickness 8 mm and electrode wire 0.25mm^2 of size with de-ionized water as dielectric medium by varying the control parameters like pulse on time, pulse off time, wire tension.

The effects of process parameters on performance parameters like Machining time, MRR and surface roughness and kerf width are discussed in detail and the constituent results are obtained.

- The optimal parameters obtained by observing the SN ratio graphs are $A_1B_3C_1$ for material removal rate i.e., pulse on time at 150 μs, pulse off time at 20 μs, and wire tension at 14kg-f has high material removal rate.
- The optimal parameters obtained by observing the graphs are $A_1B_1C_1$ for surface roughness i.e., pulse on time at 150μs, pulse off time at 40μs and wire tension at 13kg-f has lower surface roughness.
- The optimal parameters obtained by observing the graphs are $A_1B_1C_3$ for kerf width i.e., pulse on time at 130μs, pulse off time at 20μs, and wire tension at 15kg-f has lower kerf width.
- On performing the Regression analysis it has been found that surface roughness is the most effected parameter as it contains highest R-Sq value i.e., 99.03%.
- Grey analysis was performed and optimal parameters for all the three process parameters were found at pulse on time at 150μs, pulse off time at 40μs and wire tension at 14kg-f i.e., at these input parameters high material rate and lower surface roughness and kerf width are

IOP Conf. Series: Materials Science and Engineering 1132 (2021) 012027 doi:10.1088/1757-899X/1132/1/012027

obtained. Machining of weldment at the welded joint has become difficult and wire breakage has happened and experiments are performed by varying different parameters but wire burn and breakage has happened at the joint.

The following are the major reasons

- Due to bad welding and welding defects.
- Angular distortion during welding.
- Insufficient parameter ranges of WEDM machine.
- Non conductive behavior of impurities at the weld joint.

References

[1] Rajkumar P,Muralidharan S,Nagendiran B, "Optimization of Micro Electrical Discharge of Machining parameters of Ti-6Al-4V" *IJSRD* **VOL.2,** issue 11,2015.

[2] Mohandas T,Satyanarayana VV and Madhusudan reddy G, "Dissimilar metal friction welding of austenitic-ferritic stainless steels",*journal of Materials processing Technology*.2005,160(2)

[3] Sadek AA,Abass M,Zaghloul B,Elrefaey A and Ushio M. "Investigation of dissimilar joints between low carbon steel and Monel 400.Transactions of *JWRI* 2000,**29(1):21-28.**

[4] John Figert, Dan Rybicki. "Developing a welding procedure for joining Monel 400 to 316L stainless steel",*NASA M&P update (NMPU),NASA M&P Update* Newsletter 1998.

[5] Alber Sadek A,Mahmoud Abass,Bahaa Zaghloul "Investigation of dissimilar joints between low carbon steel and Monel 400",Trans *JWRI,*2000;21-28.

[6] K H Ho and S T Newman, "A state of the art electrical discharge machining (EDM)", *International journal of Machine tools and Manufacture* 43(2003) 1287-1300.

[7] Debidutta Mishra,M K Vignesh,B Ganesh Raj,Pruthvi Srungavarapu,K Devendranath Ramkumar, N Arivazhagan,S Narayanan. "Mechanical charecterization of Monel 400 and 316 stainless steel weldments", *procedia Engineering* **75(2014) 24-28.**

[8] Ko-Ta Chiang and Fu-ping Chang, "Optimization of the WEDM process of particle-reinforced material with multiple performance characteristics using Grey Relational Analysis",*journal of Materials processing Technology,* **pp 96-101,**2006.

[9] G.Rajyalakshmi, "Optimization of Process parameters of Wire Electrical Discharge Machining on Inconel 825 using Grey Relational Analysis Coupled with Principle component Analysis *Article in International Journal of Applied Engineering Research, November 2013*

Mechanical, Metallurgical and Corrosion Properties of Dual Pulsed, Pulsed Gas and Shielded Metal Arc Welded AISI 310 Austenitic Stainless Steel

Arunkumar M[1], Sudhy S Panicker[2*], Aneesh Kumar P[1], K Devakumaran[3]

[1]Department of Mechanical Engineering, NSS College of Engineering, Palakkad-678008, Kerala, India

[2]Department of Mechanical Engineering, Adi Shankara Institute of Engineering and Technology, Kalady-683574, Kerala, India

[3]Welding Research Institute, Bharath Heavy Electricals Limited (BHEL), Trichy-620014, Tamil Nadu, India

[*]email: sudhyspanicker@gmail.com

Abstract. AISI 310 is a highly alloyed austenitic stainless steel used for high-temperature applications. The high chromium and nickel contents give this steel excellent oxidation resistance as well as high strength. In the present work, the material was welded using shielded metal arc welding (SMAW), Pulsed Gas Metal Arc welding (P-GMAW) and Double Pulsed Gas Metal Arc welding (DP-GMAW). A comparative study has been carried out on the basis of chemical, mechanical, metallurgical and corrosion properties of base metal and all other welded samples. SMAW process has high heat input and hence dilution effect is more prominent in this process. The mechanical properties evaluated in terms of maximum tensile strength, impact energy and hardness was obtained for DP-GMAW sample. This superior mechanical behaviour of DP-GMA welded sample was attributed to the better grain refinement occurred due to double pulsation effect. However, the DP-GMAW process showed highest corrosion rate when compared to other two welded samples.

Keywords: AISI 310, Double Pulsed Gas Metal Arc Welding, Micrographs, Corrosion rate

1. Introduction

AISI 310 is a high chromium and nickel alloyed austenitic stainless steel used for high-temperature applications. The combination of good ductility and weldability enabled its widespread usage in applications such as furnace parts, heat treatment equipment, boiler baffles, kilns, lead pots, etc. The high chromium and nickel contents give the steel excellent oxidation resistance as well as high strength at high temperatures. As the AISI 310 material has numerous applications, it is important to study the weld behaviour and corrosiveness of the material. Gas Metal Arc Welding (GMAW) is a common welding process where an externally supplied shielding gas is used to protect the arc and molten metal from oxygen and other gases in the air. A filler wire is continuously fed in to the welding through the wire feeder and welding gun. The GMAW process can be performed continuously without changing the electrode and has the potential to improve the productivity over Shielded Metal Arc Welding (SMAW). An advancement of conventional GMAW process, named as Pulsed GMAW (P-GMAW), produces stable spray modes of metal transfer even at low average currents. All these GMAW uses constant Wire Feed Rate (WFR).

Further, Dual Pulse GMAW (DP-GMAW) was developed to avoid many problems associated with GMAW of aluminium material [1]. This welding process is a derivation of P-GMAW where a low frequency current pulsation is superimposed over the pulsed current for additional control of metal transfer and weld pool [1, 2]. The DP-GMAW process requires less operator skill as it controls the arc length and heat input together to get excellent penetration [3]. The process also ensured a better surface oxide cleaning property and bead profile with good mechanical properties [1]. DP-GMAW significantly influence the mechanical and microstructural properties of ferritic stainless steel [4].

However, it was found that very few studies reported the DP-GMAW of austenitic stainless steel [4, 5]. Also, no previous work dealt with the corrosion properties of DP-GMA welded AISI 310 steel. Hence, in this study, an attempt was made to understand the welding strength and corrosiveness of AISI 310 steel welded using DP-GMAW process. The mechanical characterisation in terms of tensile strength, Charpy impact energy, and hardness, metallographic analysis and corrosiveness were investigated. The corresponding data of AISI 310 steels welded using SMAW and P-GMAW processes were also evaluated and their comparative investigation were also performed.

2. Experimental setup

2.1. Material selection

In this study, austenitic stainless steel AISI 310 plate was used as base material. The welding was done on plate having dimensions, 10 mm thickness and 100 mm length. The important mechanical properties of AISI 310 at room temperature is shown in Table 1. The filler wires used for SMAW was E 310 and that for GMAW was E 309. The schematic of the groove preparation is shown in Fig. 1. Conventional V-groove arrangement with 70° groove angle and 1.5 mm root face and root gap was prepared for welding. This groove was designed according to the thickness of the material. The edges were prepared and cleaned thoroughly to remove oil, dust, paint and other impurities. Copper plate was used as backing plate for all the welding experiments to protect the molten metal and to avoid oxidation at root side. The multi pass deposition technique was adopted in this study and welding experiments were performed using SMAW, P-GMAW and DP-GMAW processes. The current, voltage and heat input were measured in each pass.

Table 1 Mechanical properties of AISI 310 at room temperature

Properties	Typical	Minimum
Tensile Strength, MPa	575	515
Yield Stress (0.2 % offset), MPa	290	205
Elongation (50mm gauge length), %	50	40
Hardness (Brinell)	156	
Endurance (fatigue) limit, MPa	260	

2.2. Power source

The inverted type KEMPI ProMig 530 welding machine, capable of doing GMAW, P-GMAW and DPGMAW under synergic mode was used as the power source for this experimental study. The machine had a duty cycle of 60% at rated current 520A. The actual welding arrangement with different equipment used in this study is shown in Fig. 2. Commercially pure Argon shielding gas used for GMAW process. A 100% flow of shielding gas with a flow rate 15 l/min was used for all the welding process. The gas flow was measured by a flow meter connected on the gas cylinder. During the welding process the welding current was controlled by adjusting the Wire Feed Rate (WFR). Even though the WFR is set on the machine, it was found to vary during the DP-GMAW process. In DP-GMAW the main role of the

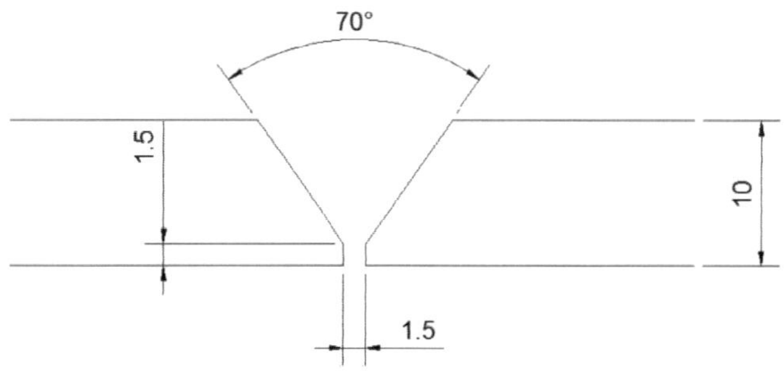

Figure 1 Schematic of the groove designed for the present study

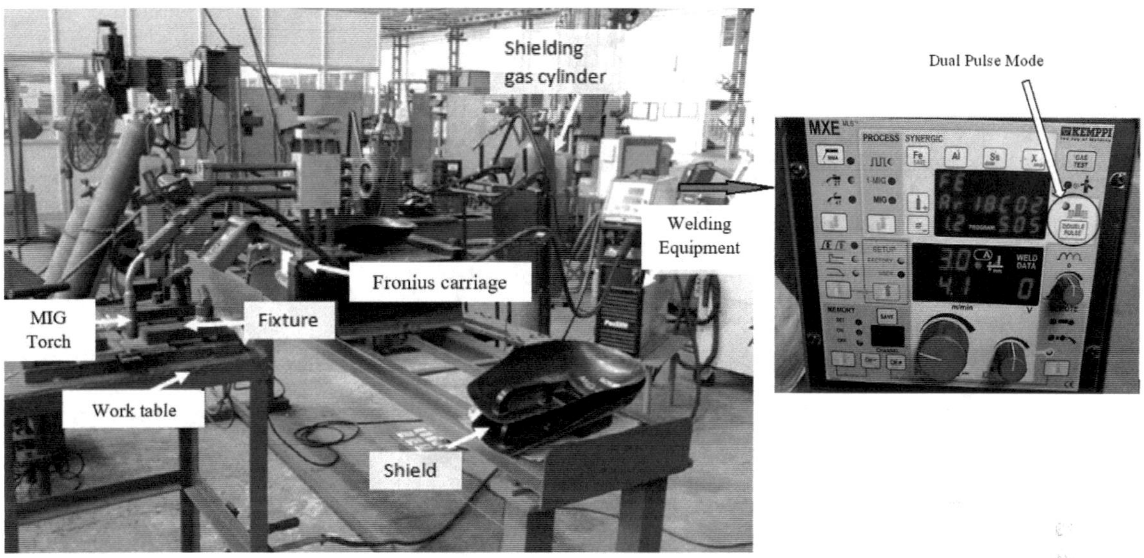

Figure 2 Welding setup showing different parts used in this study

low frequency thermal pulsation (current pulsation, also referred as Inpulse, Alu-Plus, etc [1]) was to control the weld pool. The variation of the mean current between thermal base and thermal pulse was synchronized with WFR to maintain constant arc length [1]. A large variation of WFR in DP-GMAW process significantly influences the mean current and arc voltage during welding. This variation will result in arc instability leading to porosity in the weld pool. Hence, the filler wire was fed through special arrangements of rollers and torch. The carriage FRONIUS-FCU-4-RC that moves automatically adjusting to the variation in WFR was used to minimize the error in feed rate. The current, voltage and the variation of the WFR was measured directly through the digital display on the power source. The characterisation of these power source data along with standoff distance variations of the welding processes were also performed. Welding parameters in each welding passes for all the three welding processes is enlisted in Tables 2-4. WFR value was set for P-GMAW and DP-GMAW processes, and it was observed that the heat input value was lower in DP-GMAW process for same travel speed and higher wire rate.

2.3. Metallographic examination
After welding, the welded plates were cut 25mm away from the edge with help of a power hack saw machine to prepare the samples for metallurgical analysis. The specimens were then polished using the SiC abrasive papers of grade 120, 240, 320, 400, 600, 800 and 1200 grits, one after another. They were further polished with alumina and finally finished with 9μm to 0.5 μm diamond paste. After proper polishing, the specimens were deeply etched using electrolytic etching with 10 % Oxalic acid to reveal

Table 2 Welding parameters in each welding pass for SMAW process

S. No.	Weld layer	Weld pass	Heat input (kJ/cm^2)	Welding parameters		Remarks
				Current, A	Arc voltage, V	
1	1	1	3.7	121+/-2	10+/-1	Root pass, GTAW Filler wire ER309
2	2	2	8.51	120+/-3	26+/-1	SMAW E310 ϕ 3.15 mm
3	3	3	12.97	160+/-2	27+/-1	SMAW E310 ϕ 3.15 mm
4	4	4	9.86	120+/-2	26+/-1	SMAW E310 ϕ 3.15 mm
5	4	5	8.51	120+/-2	26+/-1	SMAW E310 ϕ 3.15 mm
6	4	6	8.51	120+/-2	26+/-1	SMAW E310 ϕ 3.15 mm

Table 3 Welding parameters in each welding pass for P-GMAW process

S. No	Weld layer	Weld pass	Heat Input (kJ/cm^2)	Welding parameters				
				WFR m/min	Mean current, A	Arc voltage, V	Travel speed, S cm/min	Stand of distance, mm
1	1	1	7.41	4	183+/-2	27.0+/-1	40	15
2	2	2	7.5	5	168+/-2	30.3+/-1	40	15
3	3	3	7.73	5	184+/-1	28.0+/-1	40	17
4	3	4	7.15	5	182+/-2	26.2+/-1	40	13
5	4	5	7.02	5	180+/-2	26.0+/-1	40	15
6	4	6	7.4	5	175+/-3	28.2+/-1	40	15
7	4	7	7.18	5	182+/-3	26.3+/-1	40	15

Table 4 Welding parameters in each welding pass for DP-GMAW process

S. No	Weld layer	Weld pass	Heat input (kJ/cm^2)	Welding parameters				
				WFR m/min	Mean current, A	Arc voltage, V	Travel speed, S cm/min	Stand of distance, mm
1	1	1	5.431	6	170+/-2	21.3+/-1	40	15
2	2	2	6.3	6.5	190+/-2	22 .1+/-1	40	15
3	3	3	5.7	6.5	188+/-2	20.2+/-1	40	15
4	3	4	5.6	6.5	186+/-3	20.1+/-1	40	15
5	4	5	5.64	6.5	190+/-2	19.8+/-1	40	15
6	4	6	5.4	6.5	187+/-3	19.3+/-1	40	15
7	4	7	5.67	6.5	188+/-2	20.1+/-1	40	15

the fused metal zone. The macro images were taken at a 6.7X magnification, and microstructures are taken using optical microscope.

2.4. Mechanical characterization tests

The mechanical characterisation of the welded samples were performed through transverse tensile testing, charpy test, and Vickers hardness test. Transverse tensile testing was performed on all the welded samples prepared as per ASTM E8 standards [6] with weld bead at the middle of parallel length. The strain rate for the tensile testing was set to at 1 mm/min. The charpy V notch (CVN) impact testing was done after the completion of the welding. The CVN test provides information about behaviour of metal when subjected to a single application of a load resulting in multiaxial stresses associated with a notch coupled with high rates of loading. The specimen preparation and test procedure for the CVN test was done following ASTM E23 standards [7]. Further, the hardness values of base metal, heat affected zone and weld metal was calculated using Vickers hardness method at load 10 kg.

2.5. Corrosion testing

Corrosion resistance of the austenitic stainless steel occurs due to the presence of thin chromium oxide layer on the surface of the material [8]. The corrosion testing of the welded samples were performed by two methods, namely, immersion testing and electrochemical corrosion testing. During immersion testing method, small sections of the material are exposed to the test medium and the loss of weight of the material is measured over a period of time. In the present experiment the base metal and weld metals are cut separately and are suspended into the corrosion medium. The corrosion environments selected for the study are 3.5% NaCl, 1% KOH and 1% HCl solution due broad-spectrum existence in industrial zones [8-10]. The weight losses were measured regularly and tabulated. Further, the graphs of weight loss vs duration of experiment was plotted as per ASTM G31 standard [11]. The corrosion rates (CR) were calculated from the Eq. 1 [11], where K-constant (8.76 X 10^4 for CR in mm/year), T-time of exposure in hours, A-area in cm^2, W-mass loss in grams and D-density in $gram/cm^3$.

$$\text{Corrosion rate} = \frac{KW}{ATD} \quad .(1)$$

During electrochemical corrosion testing, the weld metal was separated from the base metal by coating paint over the base metal. The exposed area for corrosion testing is 0.36 cm^2.The rest potential selected was between -300mV to -600 mV. The specimen was then suspended over the solution exposing only the cross section of the weld metal with the solution. An electric potential was then applied across the welded specimen. An accelerating electrode was also placed in the solution to accelerate the electrochemical test. The solution considered were of 3.5 % NaCl, 1 % KOH and 1 % HCl. Further, the corrosion rate was calculated and compared from the Tafel plot [12].

3. Results and discussion

3.1. Chemical composition

The chemical composition of base metal, filler wire and weld joint prepared by SMAW, P-GMAW, DP-GMAW determined through vacuum emission spectroscopy are tabulated in Table 5. It was found that the chemical composition of SMA welded region and the E310 filler wire were similar. The loss of composition due to the vaporization effect was compensated by the diffusion effect because of high heat input in SMAW process. Even though the electrode used for P-GMAW and DP-GMAW processes were of same composition, the DP-GMAW process showed low values of Cr and Ni equivalent. This observation was also attributed to the vaporization effect during welding process. The high heat input dilution effect occurred for P-GMAW process whereas heat input was lower in case of DP-GMAW process. Hence, it was concluded that the dilution effect had insignificant role in the modifying chemical composition of the weld.

3.2. Macrographs and micrographs of weldment

Radiographic testing of all the three welded samples were taken, and corresponding macro images are shown in Fig. 3. The figures indicated that all the welded samples are free from any defects. The micro images of the base metal at different magnifications shown in the Fig. 4a shows the number of twin lines in the base metal regions. Figs 4b, c, and d shows the typical weld metal microstructure formed for SMAW, P-GMAW and DP-GMAW processes, respectively. The microstructure images with lower magnification is also shown as inset for reference. It was observed that P-GMAW and DP-GMAW process showed more refined grain structure compared to the SMAW process. The dendrite refinement

Table 5 Chemical composition test results

Materials		C	Si	Mn	Cr	Ni	P	S	Cr eq	Ni eq
Base Metal	AISI 310	0.026	1.284	1.489	25.06	19.746	0.025	0.01	26.986	21.271
Electrode/ Filler Metal	E 310 (SMAW)	0.042	0.620	1.650	25.51	19.740	0.025	0.01	26.44	22.965
	E 309 (GMAW)	0.05	0.350	1.804	24.505	13.913	0.023	0.01	25.03	16.315
Weld Joints	SMAW	0.04	0.614	1.637	25.498	19.194	0.024	0.01	26.498	22.173
	P-GMAW	0.025	0.345	1.806	24.345	13.923	0.023	0.01	24.863	15.576
	DP-GMAW	0.04	0.302	1.72	22.662	12.977	0.023	0.01	23.662	15.037

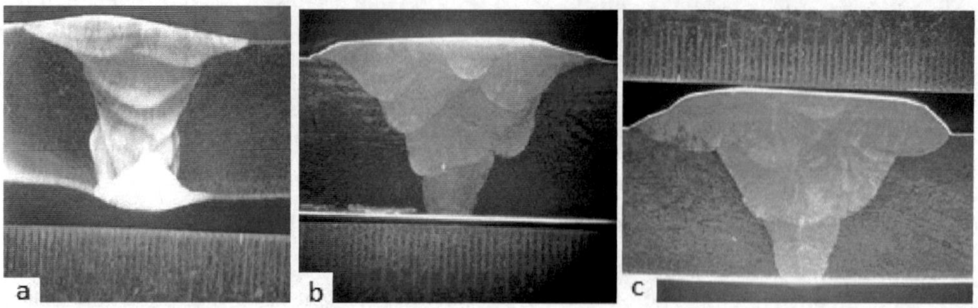

Figure 3 Macro images of (a) SMAW, (b) P-GMAW and (c) DP-GMAW weld

was uniformly distributed throughout the weld joint for DP-GMAW process, whereas dendrite refinement occurs in some areas in case of P-GMAW process. Also, the dendrite refinement was much lower for P-GMAW process. A finer grained microstructure was observed for DP-GMAW sample as the thermal pulsation induced frequent stirring of weld pool during the welding process (Fig. 4d).

3.3. Mechanical Testing

The tensile samples after testing is shown in Fig. 5 demonstrates the failure locations for each welded samples. During the tensile testing, the SMAW specimen failed in the weld metal whereas base material failure was observed for P-GMAW and DP-GMAW specimens. The ultimate tensile strength (UTS) obtained and Charpy impact energy for all the welded samples are compared in Table 6. The highest value of UTS (560 MPa), and impact energy (49 J) was recorded for DP-GMAW sample. Moreover, the Vickers hardness profile across the weld shown in Fig. 6 also revealed the better hardness trend of DP-GMAW sample. The increased values of mechanical parameters evaluated were attributed to the fine grained microstructure in DP-GMAW (Fig. 4) as compared with P-GMAW and SMAW processes [13].

3.4. Corrosion behavior of the welds

The laboratory immersion corrosion rate values using different medium, i.e. 3.5% NaCl, 1% KOH and 1% HCl solutions, are shown in the Fig. 7a, b and c respectively. The plots of potentio-dynamic polarisation (PDP) behaviours of the AISI 310 (base metal), SMAW, P-GMAW and DP-SMAW welds under different corrosion testing medium are also given in the Fig. 8. The corresponding Icorr and corrosion rate for three weld sample are tabulated in Tables 7. It was observed that the corrosion rate was maximum for DP-GMAW sample for all the three corrosion medium whereas it was almost same for both base metal and SMAW sample. In laboratory immersion testing, the corrosion rate was minimum for base metal in 3.5% NaCl solution and the maximum value was 0.32 mm/year for DP-GMAW sample. For 1% KOH solution, corrosion rate was maximum for DP-GMAW process with a maximum value of 0.15 mm/year. Minimum corrosion in this solution occurred for base metal. The

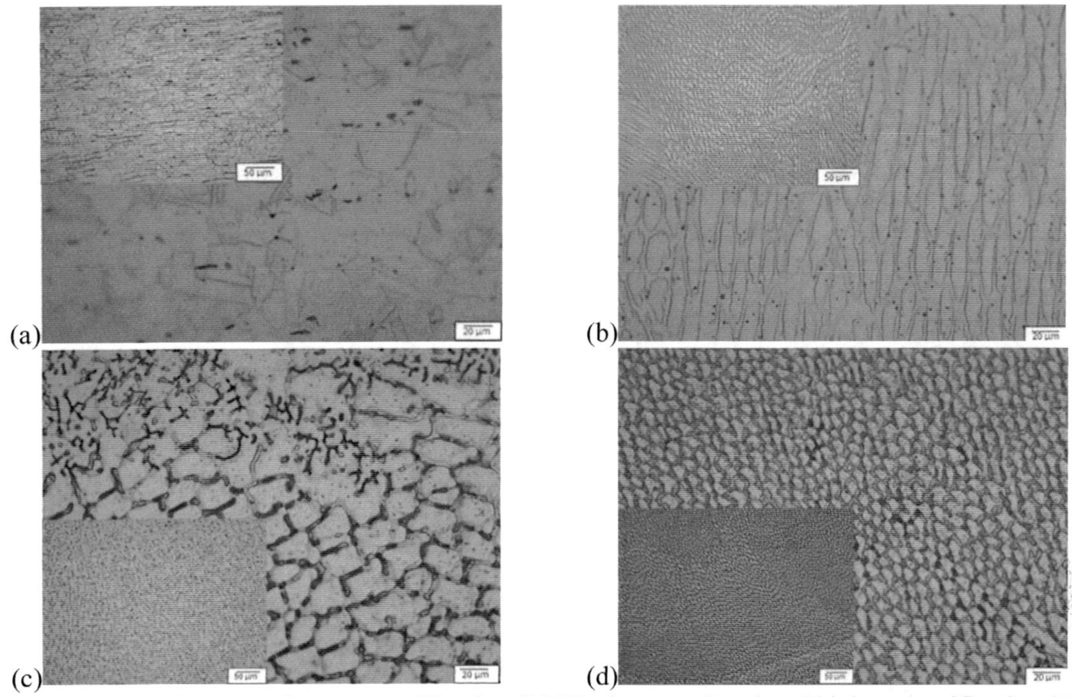

Figure 4 Micro images at lower magnification (200X, shown as inset) and higher magnification (500X, main image) of (a) base metal, and weld obtained after (b) SMAW, (c) P-GMAW, and (d) DP-GMAW processes

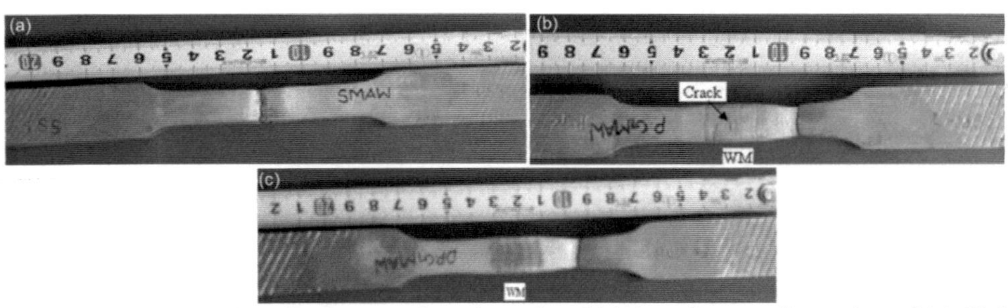

Figure 5 Photographs showing the fracture location of weld joint after tensile testing of (a) SMAW (b) P-GMAW and (c) DP-GMAW samples

Table 6 UTS and position of fracture from tensile test and Charpy impact test results

Process/Metal	UTS in MPa	Position of fracture	Impact energy in Joules
SMAW	490	WM	44
P-GMAW	516	BM	44
DP-GMAW	560	BM	49
Base metal	515- 550	----	34

corrosion behaviour under 1% HCl solution was the same as that of the 3.5% NaCl and 1% KOH solution, and the maximum value of corrosion rate (0.27 mm/year) was observed for DP-GMAW sample. As observed from Fig. 7, the corrosion rate was maximum between 48 to 72 hours in all the medium which was due to the breakage of primary layer which resists the initial corrosion. The PDP test using conventional electrochemical setup and using different electrolytes also revealed similar result as that of laboratory immersion testing. The Icorr value was minimum for base material when compared

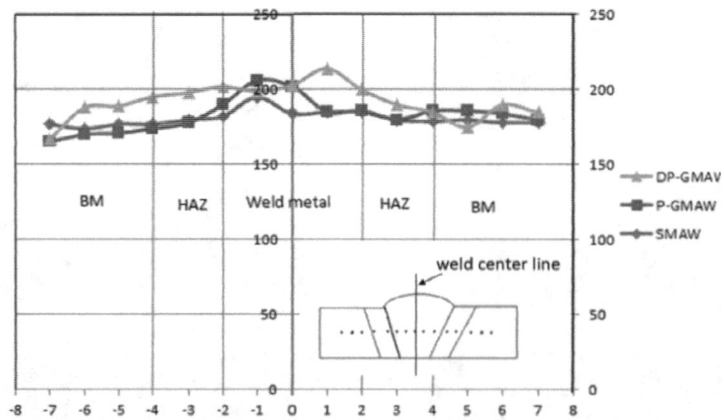

Figure 6 Vickers hardness profile across the weld for different welding processes.

Table 7 Electrochemical data for base metal and weld metal at 3.5% NaCl, 1% KOH and 1% HCl

Solution/Samples	BM	SMAW	P-GMAW	DP-GMAW
(a) 3.5 % NaCl				
Icorr (mA/cm²)	0.01992	0.03922	0.12376	0.20137
Corr.Rate (mm/year)	0.21313	0.41966	1.324	2.1547
(b) 1% KOH				
Icorr (mA/cm²)	0.33727	0.46379	0.56657	0.69313
Corr.Rate (mm/year)	3.6088	4.9625	6.0623	7.4165
(c) 1% HCl				
Icorr (mA/cm²)	0.24053	0.31719	0.59931	0.63978
Corr.Rate (mm/year)	2.5737	3.3939	6.4126	6.8457

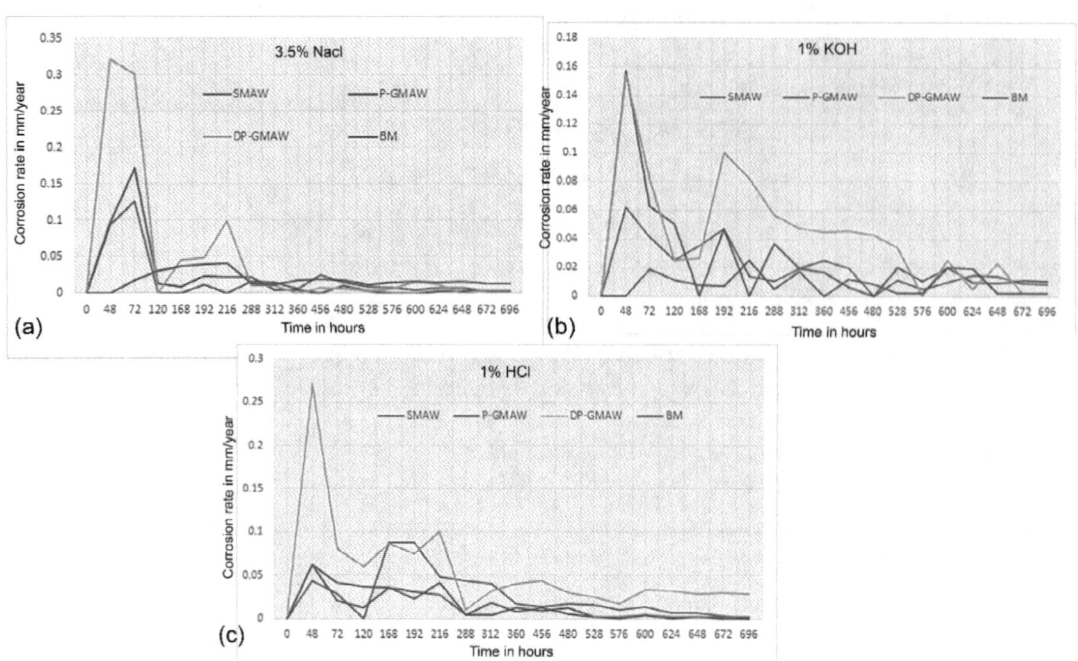

Figure 7 Graph of corrosion rate (mm/year) vs duration in (a) 3.5% NaCl, (b) 1% KOH, and (c) 1% HCl solutions

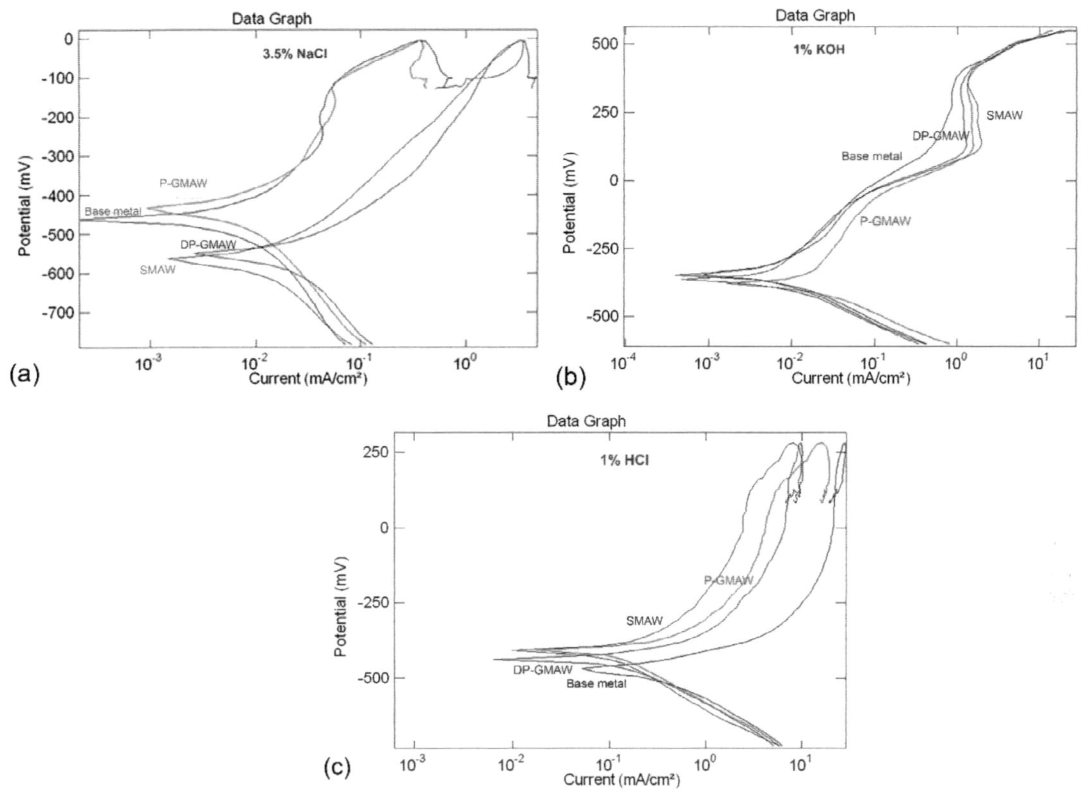

Figure 8 Potentio-dynamic polarization curves recorded in (a) 3.5 % NaCl, (b) 1% KOH, and (c) 1% HCl solutions

to the welded specimens under all three electrolytes. The maximum Icorr value was observed for DP-GMAW process in all the three electrolytes. As per the chemical analysis in Table 5, DP-GMAW process has the minimum Chromium and Nickel content. There was a reduction 9.56 % chromium content DP-GMAW welded sample as compared to base material. Also, this process has the lower heat input as compared to P-GMAW process. Both these effects resulted in higher corrosion rate of DP-GMAW sample than P-GMAW and SMAW samples.

4. Conclusion

In this work, AISI 310 Plate were welded using SMAW, P-GMAW and DP-GMAW welding processes. The mechanical, metallurgical and corrosion properties of the base material and all the three welded samples. The major conclusions derived from are listed below.

a. Chemical analysis of the welded samples shows that there is 9.56% lowering of Chromium content in DP-GMAW process in compared to base material because of the thermal pulsation and vaporization effect.
b. The metallurgical analysis using optical microscopy shows that the DP-GMAW process has the fine grained microstructure compared to other welded samples and base material because of the thermal pulsation that induced a stirring action on the molten weld pool.
c. Hardness profile and Charpy impact testing results showed that the DP-GMAW process was superior to other welding processes because of the development of finer grains.
d. Even though DP-GMAW has excellent in mechanical properties, laboratory immersion testing and PDP results shows that it has a higher corrosion rate when compared to other two welding processes because of relatively low Chromium content and heat input.

References

[1] Da Silva C L M and Scotti A 2006 The influence of double pulse on porosity formation in aluminum GMAW *Journal of Materials Processing Technology* **171**(3) 366-372

[2] Jie Y I, Cao S F, Li L X, Guo P C and Liu K Y 2015 Effect of welding current on morphology and microstructure of Al alloy T-joint in double-pulsed MIG welding *Transactions of Nonferrous Metals Society of China* **25**(10) 3204-3211

[3] Liu A, Tang X and Lu F 2013 Arc profile characteristics of Al alloy in double-pulsed GMAW *The International Journal of Advanced Manufacturing Technology* **65**(1-4) 1-7

[4] Zhang H, Hu S, Shen J, Ma L and Yin F 2015 Microstructures and mechanical properties of 30Cr-4Mo ferritic stainless steel joints produced by double-pulsed gas metal arc welding *The International Journal of Advanced Manufacturing Technology*, **80**(9) 1975-1983

[5] Hosseini H S, Shamanian M and Kermanpur A 2011 Characterization of microstructures and mechanical properties of Inconel 617/310 stainless steel dissimilar welds *Materials characterization* **62**(4) 425-431

[6] ASTM, E8/E8M Standard Test Methods for Tension Testing of Metallic Materials, 2010 *Annu. B. ASTM Stand., 2010* **4** 1-27

[7] ASTM E23-18 Standard Test Methods for Notched Bar Impact Testing of Metallic Materials, 2018 *ASTM International*, West Conshohocken, PA, 2018

[8] Hamidah I, Solehudin A, Setiawan A, Hamdani A, Hidayat M A S, Adityawarman F, Khoirunnisa F and Nandiyanto A B D 2018 Corrosion study of AISI 304 on KOH, NaOH, and NaCl solution as an electrode on electrolysis process *Journal of Engineering Science and Technology* **13** (5) 1345-1351.

[9] El-Shamy A M, El-Hadek M A, Nassef A E and El-Bindary R A 2020 Optimization of the influencing variables on the corrosion property of steel alloy 4130 in 3.5 wt.% NaCl solution *Journal of Chemistry*, *2020*

[10] Corrosion resistance of the austenitic chromium-nickel stainless steels in chemical environments, 1963 INCO, *The International Nickel Company, Inc. One New York Plaza,* New York, N.Y. 10004

[11] ASTM NACE / ASTMG31-12a, Standard Guide for Laboratory Immersion Corrosion Testing of Metals, *ASTM International*, West Conshohocken, PA, 2012

[12] Electrochemical Corrosion, *Handbook of Environmental Degradation of Materials (Second Edition)* 2012 87-125

[13] Sen M, Mukherjee M, Singh S K and Pal T K 2018 Effect of double-pulsed gas metal arc welding (DP-GMAW) process variables on microstructural constituents and hardness of low carbon steel weld deposits *Journal of manufacturing processes* **31** 424-439.

Sustainable Processing of Dissimilar Aluminium Alloy Joints by Friction Stir Welding

K Ramesha[1,2], P D Sudersanan[2], Santosh N[3] and Sasidhar Jangam[1*]
[1]Department of Mechanical and Automobile Engineering, CHRIST (Deemed to be University), Bengaluru, Karnataka, India
[2]Department of Mechanical Engineering, Dr. T Thimmiah Institute of Technology, Kolar Gold Fields, Karnataka, India
[3]Department of Mechanical Engineering, MVJ College of Engineering, Bengaluru, Karnataka, India

[*]E-mail: sasidhar.jangam@gmail.com (Sasidhar Jangam)

Abstract. Friction stir welding is one of the unique solid state welding technique that is fast gaining importance because of its excellent properties pertaining to the strength, wear, hardness and better joining capabilities. The friction stir welding technique is effectively used in this research to join 5 mm thick dissimilar aluminium alloys of AA 7075 and AA 5052 grade. The friction stir welding joints are obtained by using three different tool profiles viz., straight cylinder, taper cylinder, and straight triangular at a speed of 800, 1000, 1200 rpm and a feed rate of 100, 120, 140 mm/min. The experiments are designed based on L9 orthogonal array considering Taguchi techniques. From the experimental results, the process methodology is validated and the outcomes of the experiment are considered for fabrication of strong joints. The design and optimization of the process has led to better process control and thereby extended its scope for the use of the optimized process parameters for joining dissimilar aluminium alloys in real time engineering applications with sustained methodologies involving reduced dwell time and optimized friction forces and torque during the friction stir welding of dissimilar aluminium alloys.

Keywords: Friction stir welding, Dissimilar aluminum alloys, AA 5052, AA 7075

1. Introduction:

Friction stir welding (FSW) is the solid-state welding process, in which the joining of work pieces can be done by using non consumable tool, which helps to soften the metal by generating a heat due to friction between rotating tool and work piece. When the tool is rotating on the work piece, it will plastically deform the material and result in a strong joint. FSW process is fast gaining significance in welding of Aluminium and its alloys, especially in aerospace and automobile components [1–4].

AA 7xxx and AA 5xxx are light metal alloys which are most commonly used in various structural applications, especially in the field of aerospace, marine and automobile industries. When these materials are used to fabricate different structural components, the total weight will be decreased. The joining of dissimilar aluminum alloys by using conventional fusion welding technique results in the formation of intermetallic compounds and various welding related defects [5] due to the generation of high temperature. Formation of solidification cracks in the aluminum alloy is a common defect in fusion welding of aluminium alloys due to variation in the weld metal composition. The composition of base metal, filler metal and amount of dilution is very important to reduce the formation of solidification cracking, Henceforth, FSW process can provide the solutions to these types of defects [6, 7]. Further, the process parameters have a major effect on the outcomes of the results in the FSW process.

2. Materials and Methods

2.1. Work Piece Material

The process of friction stir welding is carried out on dissimilar alloys of aluminium, viz., AA 7075 and AA 5052, the composition of the alloys selected for the present work is given in Table 1. The work piece is cut into a required size of 170 x 60 x 5 mm, depending upon the available fixture dimensions to hold the work piece.

Table 1. Composition of AA 7075 and AA 5052 alloy

Elements (Wt. %)	Al	Zn	Mn	Si	Cu	Fe	Mg	Ti	Cr	Others
Al7075	Bal.	5.8	0.2	0.4	1.6	0.4	2.4	0.1	0.25	0.14
Al5052	Bal.	0.1	0.1	0.45	0.1	0.35	2.6	-	0.2	0.15

AA 7075, aluminium alloy is selected for the present work, since it exhibits high strength, fracture toughness and resistance to corrosion which is the basic attribute for its use in aerospace structures, while AA 5052 is an aluminium alloy having magnesium as the major alloying element with better strength, corrosion resistance and weldability that is the major factor for its selection in present work to optimize the process parameters of the FSW for plate structures especially having its wide scope in aerospace components.

2.2. Tool Material

The friction stir welding is carried out by using a typical non consumable tool made up of H13 tool steel with the hardness 55 HRC. The specification of tool is shown in Table 2. In the present research work, three pin profiles viz., cylindrical, cylindrical taper, and triangular have been considered.

Table 2. Tool Specifications

Tool shoulder	**20 mm (flat surface)**	
Tool pin conftguration	Cylindrical, Cylindrical tper, and Triangular	
Pin length	4.8 mm	
Pin Size	Cylindrical	D = 5 mm
	Triangular	L = 5 mm
	Cylindrical (Taper)	D = 5 mm d = 4 mm

2.3. Experimental Methods

Aluminium AA 7075 and Aluminium AA 5052 alloys are welded together by friction stir welding process on an ETA make 10T model horizontal FSW machine as shown in Figure 1. The process is carried out in accordance with the Design of Experiments (DOE) table framed in accordance with L9 orthogonal array considering Taguchi Design model from Minitab Software. The process parameters considered in present work involved design specific parameters like tool rotational speed (800, 1000, and 1200 rpm), traverse feed (80, 100, and 120 mm/min) and tool pin configuration (cylindrical, cylindrical taper, and triangular). The weld joint of trial 1 is as shown in figure 2. The details of welding parameters and its levels are given in Table 3. During the process of welding, Aluminium AA 7075 is located on retreating side while Aluminium AA 5052 is located on advancing side. The tool shoulder is plunged into the plates fastened on the special fixtures clamped on the table of the machine and the FSW process carried out in accordance with the Design of Experiments. The process parameters are optimized by considering the forces and torque encountered during the process.

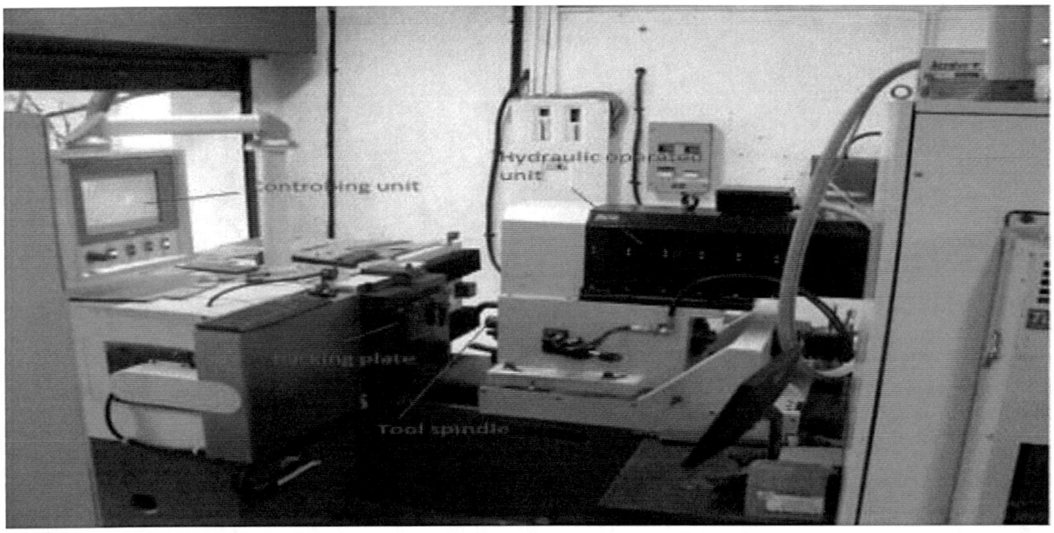

Figure 1. ETA make 10T model horizontal FSW machine

Table 3. Experimentation details

Sl. No	Parameters	Level 1	Level 2	Level 3
1.	Rotational Speed (rpm)	800	1000	1200
2.	Feed (mm/min)	100	120	140
3.	Tool Offset (mm)	-0.5	0	0.5
4.	Tool Pin Profile	Cylindrical	Cylindrical (Taper)	Triangle

3. Results and Discussions

The experimental trials are carried out in accordance with the "L9 Taguchi Orthogonal Array (OA)", which is given in the Table 4. The outcomes of the experimental trials are validated from the optimum process parameters identified from the graphs.

The "Orthogonal Array (OA)" is a scientifically disciplined frame work model, used for evaluating and optimizing the process parameters. During welding process, there are essentially two forces viz., Normal Force, Traverse Force, and Spindle Torque. The optimization of these forces are important for evolving a sustainable methodology.

3.1. S1F1T1 Condition

3.1.1. Normal Load for S1F1T1 Condition. The normal load exerted by the tool on the work piece during the course of welding for spindle speed (S1) of 800 rpm, Feed (F1) of 100 mm/min

Figure 2. Friction Stir Weld Joint of AA 7075 and AA 5052

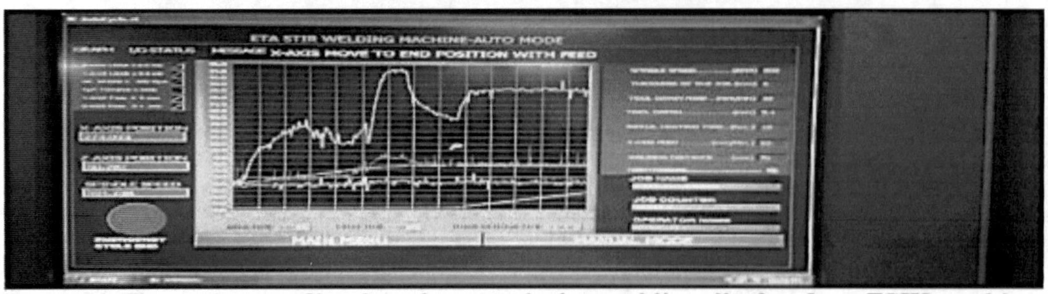

Figure 3. The display of variation of loads and torque during welding from FSW machine

Table 4. DOE– L9 Orthogonal Array

Exp. No.	Designation of the Conditions	Rotational Speed (RPM)	Feed (mm/min)	Tool Offset (mm)	Tool Pin Profile
1	S1F1T1C	800	100	(-)0.5	Cyl.
2	S1F1T1CT	800	100	(-)0.5	Cyl.(Tp)
3	S1F1T1T	800	100	(-)0.5	Tr.
4	S2F2T2C	1000	120	0	Cyl.
5	S2F2T2CT	1000	120	0	Cyl. (Tp.)
6	S2F2T2T	1000	120	0	Tr.
7	S3F3T3C	1200	140	0.5	Cyl.
8	S3F3T3CT	1200	140	0.5	Cyl.(Tp.)
9	S3F3T3T	1200	140	0.5	Tr.

and tool offset of (-) 0.5 mm for cylindrical pin profile (C), cylindrical taper (CT) pin profile and Triangular (T) pin profile is given in figure 4, it is herewith noted that the normal load exerted by the tool on workpiece with time is maximum for cylindrical tool profile and the time duration of the normal load acting on the workpiece is also found to be maximum for cylindrical tool pin profile, followed by cylindrical taper and triangular tool profile, this is because, the cylindrical pin profile offered resistance to the flow ability of the workpiece around the tool all along the length of the weld zone. The peak normal load is for S1F1T1C condition, at a magnitude of 15 k N, followed by S1F1T1CT and S1F1T1T conditions. The time period for normal loads acting on the workpiece during the course of friction stir welding is 140 seconds for S1F1T1C condition, 90 seconds for S1F1T1CT condition, and 70 seconds for S1F1T1T condition.

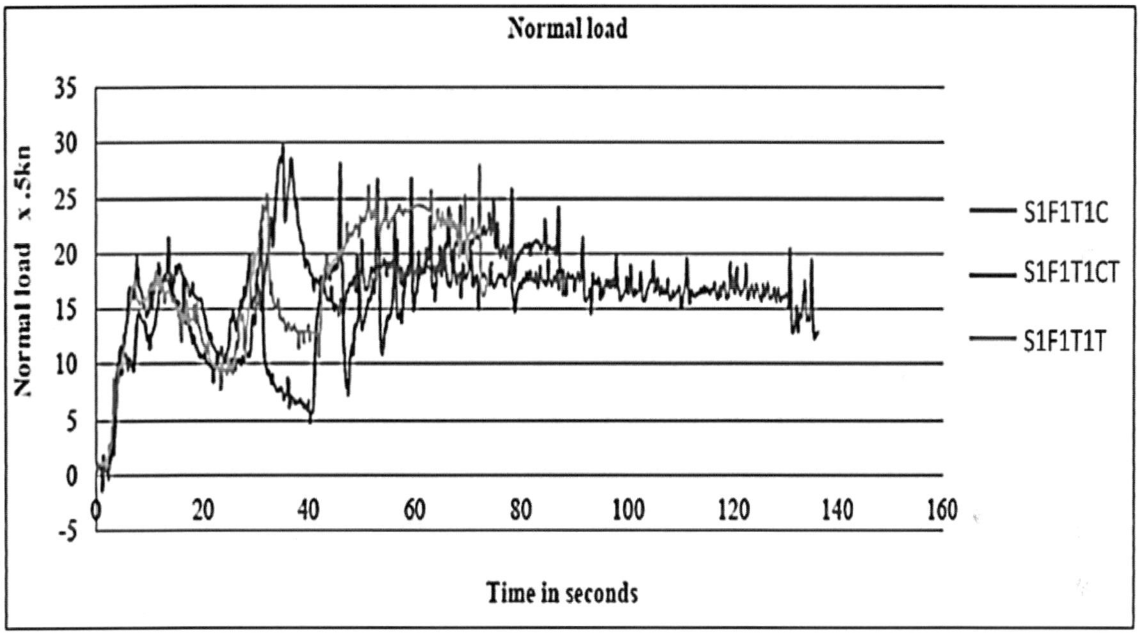

Figure 4. Normal Load for S1F1T1 conditions

3.1.2. Transverse Load for S1F1T1 Condition. The traverse load exerted by the tool on the work piece during the course of welding for spindle speed (S1) of 800 rpm, Feed (F1) of 100 mm/min and tool offset of (-) 0.5 mm for cylindrical pin profile (C), cylindrical taper (CT) pin profile and Triangular (T) pin profile is given in Figure 5, it is herewith noted that the traverse load exerted by the tool on workpiece with time is maximum for cylindrical tool profile and the time duration of the traverse load acting on the workpiece is also found to be maximum for cylindrical tool pin profile, followed by cylindrical taper and triangular tool profile. The peak traverse load is for S1F1T1C condition, at a magnitude of 1.75 k N, followed by S1F1T1CT and S1F1T1T conditions. The time period for traverse loads acting on the workpiece during the course of friction stir welding is 140 seconds for S1F1T1C condition, 90 seconds for S1F1T1CT condition, and 70 seconds for S1F1T1T condition.

3.1.3. Spindle Torque for S1F1T1 Condition. The spindle torque of the tool on the work piece during the course of welding for spindle speed (S1) of 800 rpm, Feed (F1) of 100 mm/min and tool offset of (-)0.5 mm for Cylindrical pin profile (C), Cylindrical taper (CT) pin profile and

Triangular (T) pin profile is given in figure 6, it is herewith noted that the spindle torque of the tool on workpiece with time is maximum for cylindrical tool profile and the time duration of the torque acting on the workpiece is also found to be maximum for cylindrical tool pin profile, followed by cylindrical taper and triangular tool profile. The peak spindle torque is for S1F1T1CT condition, at a magnitude of 14 N-m, followed by S1F1T1C and S1F1T1T conditions. The time period for traverse loads acting on the workpiece during the course of friction stir welding is 140 seconds for S1F1T1C condition, 90 seconds for S1F1T1CT condition, and 70 seconds for S1F1T1T condition.

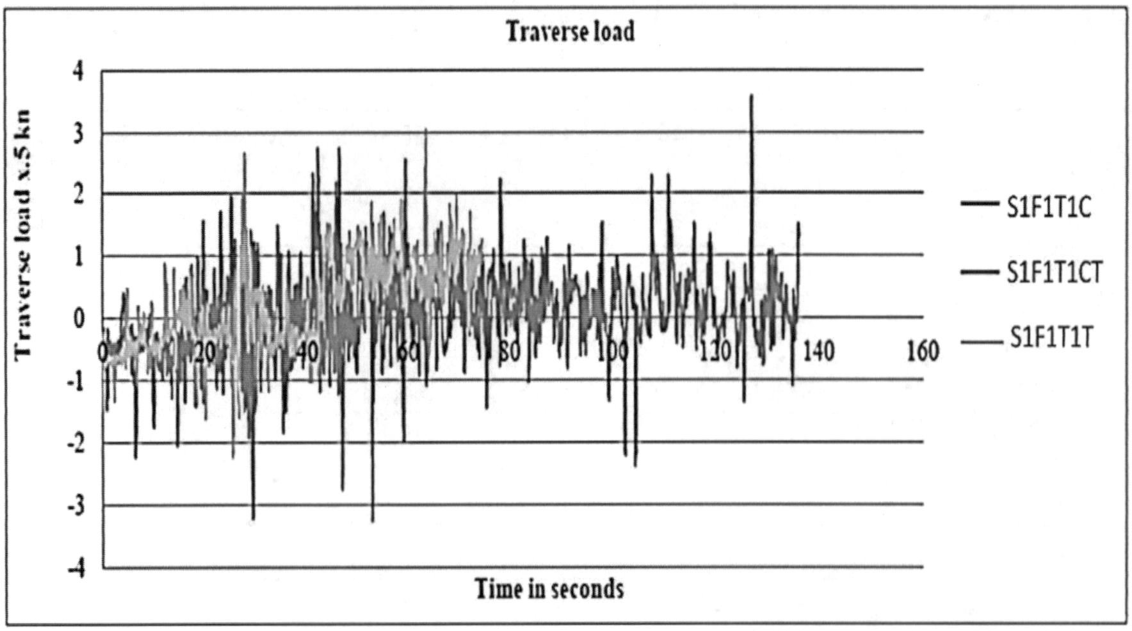

Figure 5. Traverse Load for S1F1T1 conditions

3.2. S2F2T2 Condition

3.2.1. Normal Load for S2F2T2 Condition. The normal load exerted by the tool on the work piece during the course of welding for spindle speed (S2) of 1000 rpm, Feed (F2) of 120 mm/min and tool offset (T2) of 0 mm for cylindrical pin profile(C), cylindrical taper (CT) pin profile and Triangular (T) pin profile is given in Figure7, it is here with noted that the normal loade xerted by the tool on workpiece withtime is workpiece is also found to cylindrical taper and triangular tool profile.The peak normal load is for S2F2T2CT condition, at a magnitude of S2F2T2C and S2F2T2T conditions. The time period for normal loads acting on the workpiece during the course of friction stir welding is 138 seconds for S2F2T2C condition, 91 seconds for S2F2T2CT condition, and 72 seconds for S2F2T2T condition.

3.2.2. Transverse Load for S2F2T2 Condition. The traverse load exerted by the tool on the work piece during the course of welding for spindle speed (S2) of 1000 rpm, Feed (F2) of 120 mm/min and tool offset of 0 mm for cylindrical pin profile (C), cylindrical taper (CT) pin profile and Triangular (T) pin profile is given in Figure 8, it is herewith noted that the traverse load exerted by the tool on workpiece with time is maximum for cylindrical tool profile and the time duration of the traverse load acting on the workpiece is also found to be maximum for cylindrical

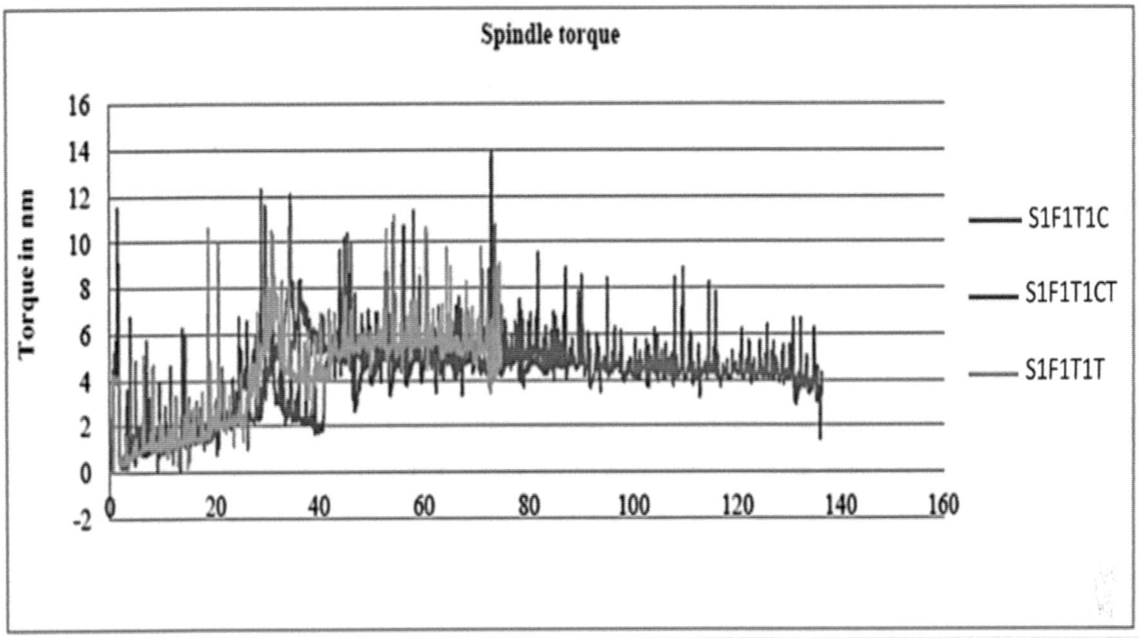

Figure 6. Spindle Torque for S1F1T1 conditions

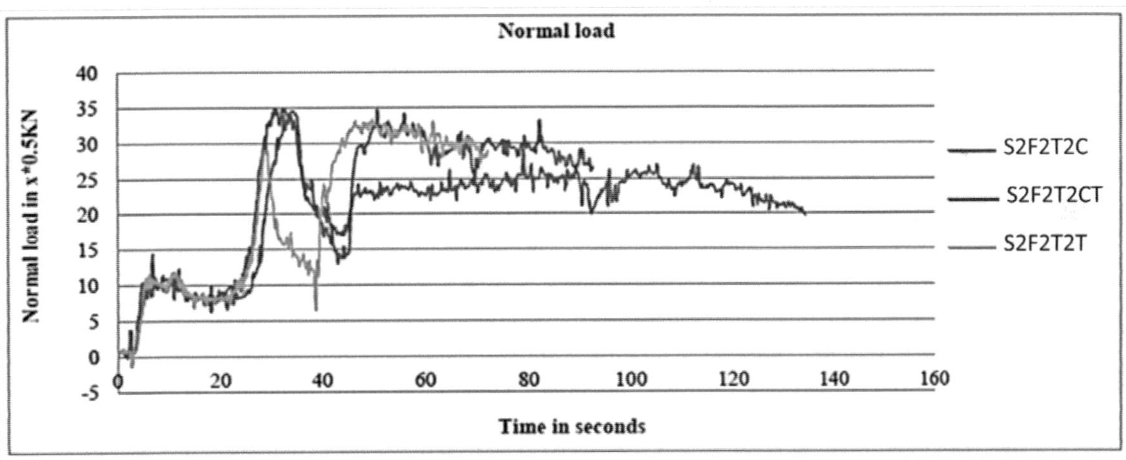

Figure 7. Normal load for S2F2T2 conditions

tool pin profile, followed by cylindrical taper and triangular tool profile. The peak traverse load is for S2F2T2C condition, at a magnitude of 2.25 k N, followed by S2F2T2CT and S2F2T2T conditions. The time period for traverse loads acting on the workpiece during the course of friction stir welding is 138 seconds for S2F2T2C condition, 91 seconds for S2F2T2CT condition, and 72 seconds for S2F2T2T condition.

3.2.3. Spindle Torque for S2F2T2 Condition. The spindle torque of the tool on the work piece during the course of welding for spindle speed (S2) of 1000 rpm, Feed (F2) of 120 mm/min and tool offset of 0 mm for Cylindrical pin profile (C), Cylindrical taper (CT) pin profile and

Triangular (T) pin profile is given in Figure 9, it is herewith noted that the spindle torque of the tool on workpiece with time is maximum for cylindrical tool profile and the time duration of the torque acting on the workpiece is also found to be maximum for cylindrical tool pin profile, followed by cylindrical taper and triangular tool profile. The peak spindle torque is for S2F2T2CT condition, at a magnitude of 6.2 N-m, followed by S2F2T2C and S2F2T2T conditions. The time period for traverse loads acting on the workpiece during the course of friction stir welding is 138 seconds for S2F2T2C condition, 91 seconds for S2F2T2CT condition, and 72 seconds for S2F2T2T condition.

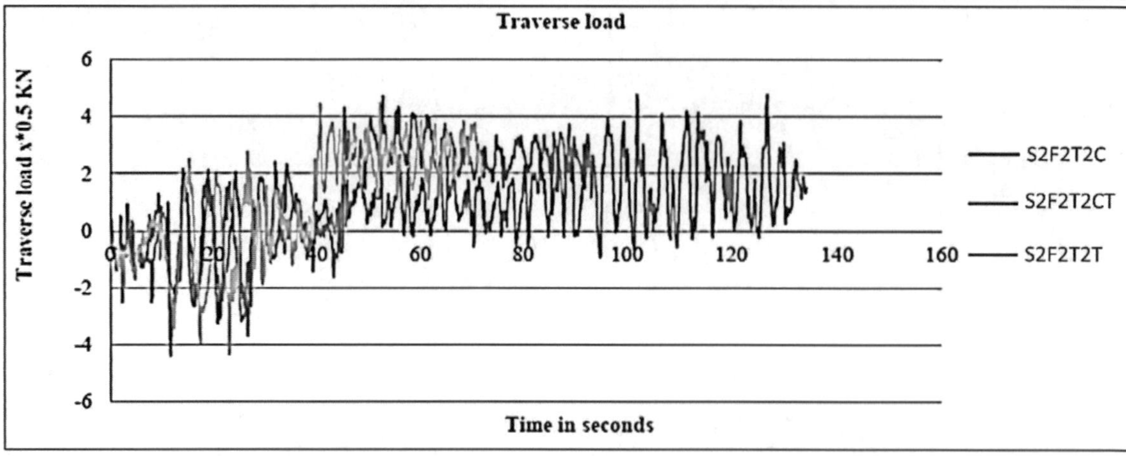

Figure 8. Traverse load for S2F2T2 conditions

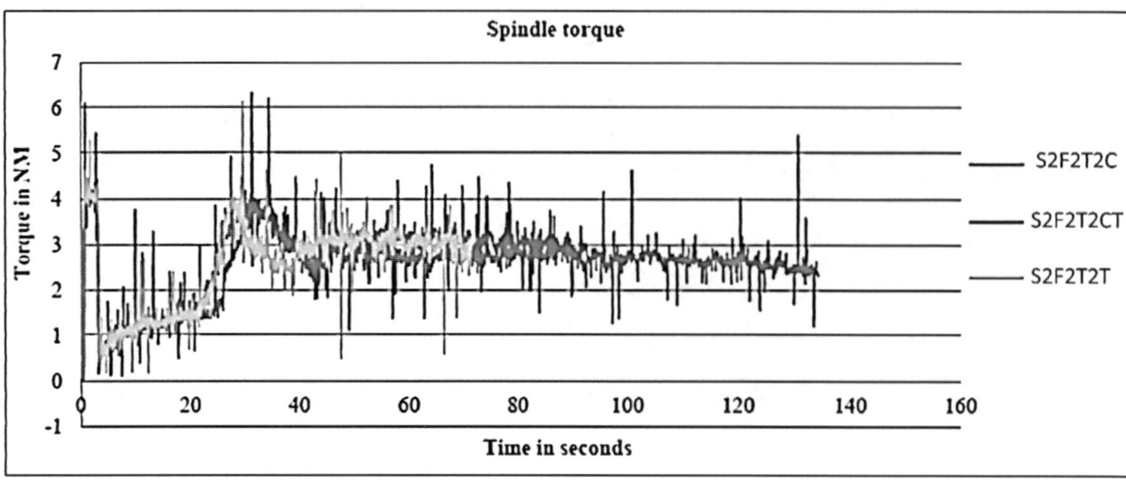

Figure 9. Spindle Torque for S2F2T2 conditions

3.3. S3F3T3 Condition

3.3.1. Normal Load for S3F3T3 Condition. The normal load exerted by the tool on the work piece during the course of welding for spindle speed (S3) of 1200 rpm, Feed (F3) of 140 mm/min and tool offset of 0.5 mm for cylindrical pin profile (C), cylindrical taper (CT) pin profile and Triangular (T) pin profile is given in Figure 10, it is herewith noted that the normal load exerted by the tool on workpiece with time is maximum for cylindrical tool profile and the time duration of the normal load acting on the workpiece is also found to be maximum for cylindrical tool pin profile, followed by cylindrical taper and triangular tool profile, this is because of the fact that the cylindrical tool profile offers more resistance to the flow of the work piece material during the due course of friction stir welding process. The peak normal load is for S3F3T3C condition, at a magnitude of 18 k N, followed by S3F3T3CT and S3F3T3T conditions. The time period for normal loads acting on the workpiece during the course of friction stir welding is 136 seconds for S3F3T3C condition, 87 seconds for S3F3T3CT condition, and 69 seconds for S3F3T3T condition.

3.3.2. Transverse Load for S3F3T3 Condition. The traverse load exerted by the tool on the work piece during the course of welding for spindle speed (S3) of 1200 rpm, Feed (F3) of 140 mm/min and tool offset of 0 mm for cylindrical pin profile (C), cylindrical taper (CT) pin profile and Triangular (T) pin profile is given in Figure 11, it is herewith noted that the traverse load exerted by the tool on workpiece with time is maximum for cylindrical tool profile and the time duration of the traverse load acting on the workpiece is also found to be maximum for cylindrical tool pin profile, followed by cylindrical taper and triangular tool profile. The peak traverse load is for S3F3T3CT condition, at a magnitude of 1.45 k N, followed by S3F3T3C and S3F3T3T conditions. The time period for traverse loads acting on the workpiece during the course of friction stir welding is 136 seconds for S3F3T3C condition, 87 seconds for S3F3T3CT condition, and 69 seconds for S3F3T3T condition.

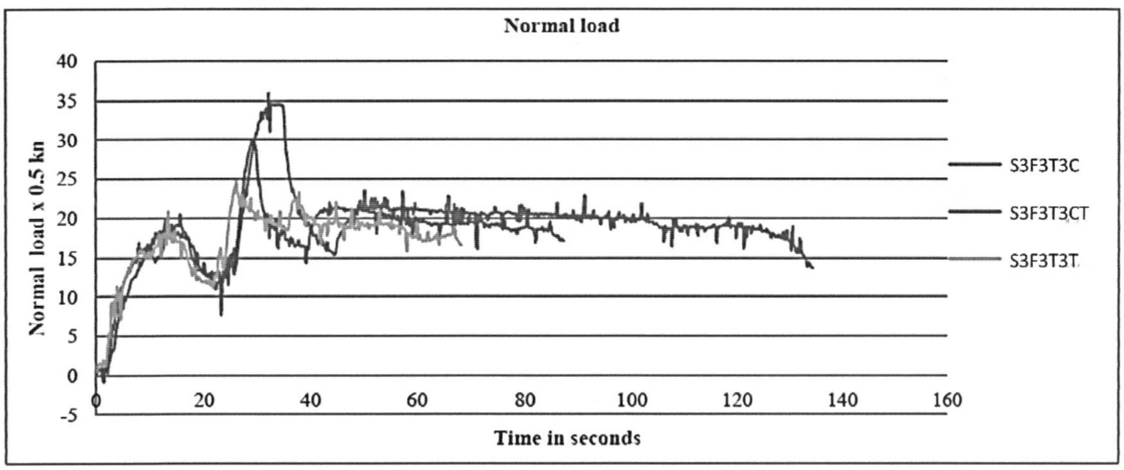

Figure 10. Normal load for S3F3T3 conditions

3.3.3. Spindle Torque for S3F3T3 Condition. The spindle torque of the tool on the work piece during the course of welding for spindle speed (S3) of 1200 rpm, Feed (F3) of 140 mm/min and tool offset of 0.5 mm for Cylindrical pin profile (C), Cylindrical taper (CT) pin profile and Triangular (T) pin profile is given in Figure 12, it is herewith noted that the spindle torque of the tool on workpiece with time is maximum for cylindrical tool profile and the time duration of the torque acting on the workpiece is also found to be maximum for cylindrical tool pin profile, followed by cylindrical taper and S3F3T3C and S3F3T3T conditions. The time period for traverse loads acting on the workpiece during the course of friction stir welding is 136 seconds for S3F3T3C condition, 87 seconds for S3F3T3CT condition, and 69 seconds for S3F3T3T condition.

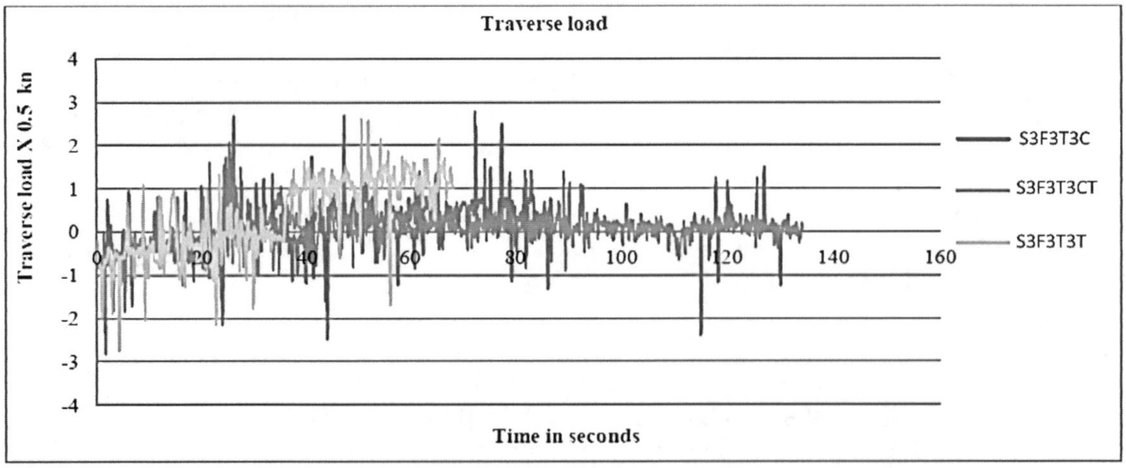

Figure 11. Transverse load for S3F3T3 conditions

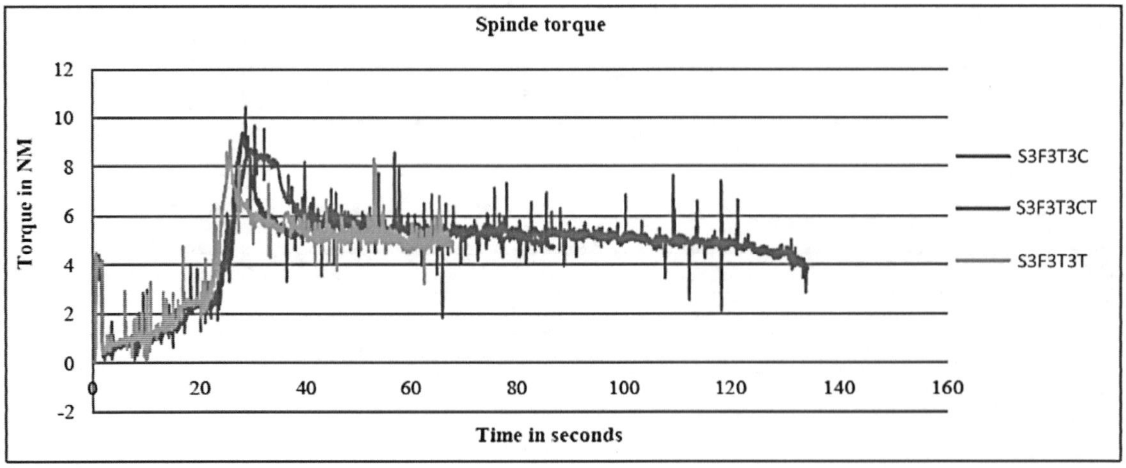

Figure 12. Spindle Torque for S3F3T3conditions

3.4. Summary of force and torque optimizations

The comprehensive review of the normal load, traverse load and torque versus time for different process conditions has provided valid base for deciding the optimum process parameters for Friction Stir Welding (FSW), the dissimilar aluminium alloys with minimum time duration and optimized forces; The Table 5 gives the time duration and the subsequent peak normal loads, peak traverse loads and spindle torque for friction stir welding at different conditions tabulated from the careful observations of the graphs, it is herewith observed that the time duration of welding and the forces acting on the tool pin profile during the process are optimized when a triangular pin profile is used for friction stir welding the plates at a spindle speed of 1200 rpm and feed rate of 140 mm/min with a positive tool offset of 0.5 mm.

William R L et al., have reported the variation of force and torque during the friction stir welding, they have investigated the relationship between axial force and torque and have plotted the variation of forces and torque with time. The results from the present work follows the same trend of peak force and torque variations and likewise when, the torque increases, so does the

Table 5. Peak Loads and Duration for FSW at different conditions

Exp. No.	Designation of the Conditions	Duration For FSW (S)	Peak FN k N	Peak FT k N	Peak ST N-m
1	S1F1T1C	140	15	1.75	12
2	S1F1T1CT	90	13.5	1.35	14
3	S1F1T1T	70	13	1.5	13
4	S2F2T2C	138	17	2.25	6.1
5	S2F2T2CT	91	17.5	2.2	6.2
6	S2F2T2T	72	15	2.1	6
7	S3F3T3C	136	18	1.4	9.6
8	S3F3T3CT	87	15	1.45	10.2
9	S3F3T3T	69	12.5	1.25	8.8

forces [8].

Maurizio B, et al., have extensively worked on the sustainability of friction stir welding of aluminium sheets, and have identified that the environmental impact of friction stir welding in terms of the dwell time and peak forces and torque encountered during the process, the findings of their work have been in parlance with the findings of the present work, with respect to the reduction of the period for friction stir welding of dissimilar aluminium alloy plates using triangular pin profile, viz., 69 seconds for the completion of the process of friction stir welding for 170 mm length among all the experimental trials design according to taguchi's L9 orthogonal array (OA) [9].

3.5. Microstructure

The microstructure of the top and transverse section of the friction stir weld zone is studied using the SEM images captured using a VEGA3 TESCAN machine.

3.5.1. Microstructure of the top surface. The morphology of the top surface of the L9 weld joints processed at 1200 rpm rotational speed, feed rate of 100 mm/min, with a tool offset of -0.5 mm for tapered cylindrical pin profile are studied using the SEM images captured using VEGA TESCAN scanning electron microscopes at various magnifications and the microstructure observed at 1.5 kx is given in the Figure 13(a). It is evident from the microstructure of the weld zone that the bonding has occurred coherently between the different zones, which eventually results in the formation of striated bands of the weldment.

3.5.2. Microstructure of the transverse surface. The morphology of the transvers surface of the L9 weld joints processed at 1200 rpm rotational speed, feed rate of 100 mm/min, with a tool offset of -0.5 mm for tapered cylindrical pin profile are studied using the SEM images captured using VEGA TESCAN scanning electron microscopes various magnifications and the microstructure observed at 1.5 kx is given in the Figure 13(b). It is evident from the microstructure of the transverse weld zone that the interstitial transverse bonding has occurred coherently between the different zones in the transverse section, resulting in the formation of striated transverse bands of the weldment in friction stir welded joints.

Further, Dhananjayalu avula, et. al., have carried out extensive work on the microstructure of the friction stir weld zone, they have critically observed that the nugget zone exhibits very fine grains, also the atomic grains depict an equi axed elongated grain structure all along the periphery of the thermo mechanically affected zone. The same observations are found in the microstructure of the friction stir weld dissimilar alloy joints observed in the present work [1].

IOP Conf. Series: Materials Science and Engineering 1132 (2021) 012029 doi:10.1088/1757-899X/1132/1/012029

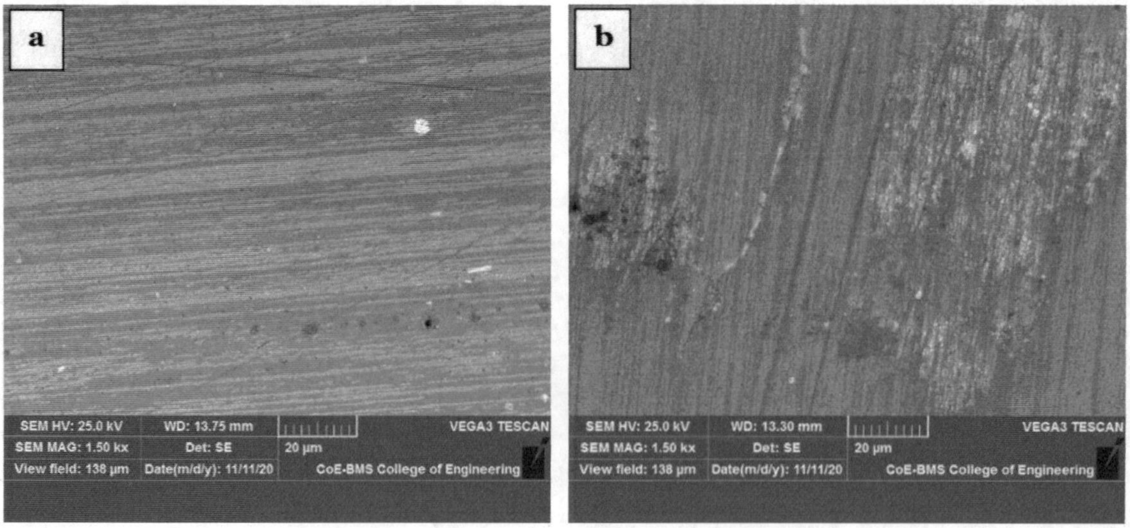

Figure 13. **(a)** Microstructure of the top surface of the weld zone **(b)** Microstructure of the transverse surface of the weld zone

4. Conclusions

The critical analysis of the outcomes of the experimental design has yielded several conclusions which are comprehensively put forth in the current section.

From, the analysis of traverse load, normal load, and spindle torque; the optimized set of process parameters of 1200 rpm tool rotational speed, 140 mm/min feed rate, tool offset of 0.5 mm and triangular pin profile, will give an effective weld of AA 7075 and AA 5052 alloy. Further, the SEM images confirm and acknowledge the formation of strong bonds between the atoms across the subsurface in the weldments. It gives an overview of the fracture type, weld zone atomic morphology, and the passivation of the corrosion samples over a period of time. The SEM images of the weld joint clearly exhibits the intrinsic strong bonding between the atoms in the interstitial zones of the weldment, further, the micro coring and segregation of the atoms leads to better solutionizing. The optimized process parameters lead to sustainable friction stir welding of dissimilar aluminium alloy joints with reduced time duration of 69 seconds for friction stir welding the 170 mm length aluminium alloy joints with optimized peak forces and torque.

References

[1] Shaysultanov D G, Stepanov N D, Vysotskiy I V and Zherebtsov S V 2019 *AIP Conf. Proc* **2167** 336–344

[2] Kumbhar N T, Sahoo S K, Samajdar I, Dey G K and Bhanumurthy K 2011 *Materials & Design* **32** 1657–1666 URL 10.1016/j.matdes.2010.10.010;https://dx.doi.org/10.1016/j.matdes.2010.10.010

[3] Carvalho G H S F L, Galvão I, Mendes R, Leal R M and Loureiro A 2019 *Materials and Manufacturing Processes* **34** 1243–1250 URL 10.1080/10426914.2019.1644452;https://dx.doi.org/10.1080/10426914.2019.1644452

[4] Kundu J and Singh H 2016 *Engineering Solid Mechanics* **4** 125–132 URL 10.5267/j.esm.2016.2.001;https://dx.doi.org/10.5267/j.esm.2016.2.001

[5] Ramesha K, Sudersanan P D, Santhosh N, Ravichandran G and Manjunath N *Advances in Structures, Systems and Materials, 2020* 293–306

[6] ZUO L, ZHAO X, LI Z, ZUO D and WANG H 2020 *Chinese Journal of Aeronautics* **33** 792–804 URL 10.1016/j.cja.2019.07.019;https://dx.doi.org/10.1016/j.cja.2019.07.019

[7] Ramesha K, Santhosh N, Bedi R and Sudersanan P D 2018 *J. Thin Film. Coat. Sci. Technol. Appl* **5** 7–14

[8] Longhurst W R, Strauss A M, Cook G E and Fleming P A 2010 *The International Journal of Advanced Manufacturing Technology* **51** 905–913 URL 10.1007/s00170-010-2678-3;https: //dx.doi.org/10.1007/s00170-010-2678-3

[9] Bevilacqua M, Ciarapica F E, D'Orazio A, Forcellese A and Simoncini M 2017 *Procedia CIRP* **62** 529–534 URL 10.1016/j.procir.2016.06.081;https://dx.doi.org/10.1016/j.procir.2016.06. 081

IOP Conf. Series: Materials Science and Engineering 1132 (2021) 012030 doi:10.1088/1757-899X/1132/1/012030

Fabrication and Characterisation of PP/TS Biocomposites

Sreenivasan S N [1*], Neethumol M David [2], Hanna J Maria [3], Sabu Thomas [3], Józef T Haponiuk [4]

[1]Department of Mechanical Engineering, Adi Shankara Institute of Engineering and Technology, Mattoor, Kalady, Ernakulam, Kerala, India-683574

[2]Department of Chemistry, Mar Ivanios College, Thiruvananthapuram, Kerala, India-695015

[3]Department of Chemical Science, Mahatma Gandhi University Kottayam, Kerala, India-686560

[4]Polymers Technology Department, Gdansk University of Technology, 80-233 Gdańsk, Poland

[*]email: sreenivasan.me@adishankara.ac.in

Abstract. Biocomposites are eco-friendly composite materials formed by the combination of any polymer matrix and the reinforcement of any natural fiber. This particular composite is gaining more attention in research field in order to reduce the utilization of inorganic fillers in the reinforcement of thermoplastics and biodegradable plastics. In this study, we used various amounts of turmeric spent (TS) viz, 5%, 7%, 10%, 15%, 20% and 25% (w/w) for the fabrication of Polypropylene/TS composites. These various composition mixtures are blended and melt on a Haake mixer, then it is placed over a compression mould of 2mm thickness and forms composite films of different composition. The developed composites were characterized using different techniques such as scanning electron microscopy, contact angle measurement and water absorption. SEM images clearly shows the presence of TS content in biocomposites. From contact angle measurement, it is clear that the 7% PP/TS composite shows better hydrophobic nature. From absorption measurement technique, the higher composition samples show better absorption capacity due to the presence of OH- group within the turmeric spent.

1. Introduction

The shortage of petroleum based resources and various environmental issues have prompted many researchers to think about biodegradable resources in substitution of petroleum based products. The recycling of composites manufactured from petroleum based non renewable resources is complicated due to the tiring effort required for the separation of their components. Due to its non biodegradable properties, several environmental issues may cause due to landfills or burning. The biocomposites exhibit superior mechanical and thermal make it to utilize in different domains such as aerospace, military, automobile, construction, naval and packaging [1]. "Toyota Raum" is a Toyota launched model where the interior parts are made up of hemp fibers and a potato derived PLA with the reinforcement of sugarcane bagasse was used in its manufacturing of springboard [1]. The biocomposites can be processed by utilizing many methods such as extrusion, thermoforming and injection moulding [2]. Many other automobile industries are investing huge amount in research and development for implementing green technology in their products. For the past several decades, scientific researches are conducting across the world on wide variety of natural fillers such as hemp, flax, kenaf, abaca, bagasse, banana etc. to reinforce this to various polymer matrices [3].

Here the polymer matrix intends to use is polypropylene (PP) and the natural filler is turmeric spent. PP have remarkable properties such as low density, high stiffness, heat resistance, good impact and stretchability [3,5]. The agro industrial waste Turmeric spent (TS) which is a byproduct of nutraceutical industry consists of soluble and insoluble dietary fibers [4]. It is more abundant, cheap, renewable and easily recycled.

The reinforced fiber PP composites shows better tensile and flexural properties compared to unreinforced PP [6,7]. PP/Banana yarn composites showed an increment of 294% in tensile strength and 72% increment in flexural strength when compared to unreinforced PP [5]. The Modulus/Cost and specific modulus of Kenaf (30% by weight) –PP composites are higher than different natural fibers such as flax, hemp, sisal, E-glass and coir [6].

The incorporation of various sizes of fillers in composite shows considerable change in mechanical properties The 3% addition of nano oil palm empty fruit fiber (OPEFB) filler into kenaf/epoxy composites reduces the presence of voids, fiber pull outs, fiber tearing and protruding on the fractured surface due to the interfacial bonding and good adhesion between fiber and matrix [8]. The addition of talc which is a hydrophobic mineral possessed stiff character to Low Density Polyethylene (LDPE) / Kenaf Core Fiber (KCF) biocomposites whereas the addition of calcium carbonate which is a hydrophilic mineral caused a tough behavior to the same biocomposites [7]. The hybrid polypropylene biocomposites with rice husk/ground nut husk wastes as fillers showed water absorption 85% lower than gypsum tiles which gives an advantage to use it as a ceiling tiles compared to gypsum tiles [9].

Recently, turmeric spent has attracted a great attention due to the increase in cellulose content after some chemical treatments. It has very little commercial value hence it can be utilized as reinforcing filler in engineering thermoplastics. The interest of using agro industrial waste has grown in recent years to promote sustainable agriculture and recycling. The specific gravity of this filler is very low and its incorporation in polymer matrix reduces the overall weight of composite and enhances their bulk property. The antimicrobial property of turmeric spent makes it possible to explore in new applications in automobile as well as in medical field.

The main objective of this work is the fabrication of a cheap thermoplastic biocomposite ie, PP/TS biocomposite and evaluation of morphological, chemical and hydrophobic properties for various applications in automobile as well as in medical field.

2. Experimental

2.1. Materials

In the present study Turmeric spent (TS) was procured from M/s. Sami Labs, India and the polymer matrix, polypropylene, REPOL H200MA obtained from M/s. Reliance Polymers, India. The density of PP was 0.92g/cc and its melt flow rate (230 ^{0}C/2.16 kg) was 35 g/10 min.

2.2. Methodology

Overnight dried TS in oven at 80^{0}C was thoroughly mixed in different ratios of PP/TS (wt/wt, %) such as 95/5, 93/7, 90/10, 85/15, 80/20 and 75/25. Torque Rheometer (Thermo Haake Rheocord 600) was used to prepare all composites. And the mixture is placed in the Haake mixer in order to blend and melt the mixture. The Haake mixer is preheated for 10 minutes in order to attain the melting temperature of polypropylene that is about 180°C. Then fed the different composition in the machine and melt it and mix it for 10 minutes with a screw speed of 60rpm. The blended composite is placed in a 2mm mould in a compression moulding machine. Before placing the composite, the compression moulding machine is preheated for 10 minutes along the mould inside the machine to achieve the melting temperature of polypropylene that is around 180°C. Then placed the composite over the mould and run the machine for 10 minutes and then cooled by water around the mould to the room temperature of 50°C to remove the prepared composite sheet from the mould. The three stages of moulding cycle were 2 min preheating, compression under a pressure of 120kg/cm^2 for 10 min, and cooling of mould by using water until it reached 50^{0}C.

2.3. Measurements

2.3.1. Scanning Electron Microscopy. The structure of PP/TS composites obtained after compression moulding process was investigated using Scanning electron microscope (Vega3Tescan, Germany). Aluminium stubs were used to place the samples by using double sided carbon tape. 10kV was used as the accelerating voltage to examine the samples.

2.3.2. Contact angle analysis. The method adopted for contact angle analysis was sessile drop method and the instrument used was Phoenix-i model from Surface Electro Optics – SEO (Florida, EUA). The micrometric dozer was used to drop distilled water on the surface of specimen and by using an interval of 200s the images captured was analysed using a software. The measurement of contact angle was done programmatically by considering the angle between the tangent of the droplet edge and the base line where the drop lies.

2.3.3. Water absorption analysis. For water absorption analysis, the specimens were prepared according to ASTM D 570-99. Two samples of each composition in the disc shape of 50mm diameter and 2mm thickness was prepared for the analysis. These all samples were immersed in distilled water at room temperature and its change in weight was recorded at the end of 48h.

3. Results and Discussion

3.1 Scanning Electron Microscopy
SEM image of Neat PP (Fig. 1(a).) shows smooth and it is free from any contaminants. On the other hand, Fig. 1(b) to Fig. 1(g) shows the presence of turmeric spent filler as white dots and white lines. Between Fig. 1(b) and Fig. 1(c), it is visible that better dispersion in Fig. 1(c). When the content of TS was enough higher, the particles were uniformly distributed in the PP matrix. TS exhibited many single disperse particles and aggregates integrated with the particles. From, Fig.1 (d) to Fig. 1(g), it is clear that the PP matrix is not enough to encapsulate the solid micro particles of TS. However, large aggregates were found, and the aggregate size increased substantially in these micrographs with higher TS loading due to the hydrogen bonding between TS particles [10].

3.2 Contact angle analysis
Inorder to find the effect of TS in biocomposites, the contact angle of the calibrated water droplets deposited on the PP/TS biocomposites measured were evaluated. There was a change in polarity on the composite surfaces were observed by analysing the contact angle results. The higher contact angle specimens show the higher hydrophobic nature.

From Fig. 2(a) to Fig. 2(g), it is clear that there is a slight increase in contact angle of 7%TS/PP composite which is 84.780, it is due to the presence of TS content in composite. However, the composites having higher content of TS displaying lower contact angle is due to the agglomeration of TS on the surface of the composite.

3.3 Water absorption analysis
By using equation (1) the predicted water absorption (PWA) can be calculated.

IOP Conf. Series: Materials Science and Engineering 1132 (2021) 012030 doi:10.1088/1757-899X/1132/1/012030

$$PWA = (W_f - W_o)/W_o \times 100 \quad (1)$$

Where W_f = weight of specimen after immersing in distilled water in g,
W_o = weight of specimen before immersing in distilled water in g.

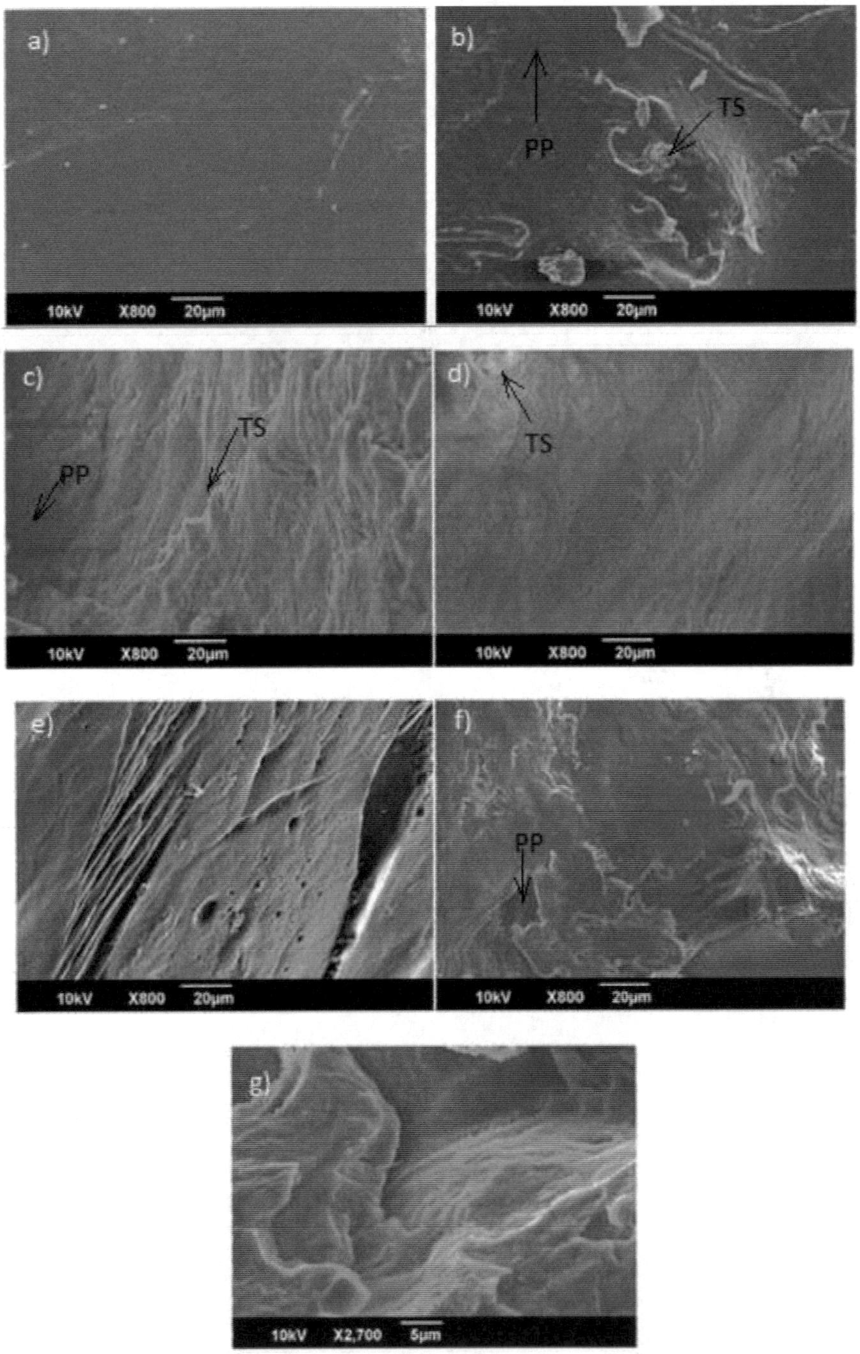

Figure.1. SEM images of (a) Neat PP, (b) 5%TS/PP, (c) 7%TS/PP, (d) 10%TS/PP, (e) 15%TS/PP, (f) 20%TS/PP & (g) 25%TS/PP

Fig. 3 shows the behaviour of composites in water absorption after 48h immersion in distilled water. From the results, it is clear that as the addition of TS to biocomposites increases, its water absorptivity increases. The overall hygroscopic nature and higher fiber content of TS makes it less enclosed structure leads to the increase in absorptivity There is an enhancement of water pathways created by the TS from surface to core due to the increase in porosity leads to the easiness of water entrance to the core.

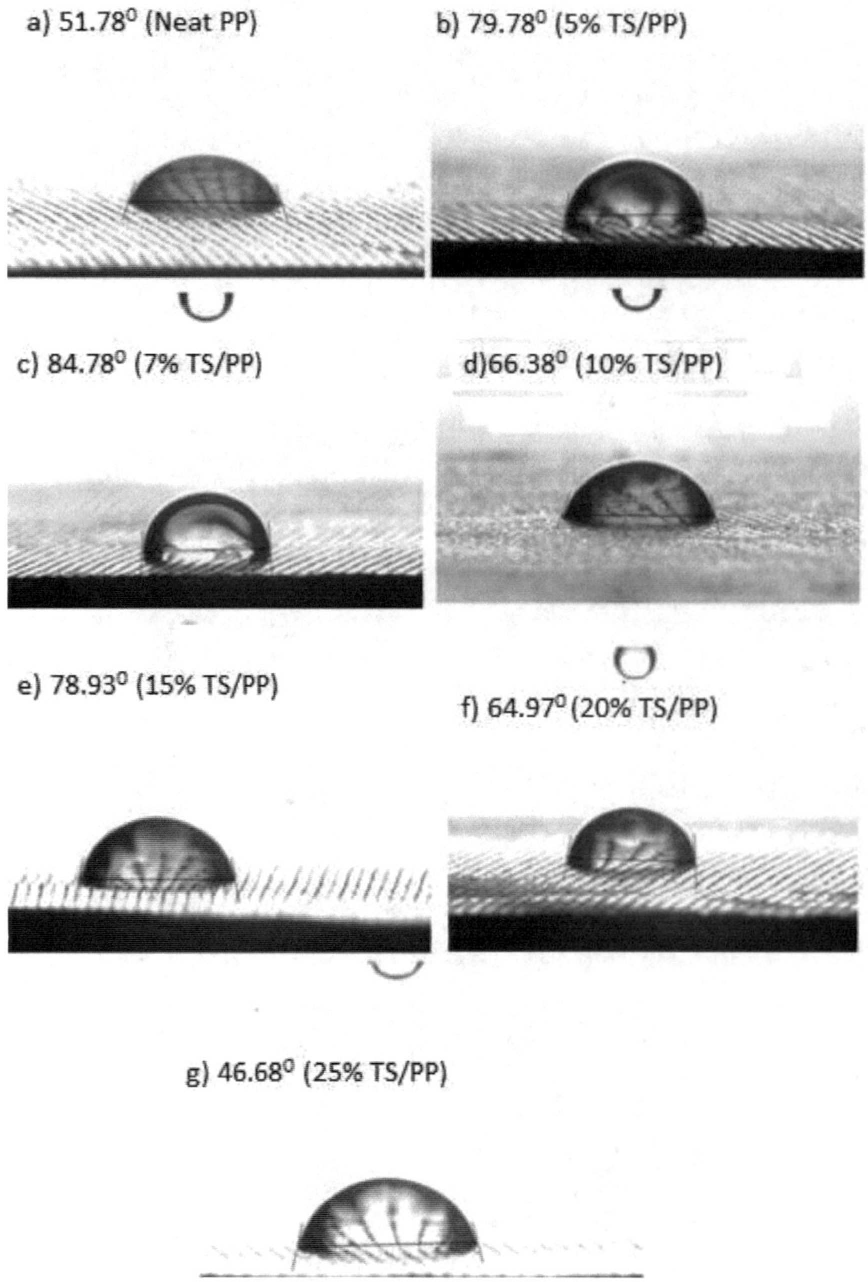

Figure. 2 Contact angle measurements of (a) Neat PP, (b) 5%TS/PP, (c) 7%TS/PP, (d) 10%TS/PP, (e) 15%TS/PP, (f) 20%TS/PP & (g) 25%TS/PP

Also there is some polarity interactions of TS fibers with the water due to the presence of hydroxyl groups in TS fiber. The water molecules are more chemically attractive to OH molecules from TS surface.

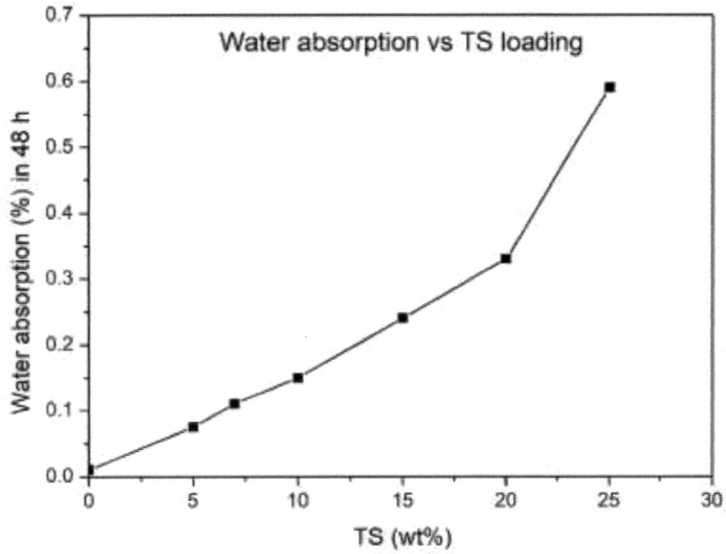

Figure.3 Graph showing variation of water absorption % corresponding to wt% of TS

4. Conclusions

Biocomposites based on turmeric spent and polypropylene were melt blended in this work. It is clear from the contact angle measurement that higher the TS content in bicomposite leads to lower the contact angle due to the agglomeration of TS particles on the surface. The water absorptivity of higher composition increased due to the presence of more hydroxyl groups on the TS surface leads to provide more opened structure which favours the water entrance. The polarity effects also provided more easiness of water entrance from surface to core and which improve the interactions between matrix and fiber respectively.

References

[1] Reddy T R K, Kim H J and Park J W 2016 *Composites from renewable and sustainable materials* InteTech Publishers 177 – 198
[2] Gallos A, Paës G, Allais F and Beaugrand J 2017 *RSC Adv.* **7** 34638
[3] Ghasemi A, Abbas E P H, Farhang L and Bagheri R 2018 *Polypropylene – Based Biocomposites and Bionanocomposite* Scrivener Publishing 247-286
[4] Gopi S, Amalraj A, Jude S, Benson K T, Balakrishnan P, Haponiuk J T & Thomas S 2019 *Journal of Macromolecular Science, Part A: Pure and Applied Chemistry* **56** 327 - 340
[5] Amir N, Abidin K A Z and Shiri F B M 2017 *Procedia Engineering* **184** 573 – 580
[6] Zampaloni M, Pourboghrat F, Yankovich S A, Rodgers B N, Moore J, Drzal L T, Mohanty A K and Misra 2007 *Composites: Part A* **38** 1569–1580.
[7] Shamsuria A A and Sumadin Z A 2018 *Composites Communications* **9** 65–69.
[8] Saba N, Paridah M T, Abdan K and Ibrahim N A 2016 *Construction and Building Materials* **123** 15–26.
[9] V Guna, M Ilangovan, M H Rather, B V Giridharan, B Prajwal, K V Krishna, K Venkatesh and N Reddy 2020 *Journal of Building engineering* **27** 100991
[10] Murtuza Ali Syed and Akheel Ahmed Syed 2012 *Materials and design* **36** 421-427.

IOP Conf. Series: Materials Science and Engineering 1132 (2021) 012031 doi:10.1088/1757-899X/1132/1/012031

Experimental Investigation on Mechanical Properties of Jute and Hemp fibre based Epoxy Composites

Ravi Kumar B[1] and Hariharan S S[2*]

[1, 2] School of Mechanical Engineering, SASTRA Deemed University, Tirumalaisamudram, Thanjavur, Tamilnadu, INDIA – 613401

[*]Corresponding author E-mail: hariharans@mech.sastra.edu

Abstract. The conventional materials are replacing via composite materials due to their special properties like high specific strength, high elastic constants, and relatively low expansion in the thermal environment. Usage of fibre reinforced composites based on natural fibres has increasingly started as viable material in low load carrying applications. In the current study, hemp and jute fibre based epoxy composites were developed and tested for assessing their suitability for possible applications in automobile and aerospace interiors. Composites were undergone tensile tests, compression tests, and flexural tests to understand their behavior under various loading conditions. The results revealed that the usage of hemp and jute fibre reinforced composites can improve mechanical properties and have shown a viable alternative to replace synthetic fibres such as glass fibres for specific applications. Hemp reinforced natural fibre composites have shown better mechanical properties as compared to jute fibre based composites.

1. Introduction

Over the years composites have been fulfilling the needs of special materials for various structural and domestic applications [1]. Today a huge awareness is being created for the usage of renewable products around the globe. The social influence, demand from consumers and various other aspects is forcing researchers and industrialists to move towards eco-friendly products. The usage of natural fibre composites would be a good solution to this issue as it has displayed a viable replacement to synthetic fibres especially glass fibre based composites in specific applications [2,3]. The various characteristics of natural fibres like biodegradability, less density, lower manufacturing costs, renewable resource, and good mechanical properties in comparison to glass fibres have shown a huge application potential in automotive, construction, aerospace, and various other applications [4–6]. The light weight of components, less fuel consumption, good recycling possibilities act as main drivers for its usage in the interior of automotive and aerospace industries.

Bharath et al. [7] classified the various types of natural fibres and explored the applications of various natural fibres with a view of the various strength parameters. They also addressed the need for emphasis on other criteria such as environmental durability, embedded energy, fire resistance.

Vaisanen et al. [8] studied the effectiveness of the hemp-epoxy composites using various modifications to the hemp fibers by treating with alkali, enzymatic, steam, and wood distillate techniques. The mechanical properties of prepared composites were observed to be reduced after modifications but it lowered significantly the water absorption characteristics.

Gopinath et al. [9] reported an experimental study on the mechanical response of jute fiber-epoxy composite and illustrated a detailed testing procedure to understand the behavior of composites under mechanical loadings. In another study, Biswas et al. examined thermal and physical behavior of unidirectional jute-bamboo-epoxy composites and have shown that jute reinforced composites show enhanced physical and thermal stability over banana fibre composites [10].

Rajesh et al. [11] studied the effect of fibre loading and alkali treatments on the tensile properties of short jute fibre reinforced composite. Alves et al. [5] reported a study on the usage of natural fibre composites in ecodesign of automotive components and addressed the effects of factors affecting the performance of the composite components. Ravi Kumar et al. [12] experimentally investigated the mechanical and surface morphological properties of natural fibre based composites and observed that natural fibre composites have the comparable specific tensile strength to that of glass fibre composites.

In the current study natural fibre based composites are fabricated using the hand lay-up technique and the properties like tensile, flexural, and compressive stability are assessed to look for possible applications of these bio-composites in Industrial sectors.

2. Material and Experimental Methods

2.1. Material

Fibres selected for composites fabrication are plant based fibres, i.e., jute and hemp due to their cheap price and rapid growth over a wide range of climatic conditions. Polymer used as matrix material is epoxy (LY556) and Araldite hardner (HY 951). Figure 1 shows the chemically treated hemp and jute fibre Mats supplied by Go green composites Chennai.

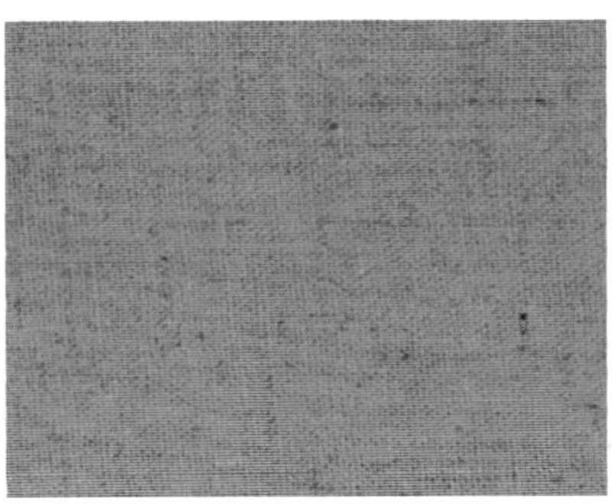

a) Hemp fibre Mat

b) Jute fibre Mat

Figure 1. (a) Hemp fibre Mat (b) Jute Fibre Mat

2.2. Composite Fabrication

The hand layup technique is the most economical method and is used for the fabrication of hemp/epoxy and jute/epoxy composite plates. Epoxy and a suitable hardener were mixed in the weight ratio of 10:1. The layup was loaded inside a square mould box of inner dimensions 300 mm x 300 mm. Four layers of

IOP Conf. Series: Materials Science and Engineering 1132 (2021) 012031 doi:10.1088/1757-899X/1132/1/012031

fibres mat in the laminate sequence of [45/0]s are placed inside the mould box and plates of 4 mm thickness with fibre weight fraction of 45% were fabricated. Curing was carried out under a load of 50kg at atmospheric temperature. Figure 2 shows the fabrication process and fabricated composite plates.

i) HFRC ii) Loaded Mould iii) JFRC

Figure 2. (i) Hemp fibre reinforced composite (HFRC), (ii) Loaded mould box, (iii) Jute fibre reinforced composite (JFRC)

2.3. Samples Preparation

Samples were prepared for different testing as per ASTM standards as follows:

2.3.1 Individual fibre Tensile Testing:

Individual fibres were tested to assess their tensile strengths before reinforcing them into the polymer matrix. Twenty single fibres of the uniform length of 30 cm were selected and tested on a 10 kN universal tensile testing machine. As fibres are having different diameters along the lengths so the average linear density was measured in Tex and fed into the UTM machine for measuring fibre breaking force and corresponding elongation as per ATMD 3822.

2.3.2 Composites Tensile Testing:

The test specimens were cut according to ASTM D3039; the most common specimen for ASTM D3039 has a constant rectangular cross-section of 250 mm long and width of 25 mm. Thickness may vary as per the fibres used.

2.3.3 Composites Flexural Testing:

Flexural test was carried out by employing the three-point bending method according to ASTM D790. Sample dimensions for the flexural test were 120mm x 12mm.

2.3.4 Composites Compression Testing:

The compression specimens were cut as per the ASTM D695 standard. In this test, samples were mounted on a test plate and were subjected to the compressive loadings until its fracture. Sample dimensions are a square prism of 12.7 mm × 12.7 mm.

3. Results and Discussion

3.1 Individual fibre Testing

Twenty fibres of uniform length and diameters were selected and tested for tensile strength. Individual fibres were subjected to tensile loading as per ASTM D 3822 in 10 kN Universal tensile testing machine with grip to grip separation at the start position as 65 mm. Cross head travel was at a speed of 10 mm/min.

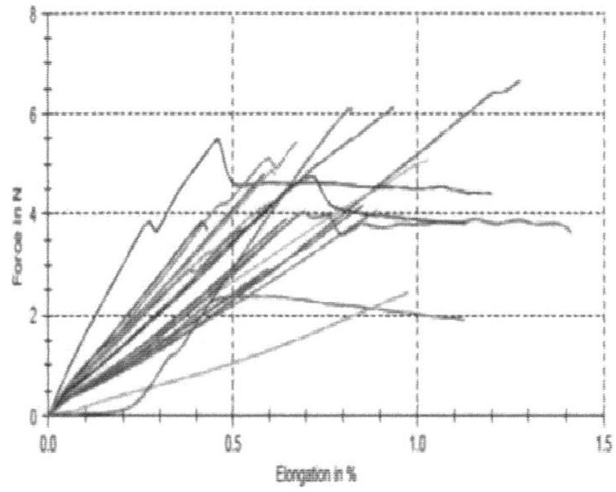

Figure 3. Force vs strain (%) for Hemp fibres

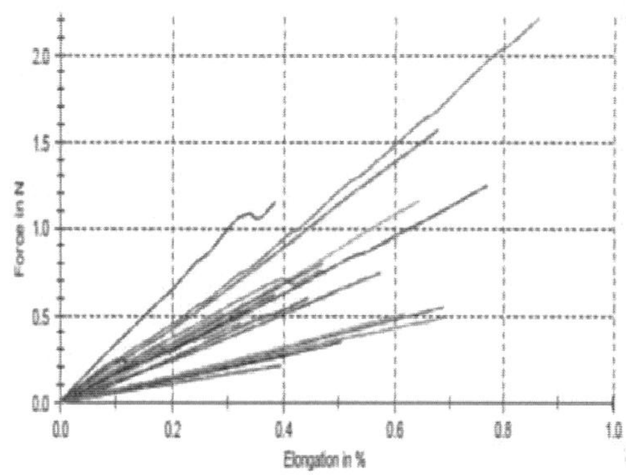

Figure 4. Force vs strain (%) for Jute fibres

Figure 3 shows the tensile results for hemp fibres, the maximum elongation of individual hemp fiber is observed at about 1.3% and the average elongation of the individual fibers is found out to be 0.7 %. The maximum and average tensile force that a single hemp fiber could withstand is 6.5 N and 4.43 N respectively.

It is observed from figure 4 that the maximum elongation of individual jute fiber is about 0.85% and the average elongation of the individual fibers is found out to be 0.52%. The maximum tensile force that the single jute fiber can withstand is 2.2 N and the average tensile force that the single jute fiber could withstand is 0.75 N.

Results revealed that in raw form hemp fibres can withstand higher tensile force as compared to jute fibres when subjected to the same tensile force environment.

3.2 Composite Tensile Test

Both the composite's cut samples were tested for finding out their tensile strengths using a 100 kN universal tensile testing machine as per ASTM D3039. Four samples for both jute fibre composite (JFRC) and hemp fibre composite (HFRC) were tested as per ASTM standards.

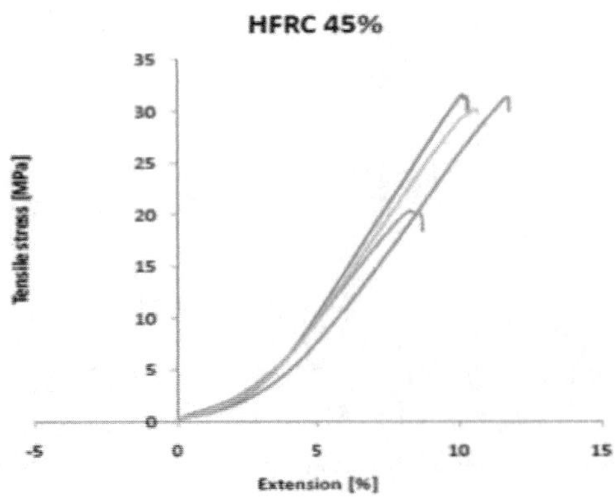

Figure 5(a). Stress vs strain for HFRC

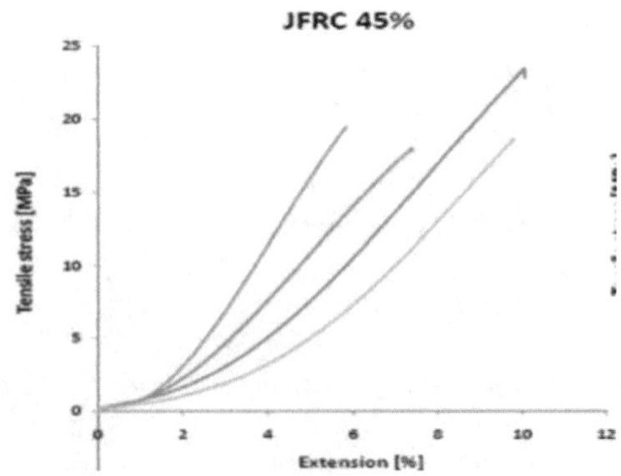

Figure 5(b). Stress vs strain for JFRC

Figure 5(a) shows the variation of tensile stress of hemp fibre composite (HFRC) with fiber weight fraction of 45%, it is clear that the average tensile strength of the composite is 28.41 MPa with an extension of 10.3 %. Figure 5(b) presents the stress variation with strain for jute fibre composite (JFRC). Average tensile strength is observed to be 19.9 MPa with an extension of 8.29%.
From results, it is clear that jute fibre reinforced composites have shown an improved tensile strength over flax reinforced composites.

3.3 Composite Flexural Testing

The test was conducted on both the composites specimens as per ASTM D790 standard. The test specimen was placed horizontally onto the three points of contact and then a force was applied on the top of the specimen through one point until the specimen fails.

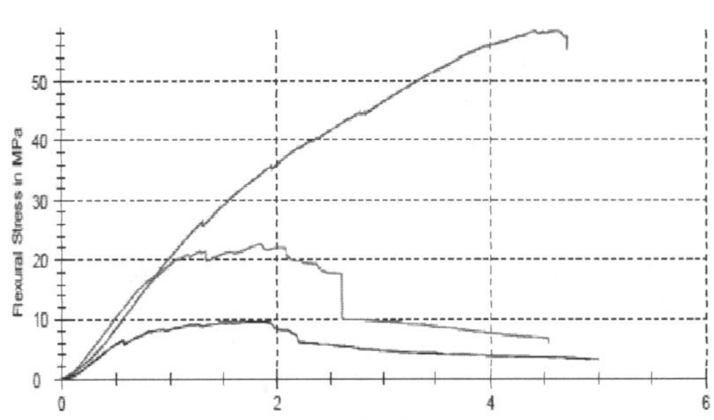

Figure 6(a). Flexural stress vs strain for HFRC

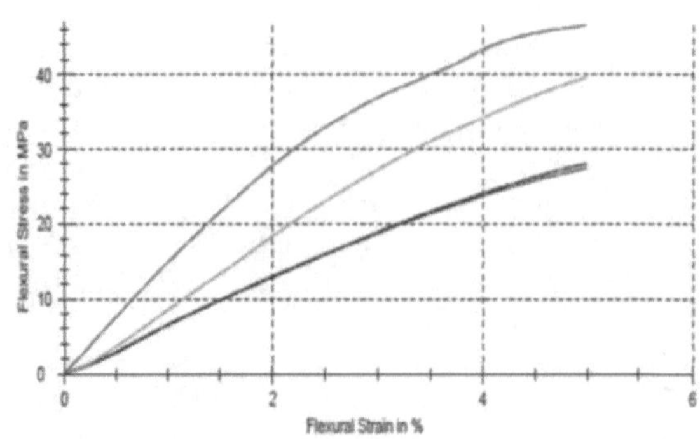

Figure 6(b). Flexural Stress vs strain for JFRC

Figure 6(a) shows the variation of flexural stress with strain for hemp reinforced bio-composite (HFRC). Three samples were tested and it is observed that maximum bending stress was found to be 53 MPa, with a maximum radius of curvature at 4.8%. The average bending stress is found to be 30.2 MPa and the average radius of curvature is 4.2%. Figure 6(b) presents the bending stress variation with strain for jute fibre reinforced bio-composite (JFRC). It is observed that maximum bending strength is at 45.5 MPa with a strain % of 5.2. The average bending strength observed is 42.9 MPa and the average radius of curvature is found to be 4.9 %. From results, it is evident that jute fibre reinforced composites have shown better bending strength over hemp reinforced composites.

3.4 Composite Compression Test

Figure 7(a) shows the variation of compressive stress with strain for hemp reinforced bio-composite (HFRC). It is observed that the maximum compressive force was found to be 42.5 kN, with a crush percentage of 18. The average compressive strength is observed as 40 kN and the average percentage crushing strain of 17.7%. Figure 7(b) represents the variation of compressive stress with strain for jute reinforced bio-composite (JFRC). It is observed that the maximum compressive force was found to be 3.4 kN, with a crush percentage of 9.9%. The average compressive strength is observed as 2.6 kN and the average percentage crushing strain of 9.2%.
From results, it is evident that hemp fibre reinforced composites have shown better compressive strength over jute reinforced composites.

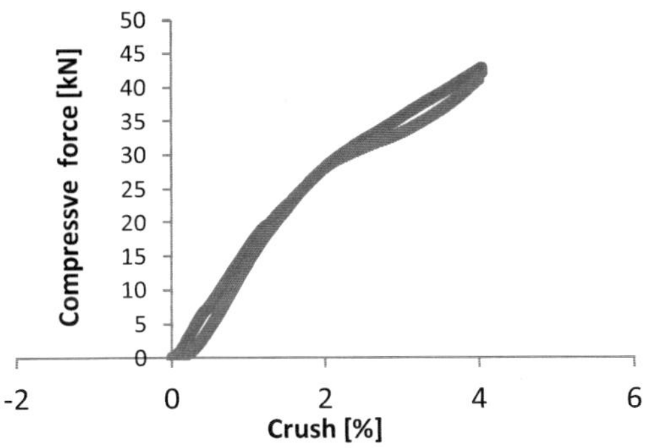

Figure 7(a). Compressive force vs strain for HFRC

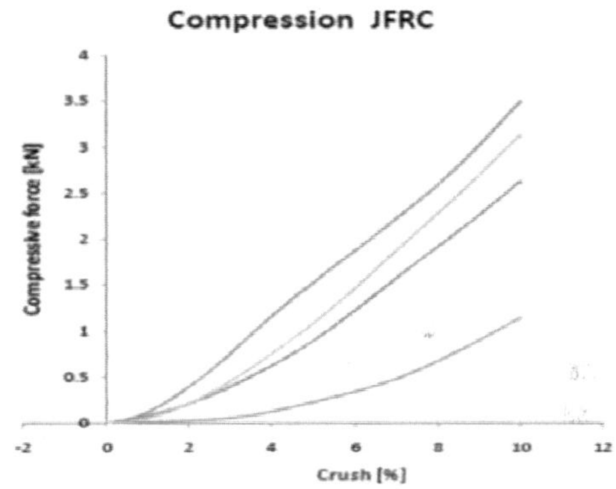

Figure 7(b). Compressive force vs strain for JFRC

4. Conclusion

In this study, the natural fibre reinforced Bio-composites were tested to assess their suitability for possible applications for interior in aircraft and Automobiles. The composites were subjected to testings such as individual fibre tensile tests, composite tensile testing, flexural, and compression test. The following conclusions are drawn:

- Individual hemp fibre in raw form can sustain higher tensile force compared to jute fibre under the same operating force range.
- Hemp fibre composites have shown better tensile properties over jute fibre composites.
- The average flexural strength in hemp fibre-based bio-composites is found to be 30.2 MPa and the average radius of curvature is 4.2% as compared to the average bending strength as 42.9 MPa and the average radius of curvature to be 4.9 % in the case of jute fibre-based composites.

- Hemp fibre composites have displayed higher compressive strength as compared to jute fibre composites.

The analysis presented in this research work would be helpful in further exploring the applicability of natural fibre composites with improved mechanical properties for possible use in Industrial, domestic and structural applications.

Acknowledgement

The authors express their sincere gratitude to *SASTRA Deemed* University for infrastructural support *(TRR Fund)* to the advanced composite fabrication lab. The authors also express deep gratitude to PSG COE Indutech, Coimbatore for providing testing facilities to carry out this research.

References

[1] R.M. Jones,1999 "Mechanics of Composite Materials", **02** ed., CRC Press.
[2] N.P. Singh, L. Aggarwal, and V.K. Gupta, 2015 *Mater. Today Proc.*, **2**, 3140–3148.
[3] T. Gurunathan, S. Mohanty, and S.K. Nayak, 2015 *Compos. Part A Appl. Sci. Manuf.*, **77**, 1–25.
[4] V.P. Arthanarieswaran, A. Kumaravel, and M. Kathirselvam, 2014 *Mater. Des.*, **64**, 194–202.
[5] C. Alves, P.M.C. Ferrão, A.J. Silva, L.G. Reis, M. Freitas, L.B. Rodrigues, and D.E. Alves, 2009 *J. Clean. Prod.*, **18**, 313–27.
[6] F. Ding, S. Sheng, Y. Yu, and Z. Yu, 2019 *Steel Compos. Struct.*, **31**, 291.
[7] K.N. Bharath and S. Basavarajappa, 2016 *Sci Eng Compos Mater*, **23**, 123–133.
[8] T. Väisänen, P. Batello, R. Lappalainen, and L. Tomppo, 2018 *Ind. Crops Prod.*, **111**, 422–429.
[9] A. Gopinath, M. Senthil Kumar, and A. Elayaperumal, 2014 Procedia Eng.,**97**, pp. 2052–2063.
[10] S. Biswas, S. Shahinur, M. Hasan, and Q. Ahsan, 2015 Procedia Eng.,**105**, pp. 933–939.
[11] G. Rajesh and A.V.R. Prasad, 2013 *Int. J. Adv. Mater. Manuf. Charact.*, **3**, 528–532.
[12] B. Ravi Kumar and S.S. Hariharan, 2019 *Steel Compos. Struct.*, **33**, 299–306.

IOP Conf. Series: Materials Science and Engineering 1132 (2021) 012032 doi:10.1088/1757-899X/1132/1/012032

Smart Reader for Visually Impaired Using Raspberry Pi

S Sarkar, G Pansare, B Patel, A Gupta, A Chauhan, R Yadav and N Battula*

School of Mechanical Engineering, Vellore Institute of Technology, Vellore, Tamil Nadu - 632014, India

*e-mail: nithin.battula@vit.ac.in

Abstract. With the humongous amount of texts and written media available today it has been increasingly necessary to make these available to people from all walks of life specially catering to the visually impaired people, thus devising a system which can assist in this task is of prime importance. The proposed device aims to solve this predicament by using Raspberry Pi module B+ to convert printed and hand written into easily accessible playable speech. It also stores speech which can be replayed at a time deemed suitable. The main focus of this paper is to develop a smart reader system which coverts hand written and printed text to speech. It has been observed that the scanned texts were converted into easily audible speech heard via the speaker with high efficiency.

1. Introduction

Today there is a huge amount of written material available everywhere, and converting that into Braille is not a suitable option due to the vastness of the content that is available and the lack of return on such a huge investment. But that would mean denying the blind an access to these huge quantities of scholarly material. A suitable and viable alternative for this predicament is designing a smart device which can convert this print media into speech format and it can in turn play it out for the visually impaired. Such a device is radical in its kind and is a huge benefit, although the initial investment may be high but in the long run it is pretty cost effective. The main objective of this paper is to make use of such a system specifically the Raspberry Pi system along with its ancillaries in unison and use it to convert text into speech and thus assist the visually impaired.

The proposed algorithm uses a camera module with which it can capture the desired text and then convert it to binary representation [1], i.e. converting the image into a gray-scale image. From this gray-scale image the individual characters are extracted and recognized all of which is carried out by the Optical Character Recognition Algorithm. Upon undergoing the processes of Scanning, Preprocessing, Segmentation and Feature Extraction, the scanned text is finally ready to give its output by means of the speaker connected to the Pi module. Even though such systems are present, most of them are in the crude forms and developing a commercially viable setup will be a huge aid for the visually impaired thus giving them access to unprecedented amounts of text and written media. Such as system which involves only a one-time investment is thus a vital assist tool. The main objective of this project is converting print and written media into playable audio with high efficiency. A unique addition in this device is to record speech in the memory and replay these audio files at a convenient time.

IOP Conf. Series: Materials Science and Engineering 1132 (2021) 012032 doi:10.1088/1757-899X/1132/1/012032

2. Methodology

The block diagram of the setup consists mainly of two input parts. The first part is the text to speech conversion and the next is where the camera module converts the scanned image file into speech. The entire project is implemented on the Raspberry Pi 3 board which consists of USB ports, GPIO pins for input and output, a dedicated Pi camera for scans, a micro SD card slot and finally the speaker which will give the output.

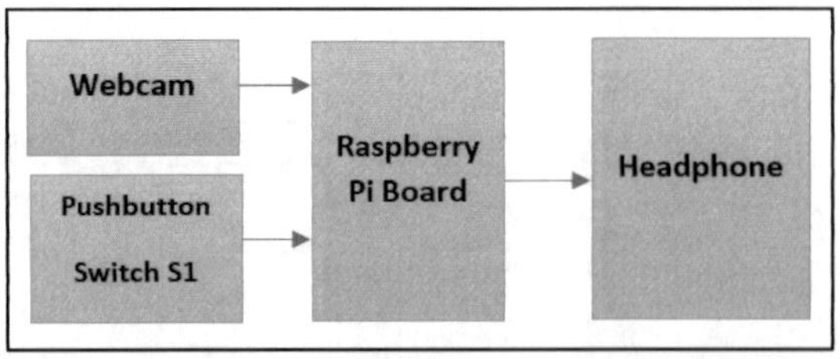

Figure 1. Block diagram of the hardware setup

Figure 1 shows the block diagram of the entire module with the input units, the processing unit and the output unit. Raspberry Pi is the control unit which is responsible for all processing and conversion. The entire module is powered by a USB cable connected to a computer. The press buttons are used to activate the program and the speaker for output [2]. The printed text to be scanned is then placed under the Pi camera to ensure a good quality image and few numbers of distortions. The acquired image is then transferred to an OCR algorithm whose task is to convert the captured image file into a playable text file. This OCR algorithm checks the scanned image for alphabets, numbers and characters and then gives a corresponding text output after comparing it with the pre-loaded database.

2.1. Hardware

The list of main hardware components used in this simulation are Pi Camera, Raspberry Pi3 Model B+, speaker/ headphone, push buttons, GPIO pins, bench Support, book, and monitor.

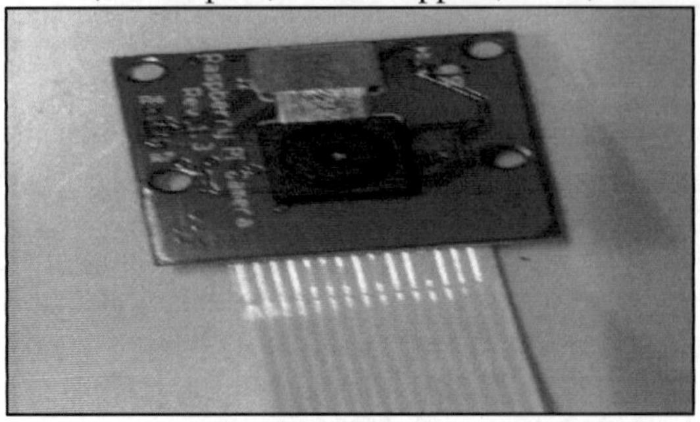

Figure 2. Camera module

The Raspberry Pi V2.1 camera is based on the Sony IMX219 silicon CMOS back lit sensor which produces 8 megapixels images that are 3280 x 2464 pixels in size. The IMX219 sensor operates in the visible spectral range [3] i.e. 400-700 nm and uses a Bayer array with a BGGR pattern. There are two versions of the Pi camera, the regular version and the No-Infrared (NoIR) version. The NoIR version has a filter on the lens which allows to capture images beyond the visible range. The scope of this study though does not involve the NoIR version.

Table 1. Features of camera module

Camera Parameter	**Specification**
Lens Focal Length	3.04 mm
f-number	2
Instantaneous field of view	0.368 mrad
Full-frame of view	59.17^{o} (H) X 58.3^{o} (V)

The Raspberry Pi V2.1 camera module has a fixed focal length of 3.04 mm [3] and a single f-number of F2.0 which is typically focused from near-field to infinity. The images are captured at ISO settings between 100-800 which can be changed manually by increments of 100, which has though not been verified above 600. Some of the most significant feature of the Raspberry Pi camera version used has been shown is table 1.

The dimensions of the entire camera module are about 25mm x 25mm x 9mm and it weighs just about 3 grams. The camera connects to the Raspberry Pi board via a 15-pin mobile industry processor interface (MIPI). This is chosen because it is accessible easily and widely. The module has small physical size and the quality of storage can be modified as per requirements, it's power-efficient and cost effective and it is possible to connect multiple cameras and process them parallelly [4].

The Raspberry Pi is a single credit card sized single board computer which has been developed by the Raspberry Pi Foundation, United Kingdom. This board has incredible computing power [5] and has the capability to do amazing projects. The device is perfect for performing all types of computing tasks and interfacing various types of devices via GPIO. The board consists of a Broadcom based ARM processor, a graphics chip, a RAM, GPIO and other connectors for external devices. It has Broadcom BCM2835 SoC processor with 700MHz ARM1176JZF-S core, 512 MB RAM, Videocore 4 GPU supports up to 1920x1200 resolution, MicroSD card slot, 10/100 Mbps Ethernet port, 4 x USB 2.0 ports, HDMI, audio/video jack, GPIO header containing 40 pins, MicroUSB power port providing 2A current supply, DSI and CSI ports.

Raspberry Pi operating system is installed in this device and Python is used as the main programming language but there is also support for BASIC, C, C++, JAVA, Perl and Ruby.

The main function of the speaker in the entire module is to give the final output of the scanned text and help in the completion of the motive behind the study. There are a few necessary conditions which have been ensured like recognition of a crowded place and modify the volume of sound accordingly and recognition of silent places and adjust the sound similarly [7].

The main function of speaker is to assist in text-to-speech conversion. gTTS module for python has is a Text-to-Speech (TTS) engine whose main function is to convert and thus create a spoken version of the text which has been scanned previously by the Pi camera. This TTS enables the reading of the text [8].

The main function of the push button is to switch something on or off depending on the necessity of the action. In this project a tangible push button is connected to a pin on the Raspberry Pi3, while the opposite side has a ground connection. When the button is pulled up the circuit is connected while when it is pushed down it is connected to the ground and the circuit is in off condition. These tactile push button switches can be easily programmed to provide a variety of output messages and switching messages.

The attaching of the tangible push button to the Raspberry Pi is an easy job, but the most important thing to remember that the GPIO pins are +3.3V DC compliant. If a voltage greater than +3.3 V DC is applied it will end up damaging the Raspberry Pi module.

An important and a very powerful feature of the Raspberry Pi is the row of GPIO (general purpose input/output) pins along the top edge of the board. The 40-pin GPIO is now found on all Raspberry Pi boards.

Any of the 40 GPIO pins can be designated as an input or output and they can be used for a wide range of applications. The board consists of two 5 V pins, three 3V pins and number of unconfigurable ground pins which have 0V.

The Optical Character Recognition (OCR) is the text recognition system which allows the hard copies and all other written text to be converted into editable soft copies [9]. An OCR facilitates the conversion of the geometric source object into digitally representable ASCII or Unicode scheme. OCR are of two types: for recognizing printed characters and for recognition of hand written text.

2.2 *Software*

The main open source programming language used in this Raspberry Pi project is Python [10]. Python is the very versatile, easily acquirable and has a simple syntax and thus the language becomes a very powerful tool in developing programs. Raspberry Pi also faces no compatibility issues while programming it via Python. Due to the wide range of libraries that are included by Python, it can be used for a variety of applications.

Python therefore comes pre-installed in on Raspbian so that it can be used upon receiving. The two main choices for writing Python on the Raspberry Pi are using MU editor and editing remotely over SSH

The MU editor is one of the pre-installed Python IDEs that can be used to write to the programs.

It often happens that a computer, keyboard and monitor are not always available to write the Python codes on Raspberry Pi. Then Raspberry Pi has a special feature which allows to connect to the module remotely over SSH.

The VLC media player is an open source cross platform media player and streaming software which has been implemented in our Raspberry Pi module; its objective being providing the platform via which the sound output is emitted.

3. Results and discussion

Figure 3 is the image of the original simulation setup which has been done during the course of this study. The assembly of the components worked in unison and gave the desired output.

Figure 3. Simulation setup

3.1 Python code

3.1.1 Text to speech conversion procedure. The text to speech conversion is mainly divided into two main modules, the image processing and thevoice processing. The "picamera" module is used to capture images via the camera and convert imagesinto text using "pytesseract" OCR image processing module of python. The voice processing module (gTTS) converts this scanned text into sound which has specific characteristics so that it can beunderstood by the intended audience.

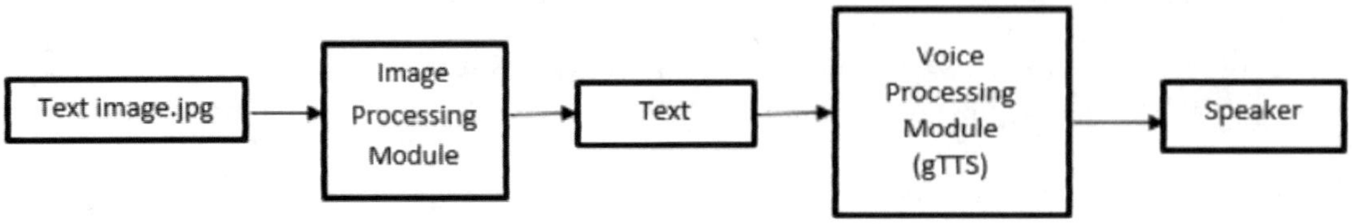

Figure 4. Block Diagram of TTS

Figure 4 shows the block diagram of a typical block diagram of the text to speech conversion module, where the first block is the image processing module. Here the image processing module converts image into texts and the next is the voice processing module which converts this text into voice.

The design of the entire module involves a fully integrated system where all devices do their tasks in harmony, the camera feeds the input text and it forms a digitized document. The image processing module processes the scanned document. The image processing module can recognise the sequence of characters present and can also indicate whether the current reading line is enabled [11]. The scanned images may vary in size some to a few idioms whereas some may be full sentences, therefore storage space management is very necessary and thus space digitization [12] of texts is very helpful. The image processing module methodology also makes use of methods like machine-learning and text-to-speech conversion to capture the image and finally this recognized speech is translated to the speaker for output.

3.1.2 Text displayed and speaker output. The hand-written text shown in Figure 5 is the actual sample of the written media which had beenscanned by the Raspberry Pi camera module. The speaker readout the following text without anymistakes. It was tested with various other type of text for different hand writings and it recognised allof them almost all the time and gave the correct output.

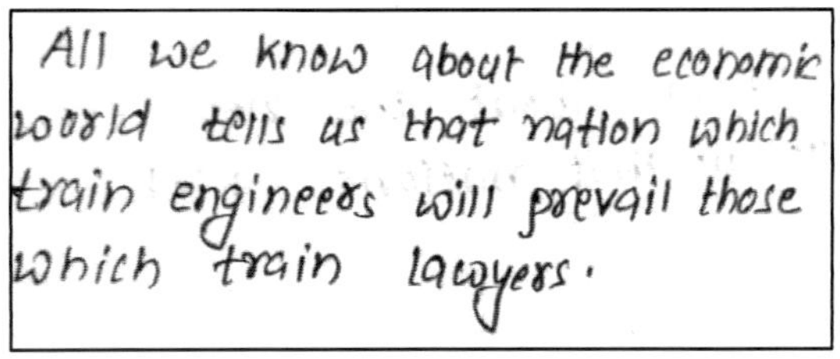

Figure 5. Hand written text shown to Pi camera module.

The Raspberry Pi Module does not have a language barrier, it can be successfully programmed to process any language which the user finds comfortable working with. We tested only for English language as modules are available for it. It has to be tested further for other languages. We observed that it works correctly 90 % of the time because of the OCR algorithm. Moreover, a special bookmark feature can be enabled such that any important information can be highlighted for future reference.

4. Conclusion

We developed a smart reader to assist the visually impaired. The device has demonstrated all the desired working criteria and thus the success of the device can be considered. Such a reading assistant has its scope limited not only to the visually impaired, but it can be widely used by people from all walks who want to have the luxury of a book reader available to them. This device thus acts both as a necessity and as a comfort supplement. The one-time cost of the entire device is around $70 (₹ 5000) and once installed successfully it has the potential to be the ideal assistant for the people it has been chiefly intended to. This device has known to find its importance in schools & libraries and also in many other fields where a voice assistant is needed.

References

[1] Aditi S, Annapoorani S P, Kanchana A 2018 Book Reader Using Raspberry Pi for Visually Impaired *Int. Res. J. Eng. Tech.* **5(03)**.

[2] George L, Rinsila S, Baby R, Thomas A, 2020 Raspberry Pi based Reader for Blind, *J. Opt. Commun. Electron.* **5(03)** 11–16.

[3] Pagnutti M A, Ryan R E, Gold M J, Harlan R, Leggett E and Pagnutti J F 2017 Laying the foundation to use Raspberry Pi 3 V2 camera module imagery for scientific and engineering purposes *J. Electron. Imaging*, ***26*(1)** 013014.

[4] Anandhalli M, Baligar V P 2018 A Novel Approach in real time vehicle detection and tracking using Raspberry Pi *Alex. Eng. J.* **57(3)** 1597–1607.

[5] Nayyar A, Puri V 2015 Raspberry Pi-A Small Powerful, Cost effective and efficient form factor computer: A review, Int. J. Adv. Res. Comput. Sci. Software. Eng. **5(12)** 720–737.

[6] Sachdeva P, Katchii S 2014 A Review Paper on Raspberry Pi *Int. J. Curr. Eng. Tech.* **4(6)** 3818-19.

[7] Nwazor N O 2019 A Raspberry Pi Based Speaker Recognition System for Access Control *Int. Res. J. Eng. Tech.* **6(3)**.

[8] Mallesh S J, Basavaraj H, Gagan L, Prateek R P 2018 Smart Speaker using Raspberry Pi *Int. J. Trend Sci. Res. Dev.* **2(4)** 1682–85.

[9] Aaron J S, Sanjana S, Monisha M 2016 OCR based automatic book reader for visually impaired using Raspberry Pi *Int. J. Innovative Res. Comp. and Commun. Eng.* **4(7)** 1111-18.

[10] Hoyo A, Guzman J L, Moreno J C 2015 Teaching Control Engineering Concepts using Open Source tools on a Raspberry Pi board *IFAC-PapersOnLine* **48(29)** 99–104.
[11] Jojie A, George A and Nayana, J D D Book Reader for Blind. *IOSR J. Eng. ISSN (e)*, 2250-3021.
[12] Mallapa D G, Shruti S S, Shruti B H, Vijayalaxmi I B 2017 *B-LIGHT:A reading aid for the blind people using OCR and OpenCV* *Int. J. Sci. Res. Eng. Tech.* **6(5)** 546–48.

Concept selection and Kinematic modeling of hand rehabilitation robotic device

Sunil, Meby Mathew, Sudheer A.P.

Mechatronics/Robotics Laboratory, Department of Mechanical Engineering, National Institute of Technology Calicut, Kerala, India

E-mail: sunil_m190167me@nitc.ac.in

Abstract. Rehabilitation using an exoskeleton is one of the widely accepted methods for the recovery of lost motor abilities. Most of the present research concentrates on either soft or rigid exoskeletons. In this paper, the design of a five-finger exoskeleton is discussed along with the advantage of soft and rigid bodied exoskeletons. Concept selection for the designs is done using the standard procedure of screening and scoring. Forward and inverse kinematic modeling of thumb and index finger is performed with Denavit-Hartenberg (DH) method.

1. Introduction

Every year, more than 15 million people are affected by stroke, which is the second leading cause of motor dysfunction after dementia. Several cerebrovascular diseases and neuromuscular disorders, such as stroke, epilepsy, dementia, paralysis, impaired motor control with symptoms ranging from spontaneous distractive gestures to inhibited movements. There are almost 9 million stroke survivors who typically face motor dysfunction. Movement disorders primarily decrease the quality of life of the patient, because the ability of a person to move is important to achieve essential daily living activities (ADL). Fortunately, there are many choices for restoring upper-hand mobility, including orthotics and physical rehabilitation. In the case of neurologically related disorders, the effectiveness of physical rehabilitation depends primarily on the onset, duration, strength, and task orientation of the training [1]. It also relies on the patient's health status and commitment to regain motor functions [2]. Intense repetitions of motor movements represent a major load on therapists during physical therapy. There seems to be a need for a robotic system that physically supports it.

The exoskeleton concept emerged in the late 19th century, but in the 1960’s, the prototype named Hardiman appeared [3]. Initially, Hardiman was designed for military purposes to increase the wearer's strength and performance. Exoskeleton research has improved dramatically since the 1990s, due to work proposed by the Berkeley Robotics Human Engineering Laboratory and other research groups [4]. Exoskeletons usually have both physical and cognitive human-robot interface (HRI) elements for physical assistive devices. It is therefore important to understand the biomechanics and neuroscience behind neuromuscular diseases and recovery. To design soft-rigid robotic based systems, understanding anatomy is also important. Robotic exoskeletons are structures used to recover the user's physical abilities [5].

There are two types of exoskeleton based on mechanical nature, i.e., rigid bodied exoskeleton and soft-bodied exoskeleton. Rigid body exoskeletons were robust and assisted the body stiffly [7]. In contrast with the soft-bodied exoskeleton, due to simpler system dynamics, the rigid bodied exoskeleton tolerates a simpler control system design [8,9]. This allows more complicated cases of deception performed with relative ease. For high force/torque transmission requirements, rigid bodied exoskeletons are useful. Rigid exoskeletons were hard, requiring high torque actuators to be combined with them and having a larger source of power [10,11]. Compared to rigid body exoskeletons, soft-bodied exoskeletons are smaller, less costly, and have less inertia. The limitation of the soft-bodied exoskeleton is that power transmission is complex and difficult. There is no rigid external frame for these systems, hence they depend on the skeleton framework of the consumer to provide the rigid frame. This makes reduced power requirements and allows for longer running times compared to rigid bodied exoskeletons with the same power source [12]. To overcome the limitation of completely soft-bodied systems, a possible middle ground having both rigid and soft parts is essential for an effective exoskeleton design as proposed by park et al. [13]. In the design of HANDEXOS [14], they had described the mechanical design of the index finger module along with dynamic modeling and some preliminary experimentation. Some misalignment related problem is also explained in the design. Hong kai et al. created a soft robotic glove and tested the efficiency of actuators as well as a glove for healthy participants in terms of kinetic and kinematic assistance. Besides, surface electromyography and radiofrequency recognition techniques were used to identify the user's intent to trigger or deactivate the glove [15]. A poly-articular tendon drive mechanism to mimic the mechanical compliance of human fingers was proposed by Hasegawa et al. They also proposed a new dual-sensing system mechanism and a new bioelectric potential-based switching control algorithm so that it could be easily used by the wearer [16]. Cempini et al. suggested a self-alignment design mechanism to absorb misplacement of human/robot joint axes and developed a novel thumb opposition mechanism. The kinematic design and actuation were addressed in the Hx exoskeleton [17] with theoretical and experimental validation of its efficiency. The PI control, which is based on the pressure value of the balloon sensor [18], is used to perform the power-assist motion. An embedded controller-dependent robotic hand module is defined by Tong et al. for each finger, it has 5 finger assemblies capable of driving 2 degrees of freedom (DOFs) at the same time. It is actuated by a linear actuator; at the metacarpophalangeal (MCP) joint, the finger assembly achieves a 55° range of motion (ROM) and 65° ROM at the proximal interphalangeal (PIP) joint. For opening-closing of hand, they have used electromyography (EMG) signals which helps in detecting the intention of the user [19]. The actuation systems in an exoskeleton are not inexpensive, particularly if the exoskeleton allows the actuator to be mounted at the user's joints, in which the actuator at one joint must endure the load of the actuator at the next joint, and so on. Researchers have solved this flaw by using cable transmission systems, which allow the exoskeleton to mount its actuator on a fixed foundation, reducing the weight of the device and lowering its price [20].

This paper introduces the design and development of a novel hand rehabilitation robotic system after considering the advantages and limitations of soft and rigid bodied exoskeletons by integrating both to achieve a positive result. This paper is structured as follows. Section 2 introduces conceptual designs for a hand rehabilitation robot. Section 3 explains concept selection and section 4 presents the design of the selected hand rehabilitation robotic device. Section 5 explains kinematic modeling, section 6 explains inverse kinematic modeling, and section 7 is the conclusion.

2. Conceptual designs for a hand rehabilitation robot

The conceptual design is a rough approximation of technology and its philosophies of operation [21]. It gives clarity in the design and helps to identify which technology will be suitable for its effective working. Some designs have been developed based on a literature survey by giving priority to the research gaps explained in the introduction. Figure1 shows the developed concepts using solid works.

Concept 1 is a webbed design in which the supports at each joint are firm and lightweight, whereas concept 2 is highlighted with a very thin design which will be compactable for almost all possible thickness of different biomechanical dimensions of fingers. Concept 3 is having less surface contact with fingers and the joints are pinned. In concept 4 the design is based on a double-slotted mechanism which will help in adjusting the lengths between each finger joint. Concept 5 is a modified version of concept 3 in which some part modification and elimination of extra materials are done. Whereas concept 6 is a slotted mechanism that helps to adjust the link lengths of each joint, and it is having fewer material requirements compared to others.

Concepts 1, 2, and 3 are having 21 joints and 23 DOF whereas concepts 4, 5, and 6 consist of 17 joints and 21 DOF. As shown in figure 1, concept 01 is bulky while concept 02 is most compact compared to others.

To achieve smooth contact and fixing, a silicon fabric strip is planned to place between the exoskeleton and the human fingers which is also used to tie it. Here soft fabrics are not shown however they will be integrated into the later stage of fabrication. This will help to achieve the advantages of a soft-bodied exoskeleton. All the concepts comprise of revolute joints with their link arrangements different for each concept. It is planned to drive the different concepts using the belt and pulley mechanism.

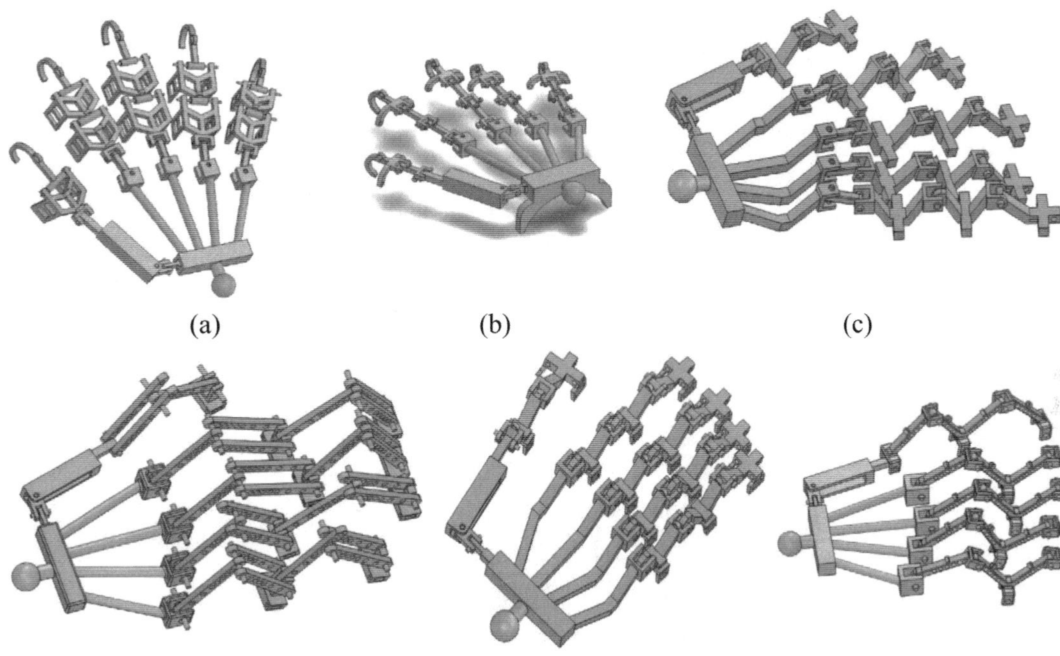

(a) (b) (c)

(d) (e) (f)

Figure 1. Conceptual designs (concept 1, 2, 3, 4,5, and 6 are represented as a, b, c, d, e, and f respectively).

3. Concept selection

Concept selection is done in two steps i.e., concept screening and concept scoring. Both concept screening and concept scoring stages follow a six-step process. The first step deals with the preparation of the selection criteria. Then rating and ranking of the concepts followed by combining and improving the concepts. After that selection and reflection on the results and processes is carried out. From the developed concepts, the most suitable concept needs to be selected. The matrices are used in the idea selection process as visual guides for finding consensus among the team members who construct the concepts. Matrices correspond to the needs and other decision criteria, based on the requirements the best concept is selected through a certain standard procedure. Criteria selected for the concept scoring and screening are ease of handling, weight, aesthetics, ergonomics, ease of manufacture, Degrees of Freedom (DOF) and ease of actuation.

3.1. Concept screening

Screening is a simple, rough assessment aimed at generating a few reasonable options. Its evaluation is done using a screening matrix as a guide. In concept screening, the concepts inferior to standard reference is given as "-" sign, and concept superior are given as " + " sign, and concept equal to the reference is given as " 0 ". Then based on the net score ranking is given, the concept with the highest net score is the best design and the lowest net score is the worst design. While analyzing the net score it is visible that concept 2 is having the highest score followed by concept 6,1,3,5,4.

Table 1. Concept screening.

Selection criteria	Concept 01	Concept 02	Concept 03	Concept 04	Concept 05	Concept 06
Ease of handling	-	+	-	-	-	0
Aesthetics	+	+	-	-	0	0
Ease of Manufacture	+	+	0	-	-	0
Weight	-	+	0	-	-	0
Ergonomics	-	0	-	-	0	0
Degrees of freedom	0	0	+	0	0	0
Ease of actuation	-	-	-	0	0	0
Sum +'s	2	4	1	0	0	0
Sum 0's	1	2	2	2	4	8
Sum –'s	4	1	4	5	3	0
Net score	-2	3	-3	-5	-3	0
Rank	III	I	IV	VI	V	II
Continue?	Yes	Yes (Modify)	Yes	No	Yes	Yes

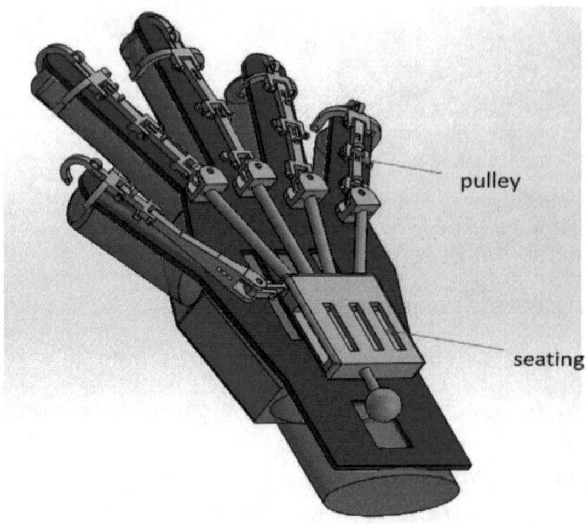

Figure 2. Modified Concept 2.

Concept 2 is modified with optimized thickness. Small pulleys are incorporated at each joint such that the actuation of the system using tendon drive will be easier and more practical. For the proper seating, the design is modified and some hallow portions are given to reduce the weight as shown in figure 2.

3.2. Concept scoring

Scoring is also similar to screening, whereas it carefully analyses few concepts and selects the best one out of them. After concept screening concept 04 is rejected due to the lowest net score and the remaining concepts were undergone concept scoring. Weightage for each criterion is given based on the literature survey. The scores for each concept is obtained by multiplying the weightage and rating. The equation for weighted score is given as $Sj = \sum_{i=1}^{n} r_{ij} w_i$ where, r_{ij} is the raw rating for concept j for the ith criterion, w_i is the weighting for ith criterion, n is the number of criteria, Sj is the total score for concept j. More weightage is given for the criteria such as ease of handling, weight, and ease of manufacture.

And rest of the criteria as given equal weightage. It is found that concept 2 having the highest net score. Hence it is screened as the best concept and selected for development.

Table 2. Concept scoring.

Concepts		Concept 01		Concept 02		Concept 03		Concept 05		Concept 06	
Selection criteria	**Weightage**	**Rating**	**Score**	**Rating**	**Score**	**Rating**	**Score**	**Rating**	**Score**	**Rating**	**Score**
Ease of handling	20%	3	0.6	3	0.6	4	0.8	3	0.6	4	0.8
Aesthetics	10%	4	0.4	5	0.5	3.5	0.35	3.5	0.35	4	0.4
Ease of Manufacture	20%	4	0.8	4.5	0.9	3	0.6	4	0.8	4	0.8
Weight	20%	3	0.6	5	1.0	2.5	0.5	2.5	0.5	3	0.6
Ergonomics	10%	3.5	0.35	3.5	0.35	4	0.4	3	0.3	4	0.4
Degrees of freedom	10%	4	0.4	4.5	0.45	3	0.3	3	0.3	4.5	0.45
Ease of actuation	10%	3	0.3	3	0.3	3.5	0.35	3.5	0.35	3.5	0.35
	Net Score	3.45		4.1		3.3		3.2		3.8	
	Rank	3		1		4		5		2	
	Continue?	NO		**Selected**		NO		NO		NO	

4. Design of selected hand rehabilitation device

This study mainly focuses on the upper limb rehabilitation of the five fingers in the right hand except for the wrist joint. The human hand consists of thumb, index, middle, ring, and little finger, in which the structure of four fingers except thumb is similar. The fingers are made up of digits, each digit is having two segments the phalangeal segment and the metacarpophalangeal segment. Thumb has one interphalangeal (IP) joint and a metacarpophalangeal (MCP) joint which has a flexion /extension motion. The selected concept 2 design is having 3 joints and 4 DOF for an index, middle, ring, and little finger. Whereas the thumb is having 3 joints and 4 DOF. Thumb differs from the other fingers due to the presence of an intermetacarpal joint. The human lower hand up to the wrist is mainly having 16 joints and 23 DOF.

In the selected model (Figure 3), the dimension of each link is done by considering the standard biomechanics data. The design of the thumb and index fingers has 4 DOF and 4 joints. Soft silicone strips are used to fix the rigid links over the fingers. The cable-driven mechanism is used to actuate the robot. Pulleys are provided at each joint to wound the cables over it, and cables are actuated by using servo motors hence they can get desired motions.

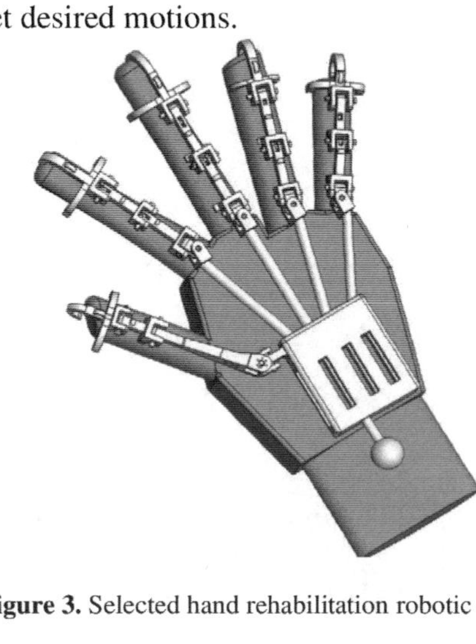

Figure 3. Selected hand rehabilitation robotic device.

IOP Conf. Series: Materials Science and Engineering 1132 (2021) 012033 doi:10.1088/1757-899X/1132/1/012033

4. Kinematic modeling

The use of the robot's kinematic equations to evaluate the end effector's position and orientation depending on the defined values of the joint parameters is known as forward kinematics. Forward kinematics is used to calculate the relationship between the pose of the end effector with respect to joint parameters. Here forward and inverse kinematics of thumb and index finger is explained in the following section.

4.1. Forward kinematic modeling of the thumb

The DH frame representation of thumb which is having the global frame {0} at the wrist joint is shown in figure 4. All the fingers are having local frames {1} and {2} at the metacarpal joint. The frame at the proximal joint is {3} and at the distal joint is {4}. Frame {1} and {2} at the metacarpal joint has the same origin as shown in figure 4. The inter metacarpal (IMCP) joint is represented as jt1 and jt2, the proximal and distal joints are represented as jt3 and jt4, respectively. Joint rotations are represented as θ whereas θ_{t2} , θ_{t3} , θ_{t4} and θ_{t5} are joint rotation at jt1, jt2, jt3, and jt4 respectively.

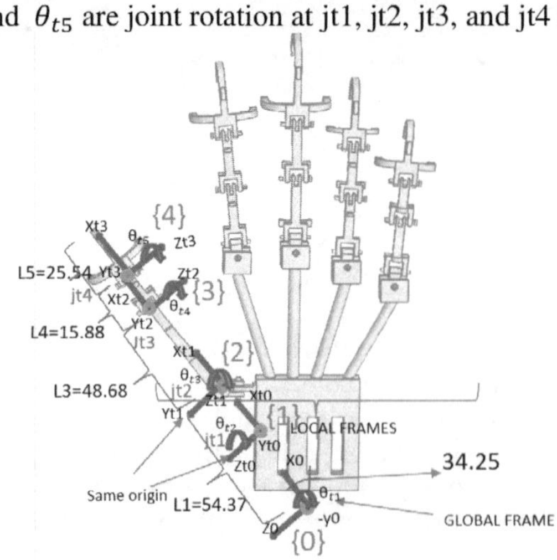

Figure 4. DH frame representation of Thumb.

Table 3. DH parameters of thumb.

Link	θ	d	a (mm)	α	
1	θ_{t1}=34.2°(fixed)	0	54.37	0°	$^{0}T_{t1}$
2	θ_{t2}	0	0	90°	$^{1}T_{t2}$
3	θ_{t3}	0	48.68	90°	$^{2}T_{t3}$
4	θ_{t4}	0	15.88	0°	$^{3}T_{t4}$
5	θ_{t5}	0	25.54	0°	$^{4}T_{t5}$

Based on biometric data [22] the distance between each joint is taken and it is as follows, the distance between the global and local frame is L1 (54.37mm) and the distance between frame {1} and {2} is L3 (48.68mm). Distance between frame {3} and {4} is L4 (15.88mm) and the length of a distal digit is L5=25.54mm. Angle about the z-axis, from x0 to x1 at global frame is fixed and is θ_{thumb1} = 34.25°

which is taken as the initial position and it is measured from the Solidworks platform. DH parameter tabulated in table 3. Where "θ" is the angle about the previous z-axis, from the previous x to next x. "d" is offset along the previous z to common normal and where "a" is the length of common normal.

The final transformation matrix of thumb is represented as $0_{T_{t5}}$. It is calculated as the product of $0_{T_{t1}}$, $1_{T_{t2}}$, $2_{T_{t3}}$, $3_{T_{t4}}$ and $4_{T_{t5}}$as shown in equation (1). $0_{T_{t1}}$represents the transformation matrix of the first link and correspondingly $1_{T_{t2}}$, $2_{T_{t3}}$, $3_{T_{t4}}$ and $4_{T_{t5}}$ represents the transformation matrix of the remaining links.

$$0_{T_{t5}} = 0_{T_{t1}} * 1_{T_{t2}} * 2_{T_{t3}} * 3_{T_{t4}} * 4_{T_{t5}} \tag{1}$$

The general format of the final transformation matrix ($0_{T_{t5}}$) of thumb is represented in equation 2.

$$0_{T_{t5}} = \begin{bmatrix} a11, & a12 & ,a13 & ,a14 \\ a21, & a22 & ,a23 & ,a24 \\ a31, & a32 & ,a33 & ,a34 \\ a41, & a42 & ,a43 & ,a44 \end{bmatrix} \tag{2}$$

Where, a11, a12,..., a44 are the elements of the final transformation matrix of thumb and their values given in appendix 1.

4.2. Forward kinematic modeling of the Index finger

DH model of the index finger is very much similar to that of the thumb. It is having one MCP joint, an IP joint, and a distal joint. MCP joint represented as ji0 and ji1, proximal and distal joints is represented as ji2 and ji3 respectively. θ_{i2} , θ_{i3} , θ_{i4} and θ_{i5} are joint rotation at ji0, ji1, ji2, and ji3 respectively. Figure 5 shows the DH frame representation of the index finger.

From the biometric data [22] the distance between each joint is taken and it is as follows, the distance between the global and local frame is L01 (106.47mm) and the distance between frame {1} and {2} is L03 (29mm). Distance between frame {3} and {4} is L04 (24mm) and the length of a distal digit is L05 (30mm). Angle about the z-axis, from x0 to x1 at global frame is fixed and is θ_{index1} = 13.98° which is calculated from the Solidworks model. DH parameters are tabulated in table 4.

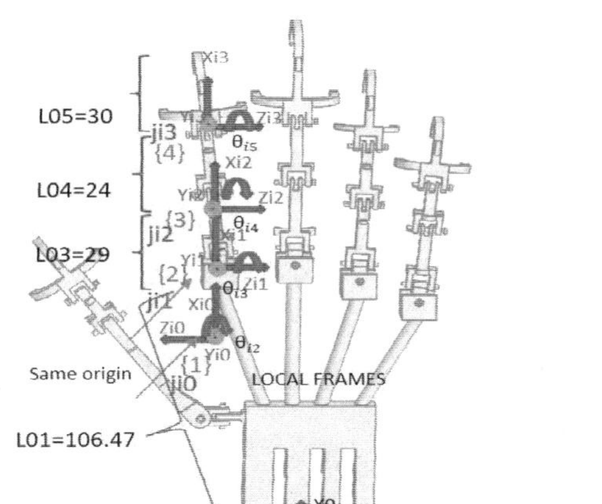

Figure 6. DH frame representation of Index finger.

Table 4. DH parameters of index finger.

Link	Θ	d	a (mm)	α	
1	θ_{index1}=13.98°(fixed)	0	106.47	0°	$0_{T_{i1}}$
2	θ_{i2}	0	0	90°	$1_{T_{i2}}$
3	θ_{i4}	0	29	0°	$2_{T_{i3}}$
4	θ_{i5}	0	24	0°	$3_{T_{i4}}$
5	θ_{i6}	0	30	0°	$4_{T_{i5}}$

Since index, middle, ring, and little finger have similar design structures the DH parameter is found to be the same. The first joints of each finger are fixed at the local frame and their angle θ_1 is measured from the Solidworks model. Angles for each finger are θ_{index1} = 13.98°, $\theta_{middle1}$ = 0°, θ_{ring1} = -12.68°, $\theta_{little1}$ = -25.68°.
The final transformation matrix for the index finger $0_{T_{i5}}$ is calculated from equation (4)

$$0_{T_{i5}} = 0_{T_{i1}} * 1_{T_{i2}} * 2_{T_{i3}} * 3_{T_{i4}} * 4_{T_{i5}} \tag{3}$$

$$0_{T_{i5}} = \begin{bmatrix} b11, & b12 & ,b13 & ,b14 \\ b21, & b22 & ,b23 & ,b24 \\ b31, & b32 & ,b33 & ,b34 \\ b41, & b42 & ,b43 & ,b44 \end{bmatrix} \tag{4}$$

Where, b11, b12, …, b44 are the elements of the final transformation matrix of an index finger, and their values are given in appendix 2.

5. Inverse Kinematic modeling

The inverse kinematics is done for finding the joint angles when the pose of the end-effector is known. The general end-effector pose is given by equation 5.

$$AE = \begin{bmatrix} nx & ox & ax & px \\ ny & oy & ay & py \\ nz & oz & az & pz \\ 0 & 0 & 0 & 1 \end{bmatrix} \tag{5}$$

The final forward transformation matrix is given by equation 5. That is equated with equation 6 to find the joint angles. For the effective mathematical solving inverse of the transformation matrix of the first link $(0_{T_1})^{-1}$ is multiplied with equation 1. The general equations for finding inverse kinematics are given as follows

$$\begin{gathered} 0_{T_5} * (0_{T_1})^{-1} = 1_{T_2} * 2_{T_3} * 3_{T_4} * 4_{T_5} \\ 0_{T_5} * (0_{T_1})^{-1} * (0_{T_2})^{-1} = 2_{T_3} * 3_{T_4} * 4_{T_5} \\ 0_{T_5} * (0_{T_1})^{-1} * (0_{T_2})^{-1} * (0_{T_3})^{-1} = 3_{T_4} * 4_{T_5} \\ 0_{T_5} * (0_{T_1})^{-1} * (0_{T_2})^{-1} * (0_{T_3})^{-1} * (0_{T_4})^{-1} = 4_{T_5} \end{gathered} \tag{6}$$

After equating RHS and LHS elements of the set of equations (6), the inverse kinematics of thumb is calculated and given in appendix 3. Since all four fingers having similar design structures the inverse formulae were found to be the same. Whereas θ_1 values are different i.e., θ_{index1} =34.25, $\theta_{middle1}$ = 0,

θ_{ring1} = -12.68, $\theta_{little1}$ = -25.68. Inverse formulae for the remaining four fingers are calculated and given in appendix 4.

6. Conclusion

A design for a hand exoskeleton robotic device for rehabilitation has been presented. The selection of the concept is done through concept screening and concept scoring for six conceptual designs. Forward and inverse kinematic analysis of the selected concept is performed by using Denavit-Hartenberg (DH) modeling. The use of lightweight materials like aluminum alloy and silicon strips will help the design framework to take the advantage of both soft and rigid structures. Actuation of the joints can be made possible with the help of tendon-driven cables which is pending future work. Several challenges are there in developing a fully equipped exoskeleton, the major challenge is to detect the intention of the user and act accordingly. Implementation of electromyography (EMG) or electroencephalogram (EEG) signal will help to detect the user intention up to a certain level.

References

[1] Platz T 2003 Evidenzbasierte Arm rehabilitation *Der Nervenarzt* vol **74(10)** pp 841-849.

[2] Patton, J, Small, S and Zev Rymer, W 2008 Functional restoration for the stroke survivor: informing the efforts of engineers *Topics in stroke rehabilitation* **15(6)** pp 521-541.

[3] Croshaw, P 1969 Hardiman I arm test-Hardiman I prototype project *General Electric Co Schenectady Ny Specialty Materials Handling Products Operation.*

[4] Bogue, R 2009 Exoskeletons and robotic prosthetics: a review of recent developments *Industrial Robot: an international journal* vol **36 (5)** pp 421- 427.

[5] Polygerinos, P, Wang, Z, Galloway, K, Wood, R and Walsh, C 2015 Soft robotic glove for combined assistance and at-home rehabilitation *Robotics and Autonomous Systems* **73** pp 135-143.

[6] Hasegawa, Y, Mikami, Y, Watanabe, K and Sankai, Y 2008 Five-fingered assistive hand with mechanical compliance of human finger In *2008 IEEE international conference on robotics and automation* pp 718-724

[7] Park, Y, Santos, J, Galloway, K, Goldfield, E and Wood, R 2014 A soft wearable robotic device for active knee motions using flat pneumatic artificial muscles In *2014 IEEE International Conference on Robotics and Automation (ICRA)* pp 4805-4810

[8] Asbeck, A, De Rossi, S, Galiana, I, Ding, Y and Walsh, C, 2014 Stronger, smarter, softer: next-generation wearable robots *IEEE Robotics & Automation Magazine*, **21(4)** pp 22-33.

[9] Varghese, R, Freer, D, Deligianni, F, Liu, J, Yang, G and Tong, R 2018 Wearable robotics for upper-limb rehabilitation and assistance: A review of the state-of-the-art challenges and future research *In Wearable Technology in Medicine and Health Care* p 340.

[10] Zoss, A, Kazerooni, H and Chu, A 2006 Biomechanical design of the Berkeley lower extremity exoskeleton (BLEEX) *IEEE/ASME Transactions on mechatronics*, **11(2)** pp 128-138.

[11] Guizzo, E and Goldstein, H, 2005 The rise of the body bots [robotic exoskeletons] *IEEE spectrum* **42(10)** pp 50-56.

[12] Asbeck, A, De Rossi, S, Galiana, I, Ding, Y and Walsh, C 2014 Stronger, smarter, softer: next-generation wearable robots *IEEE Robotics & Automation Magazine* **21(4)** pp 22-33.

[13] Park, Y, Chen, B, Pérez-Arancibia, N, Young, D, Stirling, L, Wood, R, Goldfield, E and Nagpal, R 2014 Design and control of a bio-inspired soft wearable robotic device for ankle–foot rehabilitation. *Bioinspiration & biomimetics* **9(1)** p 016007.

[14] Chiri, A, Giovacchini, F, Vitiello, N, Cattin, E, Roccella, S, Vecchi, F and Carrozza, M 2009 HANDEXOS: Towards an exoskeleton device for the rehabilitation of the hand In *2009 IEEE/RSJ International Conference on Intelligent Robots and Systems* pp 1106-1111.

[15] Yap, H, Ang, B, Lim, J, Goh, J and Yeow, C 2016 A fabric-regulated soft robotic glove with user intent detection using EMG and RFID for hand assistive application In *2016 IEEE International Conference on Robotics and Automation (ICRA)* pp 3537-3542.

[16] Hasegawa, Y, Mikami, Y, Watanabe, K and Sankai, Y 2008 Five-fingered assistive hand with mechanical compliance of human finger In *2008 IEEE international conference on robotics and automation* pp 718-724
[17] Cempini, M, Cortese, M and Vitiello, N 2014 A powered finger–thumb wearable hand exoskeleton with self-aligning joint axes *IEEE/ASME Transactions on mechatronics* **20(2)** pp 705-716.
[18] Tadano, K, Akai, M, Kadota, K and Kawashima, K 2010 Development of grip amplified glove using bi-articular mechanism with pneumatic artificial rubber muscle In *2010 IEEE International Conference on Robotics and Automation* pp 2363-2368
[19] Tong, K, Ho, S, Pang, P, Hu, X, Tam, W, Fung, K, Wei, X, Chen, P and Chen, M 2010 An intention driven hand functions task training robotic system In *2010 Annual International Conference of the IEEE Engineering in Medicine and Biology* pp 3406-3409.
[20] Sanjuan, J, Castillo, A, Padilla, M, Quintero, M, Gutierrez, E, Sampayo, I, Hernandez, J and Rahman, M 2020 Cable driven exoskeleton for upper-limb rehabilitation: A design review. *Robotics and Autonomous Systems* 126 p 103445.
[21] Ulrich, K 2003 *Product design and development* Tata McGraw-Hill Education.
[22] Garrett, J 1970 Anthropometry of the hands of male air force flight personnel *Air force aerospace medical research lab wright-patterson afb oh.*

Appendix

(1) a11= $\cos(\theta_{t3} + \theta_{t4} + \theta_{t5}) * \cos(\theta_{t1} + \theta_{t2})$;
a21= $\cos(\theta_{t3} + \theta_{t4} + \theta_{t5}) * \sin(\theta_{t1} + \theta_{t2})$;
a31= $\sin(\theta_{t3} + \theta_{t4} + \theta_{t5})$; a41=0;
a12= $-\sin(\theta_{t3} + \theta_{t4} + \theta_{t5}) * \cos(\theta_{t1} + \theta_{t2})$;
a22= $-\sin(\theta_{t3} + \theta_{t4} + \theta_{t5}) * \sin(\theta_{t1} + \theta_{t2})$;
a32= $\cos(\theta_{t3} + \theta_{t4} + \theta_{t5})$; a42=0;
a13= $\sin(\theta_{t1} + \theta_{t2})$; a23= $\cos(\theta_{t1} + \theta_{t2})$;
a33=0; a34=0;
a14= $L2*\cos(\theta_{t1} + \theta_{t2}) + L1*\cos(\theta_{t1}) + L4*\cos(\theta_{t1} + \theta_{t2})*\cos(\theta_{t3} + \theta_{t4}) + L3*\cos(\theta_{t1} + \theta_{t2})*\cos(\theta_{t3}) + L5*\cos(\theta_{t1} + \theta_{t2})*\cos(\theta_{t3} + \theta_{t4})*\cos(\theta_{t5}) - L5*\cos(\theta_{t1} + \theta_{t2})*\sin(\theta_{t3} + \theta_{t4})*\sin(\theta_{t5})$;
a24= $L2*\sin(\theta_{t1} + \theta_{t2}) + L1*\sin(\theta_{t1}) + L4*\cos(\theta_{t1} + \theta_{t2})*\sin(\theta_{t3} + \theta_{t4}) + L3*\sin(\theta_{t1} + \theta_{t2})*\cos(\theta_{t3}) + L5*\sin(\theta_{t1} + \theta_{t2})*\sin(\theta_{t3} + \theta_{t4})*\sin(\theta_{t5}) - L5*\cos(\theta_{t1} + \theta_{t2})*\sin(\theta_{t3} + \theta_{t4})*\cos(\theta_{t5})$;
a34= $L4*\sin(\theta_{t3} + \theta_{t4}) + L3*\sin(\theta_{t3}) + L5*\sin(\theta_{t3} + \theta_{t4} + \theta_{t5})$; a44=1.

Appendix 2

b11= $\cos(\theta_{i3} + \theta_{i4} + \theta_{i5}) * \cos(\theta_{i1} + \theta_{i2})$;
b21= $\cos(\theta_{i3} + \theta_{i4} + \theta_{i5}) * \sin(\theta_{i1} + \theta_{i2})$;
b31= $\sin(\theta_{i3} + \theta_{i4} + \theta_{i5})$; b41=0;
b12= $-\sin(\theta_{i3} + \theta_{i4} + \theta_{i5}) * \cos(\theta_{i1} + \theta_{i2})$;
b22= $-\sin(\theta_{i3} + \theta_{i4} + \theta_{i5}) * \sin(\theta_{i1} + \theta_{i2})$;
b32= $\cos(\theta_{i3} + \theta_{i4} + \theta_{i5})$; b42=0;
b13= $\sin(\theta_{i1} + \theta_{i2})$; b23= $\cos(\theta_{i1} + \theta_{i2})$;
b33=0; b43=0;
b14= $L2*\cos(\theta_{i1} + \theta_{i2}) + L1*\cos(\theta_{i1}) + L4*\cos(\theta_{i1} + \theta_{i2})*\cos(\theta_{i3} + \theta_{i4}) + L3*\cos(\theta_{i1} + \theta_{i2})*\cos(\theta_{i3}) + L5*\cos(\theta_{i1} + \theta_{i2})*\cos(\theta_{i3} + \theta_{i4})*\cos(\theta_{i5})$;
b24= $L2*\sin(\theta_{i1} + \theta_{i2}) + L1*\sin(\theta_{i1}) + L4*\cos(\theta_{i3} + \theta_{i4})*\sin(\theta_{i1} + \theta_{i2}) + L3*\sin(\theta_{i1} + \theta_{i2})*\cos(th3) - L5*\sin(\theta_{i1} + \theta_{i2})*\sin(\theta_{i3} + \theta_{i4})*\sin(\theta_{i5}) + L5*\cos(\theta_{i3} + \theta_{i4})*\sin(\theta_{i1} + \theta_{i2})*\cos(\theta_{i5})$;
b34= $L4*\sin(\theta_{i1} + \theta_{i2}) + L3*\sin(\theta_{i3}) + L5*\sin(\theta_{i3} + \theta_{i4} + \theta_{i5})$; b44 = 1.

Appendix 3

θ_{t1}=34.25 (Known)

$$\theta_{t2} = \tan^{-1}\left(\frac{ay}{ax}\right); \quad (1)$$

IOP Conf. Series: Materials Science and Engineering 1132 (2021) 012033 doi:10.1088/1757-899X/1132/1/012033

$$\theta_{t3} = \tan^{-1}\left(\frac{ax}{\cos(\theta_{t1}+\theta_{t2})*az}\right); \quad (2)$$

$$\theta_{t34} = \theta_{t3}+\theta_{t4} = \tan^{-1}\left(\frac{[(-ax*\cos(\theta_{t1}+\theta_{t2}))-(ay*\sin(\theta_{t1}+\theta_{t2}))]}{az}\right); \quad (3)$$

$$\theta_{t45} = \theta_{t4}+\theta_{t5} = \tan^{-1}\left(\frac{(ny*\cos(\theta_{t1}-\theta_{t2}))-(nx*\sin(\theta_{t1}-\theta_{t2}))}{(oy*\cos(\theta_{t1}-\theta_{t2}))-(ox*\sin(\theta_{t1}-\theta_{t2}))}\right); \quad (4)$$

$$\theta_{t4} = \theta_{t34} - \theta_{t3} = \tan^{-1}\left(\frac{[(-ax*\cos(\theta_{t1}+\theta_{t2}))-(ay*\sin(\theta_{t1}+\theta_{t2}))]}{az}\right) - \tan^{-1}\left(\frac{ax}{\cos(\theta_{t1}+\theta_{t2})*az}\right); \quad (5)$$

$$\theta_{t5} = \theta_{t45} - \theta_{t4} = \tan^{-1}\left(\frac{(ny*\cos(\theta_{t1}-\theta_{t2}))-(nx*\sin(\theta_{t1}-\theta_{t2}))}{(oy*\cos(\theta_{t1}-\theta_{t2}))-(ox*\sin(\theta_{t1}-\theta_{t2}))}\right) - \tan^{-1}\left(\frac{[(-ax*\cos(\theta_{t1}+\theta_{t2}))-(ay*\sin(\theta_{t1}+\theta_{t2}))]}{az}\right) - \tan^{-1}\left(\frac{ax}{\cos(\theta_{t1}+\theta_{t2})*az}\right); \quad (6)$$

Appendix 4

$$\theta_2 = \tan^{-1}\left(\frac{-ax}{ay}\right); \quad (7)$$

$$\theta_3 = \tan^{-1}\left(\frac{nz}{oz}\right); \quad (8)$$

$$\theta_{34} = \theta_{t3}+\theta_{t4} = \tan^{-1}\left(\frac{[(-ay*\sin(\theta_1+\theta_2))-(ax*\cos(\theta_1+\theta_2))]}{az}\right); \quad (9)$$

$$\theta_{345} = \theta_{t3}+\theta_{t4}+\theta_{t5} = \tan^{-1}\left(\frac{-(ox*\cos(\theta_1))-(nx*\sin(\theta_1))}{(nx*\cos(\theta_1))-(ox*\sin(\theta_1))}\right); \quad (10)$$

$$\theta_4 = \theta_{34} - \theta_3 = \tan^{-1}\left(\frac{[(-ay*\sin(\theta_1+\theta_2))-(ax*\cos(\theta_1+\theta_2))]}{az}\right) - \tan^{-1}\left(\frac{nz}{oz}\right); \quad (11)$$

$$\theta_5 = \theta_{345} - \theta_{34} = \tan^{-1}\left(\frac{-(ox*\cos(\theta_1))-(nx*\sin(\theta_1))}{(nx*\cos(\theta_1))-(ox*\sin(\theta_1))}\right) - \tan^{-1}\left(\frac{[(-ay*\sin(\theta_1+\theta_2))-(ax*\cos(\theta_1+\theta_2))]}{az}\right); \quad (12)$$

Kinematic Modelling and Structural Analysis of a Spherical Robot: BALL-E

Satwik Tholapu, Sudheer A P* and Joy M L
Mechatronics/Robotics Lab, Department of Mechanical Engineering, National Institute of Technology, Calicut, Kerala, India-673601

*apsudheer@nitc.ac.in

Abstract: This paper presents a spherical mobile robot which is capable of executing rolling and walking motions. The proposed robot can also transform into a quadruped based on the terrain conditions. The advantage of combining these motions in the robot is that it improves the collision recovery and manoeuvrability over multiple terrains. Various conceptual designs are developed, and concept scoring is carried out to find the best design. The kinematic analysis is presented for the chosen design using the Denavit-Hartenberg (D-H) method. The inverse kinematics is solved using MATLAB and the corresponding robot configurations are plotted. The structural analysis of the robot is also completed using ANSYS Workbench.
Keywords: Spherical Robot, Kinematic Modelling, Dimensional Synthesis, Quadruped, structural analysis.

1. Introduction

Spherical mobile robots are becoming popular these days due to their functionality and unique features. The rolling motion makes it to be usable over a wide range of terrains, for unmanned operations in unpredictable environments and has better collision avoidance and recovery. There are many types of Spherical Robots available in the market. These spherical robots differ by the type of internal driving mechanisms to actuate the robot in the required direction. Many research works are going on in developing these internal driving mechanisms and their control strategies.

A spherical robot having a pillar-like structure with driving wheel and supporting wheels at the ends are discussed in [1]. In this spherical, robot the pillar will move inside the spherical shell and in turn, moves the sphere in the required direction. Some spherical robots are having cart like structure inside the shell which drives the outer shell [2][3]. Spherical robot having a ball inside the spherical shell that can be rotated by a motor with the help of rollers in a designated configuration is presented in [4]. This design makes the robot to be omnidirectional as the ball rotation drives the robot. A design that uses pendulum mass as its driving force is developed in [5] and [6]. The imbalance created by the pendulum moves the robot in the intended direction. Furthermore, a compound pendulum design is implemented in [7], which improves the stability and control of the spherical robot. A concept using conservation of momentum is proposed in [8], where the authors used high-speed rotors attached to two identical hemispheres that actuate the robot. Shifting the centre of mass of the robot is used as an actuation method in [9]. The controlled motion of the internal masses causes the locomotion of the robot. As an alternative, Shape Memory Alloy (SMA) wires are used to configure the outer shell by developing a suitable control strategy [10]. Similarly, an outer shell comprising of modular Lego-like blocks are designed in [11]. This modular design helps it to shapeshift and move in the desired direction. For all the spherical robots discussed above, controlling the path followed is difficult and motions are mostly unsteady due to the actuation mechanism. Power consumption is also a limitation in these designs and

IOP Conf. Series: Materials Science and Engineering 1132 (2021) 012034 doi:10.1088/1757-899X/1132/1/012034

the use of renewable energy like solar/wind can also be explored. Placement of the components such as sensors, electronic controllers and the battery is also a challenge.

In this paper, a novel spherical robot mechanism is designed which can transform itself from sphere to a quadruped and vice versa as per need. In the spherical form, the robot can move by rolling over the surfaces and in the quadruped form, it follows legged locomotion. The conceptual design is made by type, number and dimensional synthesis. Furthermore, the kinematic modelling is carried out using the Denavit-Hartenberg (D-H) convention and is verified using MATLAB simulations. Structural analysis of the robot components is also presented.

2. Conceptual Design

This section details the conceptual designs developed for the spherical robot. All the conceptual designs are focused on novelty and to overcome the limitations of the predecessors.

2.1. Concept A

In this design spoke arms protrude from the surface of the spherical robot as shown in Figure 1. All electronics are inside the casing at the centre and spoke arms are actuated using a suitable mechanism.

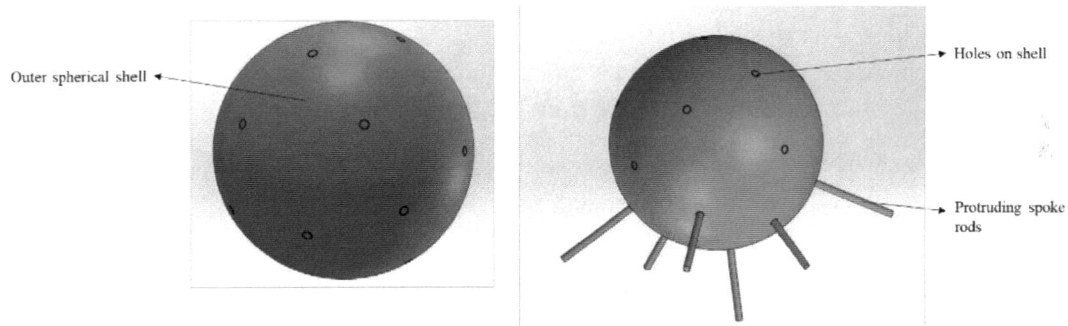

Figure 1. Spoke sphere

2.2. Concept B

It uses a combination of a pendulum and wheel-based spherical robot and also the design can be modified to have a head placed on top of it with sensors as seen in Figure 2.

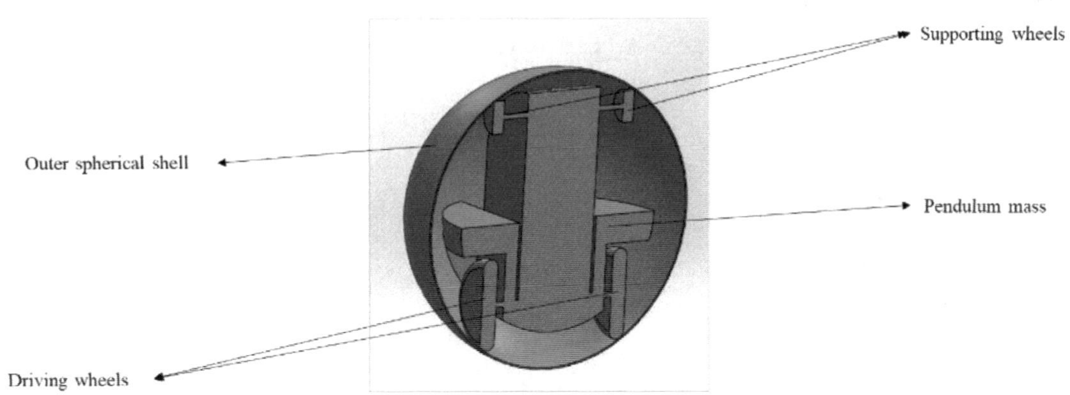

Figure 2. Wheelie robot

2.3. Concept C

The outer shell of the spherical robot is comprised of identical triangular curved panels which can be moved to facilitate the movement of the robot as in Figure 3.

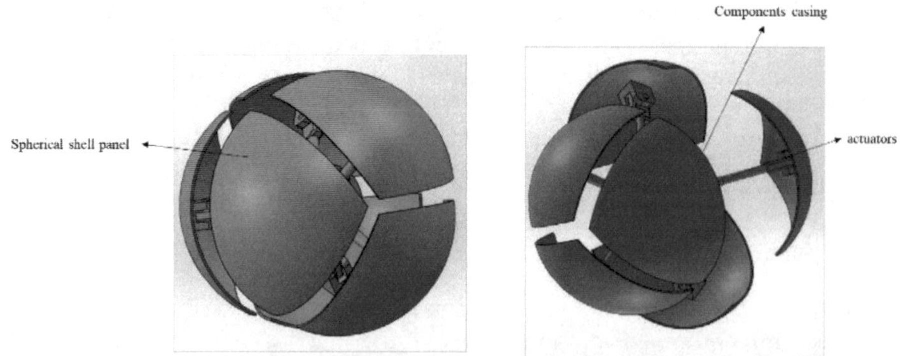

Figure 3. Trisphere robot

2.4. Concept D

In this design, the mass can be moved inside the spherical robot by sliding mechanisms, which in turn facilitate the movement of the robot as shown in Figure 4.

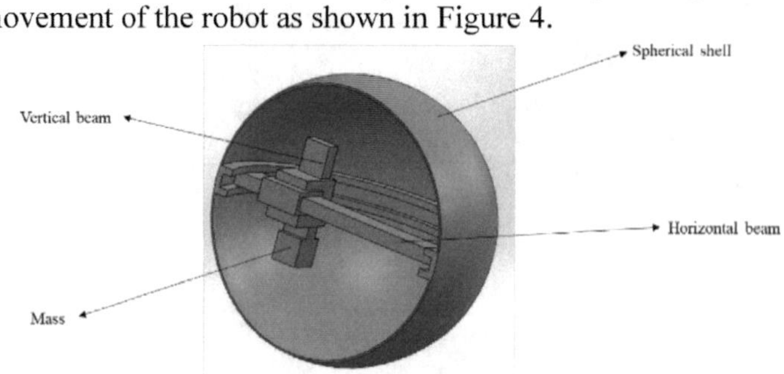

Figure 4. 3dsphere

2.5. Concept E

This concept is transformable spherical/legged robot by moving the robot legs which are attached to curved contours we can move the robot in a spherical form and when transformed it can act as a legged robot as represented in Figure 5.

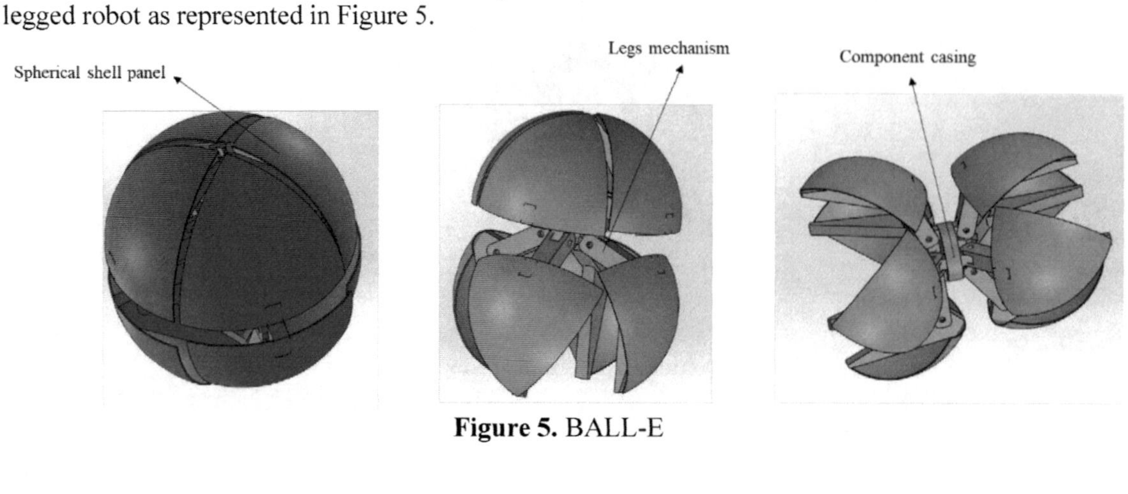

Figure 5. BALL-E

2.6. Concept F

In this concept, the outer shell is made of ferromagnetic material and the inner core is comprised of several electromagnetic coils arranged in a designated way and can be activated as per need and by activating the coils in the required direction the inner core moves in that direction causing imbalance for the robot to move accordingly as seen in Figure 6.

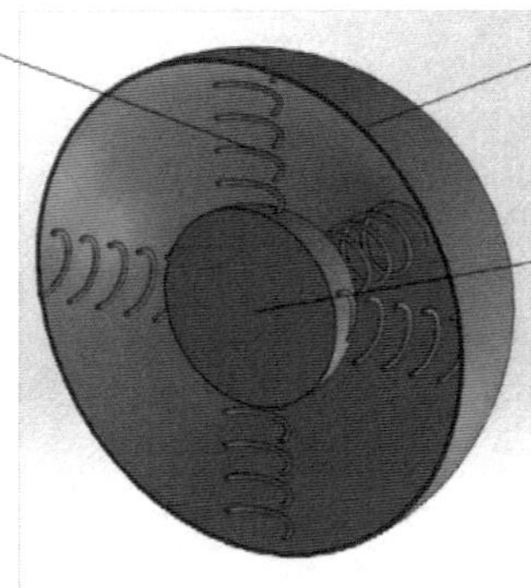

Figure 6. Springy

3. Concept Scoring and Selection

Concept scoring is a method for rating the concept designs and rank them based on a set of predetermined or user-defined criteria. It is helpful to weigh down the advantages and limitations of a design and to select the best design to satisfy the requirements. In this work, Manoeuvrability, Manufacturability, Cost, Stability, Speed, Power usage, Sensor compatibility and Ability to jump are selected as the criteria for scoring. The final scores obtained for each concept design and their rankings are listed in Table 1.

Table 1. Concept Scoring

		A		B		C		D		E		F	
Selection criteria	Weight	Rating	Weighted score	Rating	Weighted score	Rating	Weighted score	Rating	Weighted score	Rating	Weighted score	Rating	Weighted score
Manoeuvrability	25%	4	1	3	0.75	2	0.5	3	0.75	5	1.25	3	0.75
Stability	20%	3	0.6	4	0.8	3	0.6	3	0.6	4	0.8	2	0.4
Manufacturability	15%	3	0.45	4	0.6	3	0.45	4	0.6	4	0.6	2	0.3
Cost	15%	4	0.6	4	0.6	3	0.45	3	0.45	3	0.45	3	0.45
Power usage	10%	3	0.3	4	0.4	3	0.3	3	0.3	3	0.3	2	0.2
Speed	5%	4	0.2	4	0.2	3	0.15	3	0.15	3	0.15	3	0.15
Sensor compatibility	5%	3	0.15	4	0.2	2	0.1	4	0.2	4	0.2	2	0.1
Jumping	5%	4	0.2	2	0.1	4	0.2	3	0.15	4	0.2	3	0.15
Score		3.45		3.65		2.75		3.2		3.95		2.5	

IOP Conf. Series: Materials Science and Engineering 1132 (2021) 012034 doi:10.1088/1757-899X/1132/1/012034

Rank	**3**	**2**	**5**	**4**	**1**	**6**
Continue?	No	No	No	No	Yes	No

Though concept A has better control and stability, it is complex to design the internal mechanism. The cost and power usage are moderate, and the jumping motion can be accommodated with ease. However, sensor integration is difficult due to space limitation. Concept B has a good rating in all criteria except the jumping capability. Concept C is unsteady and hard to control compared to the other concepts. It has a reasonable cost and power consumption though it lacks sensor compatibility. Concept D does not excel in any criteria even though it satisfies the requirements. Concept E is innovative and fulfils the optimum requirements. Concept F is novel, complex to design, hard to control and there is a chance for the sensors to be interfered with by the magnetic field. By concept scoring, Concept E which is the transformable spherical robot is selected as the suitable design and developed further.

4. BALL-E, a Spherical Robot

BALL-E is a transformable spherical robot as shown in Figure 7 which can act as a quadruped robot as and when required. It comprises two hemispheres where the top hemisphere acts as a housing for sensors and electronics and the bottom hemisphere acts as quadruped robot when transformed. This Spherical robot contains 8 legs with 4 legs in each hemisphere and all the legs in a hemisphere are identical. All the joints in the leg mechanisms are revolute. Lower hemisphere legs have 4 links and 4 revolute joints. Upper hemisphere legs have 2 links and 2 revolute joints. Hence, there are 24 revolute joints in total and degree of freedom is equal to the total number of joints as all the legs are open chain.

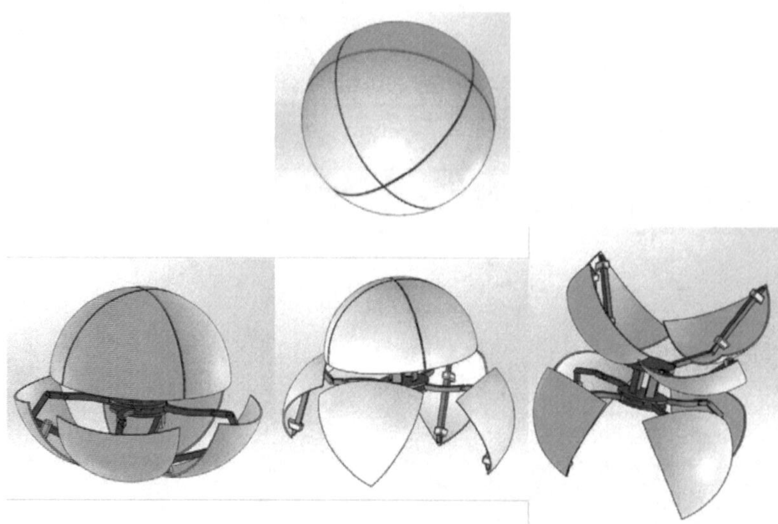

Figure 7. Concept of BALL-E

4.1. Dimensional Synthesis

The spherical robot is symmetric about a vertical axis. It has two types of leg mechanisms for each hemisphere. Based on the size of real-life actuators and electronic components, the diameter of the robot is assumed to be 340 mm. Link lengths for the legs are derived from this assumption. The size of a single shell is 1/8th of the sphere and the diameter is 340 mm.

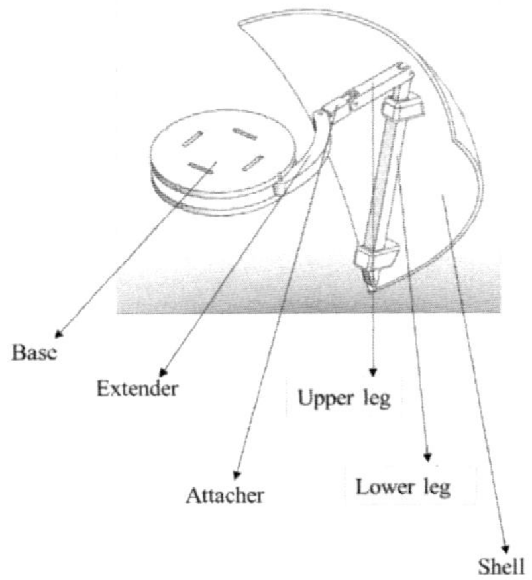

Figure 8. Lower hemisphere leg

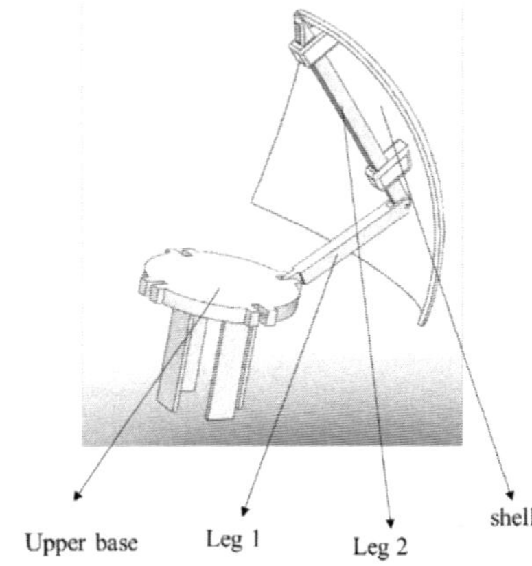

Figure 9. Upper hemisphere

The designs for leg mechanisms are presented in Figure 8 and 9. For the lower hemisphere legs, the length of the leg depends on the position of attachment to the shell. Consider the 2D view of a shell and lower leg the position as seen in Figure 10. The link length of the lower leg is found to be around 190 mm.

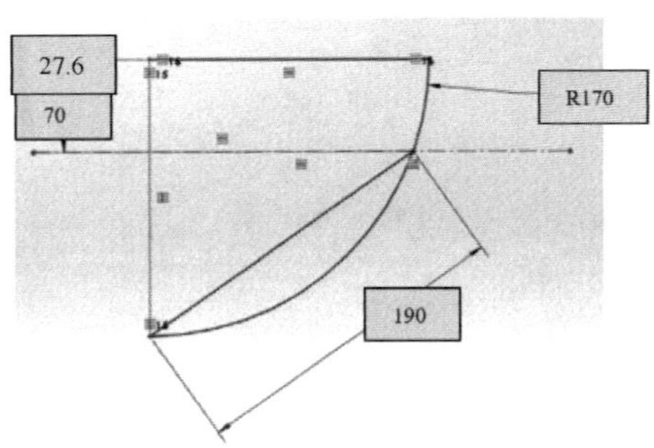

Figure 10. 2d view of leg and shell

The position of the base on the vertical axis is assumed and it affects the internal space available for components. The increase in the distance will provide more space to accommodate things but at a cost of interference between parts. Taking into account the maximum movement for lower leg and optimum space to accommodate parts, the position is assumed to be 70 mm. Upper leg link length is best if around 70 mm as it also helps in increasing the moving range of lower leg and shell. Link lengths of base, extender, attacher, the upper leg is interrelated. The exact link length is determined by obtaining link lengths for base and extender. Attacher link length is flexible as it only serves the purpose of connecting extender and upper leg. Hence, its dimension is fixed in the later stage of synthesis and the base diameter dimension is inversely proportional to extender link length. To avoid interference between legs, the base diameter is taken as 100mm. As the base diameter is fixed and the position of lower leg and shell is estimated, the sum of the link lengths of extender, attacher, the upper leg is equal to 110.8 mm. For maximum gait walk step of 300 mm, the attacher link length is assumed as 30 mm by the possible actuator placement. As discussed earlier upper leg link length is suitable around 70 mm and for ease of placement of extender, the upper link length is taken as 65.8 mm and extender link length becomes 82 mm. For upper hemisphere, the base diameter is taken as 190 mm, link length of leg 1 and leg 2 are equal to upper leg and lower leg lengths respectively for symmetry and functionality.

IOP Conf. Series: Materials Science and Engineering 1132 (2021) 012034 doi:10.1088/1757-899X/1132/1/012034

4.2. Kinematic Modelling

Kinematic modelling is used to relate the joint angles to the final pose of the robot. It helps in finding out the position of the end effector concerning base in terms of joint variables. D-H modelling is used to develop the kinematic equations of the spherical robot. The link length used for modelling is described in the previous section. Considering the lower hemisphere legs in Figure11, the reference frame is {0} and frames {1} to {4} are assigned to each joint up to the end effector. The D-H parameters obtained are given in Table 2.

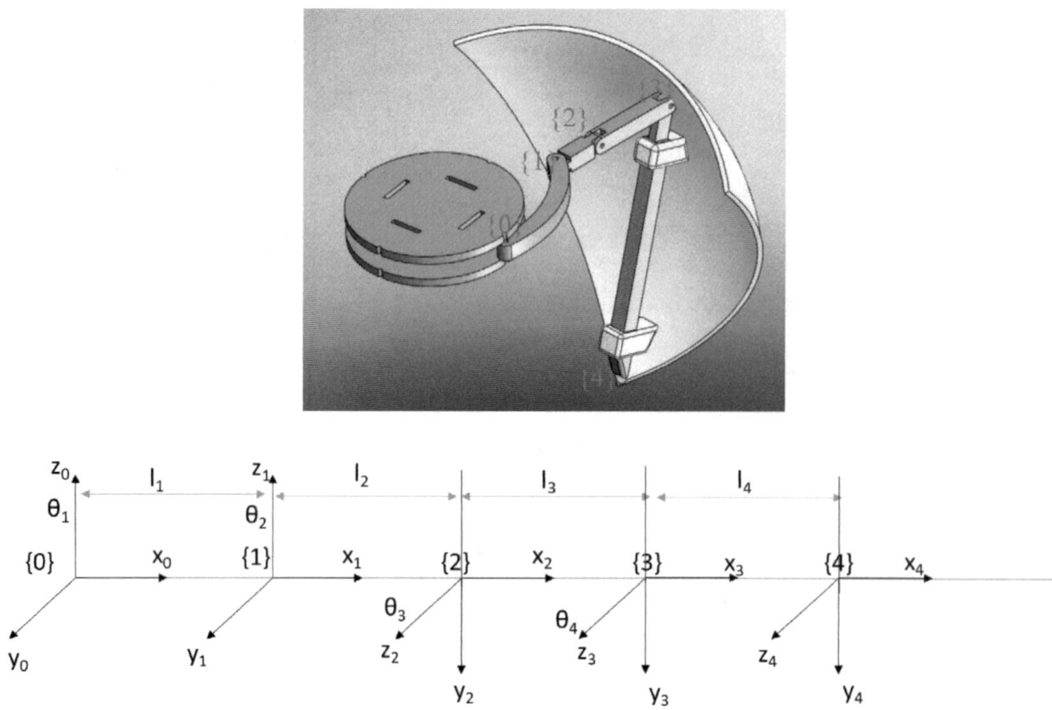

Figure 11. DH Modelling

Table 2. DH Parameters Table

Frames	Joint angle (θ_i)	Joint offset, d_i (mm)	Link length, a_i (mm)	Twist angle(α_i)
{0}-{1}	θ_1	0	L_1=82.01	0
{1}-{2}	θ_2	0	L_2=30	$\pi/2$
{2}-{3}	θ_3	0	L_3=65.8	0
{3}-{4}	θ_4	0	L_4=190	0

General transformation matrix from $i-1^{th}$ frame to i^{th} i[th] frame is

$$ {}^{i-1}_{i}T = \begin{bmatrix} c(\theta_i) & -s(\theta_i).c(\alpha_i) & s(\theta_i).s(\alpha_i) & a_i.c(\theta_i) \\ s(\theta_i) & c(\theta_i).c(\alpha_i) & -c(\theta_i).s(\alpha_i) & a_i.s(\theta_i) \\ 0 & s(\alpha_i) & c(\alpha_i) & d_i \\ 0 & 0 & 0 & 1 \end{bmatrix} \quad (1) $$

where $c(\theta)$ is cos (θ), s(θ) is sin(θ).
Transformation matrix of the end effector with respect to the base,

$$ {}^{base}_{end\ effector}T = {}^{0}_{1}T.{}^{1}_{2}T.\dots.{}^{n-1}_{n}T \quad (2) $$

Final transformation matrix for {4} wrt. {1} is,

$$ {}^{0}_{4}T = \begin{bmatrix} r_{11} & r_{12} & r_{13} & p_x \\ r_{21} & r_{22} & r_{23} & p_y \\ r_{31} & r_{32} & r_{33} & p_z \\ 0 & 0 & 0 & 1 \end{bmatrix} \quad (3) $$

where r_{ii} elements are components of the rotation matrix and p_i elements are components of the translation matrix.
The D-H parameters of a leg in quadruped and sphere form are visualized in Figure 12 using MATLAB. The joint variables can be verified from the figure whether it satisfies the desired configuration of the mechanism.

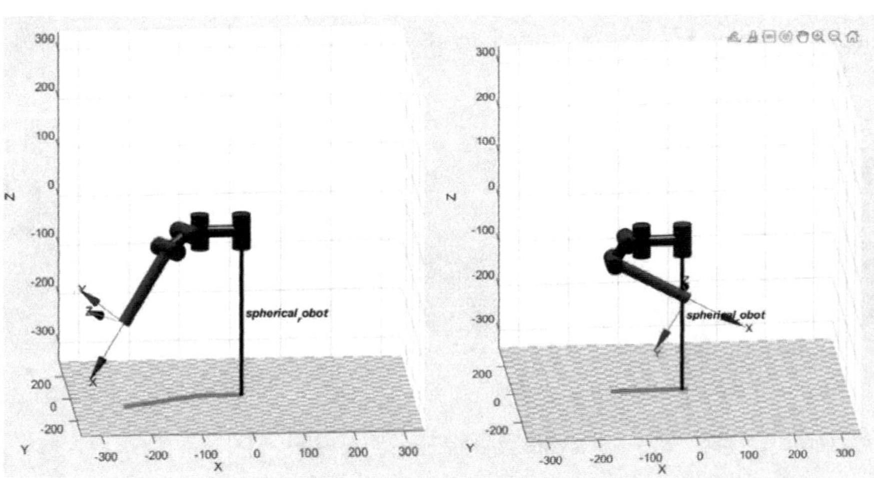

Figure 12. DH parameters

4.3. Inverse Kinematics
In this section, the inverse kinematics of the lower leg of the robot is detailed. The computational cost is higher for the geometric approach and hence the algebraic method is chosen for solving the inverse kinematics. The matrix equation to be solved is depicted in equation (4).

$$ {}_4^0T = \begin{bmatrix} r_{11} & r_{12} & r_{13} & p_x \\ r_{21} & r_{22} & r_{23} & p_y \\ r_{31} & r_{32} & r_{33} & p_z \\ 0 & 0 & 0 & 1 \end{bmatrix} = {}_1^0T(q_1).{}_2^1T(q_2).{}_3^2T(q_3).{}_4^3T(q_4) \tag{4} $$

where the left-hand side is the final transformation matrix and q_i are joint variables.
The equations (5) to (7) are nonlinear simultaneous equations derived from the matrix equation above.

$$ [{}_1^0T(q_1)]^{-1}.{}_4^0T = {}_2^1T(q_2).{}_3^2T(q_3).{}_4^3T(q_4) \tag{5} $$

$$ [{}_1^0T(q_1).{}_2^1T(q_2)]^{-1}.{}_4^0T = {}_3^2T(q_3).{}_4^3T(q_4) \tag{6} $$

$$ [{}_1^0T(q_1).{}_2^1T(q_2).{}_3^2T(q_3)]^{-1}.{}_4^0T = {}_4^3T(q_4) \tag{7} $$

By solving the set of simultaneous equations, the joint variables q_1, q_2, q_3 and q_4 are obtained.

4.4. Structural Analysis

The prototype design has been developed in SOLIDWORKS and the total mass of the robot is estimated to be around 5 kg. The main components of the lower leg mechanism are shown in Fig. 8 which are base, extender, upper leg and lower leg. The structural analysis is carried out assuming that in a given instant, the robot's weight acts on the component in a specified direction during its motion. Material is taken as aluminium alloy due to the high strength to weight ratio and the affordable cost. The load acting on the components is estimated as 50 N considering the loads that may act on the robot while executing the motion. Loading scenarios are assumed based on motion and the structure of the leg mechanism.

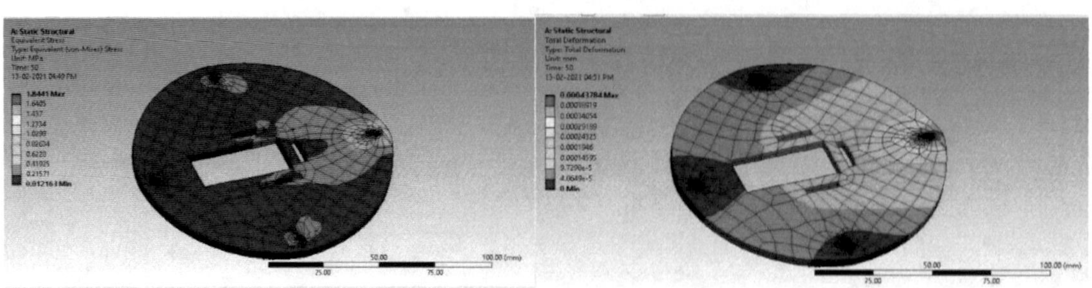

Figure 13. Base under load

As shown in Figure 13, the three holes of the spherical base plate are fixed and load is applied to the other hole radially. The maximum deflection of 4.378×10^{-4} mm is occurring at the hole on which the load has been applied and maximum equivalent stress of 1.844 MPa can be seen at the adjacent holes.

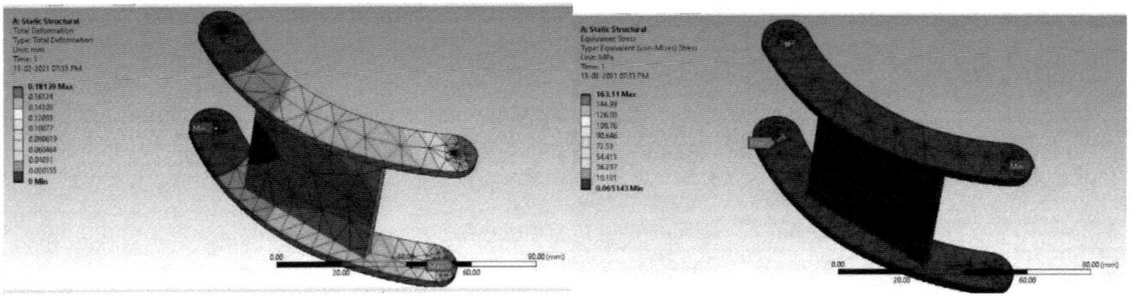

Figure 14. Extender under load

In the case of extender, it may undergo both bending and twisting forces. It was noted that stresses induced due to bending are more compared to others. The maximum deformation of 0.18 mm takes place at the load acting end and maximum stress of 163 MPa at the opposite end as of load as seen in Figure 14.

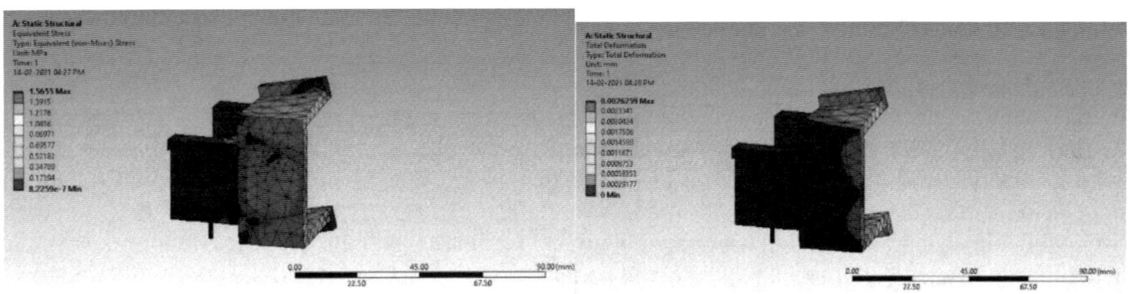

Figure 15. Attacher under load

Attacher acts as a bridge between the leg and base, extender assembly. Load acts on the walls of holders of actuators and causes a deformation of 2.629×10^{-3} mm and maximum stress of 1.565 MPa at the hinges of the top wall that can be observed in Figure 15.

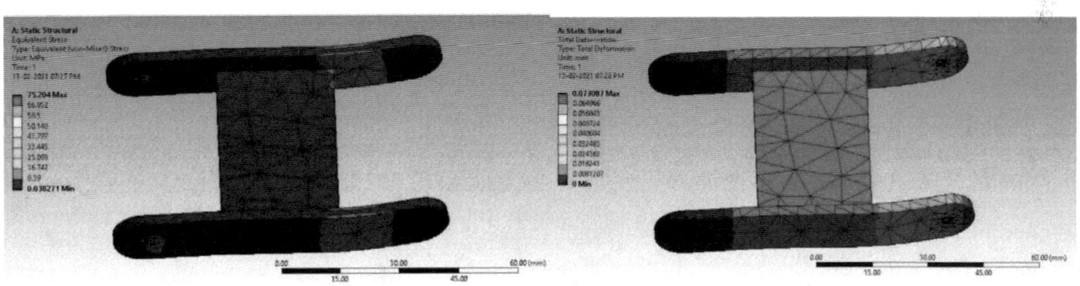

Figure 16. upper leg under load

The upper leg has a similar scenario as of extender in loading conditions. However, the stresses formed are less compared to the extender. The maximum deflection of 7.38×10^{-2} mm is observed at the load acting end and maximum stress of 75.204 MPa is seen at the opposite end of the load as depicted in Figure 16.

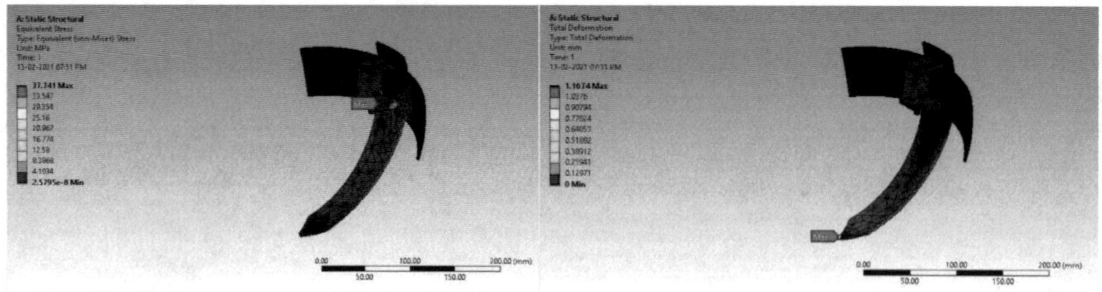

Figure 17. lower leg under load

The loading conditions of the lower leg varies based on the form of a robot that is sphere or quadruped. In sphere form, load acts on the circular section of the leg whereas in the other it acts on the end of the leg. It is found out that the stresses developed are more in quadruped state with a maximum deflection of 1.16mm at the tip of the leg and maximum stress of 37.741 MPa at the mid part of the circular section of the leg as portrayed in Figure 17.

5. Conclusion

In this paper, a novel spherical robot design is proposed. Conceptual designs based on the objectives are formulated and the best design is selected through concept scoring. Moreover, the type, number and dimensional synthesis are carried out for the chosen design. The kinematic model of the spherical robot is developed using the Denavit-Hartenberg (D-H) method and verified through MATLAB simulations. Inverse kinematic solutions are calculated using an algebraic method. Structural analysis under chosen loading conditions are performed using ANSYS Workbench and the results obtained are found to be satisfactory. The extender link in the mechanism bears more stress when compared to the other links and maximum deformation is observed in the lower leg. The dynamic analysis of the robot needs to be completed to obtain the equations of motion. The motion planning for both spherical and quadruped configurations are to be developed for generating the robot locomotion by inverse kinematics. The potential applications of the proposed spherical robot may include automated surveying, exploration in unknown environments, surveillance and security, light material transportation, etc.

References

[1] Halme A, Schonberg T, Wang Y 1996 Motion control of a spherical mobile robot *Int., Workshop on Advanced Motion Control, AMC* **1** pp 259-264

[2] Xuelei N, Suherlan A, Soh G, Foong S, Wood K and Otto K 2014 Mechanical Development and Control of a Miniature Spherical Rolling Robot *The 13th Int., Conf., on Control, Automation, Robotics and Vision, ICARCV 2014* (Marina Bay Sands, Singapore) pp 1923-1928

[3] Bicchi A, Balluchi A, Tichizzo D P and GoreUjt A 1997 Introducing the "SPHERICLE": An Experimental Testbed for Research and Teaching in Nonholonomy Proceedings *of Int., Conf., on Robotics and Automation* **3** pp 2620-2625

[4] Chen W, Chen C, Tsai J, Yang J, Lin P 2013 Design and implementation of a ball-driven omnidirectional spherical robot *Mechanism and Machine Theory* **68** pp 35-48

[5] Bhattacharya S, Agarwal S K 2000 Design, Experiments and Motion Planning of a Spherical Rolling Robot *Int., Conf., on Robotics & Automation* (San Francisco, CA) **2** pp 1207-1212

[6] Zheng M, Zhan Q, Liu J, Cai Y 2010 Control of a Spherical Robot: Path Following Based on Nonholonomic Kinematics and Dynamics *Chinese Journal of Aeronautics* **24(3)** pp 337-345

[7] Ahn S, Lee Y 2014 Novel Spherical Robot with Hybrid Pendulum Driving Mechanism *Advances in Mechanical Engineering* pp 1-14

IOP Conf. Series: Materials Science and Engineering 1132 (2021) 012034 doi:10.1088/1757-899X/1132/1/012034

[8] Jaeyeon Lee and Wooram Park 2013 Design and Path Planning for a Spherical Rolling Robot *ASME Int., Mechanical Engineering Congress and Exposition, Proc.,* (San Diego, California, USA) IMECE2013-64994

[9] Homayoun Javadi A 2002 Introducing August: A Novel Strategy for An Omnidirectional Spherical Rolling Robot *Proc., 2002 IEEE International Conference on Robotics and Automation* (Washington, DC) **4** pp 3527-3533

[10] Sugiyama Y, Shiotsu A, Yamanaka M, and Hirai S 2005 Circular/Spherical Robots for Crawling and Jumping *Proceedings of the 2005 IEEE International Conference on Robotics and Automation* (Barcelona, Spain) pp 3595-3600

[11] Wei X, Tian Y, Wen S 2018 Design and locomotion analysis of a novel modular rolling robot *Mechanism and Machine Theory* **133** pp 23-43

Kinematic modelling and analysis of single leg in hybrid wheel-legged mobile robot

Arunkumar P, Sudheer A P* and Joy M L

Mechatronics/Robotics Lab, Department of Mechanical Engineering, National Institute of Technology, Calicut, India-673601

*apsudheer@nitc.ac.in

Abstract. Mobile manipulators are extensively used in various applications. Most of the mobile manipulators are built on a wheeled mobile platform. A wheeled mobile platform performs well on flat terrains but is not suitable for navigation in rough terrains. The legged robots are commonly used for rough terrain applications. However, the navigation speed of legged robots is lesser than wheeled mobile robots. This paper focuses on the development of hybrid wheel-legged robots to overcome the above difficulties. The proposed model is quadruped with five degrees of freedom in each leg. The development of the kinematic model, its dimensional synthesis, and motion study for a single leg is presented in detail. DH-conventions are used for kinematic modelling, manipulability index is used for dimension synthesis, and motion study is carried out in ADAMS.

Keywords. Hybrid wheel-legged mechanism, rough terrains, kinematic modelling, Dimensional synthesis

1. Introduction

A mobile manipulator refers to the mounting of the manipulator over a mobile robot platform. The mobile manipulator's main advantage is its increased workspace compared to the industrial manipulators. This improvement in the workspace makes them more viable and gives greater flexibility for manipulation. A mobile manipulator also can be used as a service robot and has domestic employment potential. They can also be used in a military operation like surveillance, rescue, and combat. Legged robotic systems mounted with manipulators have been drawing much attention from both the industrial and academic community for their advantages of combining manipulation and locomotion in uneven terrains. The remaining of the paper is structured as follows. Section 2 describes the proposed mechanism and its mobility, whereas Section 3 deals with forwarding kinematic modelling in a single leg. Section 4 gives the inverse kinematic analysis of the proposed mechanism in a single leg, Section 5 discusses the dimension synthesis and then in Section 6, a motion study was carried out. Results were discussed in Section 7 and the paper was concluded in Section 8.

Very few works are carried out in the past based on Mobile manipulators. The major complexity in the development of the mobile manipulators is the design of controllers to simultaneously track the mobile base as well as the end effector's trajectory. It is to be noted that most of these works involve the use of the wheeled mobile robot for mobile manipulators [1-6]. These wheeled mobile platforms are suitable mainly for flat terrains. In order to overcome the complexity in control, various techniques were used, such as impendence with sliding mode control [1], task space formulation [2], extended Cartesian coordinates [3]. Few of the works [4-7] are focused on unknown terrains, which makes control more challenging. In spite of the above challenges, the wheeled robots are not suitable for navigation in rough terrains. Therefore, legged robots were developed to overcome the above disadvantages. Compared to legged robots, the wheeled legged robots combine the advantage of both wheel and legged robots. Obtaining the mathematical model of the wheeled legged robot has various difficulties. An external camera based approach was used for development of inverse kinematics by Teka et *al.* [8]. Grand et *al.* [9] assumed frames arbitrarily for the formulation of kinematic models. Du et *al.* [10] proposed a centroidal momentum model for the analyses and control of the wheel-legged mobile robot. Niu et *al.* [11] developed a wheel-legged robot that mimics the human leg. However, the developed model was very heavy and resulted in bad dynamic behaviour. A cost-efficient design was proposed by Nakajima [12], yet the robot lacks the control along its yaw axis in rough terrains. Ding et al. [13] studied the effects of slip ratio of the wheels with the change in the vertical height of the wheel-legged robotic platform.

It is clear from the above literature that most of the works are carried out for flat terrains and are focused on using a wheeled mobile platform. This restricts the use of mobile manipulators only for indoor environments. Very few works are carried out on wheel-legged type mobile robots and are currently being explored. This work discusses a novel design of a hybrid wheel-legged mechanism to overcome the above difficulties.

2. Model description

Figure 1 shows the conceptual model of the proposed wheel-legged hybrid robot. This model consists of five revolute joints, four links and a wheel on each leg and a base platform. This arrangement allows the robot to navigate in a flat as well in rough terrain. The focus of this study is to achieve the walking motion without undergoing any change in the angle of the platform. The motion of this robot is divided into two cases, namely 1. The motion of the leg 2. The motion of platform/base

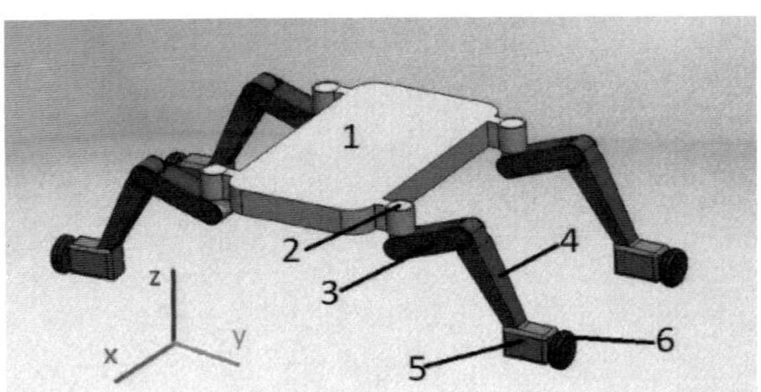

1. Platform / base
2. Link 1
3. Link 2
4. Link 3
5. Link 4 with two revolute joints
6. Wheels

Figure 1. Proposed hybrid wheel legged mechanism

2.1. Mobility Analysis

Mobility analysis is done to obtain the number of independent parameters required to control the mechanism. Kutzbach-Gŕubler criterion is used to determine these independent parameters (M) and is shown in equations (1) and (2).

$$M = 6\,(N - J - 1) + \sum f_i \text{ For mechanism} \quad (1)$$
$$M = 6\,(N - J) + \sum f_i \text{ For chain} \quad (2)$$

where, N denotes the number of links, J denotes the number of joints, and f_i denotes Degrees of freedom (DOF) of the ith joint.

The DOF analysis was done for three different scenarios:

1. Considering single leg of the mechanism with N = 6; J = 5; $\sum f_i$ = 5 yields M = 5. This shows all the five revolute joints are used in control of a single leg.
2. Considering the whole mechanism including the wheel yields M = 24, where N = 25; J = 24; $\sum f_i$ = 24. This shows twenty revolute joints are used for the control of all four leg and remaining four for the wheels.
3. For the Legged model without wheels with N = 13; J = 12; $\sum f_i$ = 12 yields M = 12 that denotes three joints each leg are used in control.

3. Forward kinematic modelling

This section deals with kinematic modelling of the mechanism. Compared to the conventional use of forward kinematics in Serial manipulators, two different forward kinematic equations are developed for the proposed robot. In both the cases, Denavit and Hartenberg modelling principles were used. The kinematic model was carried out on one leg and the same can be extended to all the four legs due to their identical structure. Figure 2 shows the frame assignment to the joints on the proposed model while Figure 3 shows the axis assignment to these frame as per the DH convention. In Figure three, numbers within the { } refers to the frame number.

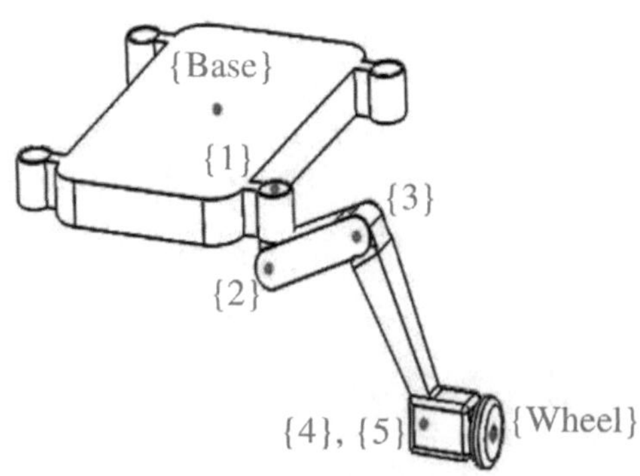

Figure 2. Frame assignment for proposed model in single leg

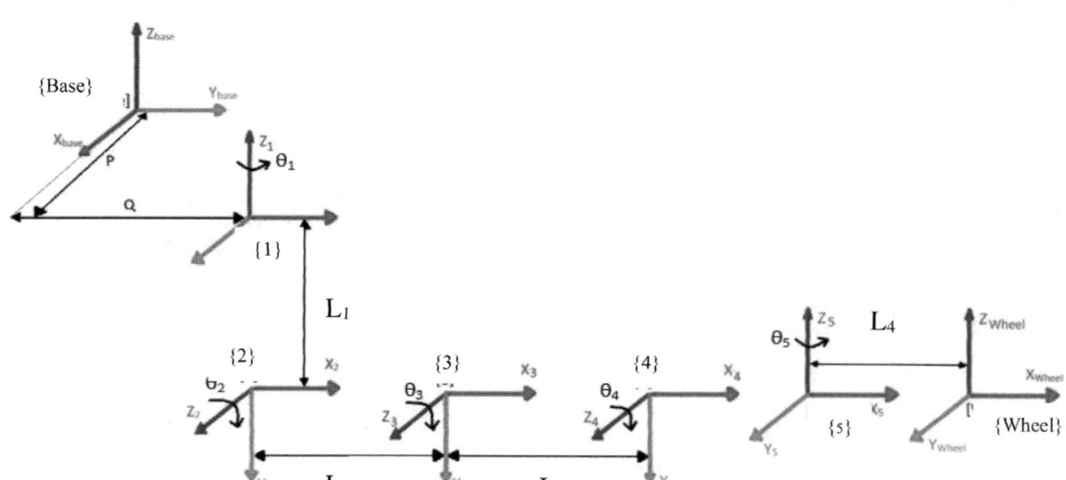

Figure 3. DH frame assignment for proposed model in single leg

IOP Conf. Series: Materials Science and Engineering 1132 (2021) 012035 doi:10.1088/1757-899X/1132/1/012035

Here L_1, L_2, L_3, L_4 denotes the link lengths between the joints θ_1, θ_2, θ_3, θ_4, θ_5 are the joint variables representing the five revolute joints, 2P and 2Q are the length and width of the platform respectively. The determination of theses link lengths are discussed in section 5.

Two types of forward kinematic analysis were performed based on the above frame assignments. In case 1, frame {1} remains fixed and wheel frame {wheel} is the end effector. Thus, case 1 is used for motion of leg in which the wheel acts as end effector and {1} acts as fixed frame. This equation is used to obtain the desired trajectory of the wheel. Table 1 shows the DH parameters for case 1 and the final transformation matrix can be obtained by post multiplying the transformation matrix of each joint.

Table 1. DH parameters for case 1

Parameter	θ (deg)	d (mm)	a (mm)	α (deg)
1	θ_1	$-L_1$	0	90
2	θ_2	0	L_2	0
3	θ_3	0	L_3	0
4	θ_4	0	0	-90
5	θ_5	0	L_4	0

The transformation matrices for various joints were obtained by substituting the DH parameters in the equation (3).

$$ {}^{i-1}_{\ \ i}T = \begin{bmatrix} cos\theta_i & -sin\theta_i cos\alpha_i & sin\theta_i sin\alpha_i & a_i cos\theta_i \\ sin\theta_i & cos\theta_i cos\alpha_i & -cos\theta_i sin\alpha_i & a_i sin\theta_i \\ 0 & sin\alpha_i & cos\alpha_i & d_i \\ 0 & 0 & 0 & 1 \end{bmatrix} \quad (3) $$

Then resulting final transformation matrix are shown in equations (4) and (5).

$$ {}^{\ \ \ \ \ \ 1}_{Wheel}T = {}^1_2T * {}^2_3T * {}^3_4T * {}^4_5T * {}^{\ \ \ \ \ \ 5}_{Wheel}T \quad (4) $$

$$ {}^{\ \ \ \ \ \ 1}_{Wheel}T = \begin{bmatrix} C_1C_5C_{234} - S_1S_5 & -C_1S_5C_{234} - S_1C_5 & -C_1S_{234} & L_2C_1C_2 - L_4S_1S_5 - L_3C_1S_2S_3 + L_3C_1C_2C_3 + L_4C_1C_5C_{234} \\ S_1C_5C_{234} - C_1S_5 & S_1S_5C_{234} - C_1C_5 & -S_1S_{234} & L_2S_1C_2 + L_4C_1S_5 - L_3S_1S_2S_3 + L_3S_1C_2C_3 + L_4S_1C_5C_{234} \\ C_5S_{234} & -S_5S_{234} & C_{234} & L_2S_3 - L_1 + L_3S_{23} + L_4S_4C_5C_{23} + L_4C_4C_5S_{23} \\ 0 & 0 & 0 & 1 \end{bmatrix} \quad (5) $$

For case 2, wheel frame {Wheel} remains fixed and frame {1} is the end effector. Table 2 shows the DH parameters for case 2 and the corresponding final transformation matrices are shown in equation (6) and (7). This equation is used to obtain the desired pose of the platform.

Table 2. DH parameters for case 2

Parameter	θ (deg)	d (mm)	a (mm)	α (deg)
1	θ_5	0	$-L_4$	90
2	θ_4	0	$-L_3$	0
3	θ_3	0	$-L_2$	0
4	θ_2	0	0	-90
5	θ_1	L_1	0	0

Similarly, from equation (3), the final transformation matrix is given by

$$ {}^{Wheel}_{1}T = {}^{Wheel}_{5}T * {}^{5}_{4}T * {}^{4}_{3}T * {}^{3}_{2}T * {}^{2}_{1}T \tag{6}$$

$$ {}^{1}_{Wheel}T = \begin{bmatrix} C_1C_5C_{234} - S_1S_5 & -C_5S_1C_{234} - C_1S_5 & -C_5S_{234} & -C_5 * (L_4 + L_3C_4 + L_1S_{234} + L_2C_{34}) \\ S_5C_1C_{234} - c_1s_5 & S_1S_5C_{234} - C_1C_5 & -S_5S_{234} & -S_5 * (L_4 + L_3C_4 + L_1S_{234} + L_2C_{34}) \\ C_1S_{234} & -S_1S_{234} & C_{234} & -L_3S_4 + L_1C_{234} + L_2S_{34} \\ 0 & 0 & 0 & 1 \end{bmatrix} \tag{7}$$

4. Inverse kinematic analysis

This section deals with formulation of inverse kinematic equations which are needed for the control of individual joints. In this work, an alternative approach is used to find the joint angles compared to conventional approach followed for serial manipulator. In order to ensure the undisturbed wheel position for the robot to execute wheeled motion at any instant of time, wheel frame needs to be parallel to frame {1}. This constraint reduces the complexity in deriving the inverse kinematic equation. Based on this constraint, the obtained transformation matrix between the frame {1} and wheel frame {wheel} is:

$$ {}^{1}_{Wheel}T = \begin{bmatrix} 1 & 0 & 0 & x \\ 0 & 1 & 0 & y \\ 0 & 0 & 1 & z \\ 0 & 0 & 0 & 1 \end{bmatrix} \tag{8}$$

Here (x, y, z) represents the wheel position coordinates.
For case 1, equation (4) gives

$$ {}^{1}_{Wheel}T * \left[{}^{5}_{Wheel}T\right]^{-1} = {}^{1}_{2}T * {}^{2}_{3}T * {}^{3}_{4}T * {}^{4}_{5}T \tag{9}$$

From (9), the inverse kinematic equations are obtained and are shown in equations (10-14).

$$\theta_1 = tan^{-1}\left(\frac{y}{x-L_4}\right) \tag{10}$$

$$\theta_2 = tan^{-1}\left(\frac{(L_1+z)*(L_3*c_3+L_2)-(L_3*s_3)*(x * c_1+y*s_1-L_4*c_5)}{(L_3*c_3+L_2)*(x * c_1+y*s_1-L_4*c_5)+(L_1+z)*(L_3*s_3)}\right) \tag{11}$$

$$\theta_3 = -cos^{-1}\left(\frac{(x * c_1+y*s_1-L_4*c_5)^2+(L_1+z)^2-L_3^2-L_2^2}{2* L_2*L_3}\right) \tag{12}$$

$$\theta_4 = -(\theta_2 + \theta_3) \tag{13}$$

$$\theta_5 = -\theta_1 \tag{14}$$

In case 2, (x, y, z) represents the platform coordinates. By using similar decoupling method the equation (6) yields

$$ {}^{Wheel}_{1}T * [{}^{2}_{1}T]^{-1} = {}^{Wheel}_{5}T * {}^{5}_{4}T * {}^{4}_{3}T * {}^{3}_{2}T \tag{15}$$

From equation (15) solved inverse kinematic are shown in equation (16-20)

$$\theta_1 = -\theta_5 \tag{16}$$

$$\theta_2 = -(\theta_3 + \theta_4) \tag{17}$$

$$\theta_3 = -cos^{-1}\left(\frac{(x * c_5 + y * s_5 + L_4 * c_1)^2 + (z - L_1)^2 - L_3^2 - L_2^2}{2 * L_2 * L_3}\right) \tag{18}$$

$$\theta_4 = tan^{-1}\left(\frac{(z - L_1) * (-L_2 * c_3 - L_3) - (-L_2 * s_3) * (x * c_5 + y * s_5 + L_4 * c_1)}{(-L_2 * c_3 - L_3) * (x * c_5 + y * s_5 + L_4 * c_1) + (z - L_1) * (-L_2 * s_3)}\right) \tag{19}$$

$$\theta_5 = tan^{-1}\left(\frac{y}{x}\right) \tag{20}$$

5. Dimensional synthesis

The determination of the link lengths L_1, L_2, L_3, L_4 and the dimension of the platform are discussed in this section. The desired step length of the proposed mechanism was considered to be 600mm [19]. Maximum step height was limited to 280 mm [21].

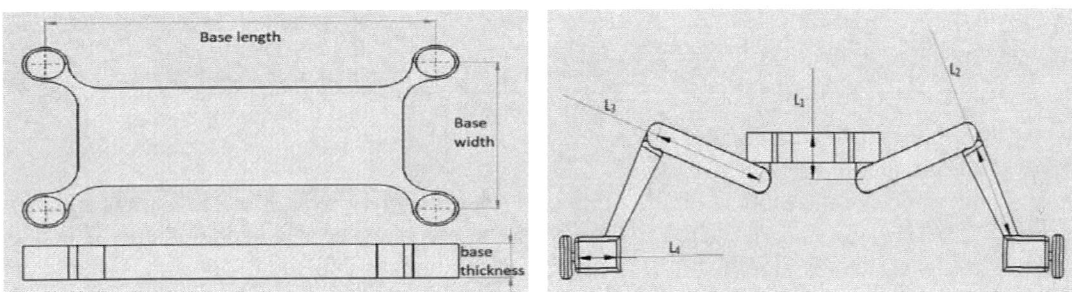

Figure 4. Shows the undefined dimensions of mechanism

Following are the constraints used in the dimensional synthesis of the proposed mechanism.

i. Initial clearance between the ground and platform is 450 mm.
ii. Initially assumed platform width = 300 mm and L_1 = 50 mm, L_4 = 50 mm. It is to be noted that L_1 and L_4 does not affect L_2 and L_3 dimensions.
iii. Step length is considered as 600 mm and its coordinates are shown Figure 5

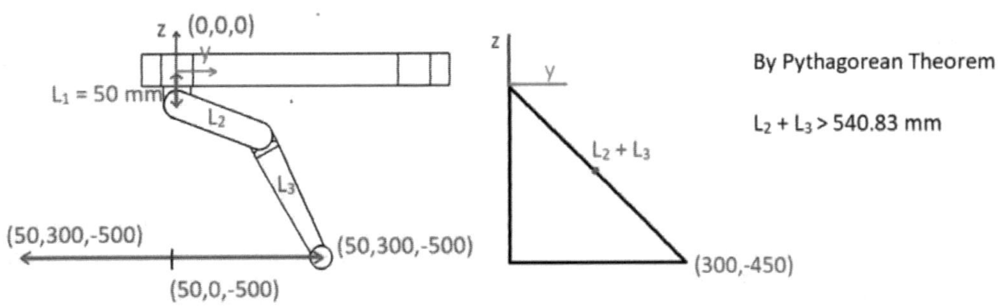

Figure 5. Step length of the proposed model

iv. In order to avoid collision between link 2 and obstacle, L_3 should be greater than obstacle height. Hence, L_3 > 280 mm.
v. Length of links L_2 and L_3 were limited within 300 mm in order to reduce the motor torque
vi. Manipulability Index (w) is a quality measure which describes the distance to singular configuration. It can be analyzed by measuring manipulability ellipsoid which is spanned by the singular vectors of the Jacobian. It should be maximum because it is proportional to the volume of ellipsoid in which no singularity is possible [18]

$$w = \sqrt{|J(q) * J(q)^T|} \tag{21}$$

where J is Jacobian matrix and q is joint variable

By using these constraints and inverse kinematic equations, a program was simulated in MATLAB software for some length combination. Then final possible dimensions are shown in Table 3.

Table 3. Final dimension of the mechanism

Parameter name	Range (mm)
Platform length	700
Platform width	300
Platform thickness	15
L_1	50
L_2	300
L_3	290
L_4	50

6. Motion study

The successful motion of the robot is divided into two stages, namely motion of the leg and motion of the platform. These motions are discussed in detail in the subsequent section. Motion study begins with obtaining the end effector's coordinates as per the desired motion and then solving for the inverse kinematic to obtain the necessary joint angles to trace the desired end-effector trajectory. The desired motion's joint trajectories were generated, and the torque required in each joint was determined.

6.1. Motion of the legs

This section defines the joint torque for all the revolute joints when the platform is fixed. The maximum walking speed of the robot is considered to be 1km/hr and the robot can cover a distance of step length of 600mm in 2.16 seconds. The clearance between the platform and ground is maintained at 500 mm, the distance between wheel and platform is 200mm. In Figure 6, coordinates for the motion of the legs are shown. A parabolic path is considered for the motion of the legs and this also ensures that the wheel does not come in contact with the ground during the motion of the legs. Figure 7 shows the three-dimensional trajectory of the wheel frame.

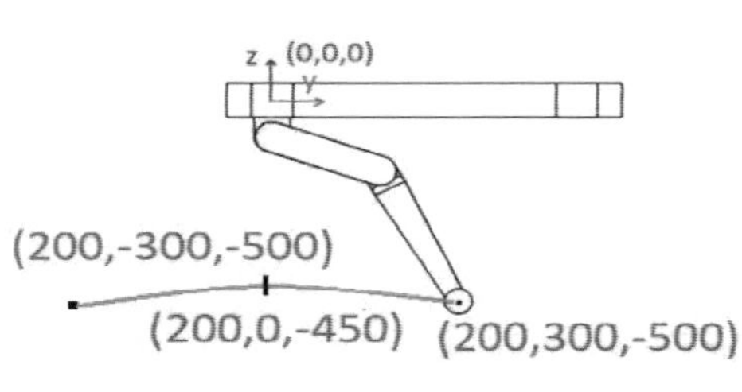

Figure 6. Displays the coordinates for this motion

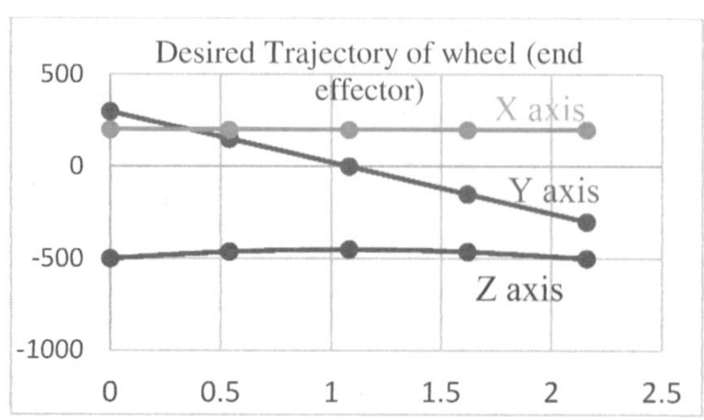

Figure 7. Desired trajectory of wheel

6.1.1. Position based Joint Trajectory

A suitable trajectory is designed for each joint to ensure the safe movement of each links without affecting the end effector trajectory. Based on the coordinates defined in Figure 6, the joint angles are obtained from inverse kinematic equations (10-14) and are shown in Table 3. Fifth order polynomial equation (22) is used to generate trajectory for joint control in every joint.

$$\theta(t) = c_0 + c_1 * t + c_2 * t^2 + c_3 * t^3 + c_4 * t^4 + c_5 * t^5 \quad (22)$$

where C_0, C_1, C_2, C_3, C_4, C_5 are constants of the polynomial equation and t refers to time.

By using the values from the Table 3 and polynomial equation (22), the values of the constants are calculated and desired fifth order polynomial equation for each joint were obtained.

Table 3. Data required for joint control

Time	x	y	z	θ_1	θ_2	θ_3	θ_4	θ_5
0	200	300	-500	63.4349	-35.6524	-35.9265	71.5789	-63.4349
0.54	200	150	-462.5	45	-25.3725	-76.352	101.7246	-45
1.08	200	0	-450	0	-26.7527	-87.2329	113.9856	0
1.62	200	-150	-462.5	-45	-25.3725	-76.352	101.7246	45
2.16	200	-300	-500	-63.4349	-35.6524	-35.9265	71.5789	63.4349

6.2. Motion of platform

This section analyses the ability of the single leg to displace the platform in vertical direction. As discussed earlier, wheel frame is fixed frame and frame {1} is the end effector frame. The displacement in the vertical direction (Z axis of wheel frame) is considered because the torque required for vertical motion comparatively higher than the torque required along other axis.

Figure 8 displays the planar motion coordinates in ZX plane such that the movement is along the Z direction. It is clear from Figure 8 that clearance of the platform from the ground is 500 mm and X direction it is maintained in 350 mm for maximum walkable position.

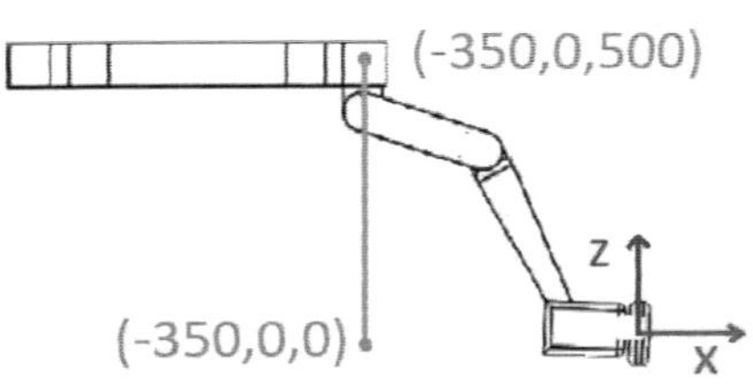

Figure 8. Coordinates for the motion

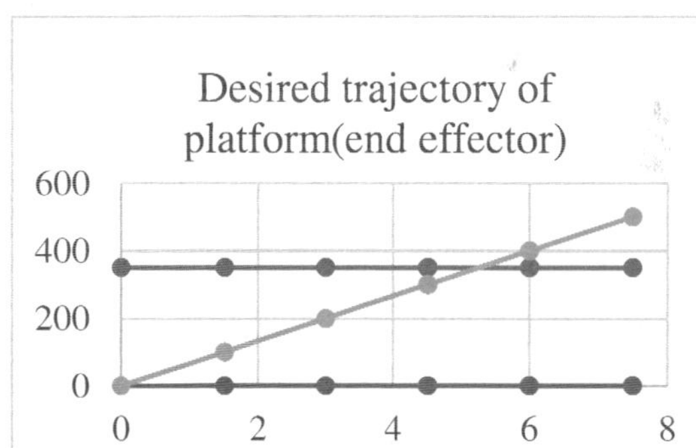

Figure 9. Desired trajectory of platform

6.2.1. Position based trajectory for the joints

In order to achieve the linear motion of the platform, a sixth order polynomial equation (23) is considered as the joint trajectory. Based on the coordinates defined in Figure 8, the joint angles are obtained from inverse kinematic equations (16-20) and are shown in the Table 4

$$\theta(t) = c_0 + c_1 * t + c_2 * t^2 + c_3 * t^3 + c_4 * t^4 + c_5 * t^5 + c_6 * t^6 \quad (23)$$

where C_0, C_1, C_2, C_3, C_4, C_5, C_6 are constants of the polynomial equation and t refers to time.

These constant are calculated using the values from Table 4 and polynomial equation (23).This yields the desired sixth order polynomial equation for each joints.

Table 4. Joint value for this motion.

Time	x	y	z	θ_1	θ_2	θ_3	θ_4	θ_5
0	350	0	0	0	-60.2274	118.9033	-58.6759	0
1.5	350	0	100	0	-39.804	115.2052	-75.4012	0
3	350	0	200	0	-18.963	104.6817	-85.7186	0
4.5	350	0	300	0	1	88.0568	-89.0597	0
6	350	0	400	0	21.3241	64.1379	-85.4619	0
7.5	350	0	500	0	50.4035	17.5511	-67.9546	0

7. Results and discussion

In this work, MSC ADAMS VIEW software is used for the motion study of the single leg of the proposed hybrid wheel-legged mechanism. The obtained equations (22 and 23) for both cases were used to analyse the motion of joints and end effector.

Figures 10 - 14 shows the results of motion of legs (case 1). It is clear from Figure 7 and 10 the desired parabolic motion of wheel is achieved. Therefore, the obtained torque values of each joint are noted from the simulation study.

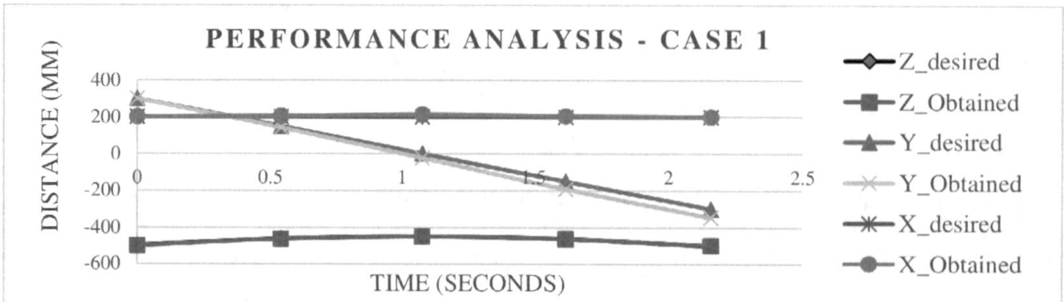

Figure 10. Performance Analysis for case 1

The wheel velocity in Figure 11 shows that the robot is able to achieve walking speed of 1.08km/hr which is close to the desired velocity of 1 km/hr. The negative sign in the wheel velocity shown in Figure 11 leg moving in negative direction of the assigned wheel frame.

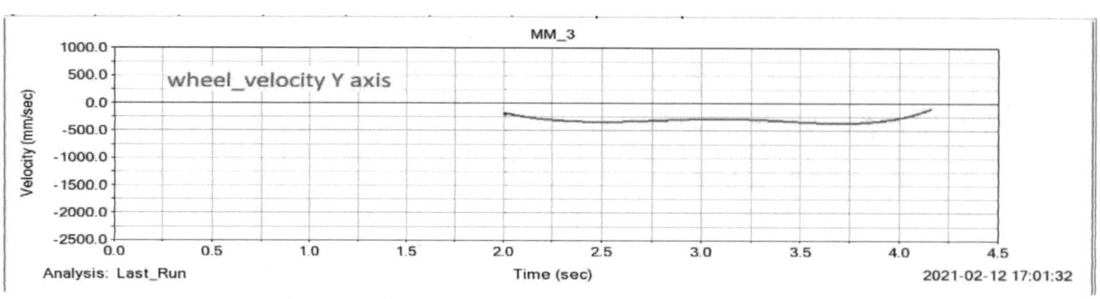

Figure 11. Achieved walking speed form simulation

Figure 12 – 14 shows the torque values obtained for the motion of the leg. Joint 2 demands a heavy torque in the range of 1500 Nmm to 2500 Nmm while joint 5 demands a torque of 7.5 Nmm. Other joints 1, 3 and 4 demand torque less than 500 Nmm. The larger torque value of joint 2 compare to joint 1 is because joint 2 lifts subsequent joints against the gravity whereas joint 1 performs a yaw motion and demands less torque.

IOP Conf. Series: Materials Science and Engineering 1132 (2021) 012035 doi:10.1088/1757-899X/1132/1/012035

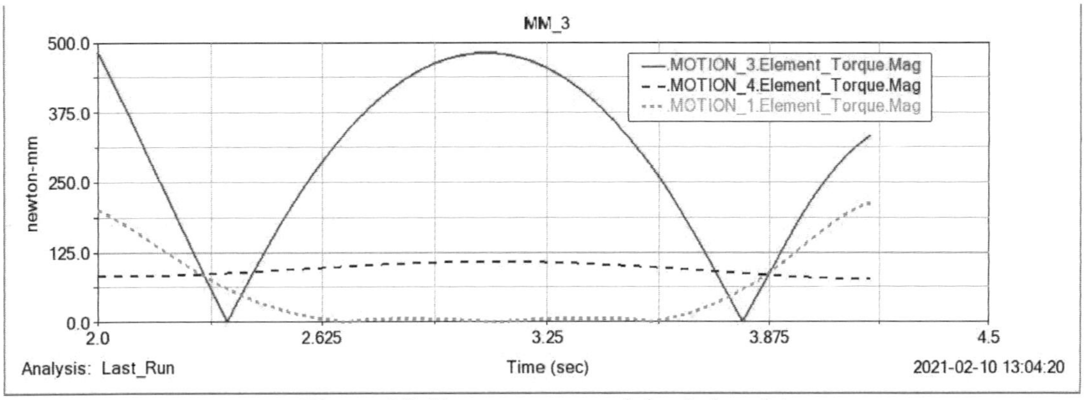

Figure 12. Torques curve at joint 1, 3 and 4

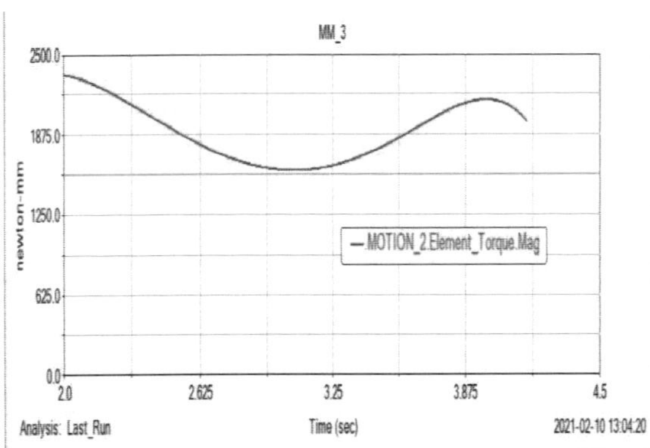

Figure 13. Torque curve at joint 2

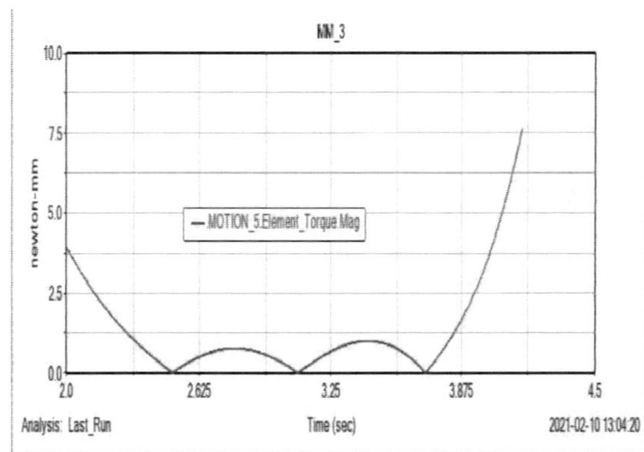

Figure 14. Torque curve at joint 5

Figure 15 – 17 represents the results of platform motion (case 2). It is clear from Figure 8 and 15 that the desired linear motion for the platform is achieved. Therefore, obtained torque values are noted from simulated results.

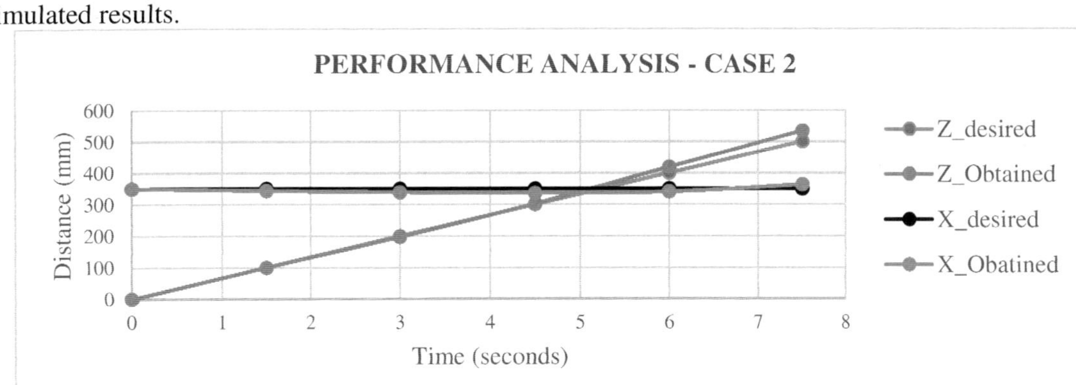

Figure 15. Performance Analysis for case 2

Figure 16 and 17 displays the torque values obtained for the platform motion. Joint 2 needs 30 Nmm while joint 3 and 4 demands heavy torque of 5100 Nmm. This is because joint 3 and 4 lifts subsequent joints as well as platform against the gravity.

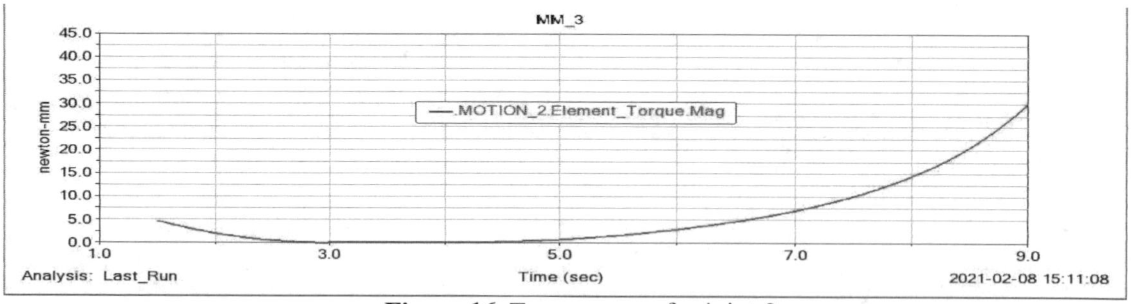

Figure 16. Torque curve for joint 2

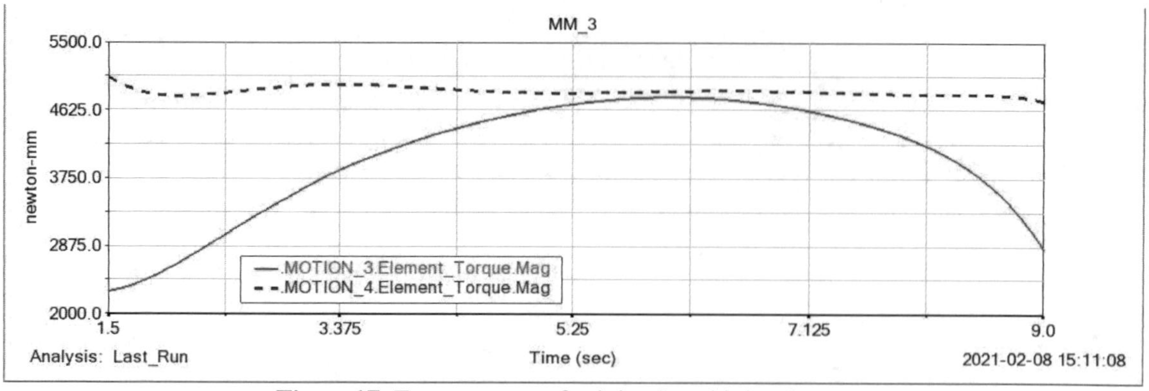

Figure17. Torque curves for joint 3 and joint 4

8. Conclusion

In this work, the hybrid wheel-legged robot was designed by considering the advantage of both wheel and legged locomotion. The motion study was successfully performed on a single leg, and the same can be extended to all four legs. The MSC ADAMS software is used for the simulation study. The resulting trajectories from the simulation study are in line with desired trajectories. Therefore the obtained torque values of each joint are considered for future work. Further, the torques are to be determined analytically and compared with the simulation study's torques. Also, the obtained dynamic model shall be used for the implementation of control strategies.

References

[1] Salehi M and Vossoughi G 2008 Impedance control of flexible base mobile manipulator using singular perturbation method and sliding mode control law *International Journal of Control Automation and Systems* **6(5)** pp 677-688

[2] Oetomo D, Ang M H, Jamisola, R and Khatib O 2002 Integration of torque controlled arm with velocity controlled base for mobile manipulation *In. Romansy* **14** pp 189-199

[3] Li M, Yang Z, Zha F, Wang X, Wang P, Li P, Ren Q and Chen F 2020 Design and analysis of a whole-body controller for a velocity controlled robot mobile manipulator *Science China Information Sciences* **63(7)** pp 1-15

[4] Carius J, Wermelinger M, Rajasekaran B, Holtmann K and Hutter M 2018 Deployment of an autonomous mobile manipulator at MBZIRC *Journal of Field Robotics* **35(8)** pp 1342-1357

[5] Suzuki T and Sekiyama K 2020 A coordinated control system for dual-arm mobile manipulator balancing grasping and viewpoint selection *IEEE/SICE International Symposium on System Integration* pp 176-181

[6] Ram R V, Pathak P M and Junco S J 2019 Trajectory control of a mobile manipulator in the presence of base disturbance *Simulation* **95(6)** pp 529-543

[7] Domski W, Mazur A and Kaczmarek M 2019 Extended factitious force approach for control of a

mobile manipulator moving on unknown terrain *Journal of Intelligent & Robotic Systems* **93(3)** pp 699-712

[8] Teka B, Raja R and Dutta A 2019 Learning based end effector tracking control of a mobile manipulator for performing tasks on an uneven terrain *International Journal of Intelligent Robotics and Applications* **3(2)** pp 102-114

[9] Grand C, Benamar F and Plumet F 2010 Motion kinematics analysis of wheeled–legged rover over 3D surface with posture adaptation *Mechanism and Machine Theory* **45(3)** pp 477-495

[10] Du W, Fnadi M and Benamar F 2020 Rolling based locomotion on rough terrain for a wheeled quadruped using centroidal dynamics *Mechanism and Machine Theory 153* p 103984

[11] Niu J, Wang H, Shi H, Pop N, Li D, Li S and Wu S 2018 Study on structural modeling and kinematics analysis of a novel wheel-legged rescue robot *International Journal of Advanced Robotic Systems* **15(1)**

[12] Nakajima S 2011 RT-Mover: a rough terrain mobile robot with a simple leg–wheel hybrid mechanism *The International Journal of Robotics Research* **30(13)** pp 1609-1626

[13] Ding X, Li K and Xu K 2012 Dynamics and wheel's slip ratio of a wheel-legged robot in wheeled motion considering the change of height *Chinese Journal of Mechanical Engineering* **25(5)** pp 1060-1067

[14] Vahrenkamp N, Asfour T, Metta G, Sandini G and Dillmann R 2012 Manipulability analysis *In 2012 12th ieee-ras international conference on humanoid robots (humanoids 2012)* pp 568-573

[15] http://fwp.mt.gov Accessed on 15-02-2021

[16] https://www.hunker.com/12377641/the-average-height-of-a-stair-step Accessed on 15-02-2021

IOP Conf. Series: Materials Science and Engineering 1132 (2021) 012036 doi:10.1088/1757-899X/1132/1/012036

Modelling of torso and dual arms for a humanoid robot with fixed base by using screw theory for dexterous applications

Shifa Sulaiman* **and Sudheer A P**

Mechatronics/Robotics Laboratory, Department of Mechanical Engineering, National Institute of technology, Calicut, Kerala, India

*Email: shifa_p170114me@nitc.ac.in

Abstract. Most of the humanoid robots are facing problems due to non-dexterous arms while working in complex environments. Kinematic and dexterity analyses help to study and solve the manipulation difficulties of a robot. This paper presents the kinematic and dexterity analyses of torso and dual arms for a tree type upper body humanoid robot with fixed base using screw theory formulation. Orientation workspace and reachability analyses of the dual arms are done for determining the dexterity of the robot. In this paper, orientation workspace of the tree type upper body is obtained by converting the rotation matrix to Euler angle set representation.The upper body of the humanoid robot is designed with three Degrees of freedom (DoF) in the torso, two DoF in the neck, five DoF in each arm.

1. Introduction

There has been a rapid rise in the development of humanoid robots for ages because of their adaptableness and interaction to the human environment. The humanoid robots are mainly used for interaction with humans and other systems.These robots are used for training purposes, personal assistance, understanding the human body structure and behavior, health care field, entertainment sector, military purposes, space explorations, etc. Humanoid robots with dexterous arms are needed for completing complex tasks. Humanoid robots employed in real life working environments require motions similar to human hand to complete given tasks. Human hands are dexterous in nature. Hence manipulators of humanoid robots are also to be dexterous for imitating human like motion capabilities. Workspace and reachability analyses help to solve the dexterity problems in robotic arms.

Screw theory formulations are used in various applications for deriving the kinematic model. Muller [1] explained the advantages of using screw theory formulations for modelling tree type systems. Yi *et al.* [2] determined the kinematic model of a planar mobile robot using screw theory approach. The computational time for deriving the jacobian matrix is less compared to other methods. Man *et al.* [3] derived the kinematic model of a ten DoF humanoid robot arm using screw theory formulation. The kinematic modelling of a spatial humanoid robot with floating base using screw theory is depicted in [4]. Toscano *et al.* [5] presented a kinetostatic model for humanoid robots using screw theory approach. The kinetostatic model also determines the forces and moments at various joints of humanoid robot. The mathematical modelling of a rehabilitation robot using screw theory approach is carried out by Liao *et al.* [6]. The comparative study of kinematic modelling for a stanford manipulator using Denavit- Hartenberg method and screw theory approach is done by Rocha *et al.* [7]. The screw theory approach is proved to be more efficient for complex systems with higher DoF as compared to DH method. A review of human like arm motion generation robots are depicted in [8]. Patel *et al.* [9] developed a manipulator to satisfy the performance requirements within given joint constraints and with least energy consumption. The optimum geometry of the orientation structure is given in [10]. The method for determining dexterous workspace of a manipulator is clearly discussed. Guenter *et al.* [11] introduced a humanoid robot which can exhibit human like torso motions.The proposed model offered enough flexibility for manipulation of objects. The main disadvantage is the slow movement of the joints due to increased number of DOF. Mizuuchi *et al.* [12] proposed the design and control of a tendon-driven flexible-spine humanoid named "Kenta". The joints are controlled by using eight muscles tendons. However, the movements exhibited are not flexible compared to human spine. Gini *et al.* [13] suggested a control system strategy and mechanics for a humanoid upper body. The developed actuation system imitated the arrangement of the human muscle structure. Dynamic behavior of pneumatic actuators used in this work is not linear and subjected to hysteresis.

Kinematic modelling with workspace analysis plays an important role in determining the reachability and dexterity of the upper body robot by obtaining the link dimensions. The performance of the humanoid robots can be increased by analyzing the shape and volume of workspace. Umetani *et al.* [14] studied the performance of a two-dimensional space robot by analyzing the workspace and manipulability of different configurations of the robot. A method for obtaining the workspace of three degrees of freedom planar robotic manipulators using limiting curves is described in [15]. The singular behavior of a robotic manipulator by analyzing the workspace and jacobian matrix is explained in [16]. Most of the works related to humanoid robots are based only on position workspace. However, orientation workspace also plays a crucial role in evaluating the dexterity of the end effectors of the humanoid robots. The method for determining the orientation workspace of a 6 DoF parallel robot is depicted in [17].

Mathematical modelling of humanoid robot is a difficult task as far as the stability is concerned. Kinematic and dexterity analyses are also difficult due to the complexity of higher degree of freedom mechanisms. This paper focuses on kinematic, workspaceand reachability of a tree type upper body humanoid robot with fixed base using screw theory formulation. Analysis of orientation workspace along with position workspace increases the accuracy of determination of dexterity and manipulability of the robotic system.

2. Kinematic analysis of humanoid upper body

The upperbody of humanoid robot with fixed base is considered as a tree type mechanism for kinematic analysis using screw theory approach. The design of the humanoid robot is done by considering the standard humanbody ratio [18]. The upper body of the humanoid robot consists of three DoF torso, two DoF neck and five DoF arms as shown in figure 1.

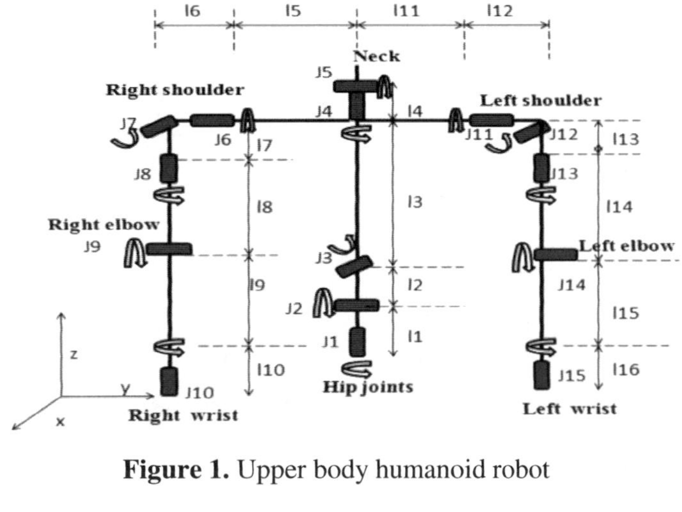

Figure 1. Upper body humanoid robot

Ji represents the joints and l_i represents the corresponding link dimensions and distances. The two arms of tree type upper body robot are designed with identical joints. The three joints in hip are assumed to be in same origin. The three shoulder joints are also assumed to be in same origin. The joint limits of various upper body joints are given in table 1.

Table 1. Joints limits for various upper body joints

Joints	Joint limits (in degrees)
1	-90 to +90
2	-60 to +60
3	-75 to +75
4	-90 to +90
5	-60 to +60
6 & 11	-180 to +180
7 & 12	-20 to +135
8 & 13	-180 to+180
9 & 14	-90 to +90
10 & 15	-180 to+180

The Screw theory formulation [19] is used for the kinematic modeling. One frame is assigned to the base frame and other frame to the end effector. The transformation matrix obtained by using screw theory approach with screw axis, ε is given by equation (1)

$$e^{[\varepsilon]\theta} = \begin{bmatrix} e^{[\hat{\varepsilon}]\theta} & (I\theta + (1-cos\theta)[\widehat{\alpha}] + (\theta - sin\theta)[\widehat{\alpha}]^2 v \\ 0 & 1 \end{bmatrix} \quad (1)$$

Where θ is the angular rotations of respective joints, $\hat{\alpha}$ is the skew symmetric matrix of angular velocity matrix, α and v is the linear velocity. The Rodrigues formula for deriving rotation matrix is given in equation (2)

$$R(\hat{\alpha},\theta) = e^{[\hat{\varepsilon}]\theta} = I + sin\theta[\hat{\alpha}] + (1 - cos\theta)[\widehat{\alpha}]^2 \quad (2)$$

The humanoid robot upper body is considered as a tree type mechanism as shown in figure 1.The upper body is divided into three branches for kinematic analysis. The first branch consists of three hip joints and two neck joints as shown in figure 2. The position of end effector with respect to base frame is denoted as N_i, where i denotes the number of corresponding branches.

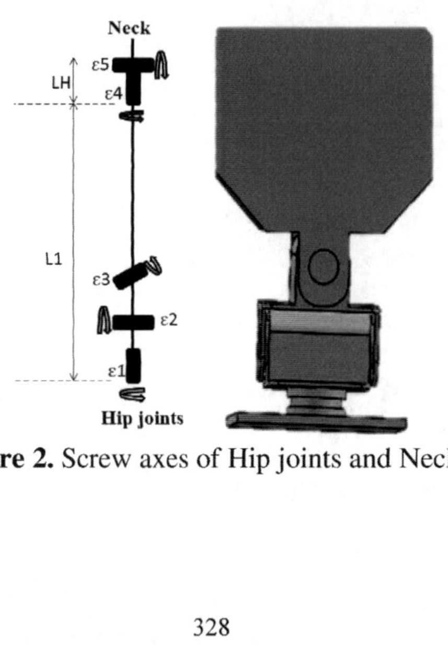

Figure 2. Screw axes of Hip joints and Neck joints

The pose of head with respect to hip, $S_1(\theta)$ is given in equation (3)

$$S_1(\theta) = e^{[\varepsilon_1]\theta_1} e^{[\varepsilon_2]\theta_2} e^{[\varepsilon_3]\theta_3} e^{[\varepsilon_4]\theta_4} e^{[\varepsilon_5]\theta_5} . N_1 \quad (3)$$

The second branch consists of five right arm joints as shown in figure 3. The pose of right hand with respect to hip, $S_2(\theta)$ is given in equation (4)

$$S_2(\theta) = e^{[\varepsilon_6]\theta_6} e^{[\varepsilon_7]\theta_7} e^{[\varepsilon_8]\theta_8} e^{[\varepsilon_9]\theta_9} e^{[\varepsilon_{10}]\theta_{10}} . N_2 \quad (4)$$

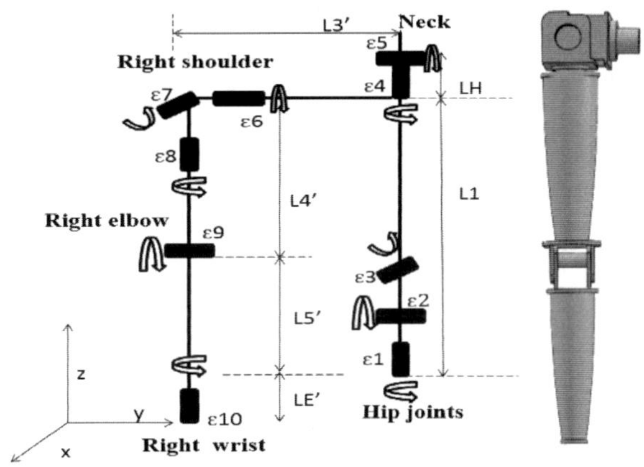

Figure 3. Screw axes of Right arm joints

The third branch consists of five left arm joints as shown in figure 4. The pose of left hand with respect to hip, $S_3(\theta)$ is given in equation (5)

$$S_3(\theta) = e^{[\varepsilon_{11}]\theta_{11}} e^{[\varepsilon_{12}]\theta_{12}} e^{[\varepsilon_{13}]\theta_{13}} e^{[\varepsilon_{14}]\theta_{14}} e^{[\varepsilon_{15}]\theta_{15}} . N_3 \quad (5)$$

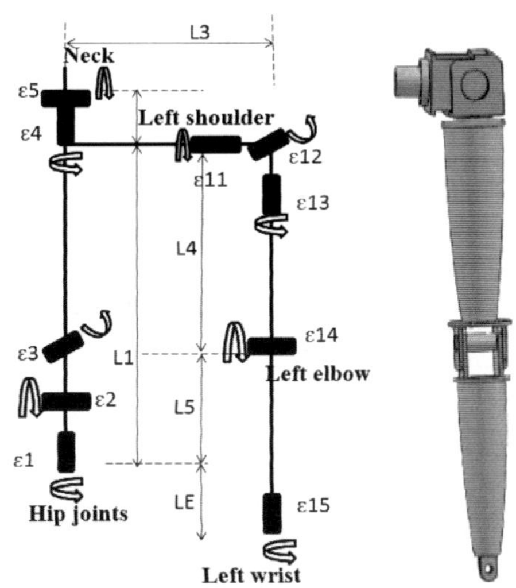

Figure 4. Screw axes of Left arm joints

3. Orientation workspace

Most of the previous works related to workspace analyses are based on constant orientation workspace or position workspace. The orientation workspace is the subset of overall workspace of the system. Determination of orientation workspace is important for obtaining the complete description of mobility of the system. The overall workspace of an end effector is the combination of position and orientation workspaces. Evaluating position workspace without considering the orientation workspace reduces the accuracy of workspace analysis. There are different alternatives for representing the orientation workspace. In this paper, Euler angle sets [17] are used for representing the orientation workspace. Euler angle sets are plotted in a spherical coordinate system.

Let φ, θ, h represents the euler angle sets. φ and θ denotes the azimuth and zenith angles whereas h denotes the length of ray as shown in figure 5.

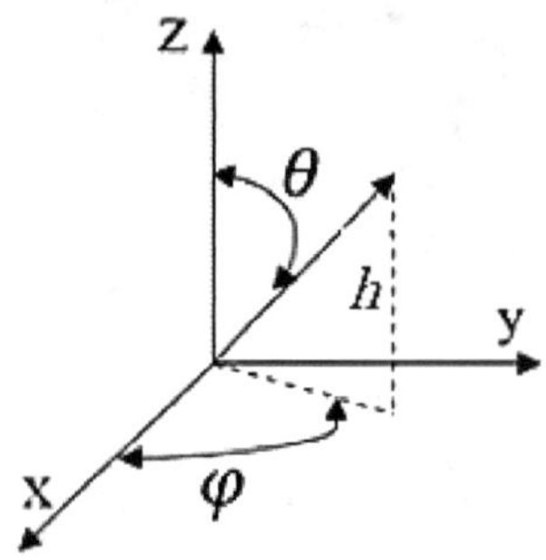

Figure 5. Representation of Euler angles in spherical coordinates

The equations for converting spherical coordinates to cartesian coordinates are given in equations (6) to (8)

$$X = h\cos(\varphi)\sin(\theta) \tag{6}$$
$$Y = h\sin(\varphi)\sin(\theta) \tag{7}$$
$$Z = h\cos(\theta) \tag{8}$$

The orientation workspace for various configurations can be analysed for determining the dexterity and manipulability of the robotic configurations. The voids inside the orientation workspace indicate the probability of existence of singular surfaces inside the workspace envelopes. The orientation workspace is plotted by considering the arm joint limits given in table 1. The left hand is taken to the corresponding positions and the orientations by giving different joint limits to elbow joint as shown in figure 6. Figure 6 shows three configurations of left hand by changing elbow joint angles to (-45,0,45) degrees.

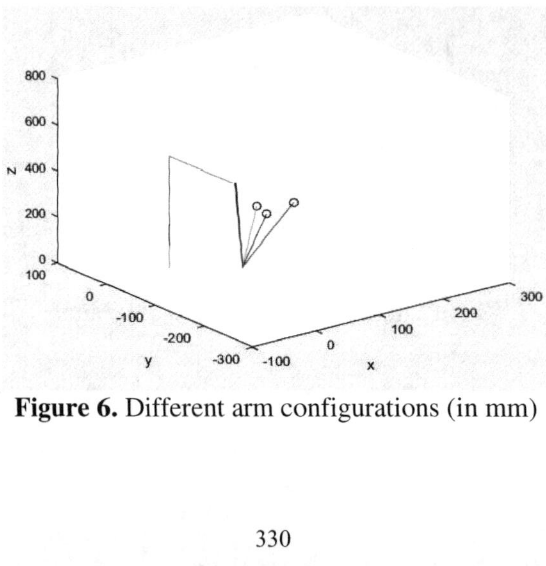

Figure 6. Different arm configurations (in mm)

IOP Conf. Series: Materials Science and Engineering 1132 (2021) 012036 doi:10.1088/1757-899X/1132/1/012036

4. Singularity analysis

The jacobian matrix of end effector with relative position, $'P_i$ and orientation, $\tilde{\omega}_i$ is given in equation (9)

$$j_i = [\,'P_i \; \tilde{\omega}_i\,]^{T} \tag{9}$$

The jacobian of each branch is derived separately for determining the singularity based on rank deficiency criteria of jacobian matrix. The jacobian matrix of first branch is given in equation (10)

$$j_{hip} = \begin{bmatrix} 'P_1 & 'P_2 & 'P_3 \\ \tilde{\omega}_1 & \tilde{\omega}_2 & \tilde{\omega}_3 \end{bmatrix} \tag{10}$$

Similarly, the singular configurations of other joints of each branch in upper body section are determined based on jacobian matrix. The non-square jacobian matrices J are converted to square matrix by using the equation given in (11)

$$J_{final} = \sqrt{det}\,(j \times j^{T}) \tag{11}$$

5. Reachability of Arms

Reachability measurement helps to determine the most optimal configurations of a robotic system for a given task. The reachability measurement of arms play an important role in determining the dexterous workspace of the upper body humanoid robot. The reachable workspace is defined as the possible set of all points accessible by the robot end effector. For most of the given task, the reachable workspace is limited as compared with the dexterous workspace due to the presence of movement constraints. Reachability function, Ʀ [9] is used for determining thereachability of a given point p_i in a task space with joint variable q_i with n DoF is given in equation (12)

$$Ʀ = max[min[\frac{(q_{i,max}-q_i)((q_i-q_{i,min})}{0.5(q_{i,max}-q_{i,min})^2}]_{i=1}^{n}]_{j=1}^{k} \tag{12}$$

Where k denotes the number of inverse kinematic solutions, j_i represents joint variables. Reachability function varies from zero to unity for a given configuration. The maximum value for a reachability function is unity and minimum is zero. The reachability function attains unity only and only if the end effector reaches the given point inside the taskspace with all joint displacements within the midrange of joint angle limits. The reachability function values for different location of point p_i is given in table 2.

Table 2. Reachability function

Location of point P_i	Value of reachability function, Ʀ
Point P_i is located inside the workspace and at least one of the solutions is within the joint limit	[0 1]
Point P_i is located inside the workspace and is reachable with joints limits midrange	1
Point P_i is located inside the workspace and is reachable with at least one of the joint angles in extreme position	0

6.Results and discussion

6.1. Determination of singular configurations of hip joints

The jacobian matrix of hip joints are given in equation (13)

IOP Conf. Series: Materials Science and Engineering 1132 (2021) 012036 doi:10.1088/1757-899X/1132/1/012036

Let $s_i = \sin\theta_i$ and $c_i = \cos\theta_i$

$$j_{\text{hip}} = \begin{bmatrix} 0 & 0 & 0 \\ 0 & 0 & 0 \\ 0 & 0 & 0 \\ 0 & -s_1 & c_1c_2 \\ 0 & c_1 & s_1c_2 \\ 1 & 1 & -s_2 \end{bmatrix} \tag{13}$$

The singular configurations of hip joints are given in equations (14) to (15)

$$\begin{vmatrix} 0 & -s_1 & c_1c_2 \\ 0 & c_1 & s_1c_2 \\ 1 & 1 & -s_2 \end{vmatrix} = 0 \tag{14}$$

$$\theta_1 = 0;\ \theta_2 = \pm\,(2n+1)\,\pi/2 \tag{15}$$

6.2. Determination of Reachability function

The reachability function for a particular point P_i (x, y, z) inside the reachable workspace indicates the reachability of the point by using different configurations of the robotic system. The reachability function can also be used for energy optimization for a given task. The reachability function, Ɽ for the three configurations of upper body movements for a point P (200,150,225), which is inside the workspace is calculated and the maximum reachability value for P_i by using equation (15) is obtained as 0.8567. The high value of reachability function for P_i indicates the reachability of the point with most of the different configurations of the end effectors. The function value approaches to zero, if the given point is along the workspace boundary.

6.3. Orientation workspace analysis

In this paper, joint limits given in table 1 are used for analyzing the orientation workspace. Euler angles sets are derived from rotation matrix obtained by inputting the joint limits. Orientation workspace of the upper body tree type robot is analysed and is plotted as shown in figure 7. The dense area indicates the dexterous region where as the void regions of the workspace is less dexterous as compared to center region. The orientation workspace with respect to orientation matrix is plotted by converting it into Euler angle sets (φ, θ, h) in radians and the coordinates are converted to cartesian coordinates with respect to position coordinates.

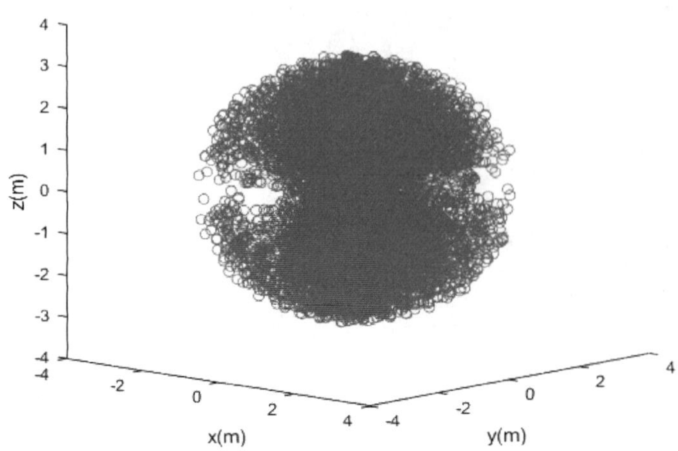

Figure 7. Orientation workspace

7. Conclusion and future scope

Screw theory based kinematic modeling of upper body humanoid robot was carried out in this paper. Singularity analysis of the hip, reachability of arms and orientation workspace are also presented. The screw theory approach is suitable for higher DoF mechanism due to the absence of intermediate frame assignments. Workspace and singularity analysis help to determine the overall workspace and voids inside the workspace. Reachability functions along with workspace and singularity analysis are used to determine the dexterity and manipulability of the upper body humanoid robot. These analyses can be used for the synthesis of robot mechanism. Modeling can be helpful for the stable humanoid locomotion and energy optimized trajectory planning. Better control strategies can also be implemented in the simulation and experimental mode of operations using a better dynamic model.

References

[1] Müller, A., 2018. Screw and Lie group theory in multibody kinematics. *Multibody System Dynamics*, ***43***(1), pp.37-70

[2] Yi, B.J., Kim, W.K. and Oh, S., 2009. Screw-based kinematic modelling and geometric analysis of planar mobile robots. *International Journal of Control, Automation and Systems*, *7*(6), pp.962-970

[3] Man, C.H., Xun, F., Li, C.R. and Zhao, Z.H., 2007. Kinematics analysis based on screw theory of a humanoid robot. *Journal of China University of Mining and Technology*, ***17***(1), pp.49- 52

[4] Toscano, g. S., simas, h., & castelan, e. B. (2014). Screw-based modeling of a humanoid biped robot. In *Anais do XX Congresso Brasileiro de Automática*(pp.944-951)

[5] Toscano, G.S., Simas, H., Castelan, E.B. and Martins, D., 2018. A new kinetostatic model for humanoid robots using screw theory. *Robotica*, ***36***(4), p.570

[6] Liao, Z., Yao, L., Lu, Z. and Zhang, J., 2018. Screw theory based mathematical modelling and kinematic analysis of a novel ankle rehabilitation robot with a constrained 3-PSPmechanismtopology. *International journal of intelligent robotics and applications*, ***2***(3),pp.351-360

[7] Rocha, C.R., Tonetto, C.P. and Dias, A., 2011. A comparison between the Denavit Hartenberg and the screw-based methods used in kinematic modeling of robot manipulators. *Robotics and Computer-Integrated Manufacturing*, ***27***(4), pp.723-728

[8] Gulletta, G., Erlhagen, W. and Bicho, E., 2020. Human-like arm motion generation: A Review. *Robotics*, ***9***(4), p.102

[9] Patel, S., & Sobh, T. (2015). Task based synthesis of serial manipulators. *Journal of advanced research*, ***6***(3), 479-492

[10] Vijaykumar, R., Waldron, K. J., & Tsai, M. J. (1986). Geometric optimization of serial chain manipulator structures for working volume and dexterity. *The International Journal of Robotics Research*, ***5***(2), 91-103

[11] Guenter, F., Roos, L., Guignard, A., & Billard, A. G. (2005). Design of a biomimetic upper body for the humanoid robot Robota. In 5th IEEE-RAS International Conference on Humanoid Robots 56-61.

[12] Mizuuchi, I., Tajima, R., Yoshikai, T., Sato, D., Nagashima, K., Inaba, M.&Inoue,H.(2002).Thedesign and control of the flexible spine of a fully tendon-driven humanoid" Kenta". In IEEE/RSJ international conference on intelligent robots and systems, 3, 2527-2532.

[13] Gini, G., Folgheraiter, M., Baroni, I., Boschetti, F., Petja, G., &Traversoni,M. (2010). A biomimetic upper body for humanoids. In ISR 2010 (41st International Symposium on Robotics) and ROBOTIK 2010 (6th German Conference on Robotics), 1-8.

[14] Umetani, Y., & Yoshida, K. (1990). Workspace and manipulability analysis of space manipulator. *Transactions of the Society of Instrument and Control Engineers*, ***26***(2), 188-195

[15] Ricard, R. M., & Gosselin, C. M. M. (1998). On the determination of the workspace of complex planar robotic manipulators, 269-278

[16] Goyal, K., & Sethi, D. (2010). An analytical method to find workspace of a robotic manipulator.*Journal of mechanical engineering*, ***41***(1), 25-30
[17] Bonev, I. A., & Ryu, J. (2001). A new approach to orientation workspace analysis of 6-DOF parallel manipulators. *Mechanism and machine theory*,***36***(1), 15-28.
[18] Arbulú, M., Kaynov, D., & Balaguer, C. (2010). The Rh-1 full-size humanoid robot: Control system design and walking pattern generation in climbing
[19] Kevin M Lynch and Frank C Park 2017 Modern Robotics-Mechanics, Planning and Control (1stedition), Cambridge University press, Chapter 4.

IOP Conf. Series: Materials Science and Engineering 1132 (2021) 012037 doi:10.1088/1757-899X/1132/1/012037

Design and Analysis of Minimally Invasive Surgical Robot

Vasundhara Ch. V S N and Sudheer A P*

Mechatronics/ Robotics Lab, Department of Mechanical Engineering, National Institute of Technology, Calicut, Kerala, India

*E-mail: venkata_m190552me@nitc.ac.in

Abstract. Minimally Invasive Surgical robots are commonly seen in medical centres for surgical purposes. However, the major challenges seen with such robotic systems are its heavy weight, costly and lack of capability to reconfigure according to the patient-specific needs. In this paper, a robot is proposed to overcome the difficulties with existing robotic systems. Six different concepts are proposed in this paper which are modelled in SolidWorks software. The best design has been selected using the concept scoring technique. Besides, the forward and inverse kinematics for the best concept is achieved using the Denavit-Hartenberg (D-H) modelling technique. Finally, results obtained from the analytical model has been verified using Peter Corke Toolbox in MATLAB. The trajectory planning of the selected concept has be done using the third-order polynomial equation and it is plotted using MATLAB.

1. Introduction

In recent years, Minimally Invasive Surgery (MIS) is extensively seen in practice for performing different medical procedures [1]. This is because of its several advantages compared to the other traditional open surgical procedures. In traditional open surgeries, the surgeon has to make larger incisions to have a clear view of the internal organ. These incisions can be a reason for the heavy blood loss in patients. This may lead to increased recovery time and infections for the patient. Unlike Open Surgery, MIS will require smaller incisions of less than 10mm. Following this, the surgeons have to insert visual devices along with the instrument from the incision to have a clear view of organ to be operated [1]. For some of the complicated surgeries, it is difficult to perform MIS due to some limitations like (1) restricted DOFs of the instrument, (2) surgeon fatigue, (3) hand-eye incoordination and (4) long learning curve.

In order to overcome the limitations with present surgical devices, robotic system has been introduced into the field of surgery. In 1985, Shao et al. used an industrial robot for performing a neurosurgical biopsy which was the first of its kind [2, 3]. Many further developments are done to design various surgical robots. Most of the surgical robots follow master-slave relationship in which surgeon operates master manipulator sitting at the work console and visualizing the operation on the video screen. In the surgical scene, there will be two human-robot interfaces established namely Surgeon-robot interface and Patient-robot interface [4]. The robot-assisted surgery has advantages like (1) instruments used are dexterous, (2) operation performed is more precise, (3) improved hand-eye coordination and (4) reduced surgeon fatigue.

The most effective way of improving the dexterity of MIS is to design instruments with additional DOFs. One kind of the dexterous surgical device is Da Vinci surgical system by Intuitive Surgical Inc. which is a 6 DOFs robot with intuitive control and is more useful for performing complicated and delicate operations. It is also the most successful and commercially available robot-assisted surgical system [5]. There are certain

IOP Conf. Series: Materials Science and Engineering 1132 (2021) 012037 doi:10.1088/1757-899X/1132/1/012037

drawbacks of this system like (1) lack of capability to reconfigure according to the different procedures, (2) requires large workspace and (3) high cost. Due to the above factors, robot-assisted MIS has limited patient base.

Recently many slave manipulators are designed suitable to satisfy the surgical needs and suitable master manipulator is being proposed ergonomically. Establishing the control between the master and slave manipulators is the most challenging part in surgical robots. The master-slave real-time control strategy for a novel surgical robot had been adopted for solving problems of intra-operative re-mapping and kinematic transformations using consistency principle [6]. An integrated MIS robotic system was proposed and a modular design for the control system was adopted. The kinematic relations were established between the master and slave manipulators and were tested using software [7]. A tele-matched surgical robot with tendon mechanism for precise force reflection and stable, intuitive operation had been developed. A torque estimation scheme instead of physical sensors was used to estimate the external forces [8].

Some surgical procedures require two or more surgeons and additional staff like nurses. The surgical robots with multi-arms have been introduced for supporting the collaborative operation of multiple surgeons. A system resembling the Raven IV system with two robotic arms had been developed. The manipulation dexterity and common workspace of each robotic arm was optimised [9]. A master-slave robotic system with two symmetric arms was developed. The kinematic computation, shape optimization algorithm and collision avoidance algorithm for the robotic arms were constructed [10].

The surgical tools used should be more dexterous to perform the operation precisely and cost effective so that they can be available easily. A reconfigurable, partially disposable and tendon-driven robotic arm was designed for providing assistance in laparoscopic surgery [11]. A small MIS robot had been designed with disposable components to reduce the cost. Modularity for the link geometry were considered to increase the flexibility to adapt the patient-specific needs [12]. A novel and dexterous robotic surgical tool for trans-gastric natural orifice surgery was designed using 3D valve, hydraulic artificial muscles and multi-segmented flexible manipulator arms [13]. A novel miniaturized cable-actuated multi-steerable tip was developed which navigates through the complex curves. Bending stiffness and the ability of the tip to form complex shapes was also tested [14]. Nevertheless, all surgical robots are not tele-operated. A modular hand-held and dexterous surgical robot with interchangeable instruments was developed. Connection between the handle and instrument was established and tested [15].

Therefore, it is necessary to have an adaptive and inexpensive design for MIS which can overcome the inherent shortcomings of presently available robotic systems. The design of a robot for MIS with reduced cost and increase flexibility to adapt different procedures and patient specific needs has been presented in this paper. This paper discusses about the design of the surgical robot for minimally invasive surgery, kinematic modelling with forward kinematics using D-H parameters and inverse kinematics of the robot in the following sections.

2. Conceptual Designs for Surgical Robot

This section explains six different conceptual designs whose CAD models were developed in SolidWorks software. All the proposed concepts have six DOFs with three DOFs for positioning and three active joints for operation. The active joints are designed to have the roll, pitch and yaw (RPY) motions. Fig. 1 shows the first conceptual design. It has two prismatic and one revolute joints for positioning. Fig. 2 depicts the second conceptual design. It is basically a Cartesian robot with three prismatic joints. Fig. 3 displays the third conceptual design which comprises of two prismatic and one revolute joints. Fig. 4 presents the fourth conceptual design which has two prismatic and one revolute joints with a different arrangement compared to the first concept. Fig. 5 displays the fifth conceptual design, which also has three prismatic joints configured differently compared to second concept. Similarly, fig. 6 shows the last conceptual design with two prismatic and one revolute joints.

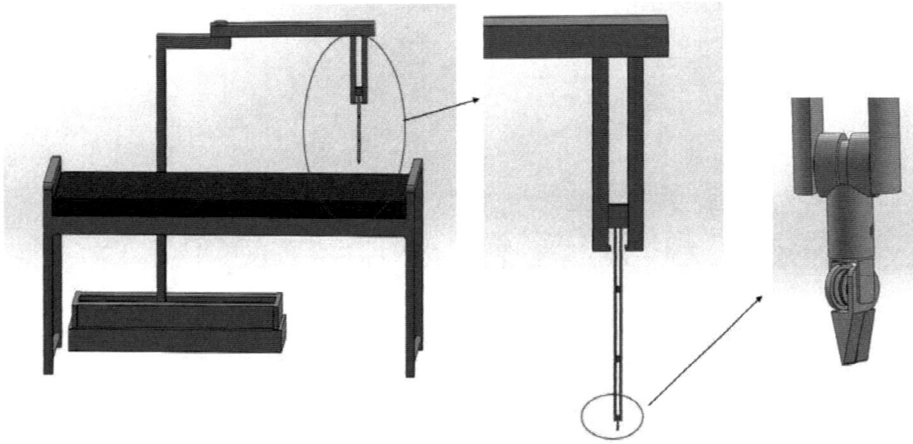

Figure 1. Conceptual Design 1

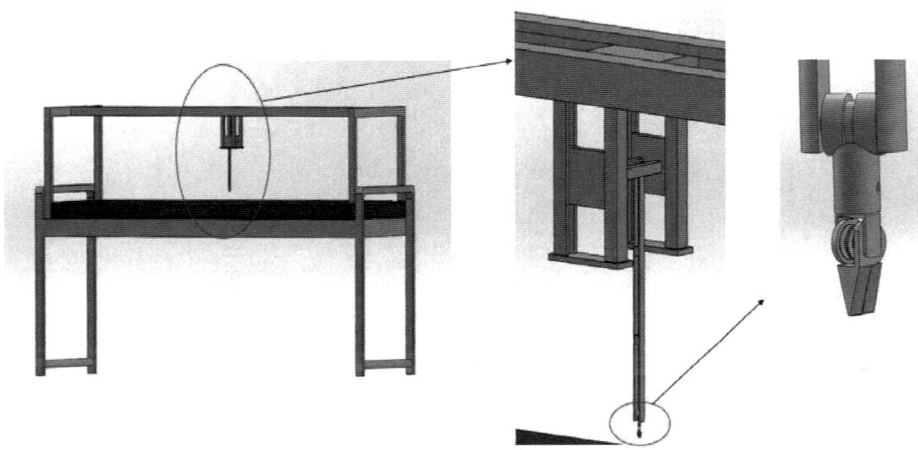

Figure 2. Conceptual Design 2

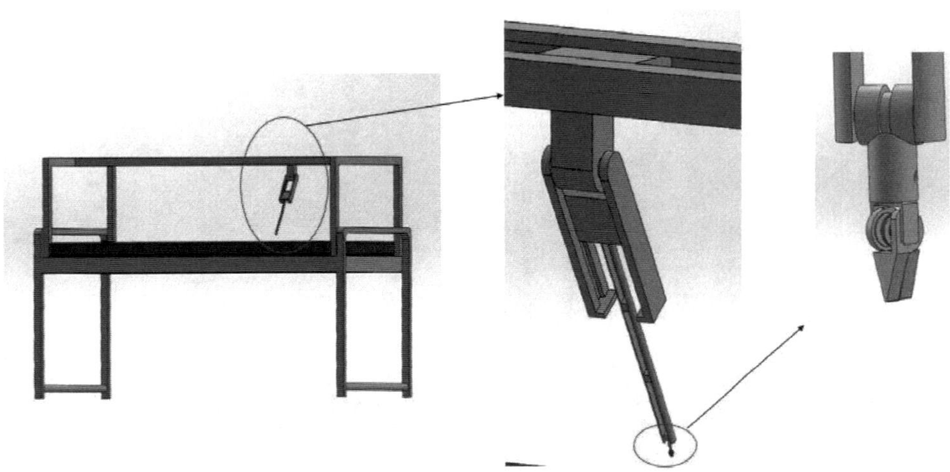

Figure 3. Conceptual Design 3

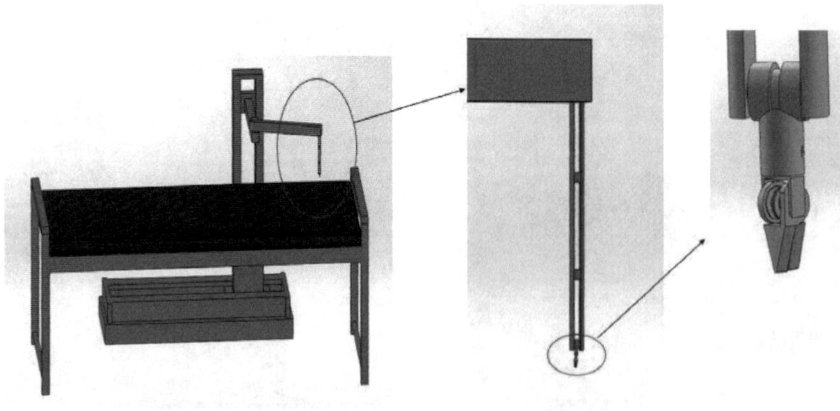

Figure 4. Conceptual Design 4

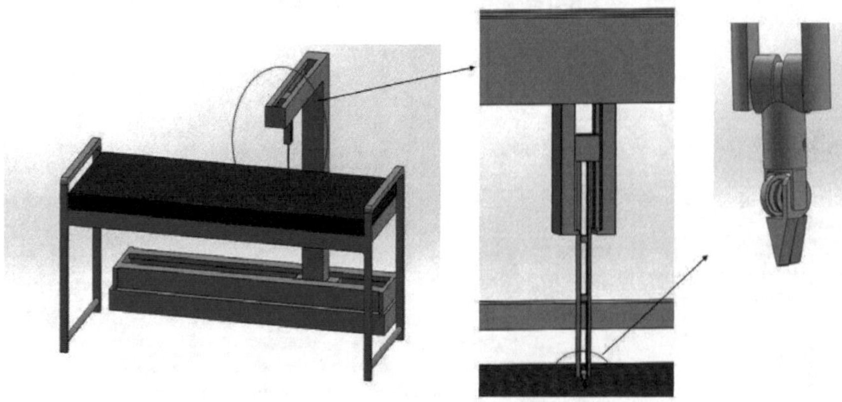

Figure 5. Conceptual Design 5

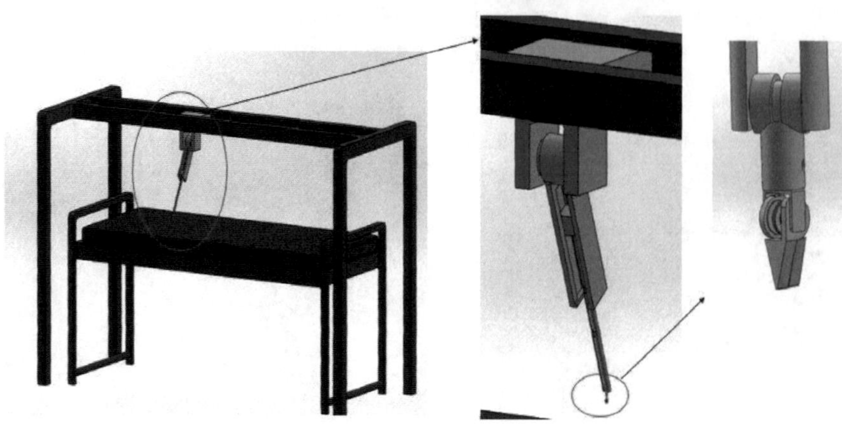

Figure 6. Conceptual Design

3. Concept Scoring

This section explains the conceptual scoring for all the designs shown in the previous section by considering the important characteristics necessary for medical robots. The scoring has been done depending upon the importance of the characteristics and the corresponding rating of the design. The desired characteristics weightage is expressed in percentage and the corresponding rating of each design is scored out of five, as shown in table 1.

Table 1. Concept Scoring

Characteristics	Weightage	Design 1		Design 2		Design 3		Design 4		Design 5		Design 6	
		R	S	R	S	R	S	R	S	R	S	R	S
Precision	20%	3	0.6	5	1	4	0.8	3	0.6	3	0.6	5	1
Cost	20%	4	0.8	5	1	5	1	4	0.8	4	0.8	5	1
Manufacturability	25%	3	0.75	4	1	5	1.25	3	0.75	3	0.75	5	1.25
Weight	10%	3	0.3	5	0.5	5	0.5	3	0.3	3	0.3	4	0.4
Reach	5%	5	0.25	4	0.2	4	0.2	5	0.25	5	0.25	5	0.25
DOF	10%	5	0.5	5	0.5	5	0.5	5	0.5	5	0.5	5	0.5
Ease of control	5%	5	0.25	4	0.2	5	0.25	5	0.25	4	0.2	5	0.25
Ease of use	5%	4	0.2	3	0.15	4	0.2	4	0.2	4	0.2	4	0.2
	TOTAL	3.65		4.55		4.7		3.65		3.6		4.85	
	RANK	**4**		**3**		**2**		**4**		**6**		**1**	

Where R is rating and S is weighted score.

Fig. 1, fig. 4 and fig. 5 resemble the model of a SCARA robot. The main disadvantage of these models are reduced precision and high deflection since it acts more like a cantilever beam. Due to these reasons, the concepts 1, 4 and 5 scores less compared to concepts 2, 3 and 6. However, the major drawback with concepts 2 and 3 designs is its manufacturing difficulty as they need to be welded to the patient bed. Therefore, sixth concept qualifies to be the best of the six concepts considered in this paper. This can be confirmed from the maximum score of the concept 6 compared to the other concepts as is evident from table 1. As shown in the fig. 6, the prismatic joint attached to the frame allows to have reach all over the patient bed. The frame is designed in such a way that the patient bed can be moved flexibly. Each link is connected using cables and the motion of every joint is given by using servomotors.

4. Kinematic Modelling

4.1. Forward Kinematics

The kinematic model has been developed for the considered six DOFs mechanism based on the Denavit-Hartenberg approach. Fig. 7 shows the frame assignment corresponding to the D-H convention. Following this, the D-H parameter table is filled as stated by the D-H convention as shown in table 2.

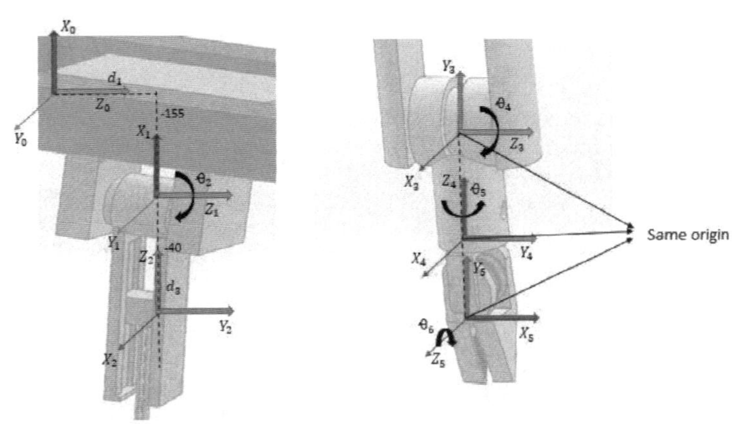

Figure 7. Frame assignment of the robot

Table 2. D-H parameters of robot

Link	θ	d	a	α
1	0	d_1	-155	0
2	$\theta_2 + \pi/2$	0	-50	$\pi/2$
3	0	$-(d_3 + 262)$	25	$-\pi/2$
4	θ_4	0	0	$\pi/2$
5	$\theta_5 + \pi/2$	0	0	$\pi/2$
6	θ_6	0	0	0

Then the D-H transformation matrices are filled for all the links as depicted in eq. 1, 2, 3, 4, 5 and 6.
For first link,

$$T_1^0 = \begin{bmatrix} 1 & 0 & 0 & -a_1 \\ 0 & 1 & 0 & 0 \\ 0 & 0 & 1 & d_1 \\ 0 & 0 & 0 & 1 \end{bmatrix} \tag{1}$$

For second link,

$$T_2^1 = \begin{bmatrix} -\sin\theta_2 & 0 & \cos\theta_2 & a_2\sin\theta_2 \\ \cos\theta_2 & 0 & \sin\theta_2 & -a_2\cos\theta_2 \\ 0 & 1 & 0 & 0 \\ 0 & 0 & 0 & 1 \end{bmatrix} \tag{2}$$

For third link,

$$T_3^2 = \begin{bmatrix} 1 & 0 & 0 & a_4 \\ 0 & 0 & 1 & 0 \\ 0 & -1 & 0 & -d_3 - a_3 \\ 0 & 0 & 0 & 1 \end{bmatrix} \tag{3}$$

For fourth link,

$$T_4^3 = \begin{bmatrix} \cos\theta_4 & 0 & -\sin\theta_4 & 0 \\ \sin\theta_4 & 0 & \cos\theta_4 & 0 \\ 0 & 1 & 0 & 0 \\ 0 & 0 & 0 & 1 \end{bmatrix} \tag{4}$$

For fifth link,

$$T_5^4 = \begin{bmatrix} -\sin\theta_5 & 0 & \cos\theta_5 & 0 \\ \cos\theta_5 & 0 & \sin\theta_5 & 0 \\ 0 & 1 & 0 & 0 \\ 0 & 0 & 0 & 1 \end{bmatrix} \tag{5}$$

For sixth link,

$$T_6^5 = \begin{bmatrix} \cos\theta_6 & -\sin\theta_6 & 0 & 0 \\ \sin\theta_6 & \cos\theta_6 & 0 & 0 \\ 0 & 0 & 1 & 0 \\ 0 & 0 & 0 & 1 \end{bmatrix} \tag{6}$$

Where $a_1 = 155$, $a_2 = 50$, $a_3 = 262$, $a_4 = 25$
The final transformation matrix for the robot is obtained from the eq. 7.

$$ {}^{0}_{6}T = {}^{0}_{1}T * {}^{1}_{2}T * {}^{2}_{3}T * {}^{3}_{4}T * {}^{4}_{5}T * {}^{5}_{6}T \quad (7) $$

The movements of the robot are analysed in MATLAB corresponding to the D-H parameters mentioned above. The motion of the robot can be analysed by specifying the values of the parameters with code 'teach'. Fig. 8 displays the posture of the robot for given end effector pose.

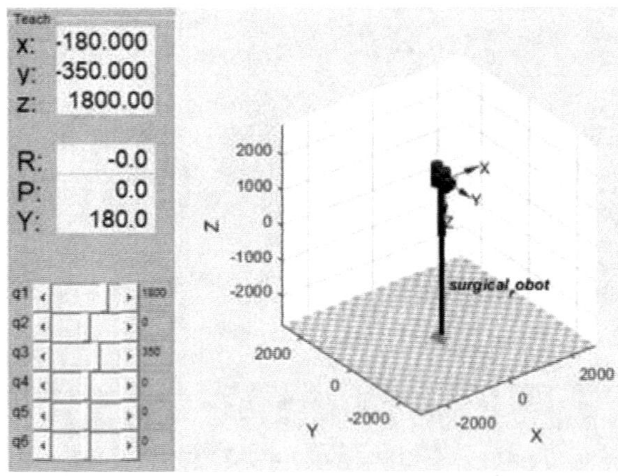

Figure 8. Posture of the robot using MATLAB

4.2. Inverse Kinematics

Inverse kinematics determines the joint variables for a given pose of end effector. Assuming a generalised equation for the end effector pose as given in eq. 8.

$$ X = \begin{bmatrix} n_x & o_x & a_x & p_x \\ n_y & o_y & a_y & p_y \\ n_z & o_z & a_z & p_z \\ 0 & 0 & 0 & 1 \end{bmatrix} \quad (8) $$

The final transformation matrix defined in eq. 7 is equated with the matrix in eq. 8.

$$ X * {}^{5}_{6}T^{-1} = {}^{0}_{1}T * {}^{1}_{2}T * {}^{2}_{3}T * {}^{3}_{4}T * {}^{4}_{5}T \quad (9) $$

Therefore, by solving the above equation we can compute all the parameters i.e., d_1, θ_2, d_3, θ_4, θ_5, θ_6 as given in eq. 10, 11, 12, 13, 14 and 15.

$$ d_1 = p_z \quad (10) $$

$$ \theta_2 = \tan^{-1}\left[\frac{(25*p_x)+3875+(-p_y)*(\pm a)}{(-155-p_x)*(\pm a)-(25*p_y)}\right] \quad (11) $$

Where a = $\sqrt{(p_x + 155)^2 + p_y{}^2 - 625)}$

$$ d_3 = -262 \pm \sqrt{(p_x + 155)^2 + p_y{}^2 - 625} \quad (12) $$

$$ \theta_4 = \tan^{-1}\left(\frac{-a_x}{a_y}\right) - \theta_2 \quad (13) $$

$$ \theta_5 = \tan^{-1}\left(\frac{a_z}{\sqrt{a_x{}^2+a_y{}^2}}\right) \quad (14) $$

$$\theta_6 = \tan^{-1}\left(\frac{-o_z}{n_z}\right) \tag{15}$$

5. Trajectory Planning

The series of displacements in the robot with respect to time are analysed using trajectories. It is mainly the trace of the path followed by the end effector. Trajectories of joint variables are defined using the polynomial equation as shown in eq. 16.

$$\theta(t) = c_0 + c_1 t + c_2 t^2 + \cdots + c_{n-1} t^{n-1} + c_n t^n \tag{16}$$

Here t is the time, θ(t) is the joint variable with reference to time, c_n are the coefficients of n-order polynomial. In this paper, a cubic polynomial equation is used to determine the trajectory of the end effector as defined in eq. 17. Eq. 18 and eq. 19 defines the velocity and acceleration equations of the joint variables respectively.

$$\theta(t) = c_0 + c_1 t + c_2 t^2 + c_3 t^3 \tag{17}$$
$$\dot{\theta}(t) = c_1 + 2c_2 t + 3c_3 t^2 \tag{18}$$
$$\ddot{\theta}(t) = 2c_2 + 6c_3 t \tag{19}$$

Using inverse kinematic equations and initial positions and orientations of the robot, the desired final positions and orientations are calculated. The movements of each joint in the robot must be considered separately as a function of time. By considering the necessary conditions, the eq. 17, 18 and 19 can be solved to obtain the constant values. Therefore, the joint positions are computed at different intervals of time. All the joint variable trajectory equations are used together to get the trajectory plot of the end effector. The Cartesian co-ordinates of the end effector are calculated with the help of joint position values and forward kinematic equations. The co-ordinate values obtained are plotted using MATLAB software as shown in fig. 9.

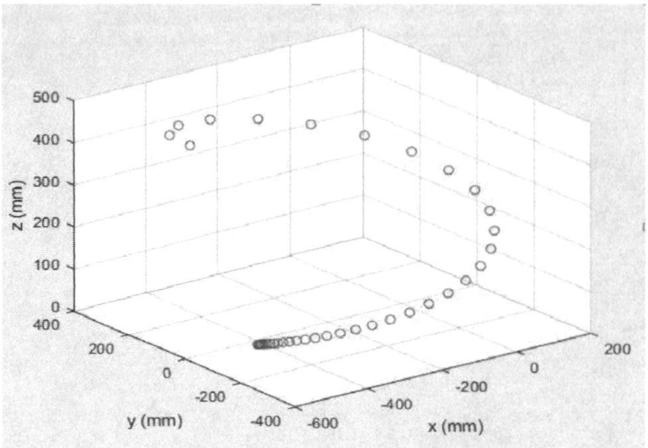

Figure 9. Trajectory planning

6. Conclusion

This paper focuses on proposing different conceptual designs for the robot-assisted minimally invasive surgery which can be altered according to the patient specific needs. Six different concepts are shown which were modelled using the SolidWorks software. The best concept was chosen based on eight important characteristics that are required for surgical robots using the technique of concept scoring. The forward and inverse kinematics of the selected conceptual design is done using Denavit-Hartenberg modelling technique and the results are finally verified using Peter Corke toolbox in MATLAB software. The trajectory planning of the end effector has been plotted by using third-order polynomial equation. The Cartesian co-ordinate values of the end effector are calculated and plotted using MATLAB software. The scope for future work includes the development of dynamic model for

IOP Conf. Series: Materials Science and Engineering 1132 (2021) 012037 doi:10.1088/1757-899X/1132/1/012037

proposed mechanism, fabrication, implementation of suitable controller and finally performance evaluation.

References

[1] Low S C and Phee L 2006 A Review of Master–slave Robotic Systems for Surgery *Int. J. Humanoid Robo.* pp 547-567.

[2] Shao H, Chen J, Truong T, Reed I and Kwoh Y 1985 A New CT-Aided Robotic Stereo Axis System *In Proceedings of the Annual Symposium on Computer Application in Medical Care American Medical Informatics Association* p. 668.

[3] Kwoh Y, Hou J, Jonckheere E and Hayati S 1988 A Robot with Improved Absolute Positioning Accuracy for CT Guided Stereotactic Brain Surgery *IEEE Transactions on Biomedical Engineering* 35(2) pp. 153–160.

[4] Jacob Rosen and Ji Ma 2015 Autonomous operations in Surgical Robots *Focus on Dynamic System and Control* pp 15-18

[5] Loi-Wah Sun, Frederick Van Meer, Yan Bailly and Chung Kwong Yeung 2007 Design and Development of a Da Vinci Surgical System *Simulator In Proceeding International Conference on Mechatronics and Automation* pp. 1050-1055.

[6] Bo Pan, Xiaofeng Qu, Yue Ai, Yili Fu and Chang Liu 2016 Master–slave real-time control strategy in Cartesian space for a novel surgical robot for minimally invasive surgery *Comput. Assist. Surg.* pp 69-77.

[7] Jianmin Li, Ningxin Zhou, Shuxin Wang, Yuanqian Gao and Dongchun Liu 2012 Design of an integrated master–slave robotic system for minimally invasive surgery *Int. J. Med. Robotics Comput. Assist. Surg.* pp. 77–84.

[8] Seung-Ju Lee, Sang-Chul Lee and Hyo-Sung Ahn 2014 Design and control of tele-matched surgery robot *Mechatronics* 24 pp 395–406.

[9] Zhi Li, Dejan Milutinovic and Jacob Rosen 2016 Design of a Multi-Arm Surgical Robotic System for Dexterous Manipulation *J. Mech. Robotics* 8 061017 pp 1-10.

[10] Quanquan Liu, Chunbao Wang and Bo Zhang 2017 Kinematics Analysis and Motion Planning of a Redundant Robotic Manipulator for Surgical Intervention *IEEE International Conference on Cyborg and Bionic Systems* 17 pp 1-10.

[11] John Lowery and Carl A. Nelson 2020 Design of A Modular Cost-effective Robot Arm for Increased Dexterity in Laparoscopic Surgery *Design Med. Dev. Conf.* V001T06A001 pp 1-4.

[12] Nicholas Nelson and Carl A. Nelson 2018 Design of a Modular, Partially Disposable Robot for Minimally Invasive Surgery *Proceedings of the Design of Medical Devices Conference* pp 1-3.

[13] Devin R. Berg, Perry Y. Li and Arthur G. Erdman 2012 Achieving Dexterous Manipulation for Minimally Invasive Surgical Robots Through the Use of Hydraulics *Annual Dynamic Systems and Control Conference joint with the Motion and Vibration Conference* 8685 pp 1-10.

[14] Aimee Sacks, Awaz Ali, Jovana Janjic and Paul Breedveld 2018 Novel Miniature Tip Design for Enhancing Dexterity in Minimally Invasive Surgery *J. Med. Dev.* 12 035002 pp 1-9.

[15] Yingkan Yang, Kang Kong, Jianmin Li, Shuxin Wang and Jinhua Li 2019 Design and Evaluation of a Dexterous and Modular Hand-Held Surgical Robot for Minimally Invasive Surgery *J. Med. Dev.* 13 041005 pp 1-10.

IOP Conf. Series: Materials Science and Engineering 1132 (2021) 012038 doi:10.1088/1757-899X/1132/1/012038

Experimental Study on Geopolymer Rubberized concrete using natural Zeolite

Suthari Bhavani[1], G. Nagesh kumar[2], M. Srinivasula Reddy[3] and E. Sanjeeva Rayudu[4]

Department of Civil Engineering, G. Pulla Reddy Engineerin College(Autonomous), Kurnool, A.P, India

Email: bhavanisuthari@gmail.com

Abstract: On one hand, the utilization of rubber tires is ever-increasing, expected to further increase at a rapid rate, and with that the waste rubber. When the tires become old, worn out and irreparable, it is often dumped in landfills. For every one billion tires produced, an equal amount of tires being removed from the vehicles. This waste can be utilized in large quantities when employed in the construction industry. On the other hand, cement-based concrete utilization also increasing at a faster rate and with that the CO_2 emissions. These two major issues i.e. waste rubber utilization and curbing of cement consumption can be effectively done by developing geopolymer concrete with the replacement of waste rubber chips and rubber powder in the place of natural aggregates. In the present study, the geopolymer concrete has been developed by using fly ash, GGBS, and zeolite as binder materials, sodium hydroxide, and sodium silicate as alkaline activators, rubber chips as a partial replacement to natural coarse aggregate, rubber powder as a partial replacement to natural fine aggregate. Rubberized geopolymer concretes have been developed by replacing fly ash with 5% zeolite, natural coarse aggregate with 2.5% waste rubber chips, and natural fine aggregate with 5%, 10%, 15%, and 20% by weight of waste rubber powder. The developed concrete is then assessed for its performance through mechanical properties, durability behaviour. The results showed the developed concrete has high impact resistance and good freeze-thaw resistance properties. This concrete is suitable for the members subjected to high-impact loads.

1. Introduction

Human comfort levels are increasing day-by-day and due to that the usage of vehicles. This, in turn, is leading to an increased need for rubber tires usage. With time, these tires going to unusable conditions and are getting thrown away. It is estimated that nearly a billion tires per annum are being thrown out as garbage after they becoming a spent force. It is expected to reach the disposed of tires number above five billion by 2030. Approximately 60% of end-of-life tires are estimated to be in landfills. The combustion of these tires is dangerous, as they produce a poisonous smell and releases lethal gases, and their combustion is restricted in many countries. The most profitable and efficient way of utilizing these rubber tire waste, without flaming, is by cutting the rubber tires into small pieces and also into powder form to utilize them in the construction industry.

Currently, the construction industry, especially concrete producers, is looking to integrate sustainability into production processes by incorporating many wastes in the place of conventional concrete ingredients. They are doing it mostly in the form of partially or fully replacement of the cement with industrial wastes [1, 2]. Some researchers even attempted to replace natural aggregates with inert materials like stone dust, agriculture waste, demolition wastes, ceramic tile powder [3, 4]. Further, studies are also available on the development of Portland cement-based concrete with scrap tire chips, crumb rubber, etc. [5, 6]. The literature suggests that concrete incorporated with rubber has shown less wear and tear when compared to conventional concrete [7]. Studies, further, reported that concrete developed using waste has good ductility and high impact resistance which can be used in structural elements such as bridges, railway lines, airport runways, where the structural elements are subjected to heavy impact and dynamic loads [8-9]. Even, some researchers developed geopolymer concrete, a cement-free concrete, by incorporating rubber waste in it [10]. They found that the addition of rubber wrecks the strength of the geopolymer concrete. The main reason for the decrease in the strength is found to be the poor bonding between the binder and rubber particles [11]. Jokar et al. [12] reported that the bond between the rubber particles and cement binder can be improved by treating the rubber particles with 1M NaOH solution. It is also reported that cement replacement with Zeolite in the concrete made with rubber particles has improved the compression and bending characteristics of the cement-based concrete. He et al. [13] reported that gradation plays a major role in the bonding between the rubber particles and binder.

Based on the literature study and also by keeping in view the environmental pollution that the unutilized industrial wastes such as fly ash, slag is creating, efforts have been made in this study to develop geopolymer rubberized concrete by employing fly ash, ground granulated blast furnace slag (ggbs), and zeolite powder as binder materials, and replacing coarse and fine aggregates with rubber chips and rubber powder, respectively. In this study, fine aggregate is replaced with both treated and untreated rubber powder with various percentages (5%, 10%, 15%, and 20%). After the formation of rubberized concrete with these percentages, specimens are allowed to ambient curing for gaining sufficient strength. The developed concretes are then assessed for mechanical properties through compressive strength, split tensile strength, impact resistance, and durability through freeze-thaw resistance tests.

2. Methods

2.1 Fly ash

Fly ash is a by-product of coal based thermal power plants. Fly ash is classified into two types, Class C and Class F, based on its chemical oxide composition. In this study, class F fly ash is used which is obtained from the Rayalaseema TPP located in Kadapa district of Andhra Pradesh. Table 1 show the physical properties of fly ash and ggbs, and Table 2 shows the chemical composition of all the binder materials used in this study, i.e. fly ash, ggbs and zeolite powder.

Table 1: Physical properties of fly ash and ggbs

Properties	Test results	
	Fly ash	ggbs
Colour	Grey	Whitish
Specific Gravity	2.2	2.9
Surface Gravity (m^2/kg)	420	407

2.2 Ground Granulated Blast Furnace Slag

GGBS is a by-product from iron industry. Molten slag and molten iron is formed when blast furnace is fed with proportionate lime, coke and iron-ore operated at the temperature about 1500°C. This molten slag when water quenched gives a hard clinker type material and which upon fine grinding gives a

IOP Conf. Series: Materials Science and Engineering 1132 (2021) 012038 doi:10.1088/1757-899X/1132/1/012038

material called ggbs. In this work, ggbs is obtained from JSW steel Ltd, Bellary.

2.3 Zeolite powder

Natural Zeolite is an excellent Supplemental Cementing Material (SCM). Natural zeolite has three-dimensional framework structures and is classified as an alkali and alkaline earth cations hydrated alumino silicate. Zeolite powder used in this study is obtained from Bethamcherla of Kurnool district. Specific gravity of the zeolite used is 2.47. Figure 1 shows the natural zeolite powder.

Figure 1. Zeolite

Table 2. Chemical composition of fly ash, GGBS, and zeolite

Material	Percentage of Chemical Oxides									
	SiO_2	CaO	Al_2O_3	MgO	Fe_2O_3	SO_3	TiO_2	NaO_3	K_2O	L.O.I
Fly ash	62.19	1.97	21.15	0.40	3.23	0.07	1.06	0.30	0.89	1.75
GGBS	37.73	37.34	14.42	8.71	1.11	-	-	-	-	-
Zeolite	67.79	1.68	13.66	1.20	1.44	-	-	2.04	-	12.23

2.4 Coarse aggregate

Coarse aggregate passing through 20 mm and 12.5 mm sieves are used and are taken in equal proportions on weight-basis. The coarse angular crushed granite aggregate was obtained from local crushing plants. The specific gravity of the coarse aggregate used is 2.77.

2.5 Fine aggregate

The fine aggregate used is river sand and is collected from the beds of the Tungabhadra River. From the sieve analysis, it is found that the fine aggregate confirms to zone II, as per IS: 383-1970[14]. The specific gravity of fine aggregate is 2.60.

2.6 Alkaline solutions

Sodium hydroxide (NaOH) and sodium silicate (Na2SiO3) are used as alkaline activator solutions in this study. After doing preliminary study and based on the observations from it, Na2SiO3 to NaOH ratio of 1.5 and NaOH molarity of 12M has been adopted to develop both conventional geopolymer concrete and geopolymer rubberized concretes. The sodium hydroxide solution is prepared 24 hours prior to adding in the concrete mix.

2.7 Water

Water used in this study is fresh, colourless and clean portable municipal tap water, free of organic matter.

2.8 Super plasticizer

Superplasticizer used is 1B233Glenium made by BASF Chemicals. It is used as high range water reducer. Superplasticizer is added to increase workability and setting times of the fresh concrete mixes. Specific gravity of 1B233 Glenium is 1.06.

2.9 Rubber Powder

The rubber powder is prepared by passing the tyre rubber through rotating corrugated steel drums. Figure 2 shows the rubber powder used in this study. This is replaced in place of the fine aggregate. The specific gravity of the rubber powder is 0.52.

Figure 2. Rubber powder

2.10 Chipped Rubber

Chipped rubber piece having size about 20 mm is used. The specific gravity of Chipped Rubber is found to be 1.1. Figure 3 shows the rubber chips used in the geopolymer rubberized concrete.

Figure 3. Rubber chips

2.11 Rubber Treatment

The rubber particles are soaked in a 1Molarity of NaOH solution for 24 hours. After 24 hours of treatment with NaOH solution, the rubber particles are thoroughly washed with tap water several times until the pH of the wash water was close to neutral. The treated rubber particles are then air dried by keeping at room temperature.

3. Experimental Study

3.1 Mix Proportioning, Casting and Curing of Geopolymer Rubberized Concrete

In the present study, a total of 9 mixes are synthesized. First, a conventional geopolymer concrete mix is prepared by taking fly ash and ggbs as binder materials in 40% and 60% by weight, respectively, natural crushed stone and river sand as aggregate, and NaOH and $Na2SiO3$ as alkaline activators. The remaining mixes are synthesized in two batches. The two batches consists of geopolymer concrete mixes developed using 35% fly ash, 5% zeolite powder, 60% ggbs as binders (constant for all the mixes), replacing coarse aggregate with 2.5% by weight of rubber chips (constant for all the mixes), and replacing river sand with 5%, 10%, 15% and 20% by weight. While, in the first batch, rubber particles are directly induced in mix. In the second batch, the rubber particles are treated with 1M NaOH solution. The properties studied, the age of concrete at the study, the tests conducted and their related codes are depicted in Table 3.

All the concrete mixes are developed by following the mixing sequence shown below.

Dry mixing of binder materials for 3-4 minutes → Addition of coarse and fine aggregates and dry mixing for further 3-4 minutes →Addition of NaOH and Na2SiO3 and then mixing till uniform mix is attained → Addition of SP of required quantity and further mixing for 1-2 minutes.

The fresh concrete mix is then immediately placed in the moulds and given required compaction before keeping them under ambient temperature curing.

Table 3. Testing Parameters

Properties	Tests	Codes	Test Ages	Cubes Sizes in mm
Mechanical	Compressive Strength	IS:516-1959	28,56	100 x 100 x 100
	Split Tensile Strength	IS:5816-1999	28	100 dia x 200 height
	Impact Resistance Strength	ACI committee 544.2R-89	28	100 x 100 x 500
Durability	Freeze-Thaw	ASTM C666 2008 C	28	100 X 100 X 100

4. Mix design

Mix design methodology proposed by Pavithra et al. (2016) [15] has been followed and the mix proportioning has been done accordingly. The following are the parameters considered for the mix design:

- AAS content: 200 kg/m^3
- Alkaline liquid to binder ratio: 0.40
- Sodium silicate to sodium hydroxide ratio: 2.5
- Molarity of sodium hydroxide solution: 12 M
- Admixture dosage: 1%

The mix designations and mix proportioning of the various mixes developed in this study are shown in Table 4 and Table 5.

Table 4. Designations

G0- Control/conventional geopolymer concrete	
Untreated	Treated
G1-5% rubber powder in F.A	G5-5% rubber powder in F.A
G2-10% rubber powder in F.A	G6-10% rubber powder in F.A
G3-15% rubber powder in F.A	G7-15% rubber powder in F.A
G4-20% rubber powder in F.A	G8-20% rubber powder in F.A

Table 5. Mix proportions of concrete

S. No.	Binder Content (kg/m^3)			Alkaline solution (kg/m^3)		C.A (kg/m^3)		F. A (kg/m^3)		SP
	Fly-Ash	Zeolite	GGBS	NaOH	Na_2Sio_3	C.A	Rubber Chips	F.A	Rubber powder	
G0	200	0	300	80	120	1404	0	338	0	1%
G1	190	10	300	80	120	1368	35.1	321	16.9	1%
G2	190	10	300	80	120	1368	35.1	304	33.8	1%
G3	190	10	300	80	120	1368	35.1	287	50.7	1%
G4	190	10	300	80	120	1368	35.1	270	67.6	1%
G5	190	10	300	80	120	1368	35.1	321	16.9	1%
G6	190	10	300	80	120	1368	35.1	304	33.8	1%
G7	190	10	300	80	120	1368	35.1	287	50.7	1%
G8	190	10	300	80	120	1368	35.1	270	67.6	1%

*C.A indicates coarse aggregate; F.A indicates fine aggregate; SP indicates Super plasticizer;

5. Results and Discussion

5.1 Compressive Strength

The geopolymer concretes developed are kept under ambient temperature curing till the time of testing i.e. 28 and 56 days. The specimens are then tested for compressive strength at 28 and 56 days and the compressive strength test results are shown in the figure 4 and figure 5, respectively. From figure 4 and figure 5, it can be observed that the compressive strength decreases with the increase in the percentage of rubber powder content in the mix. The same trend has been observed at 28 days as well as 56 days. About 10-12% improvement in the compressive strength has been seen in all the mixes from 28 to 56 days. The decrease in the strength, with the addition of rubber powder in the mixes, could be poor bonding between the geopolymer paste and the rubber particles. The hydrophobic nature of rubber is the main reason behind this weak bonding . Surprisingly, the compressive strength, both at 28 and 56 days, increased when the mixes are treated with 1M NaOH solution before adding in the mixes. This could be due to the enhanced hydrophilic nature of the rubber particles, when treated with NaOH solution, causing the improved adhesion between the geopolymer paste and rubber particles [16]. The compressive strength of the geopolymer concrete developed varies from 63 to 80 MPa at 28 days and 65 to 89.7 MPa at 56 days.

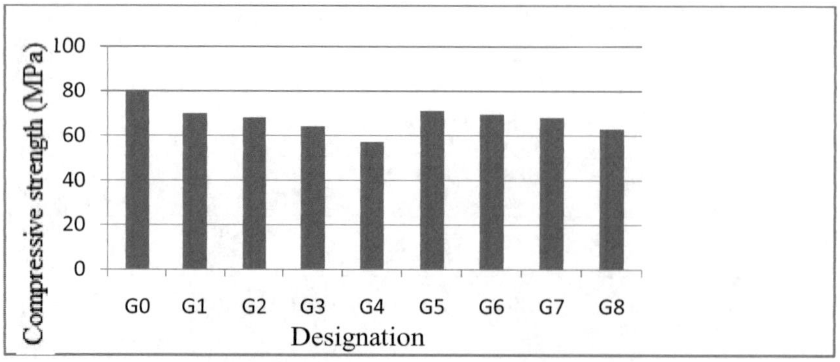

Figure 4. Compressive strength at 28 days

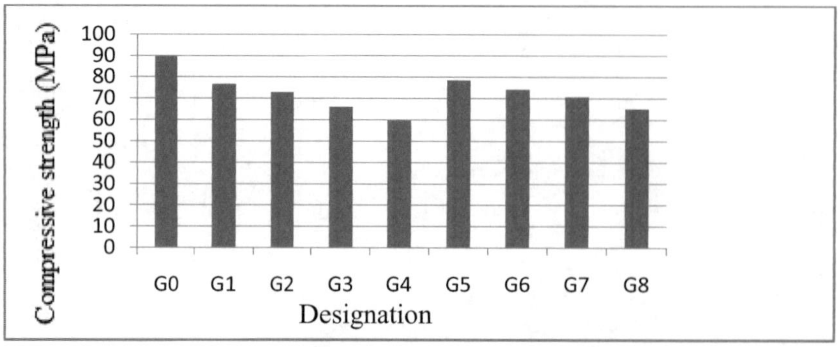

Figure 5. Compressive strength at 56days

5.2 Split Tensile Strength

Figure 6 shows the split tensile strength properties of the developed geopolymer concrete mixes after 28days. Before modification of rubber particles from G1 to G4 the split tensile strength is depleted, similar to compressive strength. The same parameters that are responsible for the decrease in the compressive strength could be the reason for similar behaviour of geopolymer concrete for tensile strength also. After modified rubber is induced in concrete spilt tensile strength is increased up to G7 and then it is decreased. Strength varies from 4.0 to 5.57 MPa. The highest and lowest tensile strength is found at G6, G7 with 5.25 MPa and G4 with 4 MPa.

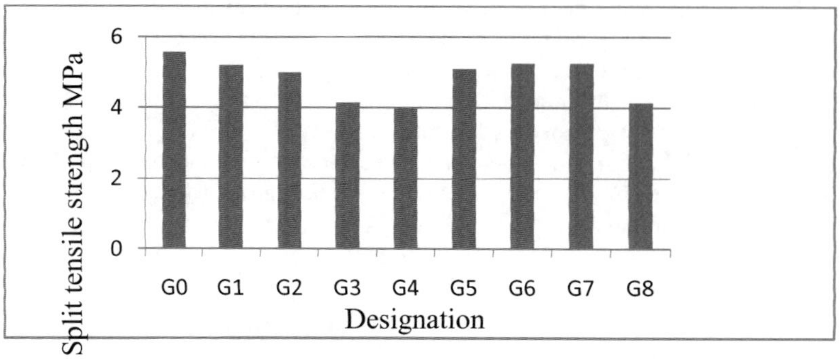

Figure 6. Split tensile strength at 28days

5.3 *Impact resistance test*

Impact resistance test is conducted on the beam specimen with size 100 x 100 x 500 mm. Impact resistance test is conducted as per ACI-544.2R-89. The weight of hammer used in this test is 4.45 kg, diameter of ball is 60.2mm, and the hammer is dropped from a height of 457mm repeatedly at the centre of the specimen. In each test "N1" number of blows at the first crack or initial crack and "N2" number of blows at the failure strength or failure crack of the specimen is recorded. The impact resistance results of all the mixes are depicted in figure 7. From the figure 7, it can be noticed that, the impact resistance of geopolymer concrete increases with the increase in the rubber content in the mix. It can be observed for both the NaOH treated and untreated mixes. Addition of the rubber particles in concrete leads to low stiffness reduced brittleness of concrete, which in turn lead to increased flexibility of rubberized concretes and considerably enhances the energy absorption of the mixture as compared to geopolymer concrete without rubber [17]. The test set-up for conducting the impact resistance test is shown in figure 8 and figure 9. Highest impact resistance is seen at treated rubber concrete with 68 blows and lowest at control mix with 12 blows.

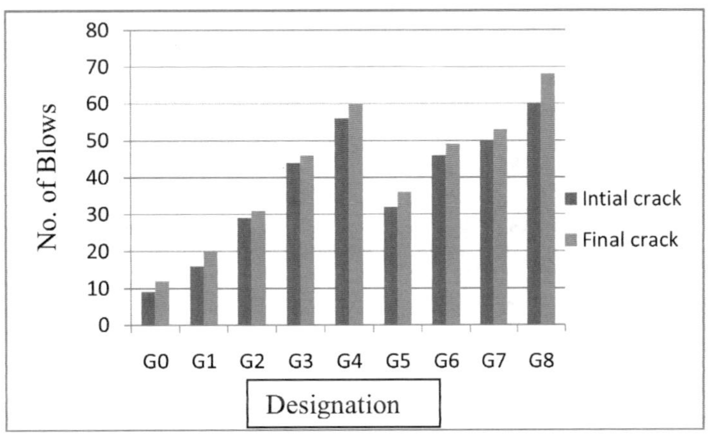

Figure 7. Impact resistance test

Figure 8. Impact resistance test equipment

Figure 9. Testing the specimen

5.4 Freeze- Thaw resistances test

The freeze-thaw resistance test is conducted as per ASTM C666 – 2008 [18] standards. In this test, after 28 days of ambient curing, the specimens are kept in freezer for 24 hours followed by removal and kept in water for another 24 hours. This is considering as 1 cycle, in the same way 3 cycles is conducted. Mass of samples is then measured throughout the freeze–thaw test. After completion of cycles, the weight loss of specimens is measured. Table 5 shows the freeze-thaw test results. From the results, it can be noticed that, treated rubberized concrete shows good freeze-thaw resistances than conventional concrete. The better freeze-thaw resistance of the rubberized concrete could be due to its flexibility to expand and contract. Figure 10 and figure 11 depicts the specimens before testing and after testing, respectively.

Table 6: Freeze-thaw test results

Designation	% of Weight loss
G0	1.44
G5	1.00

Figure 10. Before testing specimen

Figure 11. After testing specimen

6. Conclusions

Based on the results obtained from mechanical properties study and freeze-thaw resistance test, the following are the major conclusions drawn. Both the compressive strength and tensile strength decreases with an increase in the rubber content from 5 to 20%. Both the compressive strength and tensile strength decreases for concretes made with treated as well as untreated rubber particles. Maximum compressive strength of 70 MPa for geopolymer rubberized concrete made with untreated rubber particles and 71.5 MPa with treated rubber particles. The highest and lowest tensile strength is found to be 5.25 MPa and 4 MPa, respectively. Highest tensile strength is shown by geopolymer concrete developed with 15% treated rubber particles. By increasing rubber content the impact resistances also increased in both untreated and treated rubber. Among all the mixes, highest resistance shown by the mix with 20% treated rubber particles. This type of concrete can be used where the concrete is subjected to impact loads.

References

[1] Manoj Kumar Dash, Sanjaya Kumar Patro, Ashoke Kumar Rath, *Sustainable use of industrial-waste as partial replacement of fine aggregate for preparation of concrete – A review*, International Journal of Sustainable Built Environment, Volume 5(2) (2016)

[2] Lahiba Imtiaz, Sardar Kashif Ur Rehman, Shazim Ali Memon, Muhammad Khizar Khan, Muhammad Faisal Javed, *A Review of Recent Developments and Advances in EcoFriendly Geopolymer Concrete,* Applied Sciences Volume 10 (2020)

[3] K.Shyam Prakash, Ch. Hanumantha Rao, *Study on Compressive Strength of Quarry Dust as*

Fine Aggregate in Concrete, Advances in Civil Engineering (2016). Article ID 1742769

[4] Tahir Abdullah, Poloju Kiran, *Use of Ceramic Powder in Concrete -Strength & Durability Properties,* Journal of Student Research (2016)

[5] Najib N Gerges, Camille A Issa, *Samer A Fawaz, Case study Rubber concrete: Mechanical and dynamical properties*, Case Studies in Construction Materials Volume 9 (2018)

[6] Amar Ben Nakhai and Jasem M Alhumoud, *Properties of Concrete Containing Scrap (Recycled) Tire-Rubber,* Journal of Engineering and Applied Sciences 15(2) (2020)

[7] Trilok Gupta, Sandeep Chaudhary, Ravi Sharma, *Mechanical and durability properties of waste rubber fiber concrete with and without silica fume,* Journal of Cleaner Production Volume 112 (2015)

[8] Salmabanu Luhar, Sandeep Chaudhary, Ismail Luhar, *Development of rubberized geopolyme concrete: Strength and durability studies,* Construction and Building Materials volume 204 (2019)

[9] Aly Muhammed Aly, M.S. El-Feky, Mohamed Kohail, El-Sayed A.R. Nasr, *Performance of geopolymer concrete containing recycled rubber,* Construction and Building Materials volume 207 (2019)

[10] Sesha Choudary Yeluri, Nishant Yadav, *Mechanical properties of rubber aggregates based geopolymer concrete - A review,* IOP Conference Series: Materials Science and Engineering, Volume 989, International Virtual Conference on Emerging Research Trends in Structural Engineering (2020)

[11] Yeonho park, Ali Abolmaali, Young Hoon Kim, Masoud Ghahremannejad, *Compressive strength of fly ash-based geopolymer concrete with crumb rubber partially replacing sand*, Construction and Building Materials Volume 118 (2016)

[12] Farnoosh Jokar, Mohammad Khorram,Gholamreza Karimi, Nader Hataf, *Experimental investigation of mechanical properties of crumbed rubber concrete containing natural Zeolite,* Construction and Building Materials Volume 208 (2019)

[13] Liang He, Yu Ma, Quantao Liu, Yuanhua Mu, *Surface modification of crumb rubber and its influence on the mechanical properties of rubber-cement concrete,* Construction and Building Materials Volume 120 (2016)

[14] IS 383: Standard specification for Coarse and Fine Aggregates From Natural Sources For Concrete, BIS New Delhi, India (1970)

[15] P.Pavithra, M.Srinivasula Reddy, Pasla Dinakar, B. Hanumantha Rao, B. K. Satpathy, A.N. Mohanthy, *A mix design procedure for geopolymer concrete with fly ash,* Journal of Cleaner Production volume 133(2016)

[16] X.Huan-xiu, A, Ning-jian, T.Hai-sheng, *Effect of polyorganosilaxane on solvent resistance of NR surface,* China Elastomerics 6 volume 002 (2007)

[17] Farnoosh Jokara, Mohammad Khorrama, Gholamreza Karimia, Nader Hataf, *Experimental investigation of mechanical properties of crumbed rubber concrete containing natural zeolite,* Construction and Building Materials volume 208, (2019)

[18] ASTM C666 2008c: Standard test method for resistance of concrete to rapid freezing and thawing, ASTM International, United States (2008)

IOP Conf. Series: Materials Science and Engineering 1132 (2021) 012039 doi:10.1088/1757-899X/1132/1/012039

Comparative analysis of an environmentally friendly refrigerant with DI water when used in a thermosiphon heat pipe

Vivek K.S[1], Arun Kurien Reji[2], Abdulriyazdeen A[3], Abishek S[4]

[1]School of Engineering and Technology, Christ University, Bengaluru
[2,3,4] Department of Mechanical Engineering, Sri Krishna College of Engineering and Technology, Tamil Nadu

E-mail: arunkurien99@gmail.com

Abstract. The focus on building better and efficient cooling techniques for electronic gadgets are gaining importance day-by-day. Among the various techniques being studied passive cooling techniques using heat pipes are gaining momentum; the challenge being finding a suitable working fluid which would enhance the performance of these devices. With the present scenario demanding eco-friendly alternatives, the researchers have tried using R600a, a hydrocarbon refrigerant with zero Ozone Depletion Potential. The optimum fill ratio of 50% was determined experimentally at an orientation of 90^0 from horizontal. Experiments were carried out to find the performance parameters of the thermosiphon heat pipe at the said fill ratio by varying the input heat from 20 W to 140 W with an increment of 20 W each. Next, the working fluid was replaced by de-ionized water and experiments were repeated. The results suggest that the resistance of the heat pipe thermosiphon reduced by 27% when R600a was used. Also, heat transfer coefficients at the evaporator and condenser section increased by a significant margin of 19% and 30% respectively when R600a was used in comparison to DI water.

1. Introduction

Cooling has become an integral part of our lives, be it for human comfort or for increasing the working life of devices. A scaling down in terms of size of devices has contributed to an exponential increase in heat flux, which has a detrimental effect on its life and performance. Advanced, effective and economic methods that are being researched on to expel the heat that is generated, has led to the invention of heat (carrying) pipes. They are hollow evacuated tubes made of metals and partially filled with a fluid. This fluid is continuously made to undergo evaporation and condensation process to take heat away from a source and remove it to a sink without any need for any external power input [1].

Heat pipes may be classified as thermosiphon heat pipes and heat pipes with wicks. Among the two types, the complexities and difficulties in operation of wick used in heat pipes have been overcome by the use of thermosiphon heat pipes [2]. This simple system comes with a configurational disadvantage wherein the condenser always has to be on top during the operation, and not the evaporator.

Ever since its invention, attempts have been made to vary the geometrical parameters, working fluids, materials etc. to suit various applications. Researchers have come up with different working fluids which would positively influence the working of heat pipes [2-14]. Among other working fluids, good thermodynamic properties of refrigerants, especially R134a, has prompted researchers to use it as heat transfer fluid[15]. The downside of using R134a is its high Global Warming Potential [16]. In a work that was reported, the performance of a thermosiphon heat pipe with porous copper coating charged with R134a showed reduction in resistance and increase in evaporator and condenser heat transfer coefficients by 11% and 25% respectively at 45^0 inclination of the thermosiphon heat pipe. In another work conducted with HFE7000 and nano-refrigerant HFE7000/Al_2O_3 on a thermosiphon, the heat transfer coefficient in evaporator was increased by 40.48% for 0.05% concentration of nano-refrigerant [17]. Eco-friendly nature characterized by low GWP and zero Ozone Depletion Potential have led researchers to turn their attention to R600a [18]. The authors have tried to compare the performance of R600a with DI water in thermosiphon heat pipe under different operating parameters.

2. Experimental section

The experimental layout of the test rig has been represented in Figure 1. The setup consists of a 500 W capacity wattmeter, Voltmeter, Ammeter, electric heater element, variable transformer, DAQ Agilent 34972A – data acquisition system, temperature sensors, fibre glass insulation and flow meter. The heat input to the evaporator section was given using the electric heating element. The heat supplied at this end would be absorbed by the working fluid and would be transported to the condenser end of the thermosiphon heat pipe while passing through the adiabatic section in the middle. A fibre glass insulation helped reduce the heat loss around the adiabatic section. The amount of heat transported to the condenser section was determined by passing cooling water around the section and noting the inlet and outlet temperatures. A 300 mm long thermosiphon with 100 mm long evaporator section, 80 mm adiabatic section and 120 mm long condenser section was used for the experimentation purpose. The thermosiphon had a thickness of 1.6 mm and an outer diameter of 12.7 mm. Data Acquisition system was used to record the surface temperatures of the thermosiphon and the inlet and outlet temperature of cooling water, all measured using 'K' type thermocouples.

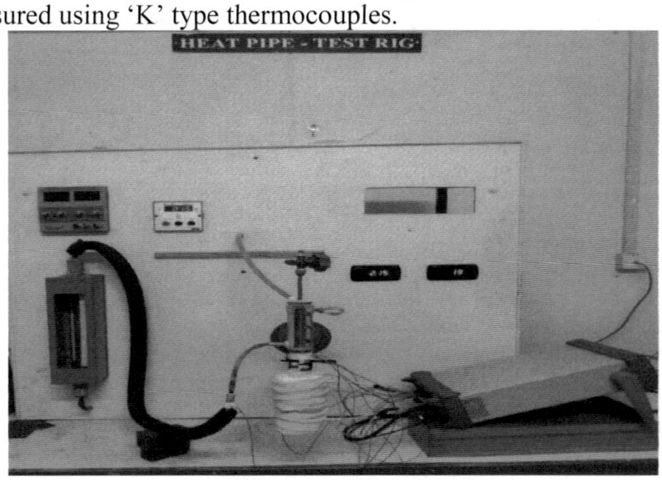

Figure 1. Test rig

The best fill ratio of a thermosiphon heat pipe is experimentally determined when charged with R600a. The performance of the thermosiphon charged at this best fill ratio is found out under varying operating conditions of heat load and has been compared to the performance of the thermosiphon charged to the same fill ratio with DI water under similar set of conditions. The experiments were run at different input values from 20 W to 140 W. Surface temperatures of various sections and inlet water temperature and outlet water temperatures of the condenser section were recorded. In order to attain steady state, the system was allowed to run for half an hour before readings were noted

3. Results and discussion

Like heat pipes, thermosiphon heat pipes also helps transport heat through a considerable distance from a source to a desirable sink, even if the temperature difference is not significantly great. Effectiveness of the device may be computed if fraction of heat transported from evaporator end to condenser end can

IOP Conf. Series: Materials Science and Engineering 1132 (2021) 012039 doi:10.1088/1757-899X/1132/1/012039

be measured. This can be found out by recording the difference in temperatures of cooling water between the outlet and inlet. The resistance (R) offered by the heat pipe to the flow of heat can be found out using the below mentioned relation:

$$R = \frac{T_e - T_c}{Q_{out}} \tag{1}$$

Here R, T_e, T_c, Q_{out} corresponds to resistance of thermosiphon, average surface temperature of evaporator, average surface temperature of condenser and heat rejected from the condenser section respectively. Heat rejected may be calculated using the relation:

$$Q_{out} = m\, C_p(T_o - T_i) \tag{2}$$

In the above equation, C_p, m, T_o and T_i represents to specific heat, mass flow rate, condenser cooling water outlet temperature and inlet temperature. The performance of the thermosiphon heat pipe was evaluated at different fill ratios of R600a starting from 40% to 70% with 10% increments while the thermosiphon heat pipe was aligned vertically. The best fill ratio was identified and other performance parameters were calculated at this fill ratio. Next, the refrigerant was replaced by de-ionized water and tests were repeated for computing performance of the thermosiphon at the best fill ratio corresponding to the refrigerant.

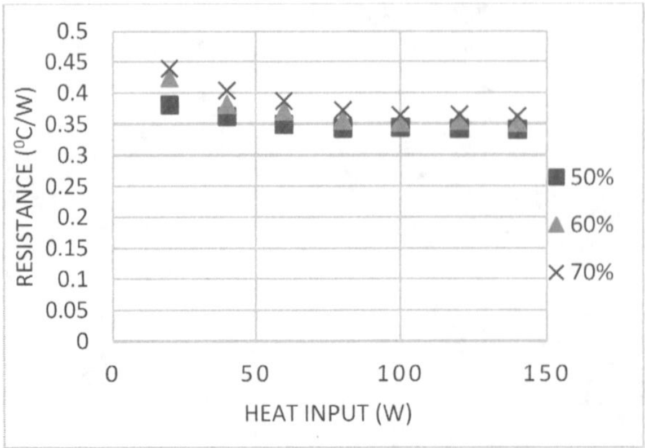

Figure 2. Variation of resistance under varying fill ratios

A 40% fill ratio resulted in dry-out condition, which suggests that the refrigerant was under filled. While comparing the resistance values at 20 W of heat input, the 60% fill ratio exhibited an increase in resistance by 9% when compared to 50% fill at the same heat input. Similarly, the resistance increased to 13% at 70% fill ratio at the same heat input. The results suggest that the 60% and 70% fill ratios might have flooded the evaporator with working fluid, which would have affected its working. The experimental results suggest that from among the various fill ratios, the thermosiphon heat pipe performed best when it was filled up to 50% of its volume. Hence, experiments were repeated with de-ionized water at 50% fill ratio. The graphical representation of the resistance at various fills have been plotted in Figure 2.

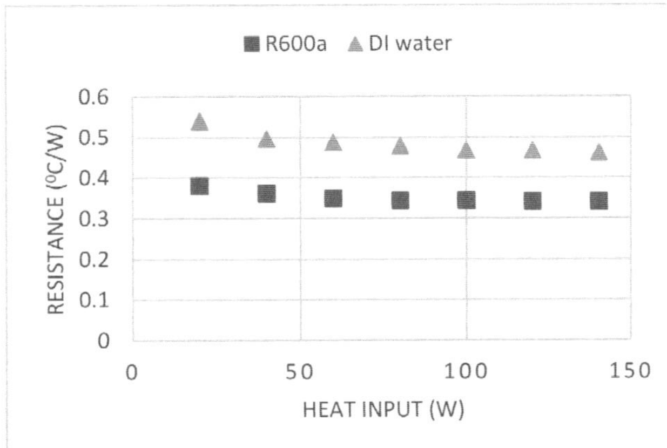

Figure 3. Resistance offered by the thermosiphon working with different fluids

Figure 3 shows the resistance offered to the flow of heat by the thermosiphon working with R600a as well as DI water at 50% fill ratio for various input heat values. From the general trend, it could be observed that the variation in resistance with the heat input exhibited by thermosiphon heat pipe working with DI water was in line with the trend exhibited by R600a. It was noted that the resistance value came down as heat input increased which complements to the findings of few other researchers [13,15]. The values suggest that the thermosiphon heat pipe exhibits better heat transfer capabilities at greater values of heat inputs.

Other parameters considered to analyse the heat transfer ability of the thermosiphon were condenser and evaporator heat transfer coefficients (h_c and h_e), which can be computed by:

$$h_c = \frac{q_c}{T_{ad} - T_c} \tag{3}$$

$$h_e = \frac{q_e}{T_e - T_{ad}} \tag{4}$$

The heat flux at condenser and evaporator sections are denoted by q_c and q_e respectively. T_e, T_{ad} and T_c refers to evaporator, adiabatic and condenser sections' surface temperature respectively.

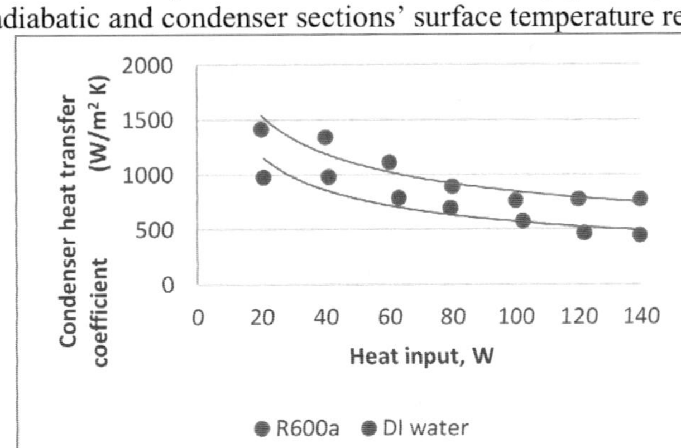

Figure 4. Variation of condenser heat transfer coefficient

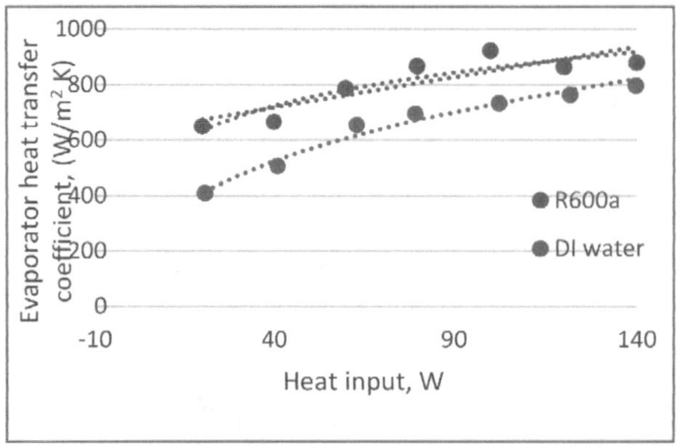

Figure 5. Variation of evaporator heat transfer coefficient

Figure 4 and Figure 5 represent the heat transfer coefficients at source as well as sink sections of thermosiphon for both R600a and de-ionized water. The observations suggest that heat transfer coefficients at sink or condenser section is more than that of the heat transfer coefficient at the evaporator section for both sets of working fluids. The smooth interior of the tube would have aided the easy recirculation of fluid back to the evaporator from condenser. Corrugations or surface roughness at the evaporator end would have enhanced the heat transfer coefficient at the evaporator end. It was also observed that as the heat input was increased, the evaporator heat transfer coefficient increased exponentially, more for R600a than de-ionized water. The shift from 'drop-wise' to 'film-wise' condensation at greater inputs would justify a gradual decrease of condenser heat transfer coefficient after an initial increase at 40 W. The graphs also suggest that the thermosiphon working with R600a exhibits superior performance in comparison to de-ionized water.

4. Conclusion

The optimum fill of R600a which delivered the best performance corresponded to 50% fill ratio. The mean value of resistance obtained at this fill was 0.352 W/m^2K which was than 5% and 9%. The working of the heat pipe was compared when the fluid was replaced by DI water. The results show that the even though the geometric specifications and the experimental conditions remain unchanged, thermosiphon heat pipe performance is remarkably affected by the changes in working fluid. Owing to better heat transfer capabilities, the mean resistance value when refrigerant was used as working fluid was 27% lesser when compared to DI water. The average evaporator heat transfer coefficient was 805.74 W/m^2 K when R600a was used. This value decreased to 652 W/ m^2 K when DI water was used marking a reduction of 19%. Similarly, the heat transfer coefficient at the condenser side was 1009.56 W/m^2K compared to 704.85 W/m^2 K when R600a and DI water were used respectively. The heat transfer coefficient also marked a decrease by 30% when DI water was used. This also focuses the strong need for using a working fluid with better thermos-physical properties in order to extract the best performance from the heat transfer device.

References

[1] Faghri Amir 2014 *Heat pipes: review, opportunities and challenges* Frontiers in Heat Pipes (FHP) **5**

[2] Mahfoud M and Emadi D 2005 *Application of heat pipe technology in thermal analysis of metals* Journal of thermal analysis and calorimetry **81** no. 1 pp 161-167

[3] Asirvatham Lazarus Godson, Rajesh Nimmagadda and Somchai Wongwises 2013 *Heat transfer performance of screen mesh wick heat pipes using silver–water nanofluid* International journal of heat and mass transfer **60** pp 201-209

[4] Senthilkumar R, Vaidyanathan S, and Sivaraman B 2012 *Experimental analysis of cylindrical*

heat pipe using copper nanofluid with an aqueous solution of n-Hexanol Frontiers in Heat Pipes *(FHP)* **2**, no. 3

[5] Kumaresan G and Venkatachalapathy S 2013 *A review on heat transfer enhancement studies of heat pipes using nanofluids* Frontiers in Heat Pipes *(FHP)* 3, no. 4

[6] Ramachandran Raghavan Nair, Kalliappan Ganesan and Lazarus Godson Asirvatham 2016 *The role of hybrid nanofluids in improving the thermal characteristics of screen mesh cylindrical heat pipes* Thermal science 20, no. 6 pp 2027-2035

[7] Hung Yi-Hsuan, Tun-Ping Teng and Bo-Gu Lin 2013 *Evaluation of the thermal performance of a heat pipe using alumina nanofluids* Experimental Thermal and Fluid Science 44 pp 504-511

[8] Kumaresan G, Venkatachalapathy S and Indraneel C N 2014 *An experimental study on improvement in thermal Efficiency of mesh wick heat pipe* Applied Mechanics and Materials, vol. 592, pp 1423-1427

[9] Ghanbarpour Morteza, Nikkam N, Rahmatollah Khodabandeh and Muhammet S T 2015 *Improvement of heat transfer characteristics of cylindrical heat pipe by using SiC nanofluids* Applied thermal engineering 90 pp: 127-135

[10] Kumaresan G, S Venkatachalapathy, Lazarus Godson Asirvatham, Somchai Wongwises 2014 *Comparative study on heat transfer characteristics of sintered and mesh wick heat pipes using CuO nanofluids* International Communications in Heat and Mass Transfer 57 pp: 208-215

[11] Reji A K, Kumaresan G, Sarathi A, Arasappa Saiganesh GP, Suriya Kumar R and Matthew M.S 2020 *Performance analysis of thermosyphon heat pipe using aluminum oxide nanofluid under various angles of inclination* Materials Today: Proceedings

[12] Solomon A B, Arul Daniel V, Ramachandran K, Pillai B C, Renjith Singh R, Sharifpur M and Meyer J P 2017 *Performance enhancement of a two-phase closed thermosiphon with a thin porous copper coating* International Communications in Heat and Mass Transfer 82 pp 9-19

[13] Alammar Ahmed A., Raya Al-Dadah K and Saad Mahmoud 2016 *M Numerical investigation of effect of fill ratio and inclination angle on a thermosiphon heat pipe thermal performance* Applied Thermal Engineering 108 pp: 1055-1065

[14] Senthilkumar C, A S Krishnan and Solomon A B *2020 Effect of thin porous copper coating on the performance of wickless heat pipe with R134a as working fluid* Journal of Thermal Analysis and Calorimetry 139, no. 2 pp: 963-973

[15] Platzer Bernhard, Gerd Maurer and Axel Polt 1990 Thermophysical properties of refrigerants

[16] Mohanraj M, Jayaraj S and Muraleedharan C 2009 *Environment friendly alternatives to halogenated refrigerants—A* review International Journal of Greenhouse Gas Control 3, no. 1 pp: 108-119

[17] Anand R. S, Jawahar C P, Solomon A B, Varghese Benson, Ashie Alan K, Vignesh Nair K P and Abraham Alan V 2020 *Experimental studies on thermosyphon using low global warming potential refrigerant HFE7000 and nanorefrigerant HFE7000/*Al_2O_3 Proceedings of the Institution of Mechanical Engineers *Part E: Journal of Process Mechanical Engineering,* 0954408919896690

[18] Reji A K, Kumaresan G, Sarathi A, Arasappa Saiganesh GP, Suriya Kumar R and Matthew M.S 2020 *Analysis of thermosiphon working with eco-friendly refrigerant under various fill ratios* Materials Today: Proceedings

IOP Conf. Series: Materials Science and Engineering 1132 (2021) 012040 doi:10.1088/1757-899X/1132/1/012040

Effect of wall-slip on natural convection in a square cavity

Kartik Suresh*, C K Arjun, Damu Murali and S Ajith Kumar

Department of Mechanical Engineering, Amrita Vishwa Vidyapeetham, Amritapuri, India.

*E-mail: sureshkartik6@gmail.com

Abstract. Numerical simulations are performed to investigate the effect of wall-slip on natural convection inside a square cavity for Rayleigh number (Ra) = 10^5. The bottom wall is heated and top wall is maintained at a cold temperature keeping the other two vertical walls adiabatic. The analysis is conducted by varying the $Slip\ factor$ (SF) from 0 (no-slip) to 1 (full-slip) in steps of 0.2. For each value of SF, the flow patterns are visualised using streamlines and heat transfer behaviour is examined using isotherms. Results are quantified using local Nusselt number (Nu_l) and average Nusselt number (Nu_{avg}) along the hot wall. It is observed that for SF=0.8 and SF=1 configurations, the natural convection patterns and heat transfer characteristics undergo notable changes.

1. Introduction

Natural convection in a square cavity is a highly researched topic because of the simple geometry and wide range of applications. Numerous studies have been conducted to understand the importance of convective heat transfer in the fields of atmospheric sciences to real world engineering problems like design of nuclear reactors and cooling of electronic devices.

One of the pioneer researchers, Vahl Davis[1] numerically investigated buoyancy-driven flow in a square cavity with differentially heated vertical walls. Wan et al.[2] employed a high-accuracy discrete singular convolution (DSC) method for simulating natural convection in a square cavity in the range $10^3 \leq \text{Ra} \leq 10^8$. These investigations are considered to be benchmarks for natural convection in a square enclosure in the laminar regime. Buoyancy driven laminar and turbulent heat transfer in a square cavity with differentially and partially heated walls was studied by Markatos and Pericleous[3], the study being important in understanding the thermal and dynamic effects of natural convection in many practical problems. Valencia and Frederick[4] performed investigations on natural convection in a square enclosure with half heated and half insulated vertical walls and observed that at higher Ra, flow features near to the unheated vertical walls exhibited significant changes owing to the absence of buoyancy forces.

Laje and Bejan[5] investigated the phenomenon of natural convection in a square cavity for a range of Prandtl number ($0.01 \leq \text{Pr} \leq 10$) and Rayleigh number ($10^2 \leq \text{Ra} \leq 10^{11}$). In this study, attention was given to find out the highest Ra where inertial fluctuations were possible in the steady laminar regime. Hasnaoui et al.[6] discussed buoyancy driven flow in an inclined cavity bounded by a porous layer. Another study by Dalal and Das[7] focused on the behaviour of flow in the laminar regime with sinusoidally heated bottom wall and other walls kept at constant cold temperature. Basak et al.[8] studied laminar natural convection flow with uniform and non-uniform heating of the bottom wall. Numerical investigation of steady natural convection in a square cavity was conducted by Sathiyamoorthy et al.[9] with insulated top wall and sinusoidally heated bottom wall and linearly heated vertical walls. They

IOP Conf. Series: Materials Science and Engineering 1132 (2021) 012040 doi:10.1088/1757-899X/1132/1/012040

Nomenclature

g	Acceleration due to gravity $[m/s^2]$
L	Enclosure length [m]
Nu_l	Local Nusselt number
Nu_{avg}	Average Nusselt number
P	Non-dimensional pressure
Pr	Prandtl number ($\frac{\nu}{\alpha}$)
Ra	Rayleigh number ($\frac{g\beta L^3(T_h-T_c)}{\nu\alpha}$)
SF	Slip factor
T	Temperature [K]
T_c	Cold wall temperature [K]
T_h	Hot wall temperature [K]
T_{ref}	Reference temperature [K]
U	Non-dimensional horizontal velocity
u	Horizontal velocity $[m/s]$
V	Non-dimensional vertical velocity
v	Vertical velocity $[m/s]$
X	Non-dimensional horizontal coordinate
x	Horizontal coordinate
Y	Non-dimensional vertical coordinate
y	Vertical coordinate

Greek Symbols

α	Thermal diffusivity $[m^2/s]$
β	Coeffcient of thermal expansion $[K^{-1}]$
λ	Slip length
ν	Kinematic viscosity $[m^2/s]$
Θ	Non-dimensional temperature($\frac{T-T_c}{T_h-T_c}$)

compared the results by varying the Prandtl number (Pr) and aspect ratio (AR) and found out that multiple secondary circulations were formed at lower Pr values whereas Pr ≥ 0.71 witnessed only a pair of secondary circulations. Kane et al.[10] studied the behaviour of boundary layer in natural convection in a square cavity. The square enclosure they considered had the central portion of the vertical walls kept at different temperatures. The remaining parts of the vertical walls and the top and bottom walls were kept insulated and the effect of Rayleigh number was studied.

Various other studies (Devaraj et al.[11], Zhang et al.[12], Cibik and Kaya[13]) have been conducted on natural convection in a square enclosure applying different boundary conditions. However, attention is not given to understand the influence of more complex boundary conditions such as slip boundary conditions, hence it is essential to carry out such a study. The application of this problem can be extended to industries that work with hydrophobic surfaces and modern-day engineering applications like MEMS devices. This paper discusses the effect of slip boundary condition on natural convection in a square cavity at Rayleigh number (based on the width of the cavity) 10^5 by varying the slip boundary condition from no-slip (SF=0) to full-slip (SF=1).

The structure of the presentation of our observations in the rest of the paper is as follows. Solution method in Section 2 describes the solver details, computational domain and the boundary conditions. The results are explained in detail in Section 3 which contains grid sensitivity studies followed by code validation. Flow and thermal patterns are visualised using streamlines and isotherms and heat transfer characteristics are quantified using local Nusselt number (Nu_l) and average Nusselt number (Nu_{avg}) for different configurations of slip conditions. Conclusions in Section 4 summarises the overall numerical findings.

2. Solution method

2.1. Numerical methodology

Numerical simulations are carried out using $buoyantBoussinesqPimpleFoam$, the generic solver available in the open source computational fluid dynamics toolbox OpenFOAM. A second order central differencing $Gauss\ linear$ scheme is used for spatial discretization of convective and diffusive terms. The pressure-velocity coupling at each iteration is handled by PIMPLE algorithm. With the Boussinesq approximation, fluid properties are assumed to be constant except the fluid density which is obtained by 1.0 -$\beta(T - T_{ref})$ where, β is the coefficient of thermal expansion, T is the temperature and T_{ref} being

IOP Conf. Series: Materials Science and Engineering 1132 (2021) 012040 doi:10.1088/1757-899X/1132/1/012040

the reference temperature.

The system is governed by the PDEs corresponding to mass, momentum and energy conservation, given as follows:

$$Continuity : \frac{\partial U}{\partial X} + \frac{\partial V}{\partial Y} = 0 \tag{1}$$

$$X - momentum : U\frac{\partial U}{\partial X} + V\frac{\partial U}{\partial Y} = -\frac{\partial P}{\partial X} + \Pr\left(\frac{\partial^2 U}{\partial X^2} + \frac{\partial^2 U}{\partial Y^2}\right) \tag{2}$$

$$Y - momentum : U\frac{\partial V}{\partial X} + V\frac{\partial V}{\partial Y} = -\frac{\partial P}{\partial Y} + \Pr\left(\frac{\partial^2 V}{\partial X^2} + \frac{\partial^2 V}{\partial Y^2}\right) + RaPr\Theta \tag{3}$$

$$Energy : U\frac{\partial \Theta}{\partial X} + V\frac{\partial \Theta}{\partial Y} = \frac{\partial^2 \Theta}{\partial X^2} + \frac{\partial^2 \Theta}{\partial Y^2} \tag{4}$$

The equations are non-dimesionalised as below:

$$X = \frac{x}{L}, \quad Y = \frac{y}{L}, \quad U = \frac{uL}{\alpha}, \quad V = \frac{vL}{\alpha}, \quad \Theta = \frac{T - T_\mathrm{c}}{T_\mathrm{h} - T_\mathrm{c}}, \quad P = \frac{pL^2}{\rho\alpha^2} \tag{5}$$

Here, X and Y represent non-dimensional horizontal and vertical coordinates. Similarly U and V are horizontal and vertical non-dimensional velocities. Θ is the non-dimensional temperature, T_c is the cold wall temperature and T_h is the temperature of the hot wall. α represents the thermal diffusivity, L denotes the length of the cavity and P is used to represent the non-dimensional pressure.

Rayleigh number based on the enclosure width is defined as $Ra = \frac{g\beta L^3(T_\mathrm{h}-T_\mathrm{c})}{\nu\alpha}$ and Prandtl number as $Pr = \frac{\nu}{\alpha}$ where, ν is the kinematic viscosity of the fluid, g is the acceleration due to gravity acting downwards in the negative y direction.

The rate of heat transfer is computed at each wall and is expressed in terms of local Nusselt number (Nu_l) and surface-averaged Nusselt number (Nu_{avg}).

$$Nu_l = \left.\frac{\partial \Theta}{\partial n}\right|_{\text{wall}}, \quad Nu_{avg} = \frac{1}{W}\int_0^W Nu_l \, \mathrm{d}S \tag{6}$$

where *n* is the unit normal to the wall, *dS* is the small elemental area and *W* is the surface area of the wall.

2.2. Computational domain and boundary conditions

Figure 1(a) illustrates the schematic of the computational domain which consists of a square enclosure with width L. The top and bottom walls are differentially heated and maintained at cold temperature (Θ=0) and hot temperature (Θ=1) respectively whereas the vertical walls are adiabatic.

Slip conditions on the walls are varied using the $partialSlip$ boundary condition (BC) available in OpenFOAM. $partialSlip$ represents a mixed boundary condition which is an implementation of the Navier slip law as expressed below. For a 2-D flow, the first order approximation for the slip-velocity is given by

$$U + \lambda\frac{dU}{dY} = 0 \tag{7}$$

here λ is the slip length as defined in Legendre et al.[14]

Let U_0 be the tangential velocity on the wall face boundary and U_1 the cell center velocity at the boundary cell. Equation (7) can be re-written in terms of U_0, U_1 and D as follows

$$U_0 + \lambda \frac{(U_0 - u_1)}{D/2} = 0 \tag{8}$$

where D is the size of the adjacent cell. The above equation after re-arranging becomes

$$U_0 = \frac{2\lambda}{D + 2\lambda} U_1 \tag{9}$$

Here, we introduce a new term $Slip\ factor\ (SF)$ defined as

$$SF = \frac{2\lambda}{D + 2\lambda} \tag{10}$$

Hence equation (9) becomes

$$U_0 = (SF)U_1 \tag{11}$$

For $\lambda \to 0$, we have $SF = 0$, which indicates no-slip condition.
For $\lambda \to \infty$, we have $SF = 1$, which indicates full-slip condition.

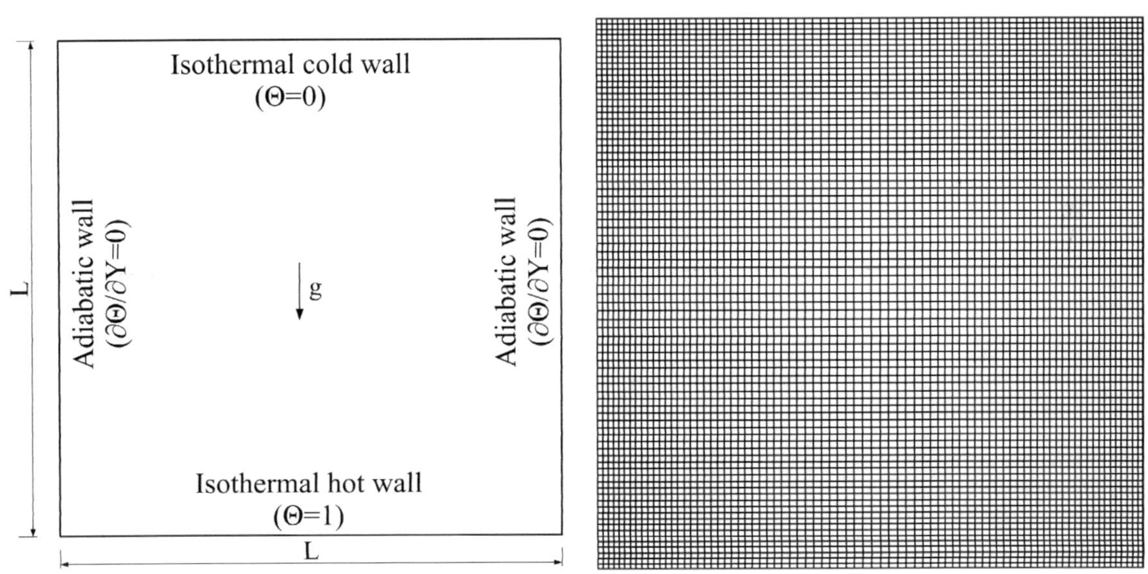

Figure 1: Schematic of (a) computational domain (b) grid structure used for the present study.

Conditions imposed at the walls other than the slip boundary conditions are as follows:

Left wall :Neumann boundary condition, $\frac{\partial \Theta}{\partial Y} = 0$
Top wall : Dirichlet boundary condition, Θ=0
Right wall :Neumann boundary condition, $\frac{\partial \Theta}{\partial Y} = 0$
Bottom wall : Dirichlet boundary condition, Θ=1

IOP Conf. Series: Materials Science and Engineering 1132 (2021) 012040 doi:10.1088/1757-899X/1132/1/012040

3. Results and discussion

3.1. Grid sensitivity

Grid sensitivity studies are performed to optimise the grid size for accurate results and to minimise the computational time. The analysis is conducted for grid sizes of 21×21, 41×41, 61×61, 81×81 and 101×101, for Rayleigh number 10^5 with no-slip boundary condition. Table 1 given below summarises the grid independence results for the average Nusselt number (Nu_{avg}) at the hot wall. The percentage deviation (shown in brackets) reduces with grid size upto 81×81, and further increase in grid size to 101×101 is accompanied by an increase in percentage difference. It can be attributed to the increase in numerical errors (truncation error, round-off error and iterative convergence error) with increase in grid size. Hence, a grid size of 81×81 is employed throughout our computations.

Table 1: Grid sensitivity study: Ra=10^5, Pr=0.71, SF=0; The percentage difference (absolute value) in successive cases are indicated in brackets.

Grid size	Nu_{avg}
21×21	4.820 (-)
41×41	4.607 (4.419%)
61×61	4.565 (0.912%)
81×81	4.559 (0.131%)
101×101	4.534 (0.548%)

3.2. Validation

The computational code used for the present study is validated using the benchmark solutions available in the existing literature.

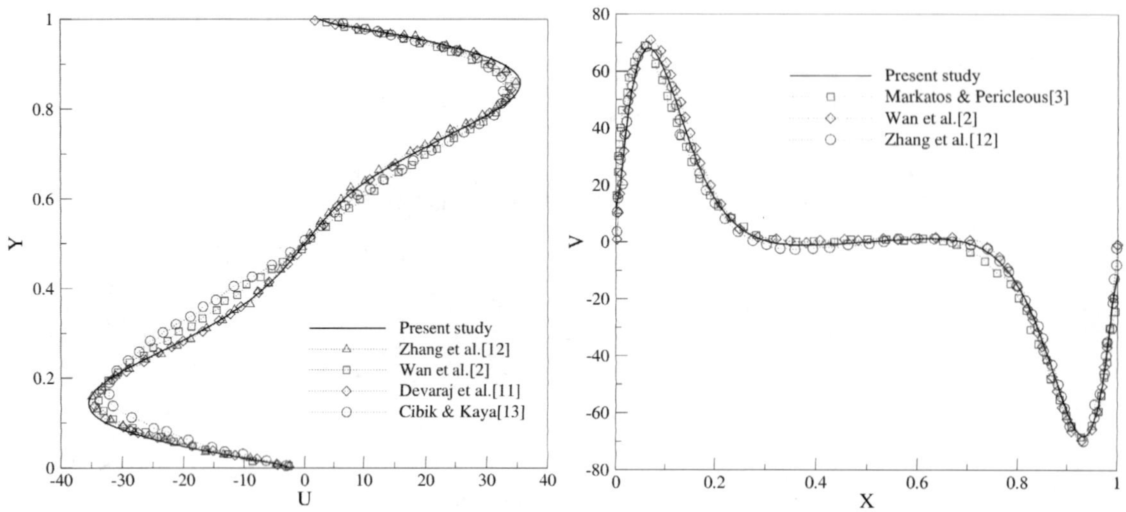

(a) (b)

Figure 2: Comparison of (a) horizontal velocity at X=L/2 (b) vertical velocity at Y=L/2.

Figures 2(a) and (b) illustrate the comparison of velocity distribution along the horizontal and vertical axes passing through the centre of the domain.

IOP Conf. Series: Materials Science and Engineering 1132 (2021) 012040 doi:10.1088/1757-899X/1132/1/012040

Table 2: Validation study: Comparison of Nu_{avg} values with the existing literature; The percentage deviation (absolute value) of calculated values with the literature is indicated in brackets.

Ra	Present Study	Wan et al.[2]	Vahl Davis[1]	Tang & Tsang[15]	le Quere & Roquefortt[16]
10^3	1.109	1.117 (0.716%)	1.12 (0.982%)	1.118 (0.805%)	1.117 (0.716%)
10^4	2.246	2.254 (0.355%)	2.243 (0.134%)	2.25 (0.178%)	2.238 (0.357%)
10^5	4.559	4.598 (0.848%)	4.519 (0.885%)	4.52 (0.863%)	4.519(0.885%)

Table 2 compares the results obtained for average Nusselt number(Nu_{avg}) on the hot wall for the present study with the data in previous literature and the maximum deviation is found to be less than 1%. The above observations conclude that the present results are in good agreement with the existing data.

3.3. Streamlines

Figure 3 illustrates the streamlines for Ra = 10^5 for varying slip conditions. At no-slip condition (SF = 0), a unicellular primary vortex rotating clockwise is seen to occupy majority of the domain. It is also observed that in addition to the primary vortex, two tiny secondary vortices inhabit the top-left and bottom-right corners of the domain. These secondary vortices are seen to be circulating in a direction opposite to the primary vortex. As the SF value is increased, the recirculating secondary vortices are seen to gradually reduce in size giving way to the dominance of the primary vortex and this continues upto SF=0.6.

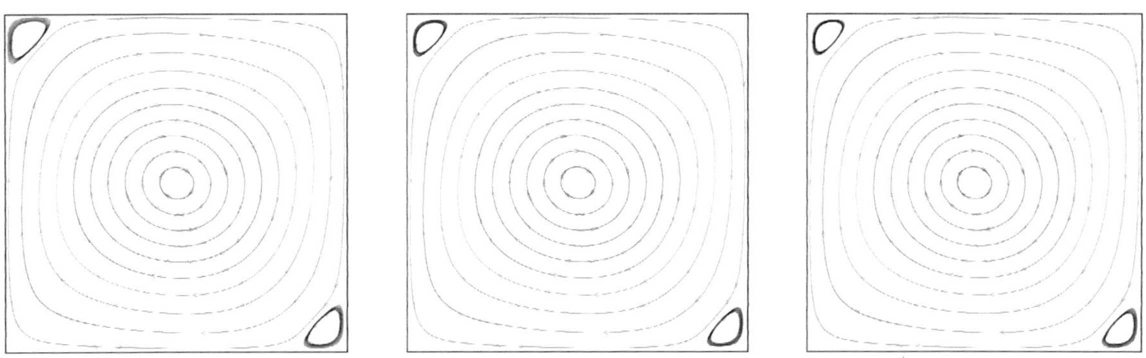

(a) SF=0.0 **(b)** SF=0.2 **(c)** SF=0.4

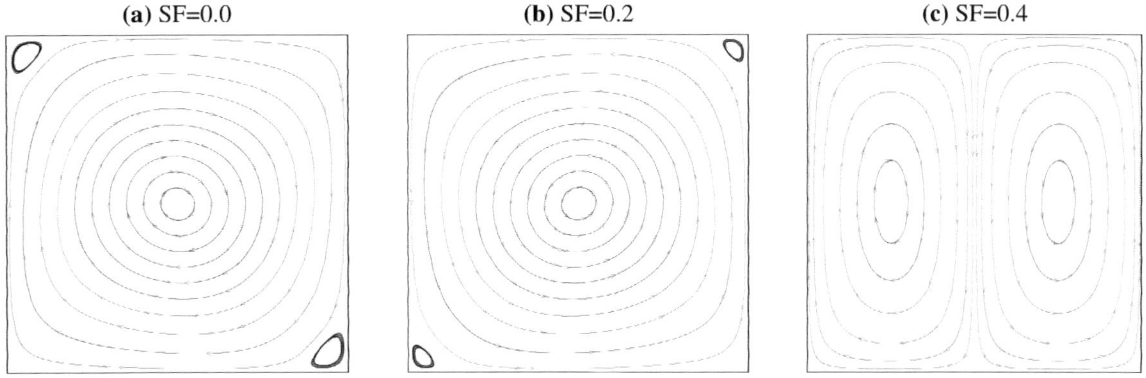

(d) SF=0.6 **(e)** SF=0.8 **(f)** SF=1.0

Figure 3: Streamlines for various slip configurations plotted at Ra =10^5.

IOP Conf. Series: Materials Science and Engineering 1132 (2021) 012040 doi:10.1088/1757-899X/1132/1/012040

At SF=0.8, the primary vortex suddenly reverses the direction of circulation and begins to rotate in the counter-clockwise direction. At the same time, the tiny secondary vortices change its position as well as direction and appear in the top-right and bottom-left corners. The size of the secondary vortices is also seen to decrease compared to the previous slip configurations. The influence of abrupt change in the flow fields bring about significant changes in the temperature fields which will be discussed in the next section.

At full-slip (SF=1), the primary vortex splits into two counter rotating vortices that is seen to occupy the entire domain and the secondary vortices cease to exist.The streamline pattern at SF=1 is similar to the pattern reported for a square enclosure by Basak et al.[17] with cold vertical walls, hot bottom wall and insulated top wall however, the case used no-slip boundary condition on the walls. The existence of similar flow fields for entirely different boundary conditions requires further exploration.

3.4. Isotherms

Existing literature report strong convection currents at Ra=10^5 and discrete heating of top and bottom walls causes strong density variations in the domain. The strong convection currents influence the separation of boundary layer on the bottom wall and subsequent development of plumes. These hot plumes ascend and impinge on the cold wall and get distorted. Visual inspection reveals that for no-slip condition (SF=0), the plumes get initialised near to the edges of the bottom wall and ascend through the vertical adiabatic wall. Impingement of these hot plumes on the cold top wall ejects the cold sub-layers

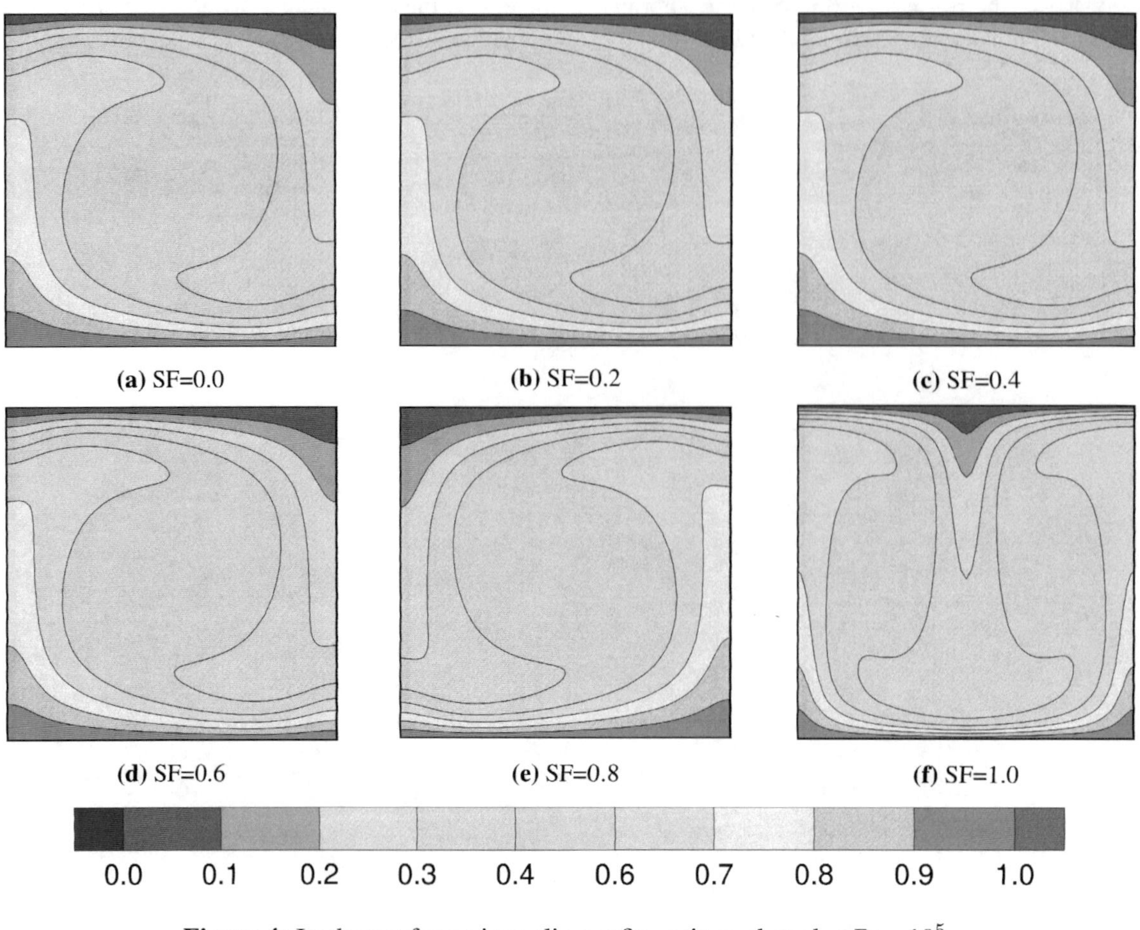

Figure 4: Isotherms for various slip configurations plotted at Ra =10^5.

IOP Conf. Series: Materials Science and Engineering 1132 (2021) 012040 doi:10.1088/1757-899X/1132/1/012040

which gives rise to a cold plume which later moves down towards the hot bottom wall. The combined influence of viscous forces and strong convection currents promote continuous advancement of the hot plume through the left adiabatic wall which pushes down the cold plume through the right adiabatic wall. The same phenomenon is seen to occur with increase in SF upto 0.6 (Figures 4(a)-(d)). At SF=0.8, similar to the flow reversal observed while visualising streamlines, the thermal fields too undergo an abrupt transition. As evidenced in Figure 4(e), the advancement of hot plume through the right adiabatic wall distorts the cold plume and paves way for its descend through the left wall. For the full-slip (SF=1) configuration, the hot plumes symmetrically advance along the vertical walls thus giving uniformity to the thermal fields.

3.5. Variation in Nusselt number

Nusselt number is an important non-dimensional parameter that describes the prominent mode of heat transfer; conductive or convective.

Figure 5(a) elucidates the variation in local Nusselt number (Nu_l) along the hot bottom wall. Moving from A to B, for SF=0, the variation in Nu_l is found to be increasing upto X=0.8 after which it decreases towards the end. This is the general trend for slip configurations upto 0.6. As evidenced in the isotherms (Figures 4(a)-(d)), the boundary layer becomes thinner from left to right and thereafter it thickens slightly towards the end. The maximum value of local Nusselt number is registered where the boundary layer thickness is minimum. The trend reverses when SF=0.8 as the boundary layer is stuffed closer to each other near to the left corner, A. The boundary layer thickness increases while moving towards B which is accompanied by a decrease in heat transfer rate as evidenced in the Nu_l curve. This variation can be related to the reversal in flow and thermal fields as illustrated in Figures 3(f) and 4(f). At full-slip condition, the Nu_l curve becomes symmetric to the vertical axis passing through the centre of the domain which is due to the uniformity in flow and thermal fields on either side of the vertical axis. Also, the peak value of is observed at mid-plane. Maximum value for local Nusselt number (Nu_l) is observed at full-slip boundary condition (SF=1) and as SF increases, the maximum value of Nu_l also increases. It is observed that the average Nusselt number (Nu_{avg}) along the hot wall of the enclosure increases with increase in slip. This may be a result of the reducing influence of viscous forces thereby creating strong

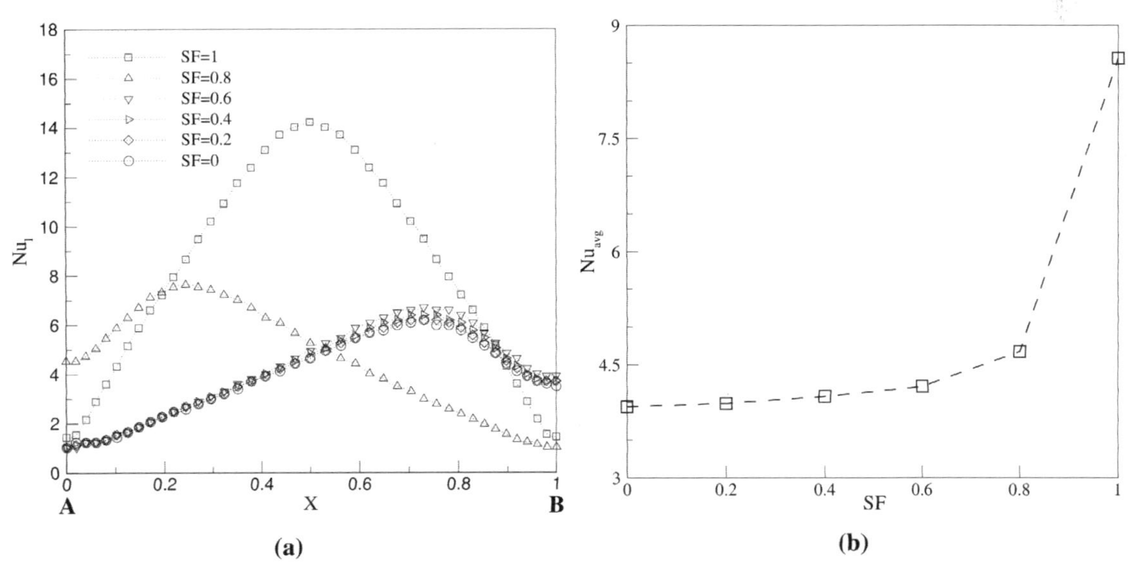

Figure 5: Nusselt number variation : (a) Nu_l variation along the hot wall and (b) Nu_{avg} variation with slip factor.

convection currents in the domain. The variation in Nu_{avg} is gradual and increasing when the slip factor is increased to 0.6. Between SF = 0.6 and 0.8, there is appreciable rise in average Nusselt number and further increase in slip (SF = 0.8 to SF = 1.0) registers a steep increase in the Nu_{avg} as the viscous effects become negligible. Between SF=0 (no-slip) and SF=1 (full-slip) there is an enhancement of nearly 117% in average Nusselt number.

4. Conclusions

Numerical investigations are conducted to analyse the effect of varying slip boundary condition on natural convection in a square cavity. It is observed that changing the slip factor (SF) from 0 to 0.6 brings marginal variation in flow and thermal fields as well as the heat transfer characteristics. For SF=0.8 and full-slip (SF=1) condition, significant changes are observed in the convection patterns. The other inferences are summarised as follows.

- For the slip configurations ranging from SF=0 (no-slip) to SF=0.6, visual inspection reveals a clockwise rotating primary vortex along with two secondary vortices at the top-left and bottom-right corners. For SF=0.8, this primary vortex reverses its direction and the secondary vortices shift to the top-right and bottom-left corners.
- At full-slip condition (SF=1), the primary vortex breaks into two counter rotating vortices and the secondary vortices disappear.
- Between SF=0 and SF=0.6, the combined influence of convection currents and viscous forces promote the advancement of hot plume along the left adiabatic wall. At SF=0.8, the thermal fields undergo a sudden transition causing the hot plume to ascend through the right adiabatic wall.
- For full-slip (SF=1) configuration, the hot plumes symmetrically advance along the vertical walls thus giving uniformity to the thermal fields.
- Between SF=0 and SF=0.8, the variation in local Nusselt number is asymmetric over the hot wall and the maximum value of Nu_l increases with increase in slip factor. Symmetric variation is observed only at SF=1 (full-slip) for which the peak value is the maximum among all the slip configurations considered.
- The results obtained for average Nusselt number (Nu_{avg}) for varying slip conditions show a marginal increase between SF=0 and SF=0.6, becomes appreciable towards SF=0.8 followed by a steep rise for the full-slip configuration.

References

[1] de Vahl Davis G 1983 *International Journal for Numerical Methods in Fluids* **3** 249–264
[2] Wan D C, Patnaik B S V and Wei G W 2001 *Numerical Heat Transfer, Part B: Fundamentals* **40** 199–228
[3] Markatos N C and Pericleous K A 1984 *International Journal of Heat and Mass Transfer* **7** 755–772
[4] Valencia A and Frederick R L 1989 *International Journal of Heat and Mass Transfer* **32** 1567–1574
[5] Lage J L and Bejan A 1991 *Numerical Heat Transfer, Part A: Applications* **19** 21–41
[6] Hasnaoui M, Vasseur P and Bilgen E 1993 *International Journal of Numerical Methods for Heat and Fluid Flow* **3** 91–105
[7] Dalal A and Das M K 2006 *Numerical Heat Transfer, Part A: Applications* **49** 301–322
[8] Basak T, Roy S and Balakrishnan A R 2006 *International Journal of Heat and Mass Transfer* **49** 4525–4535
[9] Sathiyamoorthy M, Basak T, Roy S and Mahanti N C 2007 *Journal of Heat Transfer* **129** 1723–1728
[10] Kane M K, Mbow C, Sow M L and Sarr J 2017 *Open Journal of Fluid Dynamics* **7** 623–641
[11] Devaraj C, Muthuswamy E and Kandasamy S 2015 *Journal of Applied Mathematics* **2015** 1–9
[12] Zhang T, Zhao X and Huang P 2015 *Numerical Algorithms* **68** 837–866
[13] Çıbık A and Kaya S 2011 *Journal of Mathematical Analysis and Applications* **381** 469–484
[14] Legendre D, Lauga E and Magnaudet J 2009 *Journal of Fluid Mechanics* **633** 437–447
[15] Tang L Q and Tsang T H 1993 *International Journal for Numerical Methods in Fluids* **17** 271–289
[16] Quere P L and Roquefortt T A D 1985 *Journal of Computational Physics* **57** 210–228
[17] Basak T and Roy S 2008 *International Journal of Heat and Mass Transfer* **51** 3486–3503

Effect of Baffles Position on Thermo-Hydraulic Efficiency of a Solar Air Heater

Rahul Kumar Bhatter[1,*], Ali Abbas[2] and Anant Kumar Rai[3]

[1,3]Department of Mechanical Engineering, National Institute of Technology, Warangal, India
[2]Department of Mechanical Engineering, Madanapalle Institute of Technology & Science, Madanapalle, A.P, India
[*]Email: rahul_193713@student.nitw.ac.in

Abstract. Providing artificial irregularity in the form of baffles on the bottom plate of a solar air heater improves the thermal performance of the heater, whereas it reduces the hydraulic efficiency due to friction losses. To optimize the thermo-hydraulic performance of a rectangular duct solar air heater, a thermo-hydraulic performance factor (THPF), dependent on shape, position and size of baffles, is considered in this study. The performance of the heater in various configurations related to the position of the baffles was evaluated considering Renormalization group k-Ɛ model to simulate the fully turbulent flow for a forced convection condition. At a constant value of heat flux of 1000 W/m^2 on the absorber plate and a fixed value of Reynolds number as 2350, pitch between the baffles was varied from 15mm to 75mm. Further, the distance of 1st row of baffles from the inlet was varied from 200mm to 500mm and the optimum configuration was figured out. For pitch value of 60mm between the baffles and the distance of 1st row of baffles as 200mm from the inlet, the highest value of THPF was obtained as 0.9339. The fluid behavior in presence of baffles was analyzed together with its influence on the thermo-hydraulics parameters.
Keywords: Solar air heaters, Baffles, Nusselt number, Skin friction coefficient, Reynolds number

1. Introduction

Twenty first century is looking towards the renewable sources of energy for fulfilling its energy demands sustainably. The day-to-day rising prices of energy obtained using fossil fuels, depletion in supply of fossil fuel and rising environmental health impact concern are changing the global energy scenario. Establishing renewable energy as a means to supply the growing demand of the rising population of the world is one of the greatest challenge. It is estimated that the world's population will be about 10 billion by the end of twenty-first century [1]. The increasing demand for electric vehicles in the future plays a crucial role to explore the charging of the electric vehicle batteries using renewable sources of energy such as solar energy. Solar energy is perennial, abundantly and freely available [2, 3, 4]. Harnessing the renewable solar energy is important in supplying both thermal and electricity demand as solar radiation can be converted to heat or electricity. The technology to convert solar energy into either heat or electrical energy is available and improving rapidly [5, 6]. Due to the advantages like ease of fabrication, low cost, collection of beam and diffuse radiation as well as less maintenance, the flat plate solar air collectors are one of the most commonly used technology for capturing solar radiation [2, 4, 6]. However, the major challenge with the use of flat plate solar collectors is its considerably low efficiency due to the poor heat transfer characteristics in the system.

To increase the efficiency of the collector, researchers provided several ways like use of double or multiple pass ducts instead of using single pass ducts, incorporating baffles in the duct etc. [6, 7]. In case of turbulent flow over a smooth heated duct, the rate at which heat is transferred to the surrounding air is significantly low due to viscous sub-layer present near wall of the duct. The heat transfer through this sub-layer is dominated by conduction mode whereas; convective mode of heat transfer is reduced. The viscous sub-layer is disturbed by roughening the surface using baffles, surface modifications etc. [7, 8]. In such attempt, Bensaci et al. [9] provided the numerical results of thermal and hydraulic performance of a solar air heater with different positions of the baffle in the entire length of the duct. In their study, the positions of the baffles were changed and found that the thermal as well as hydraulic performance depends significantly on the position of baffles in addition to the shape, size or other geometric parameters of the baffles. However, they only considered four configurations with respect to the baffle positions and no change in inter space between baffles were considered. Thus, some more study is required considering the position of baffles. In present research, the baffles placement is varied from the inlet and the inter baffle spacing is also varied to obtain an optimum value of thermo-hydraulic performance factor.

Menasria et al. [10] performed a 2-dimensional numerical study on solar air heater with bottom plate containing rectangular shaped baffles having upper part in an inclination position. Keeping the heat flux value constant at 1000 W/m^2, they studied the combination of four different non-dimensional values of baffle blockage ratio (B_R) between 0.7 to 0.98 and four different non-dimensional values related to baffle pitch (P_R) between 2 to 8. The Reynolds number range between 4000 and 18000 was studied and found that the arrangement with B_R = 0.7, P_R = 2 at Re = 5000 provides the highest value of thermohydraulic performance factor (THPF) – a factor representing the combined effect of thermal and fluid dynamic performance as 0.857. This optimum configuration leads to improvement in both heat transfer and friction factor by 2.16 and 15.95 times, respectively compared to a smooth duct configuration. They concluded that THPF depends on pitch, height, shape and position of the baffles. Prasad and Saini [11] observed consequences of the non-dimensional height (e/D) and non-dimensional pitch (P/e) on the frictional as well as the heat transfer characteristics of flow in the duct of a solar air heater by providing the irregularities on the bottom surface of the absorber plate. They obtained design curves pertaining to the flow parameters and the roughness for obtaining the optimum value of THPF.

Karwa et al. [12] carried out experiments on a rectangular solar air duct having rectangular cross section artificial ribs arranged in both V-shaped patterns. They observed that the thermo-hydraulic performance of V-down ribs were better than the V-up ribs considering equal amount of pumping power. Similarly, the effects of discrete W-shaped ribs for improving the performance of an absorber plate was also explored [13]. An increase in the thermal performance by a factor of 1.2-1.8 was observed for this arrangement of the duct compared to a smooth configuration of the duct for the range of parameters considered. Lanjewar et al. [14] performed experiments to study the thermal and hydrodynamic performance of a rectangular shape duct fitted with W-shaped ribs roughness on its underside on a wall, which is kept leaning to the flow path. For an angle of attack of 60^0, the highest improvement in the performance parameters due to W-shaped roughness was observed compared to smooth duct. Researchers considered different shapes of the fins like helicoidal spring for improving the THPF [15]. The characteristics of the spring fin like wire and spring diameter, and helicoidal pitch were varied and the heat transfer was found to enhance without considerable resistance to flow.

As the thermal efficiency of solar air heaters is significantly low, to increase its efficiency, roughness in form of baffles are introduced taking into consideration that thermal efficiency increases with minimum power loss because of friction. A lot of experimental study is done in this field whereas, a very few numerical studies are done. The main motive of this work is to check the variation of THPF with respect to the position of baffles and to compare the heat transfer and hydraulic parameters for smooth case and baffled case of the duct.

2. Methodology

2.1. Geometry of computational domain

The experimental data and the geometry for present study are considered from Bensaci et al. [9] as shown in Figure 1. In case of experimental arrangement, the baffles were placed on the 2nd half of the fluid domain, i.e. between the mid and outlet section of the duct. A numerical model was developed based on the experimental set up. The numerical results obtained for temperature variation from inlet to the outlet were validated with the experimental results.

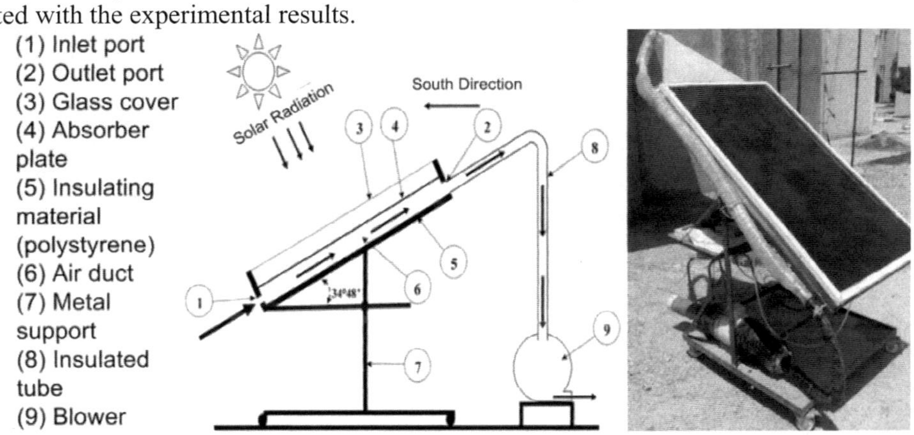

Figure 1. Experimental set-up.

A 3D geometrical model was developed in Design Modeler with the assumptions that fluid is incompressible and having turbulent flow behavior, fluid sustains single phase, negligible radiation and natural convective heat transfer between different parts, steady state situation is achieved, thermo physical properties of air and absorber plate are constant.

Further, two cases were analyzed to compare the results of the present study:

Case-1: solar air duct without baffles throughout the fluid domain.

Case-2: solar air duct with baffles on the second half of the fluid domain.

Figure 2 and 3 shows a 3D computational model of the solar air duct for Case-1 and Case-2, respectively. The geometry is created using ANSYS Design Modeler. Table 1 shows various geometrical parameters of the computational domain. The materials of absorber plate and the bottom plate are aluminum and wood (an insulating material), respectively. Air is considered as the working fluid medium which enters through inlet section and exits through outlet section. To match the simulation conditions with the experiments, the properties of air are taken from Bensaci et al. [9] as shown in Table 2.

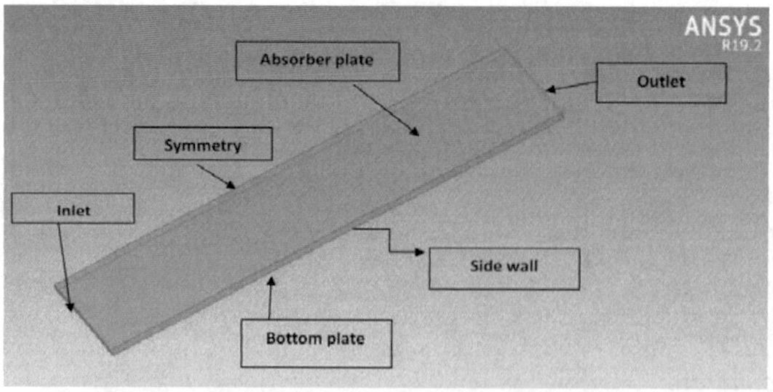

Figure 2. Computational scheme for Case-1.

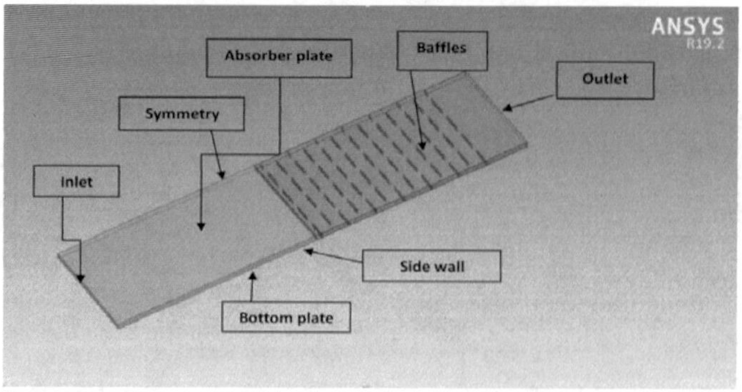

Figure 3. Computational scheme for Case-2.

Table 1. Geometrical parameters.

Duct length (mm)	1500
Duct width (mm)	750
Height of air column (mm)	25
Width of baffles (mm)	60
Height of baffles (mm)	20
Thickness of baffles (mm)	2
Distance between two adjacent baffles in a row (mm)	30
Distance between two adjacent rows of baffles (mm)	50

Table 2. Details of the properties of air and absorber plate.

Properties	Air(Fluid)	Aluminum(Absorber plate)
Viscosity (kg/m.s)	1.85 e-05	-
Density (kg/m^3)	1.167	2719
Specific heat (J/kg.k)	1006	871
Thermal conductivity (W/m.k)	0.0262	202.4

IOP Conf. Series: Materials Science and Engineering 1132 (2021) 012041 doi:10.1088/1757-899X/1132/1/012041

2.1.1. Grid generation

For Case-1, uniform structured grid consisting of only hexahedral elements were used with enough elements near the wall by keeping the y plus value less than one. The mesh metric parameters such as aspect ratio, orthogonal quality and skewness were kept within the permissible range for convergence of the solution. For Case-2, hexahedral elements were used in the first half of fluid domain, whereas tetragonal elements were used for the second half of the fluid domain where baffles were placed. The mesh metric parameters are shown in Table 3. A flow chart of the methodology used for the study is shown in Figure 4.

Table 3. Mesh metric parameters.

Mesh parameters	Case-1	Case-2
First layer thickness (mm)	0.209	2
Maximum aspect ratio	38.5	9.8
Minimum orthogonal quality	1	0.23
Maximum skewness	0.008	0.84
Number of elements	206448	709148

The Reynolds number is defined as equation (1).

$$Re = UD_h/v \quad (1)$$

where, U, v and D_h are velocity, kinematic viscosity of air and hydraulic diameter, respectively. The hydraulic diameter is calculated using equation (2).

$$D_h = 4A/P \quad (2)$$

where, A and $P$are the cross sectional area and the perimeter of the duct, respectively.

The local heat transfer coefficient is obtained using equation (3).

$$h_x = q/T_{w,x}\text{-}T_{b,x} \quad (3)$$

where, $T_{w,x}$, $T_{b,x}$ and q are the local wall temperature, bulk fluid temperature and heat flux, respectively.

The local Nusselt number, calculated near the absorber plate wall, is computed directly using equation (4).

$$Nu_x = h_x D_h/K \quad (4)$$

where, K is defined as the thermal conductivity.

The hydraulic performance is obtained by calculating the friction factor defined as equation (5).

$$f = (\Delta P/L)D_h/\ 2\rho U^2 \quad (5)$$

where, ΔP is the pressure difference observed between the inlet and the outlet sections.

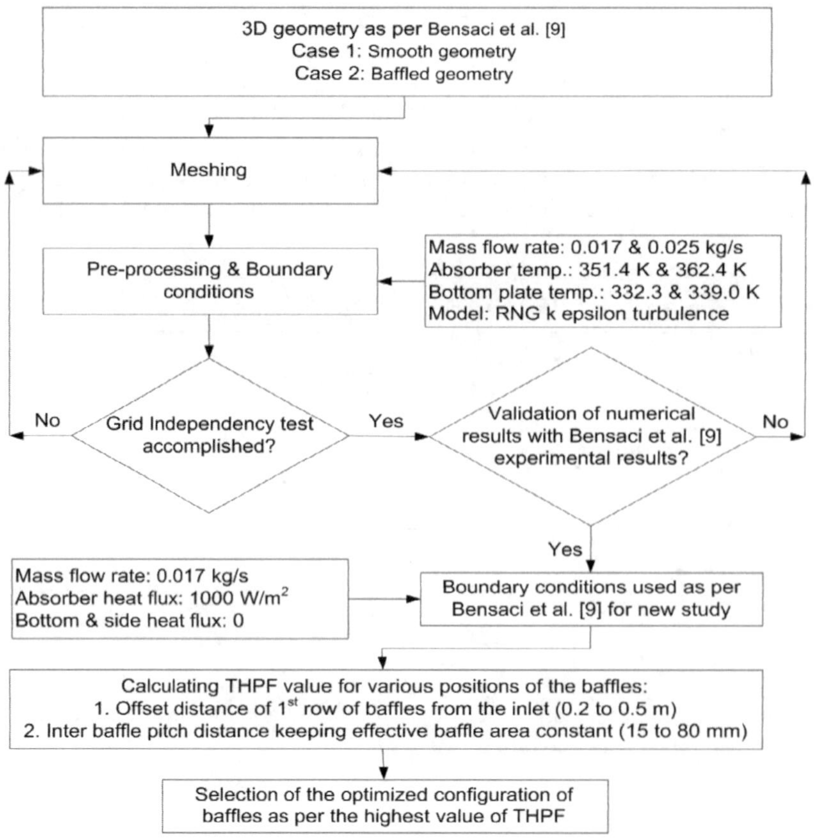

Figure 4. Methodology for numerical study.

THPF is defined as a ratio of thermal and hydraulic outputs which is expressed as equation (6).

$$THPF = (Nu/Nu_s)/\,(f/f_s)^{1/3} \qquad (6)$$

where, Nu and Nu_s are Nusselt number for baffled and smooth duct, respectively. Similarly, f and f_s are friction factors for baffled and smooth duct, respectively.

2.1.2. Boundary conditions

As shown in Table 4, the boundary conditions used for the validation are obtained from the experimental values from Bensaci et al. [9].

Table 4. Boundary conditions used for validation.

	Boundary condition 1 (BC_1)	Boundary condition 2 (BC_2)
Mass flow rate	0.017 kg/s	0.025 kg/s
Absorber temperature	351.4 K	362.4 K
Bottom temperature	332.28 K	339.61 K
Side heat flux	0 W (adiabatic)	0 W (adiabatic)

The inlet air temperature were 305K and 308K for boundary condition 1 (BC_1) and 2 (BC_2), respectively. The outlet pressure condition was considered as 0 Pascal (atmospheric pressure). A double precision pressure based solver was used to obtain the solution. For Case-1, the residual for convergence was set to the value of 10^{-6} for energy equation. Such criterion for other parameters like momentum, continuity, k and epsilon equations was set to a value of 10^{-5}. For Case-2, the residuals values were taken as 10^{-5} and 10^{-3} for respective convergence conditions. The SIMPLE numerical algorithm was used to solve the solution using RNG k-epsilon turbulence model. The turbulence intensity values were related to the Reynolds number using equation (7).

$$I = 0.16\ (Re)^{-1/8} \tag{7}$$

The mesh independency was carried out for four different mesh densities for Case-1 condition and the effect of different mesh densities on outlet temperature of air has been plotted as Figure 4.

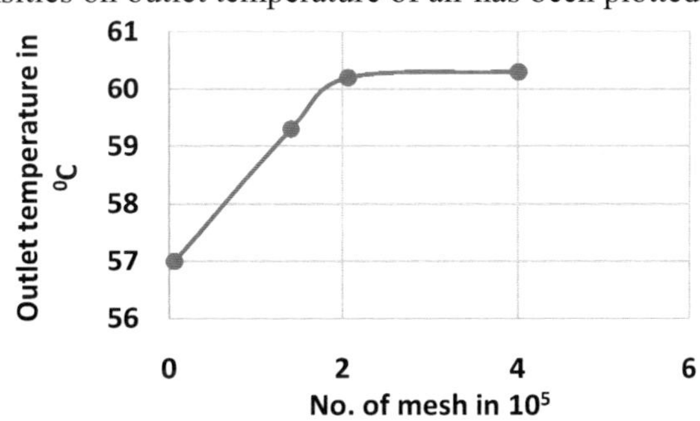

Figure 5. Mesh independency test.

From the Figure 5, no significant change of outlet temperature of air was observed from mesh density 2×10^5. Thus, for present study, mesh density corresponding to number of mesh value 2.06448×10^5 has been taken as an optimal number of mesh counts for further simulation.

3. Results and discussion

For validation of numerical results, air temperature at the outlet section of flow duct obtained from numerical simulation is plotted against the experimental data available from Bensaci et al. [9] for both BC_1 and BC_2 as shown in Figure 6. For BC_2 having mass flow rate 0.025 kg/s, the difference between the experimental and numerical values is 2°, i.e. 0.6% of the experimental value. Further, the difference in outlet temperature reduces to 0.6°, i.e. 0.1% of the experimental value for BC_1 having mass flow rate 0.017 kg/s. The minor variation in outlet temperature can be attributed to the fact that the experimental set-up offers minor unaccounted resistance to the flow which leads to higher thermal interaction of fluid within the duct leading to higher outlet temperature. However, the simulations did not account for these minor thermal interactions which lead to slightly less outlet temperature. As the outlet temperature of air obtained numerically is very close to the experimental values for both boundary conditions, the numerical scheme is considered validated and accurate enough for further study.

IOP Conf. Series: Materials Science and Engineering 1132 (2021) 012041 doi:10.1088/1757-899X/1132/1/012041

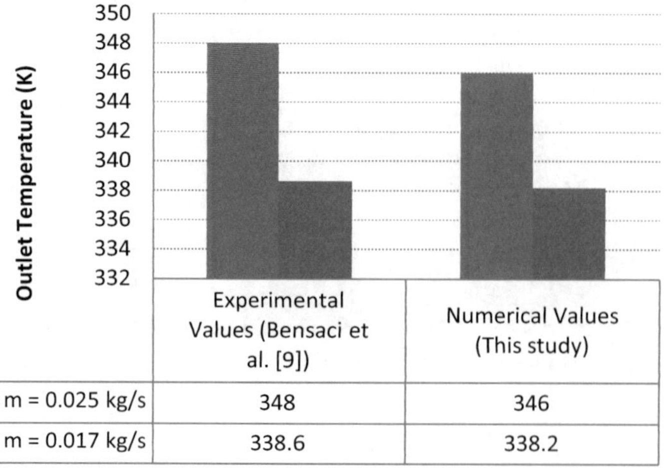

	Experimental Values (Bensaci et al. [9])	Numerical Values (This study)
m = 0.025 kg/s	348	346
m = 0.017 kg/s	338.6	338.2

Figure 6. Temperature of the air at outlet section for various flow conditions.

Figure 8 and 9 shows velocity and pressure contours respectively for the Case-2 having baffled duct. It is evident from the contour plot that the obstruction by the baffles in the flow field led to significant pressure drop from inlet to the outlet which is directly apparent in the form of higher frictional losses. These losses lead to higher power requirement by the motor. A negative pressure is observed near the wall of baffles which is responsible for creating vortices in the flow as shown in Figure 7. The produced vortices create more turbulence in the flow resulting in an increase in air velocity in the baffled region and higher interaction of air with baffles. The increase in velocity means an increase in the Reynolds number which leads to a rise in the Nusselt number (Nu); hence, an increase in heat transfer coefficient.

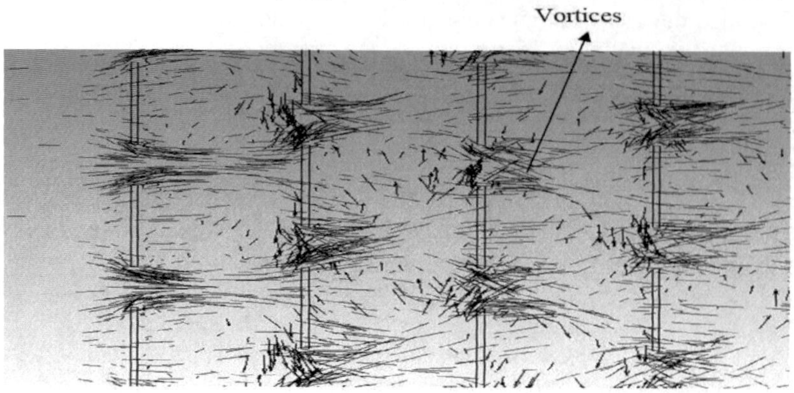

Figure 7. Velocity vectors showing the vortices generated.

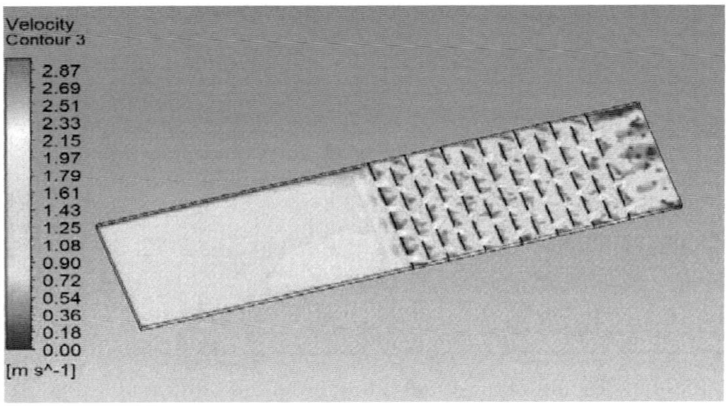

Figure 8. Velocity contours at mid plane for Case-2 and BC_1.

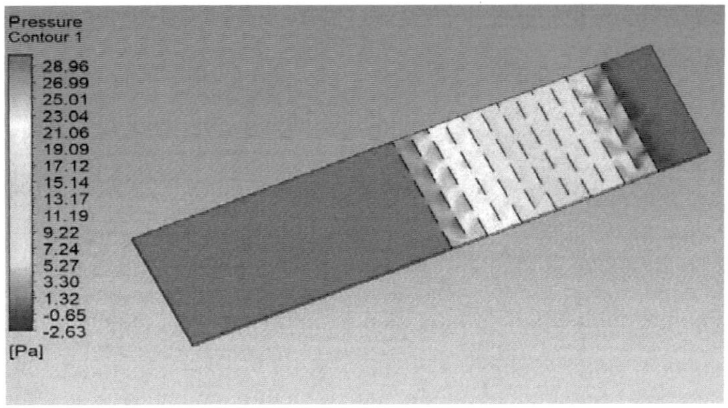

Figure 9. Pressure contours for Case-2 and BC_1.

For Case-1 and Case-2, the variation in parameters such as Nusselt number and friction factor along the length of the duct are shown in Figures 10 and 11, respectively. Though the total length of duct remains 1.5 m for both cases, the length scale for Case-1 is shown from 0 to 1.5 m whereas, for Case-2, it is from -0.75 to 0.75 m as the coordinate system for Case-2 is taken at the center of the duct. From Figure 10a showing the surface Nusselt number variation for Case-1 with smooth duct, it is observed that the Nusselt number decreased along the length of duct. However, Figure 10b for Case-2 with baffled duct showed peaks with larger amplitudes in the baffled part region corresponding to high local convective heat transfer coefficient; hence, high local Nusselt number due to high vortices. From Figure 11, showing skin friction coefficient variation for both cases, the skin friction coefficient was found to decrease along the length of flow for smooth duct. However, the introduction of baffles increased the skin friction coefficient in the second half of fluid domain for duct with baffles. Thus, it is observed that roughening the surface of duct improves the heat transfer characteristics; but, it leads to increase in friction, i.e. increase in the hydraulic power requirement for forced circulation. So, an optimum design is required to balance between the heat transfer characteristics and hydraulic characteristics.

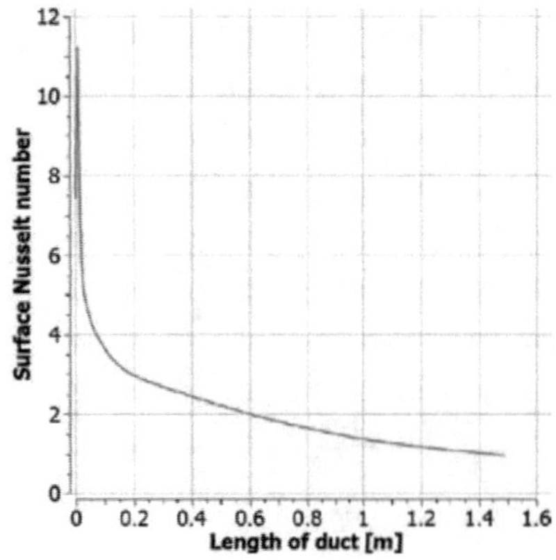

Figure 10(a). Variation of Nu along length of the duct for Case-1.

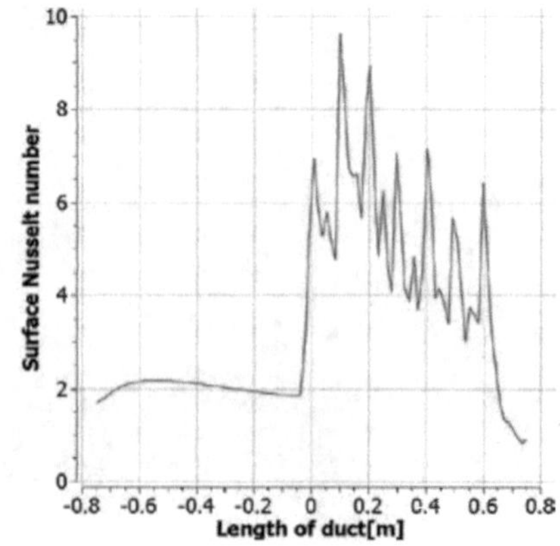

Figure 10(b). Variation of Nu along length of the duct for Case-2.

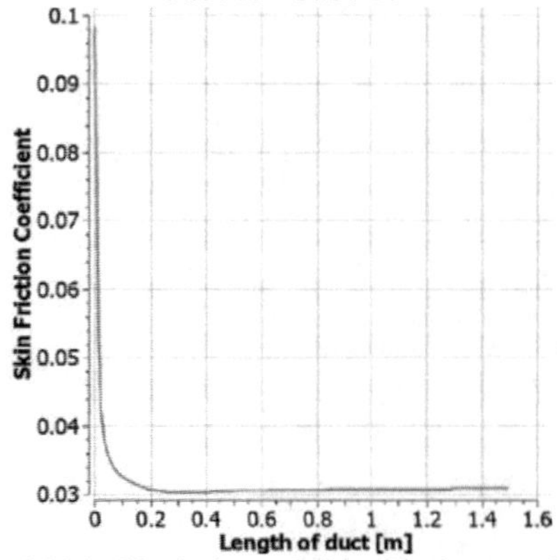

Figure 11(a). Variation of skin friction coefficient along length of the duct for Case-1.

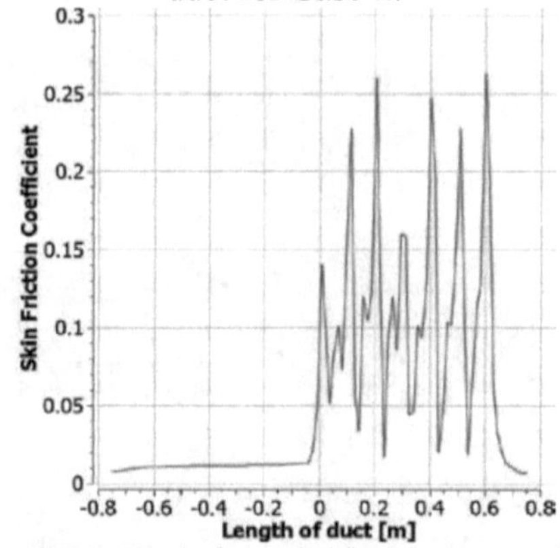

Figure 11(b). Variation of skin friction coefficient along length of the duct for Case-2.

The variation of Nusselt number and skin friction coefficient with respect to Reynolds number ranging from 2370 to 8340 is studied as shown in Figure 12a and 12b. The Nusselt number was found to increase with increase in Reynolds number and skin friction coefficient decreases with increase in the value of Reynolds number.

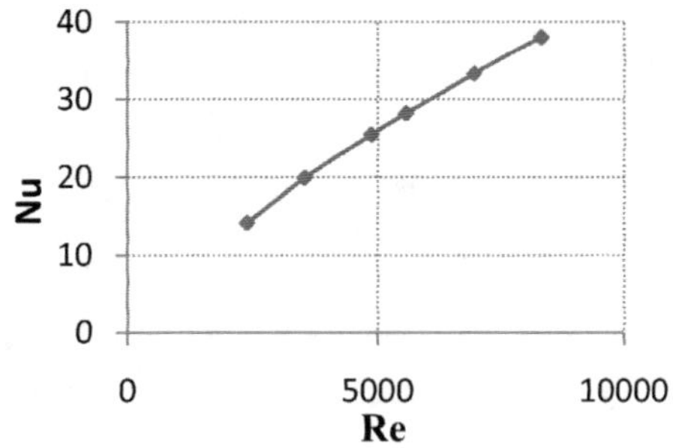

Figure 12(a). Nu versus Re.

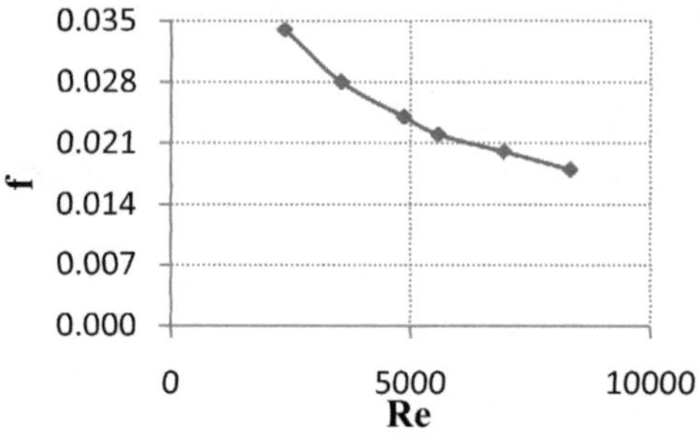

Figure 12(b). f versus Re.

The validated model obtained from the previously explained processes is further used for improving the configurations not studied in Bensaci et al. [9]. For the purpose, the boundary conditions were taken from Bensaci et al. [9] as shown in Table 5 and THPF is obtained for various positions of the baffles.

Table 5. Boundary conditions used for the model taken from Bensaci et al. [9].

	BC_3
Mass flow rate	0.017 kg/s
Absorber heat flux	1000 W/m^2
Bottom and side heat flux	0

Figure 13 shows THPF variation with pitch distance between baffles for 4 different cases of placement of 1st row of baffles from the inlet; namely, i) 1st row at 200mm offset, ii) 1st row at 300mm offset, iii) 1st row at 400mm offset, and iv) 1st row at 500mm offset. In addition to these 4 different cases of placement of 1st row of baffles from the inlet, the inter baffle pitch distance was varied from 15mm to 80mm in steps of 5mm keeping effective baffle area constant, i.e. the total number of baffles were kept fixed for all the cases considered. The highest THPF obtained is 0.9339 for the 1st row at 200mm offset for a pitch distance of 60mm.

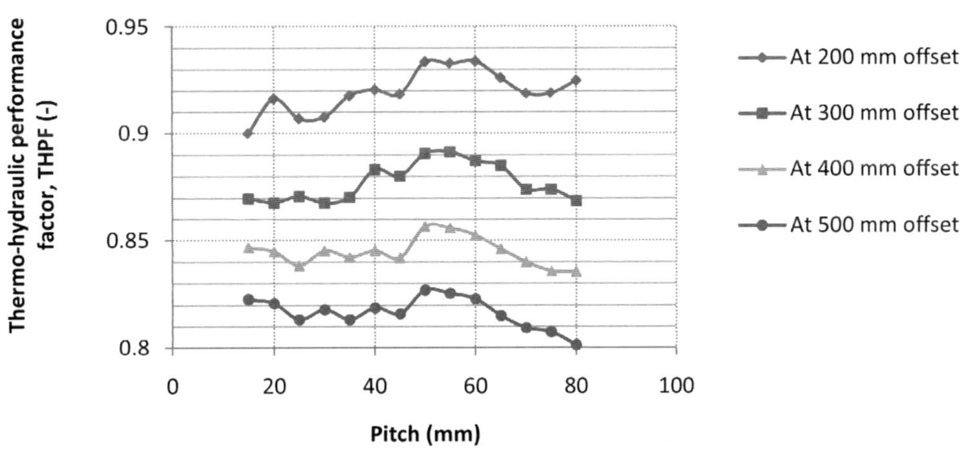

Figure 13. THPF versus pitch for different offset distance from inlet.

4. Conclusion

The effect of THPF variation for different position of baffles from inlet has been evaluated. This study also showed the comparison between thermo-hydraulic characteristics of smooth and baffled ducts considering various parameters such as Nusselt number and skin friction coefficient along the duct. The major results obtained from the research are as follows: (1) The maximum THPF was obtained for case with 1st row of baffles at 200mm offset from inlet for a pitch of 60mm between the baffles, the maximum value of THPF is 0.9339 (2) Nusselt number for a baffled duct has three different behaviors along the length of duct i) upstream of the baffled part, peaks with smaller amplitude are seen corresponding to low

IOP Conf. Series: Materials Science and Engineering 1132 (2021) 012041 doi:10.1088/1757-899X/1132/1/012041

Nusselt number because of reversed flow between first row of baffles and flow regime ii) peaks with larger amplitude are observed in baffled part corresponding to high Nusselt number because of more vortices iii) downstream of baffled part, peaks with smaller amplitude are found than upstream because of the flow regime (3) Nusselt number increases along the length of duct with the increase in Reynolds number and skin friction coefficient decreases along the length of duct as the Reynolds number increases.

References

[1] Yadav A S and Bhagoria J L 2013 *Renewable and Sustainable Energy Reviews* **23** 60-79
[2] Foster R, Ghassemi M and Cota A 2010 *Solar energy: renewable energy and the environment* (New York: CRC Press Taylor & Francis group)
[3] Li X 2011 *Green energy: basic concepts and fundamentals* (New York: Springer)
[4] Maczulak A 2011 *Renewable energy: sources and methods* (New York: Infobase Publishing)
[5] Twidell J and Weir T 2006 *Renewable energy* (New York: Taylor & Francis)
[6] Sukhatme S P and Nayak JP 2011 *Solar energy* (New Delhi: Tata McGraw Hill)
[7] Ravi R K and Saini RP 2016 *Energy* **116** 507-516
[8] Hans VS, Saini RP and Saini JS 2009 *Renewable and Sustainable Energy Reviews* **13** 1854-1869
[9] Bensaci C E, Moummi A, Flor F J S de la, Jara E A R, Casado A R and Pardo A R 2020 *Renewable Energy* **155** 1231-1244
[10] Menasria F, Zedairia M and Moummi A 2017 *Energy* **133** 593-608
[11] Prasad BN and Saini JS 1991 *Solar Energy* **47(2)** 91–6
[12] Karwa R, Bairwa RD, Jain B P and Karwa N 2005 *Journal of Enhanced Heat Transfer* **12(4)** 343–55
[13] Kumar A, Bhagoria JL and Sarviya RM 2008 *Proc. 19th National & 8th ISHMT-ASME heat and mass transfer Int. Conf.*
[14] Lanjewar A, Bhagoria JL and Sarviya RM 2011 *Energy* **36** 4531–41
[15] Kumar HSA, Kumar Sand Karanth KV 2020 *Renewable Energy* **16 0**297-311

Effect of wall-slip on natural convection in a circular annulus

Gautham S. Nair*, Harikrishna R., Damu Murali and S. Ajith Kumar

Department of Mechanical Engineering, Amrita Vishwa Vidyapeetham, Amritapuri, India

*E-mail: gsn.sudhir28@gmail.com

Abstract. Two-dimensional natural convection in an annulus enclosed by a hot inner cylinder and cold outer cylinder with air (Prandtl number,Pr=0.71) as the fluid is numerically investigated for Rayleigh number (Ra)= 5×10^4 using OpenFOAM. For a gap width to radius ratio (L/R_i) of 1.6, the slip configurations are imposed by varying the $Slip\,factor$ (SF) from 0 (no-slip) to 1 (full-slip) in steps of 0.2. The flow patterns and thermal fields for each slip factor are visualised using streamlines and isotherms and heat transfer is quantified by Nusselt number (Nu). It is observed that the natural convection patterns at full-slip boundary condition (SF=1) exhibit unique phenomena like vortex splitting, shrinkage in the thermally stagnant regions, and reduction in thermal boundary layer thickness. The heat transfer characteristics, in general, get enhanced with the increase in SF.

1. Introduction

Natural convection in an annulus has been a topic of interest due to several applications in nuclear reactors, thermal storage systems, electronic cooling, etc., to cite a few. In one of the earliest reported studies, Kuehn & Goldstein[1] experimentally and numerically investigated natural convection in the horizontal annulus between two concentric circular cylinders for a range of Rayleigh numbers (Ra). The transition of flow regime from pure-conduction to convection was discussed in detail through visual inspection of flow and thermal fields.

Boyd[2] established a correlation between mean Nusselt number and Rayleigh Number using fundamental concepts of different flow regimes and derived its governing equations in the annulus. This correlation technique was found to be applicable for the irregular annulus formed by an inner hexagonal cylinder and an outer concentric circular cylinder. The flow regimes were divided into inner and outer boundary layers, inner and outer intermediate region, plume, and stably stratified region. Rao et al.[3] investigated experimentally and numerically the natural convection and flow patterns of wide, moderate, and narrow annuli. They noticed that the number of vortices increased with an increase in Rayleigh number and for higher Rayleigh number, oscillatory flow patterns were observed. Date[4] conducted numerical investigations on natural convection heat transfer in a horizontal annulus for gap width to inner diameter ratios $(\frac{L}{D_i}) = 0.8$ and 0.15. The latter value was of interest for their study due to its application in horizontal heavy water reactors.

Numerous other studies discussed heat transfer in the annulus of horizontal circular cylinders by changing the location of the inner cylinder. Projahn & Beer[5] studied the influence

IOP Conf. Series: Materials Science and Engineering 1132 (2021) 012042 doi:10.1088/1757-899X/1132/1/012042

Nomenclature

$\overline{Nu}$	Average Nusselt number over a surface
g	Acceleration due to gravity $[m/s^2]$
L	Gap width [m]
Nu_l	Local Nusselt number
Nu_{avg}	Average Nusselt number of hot and cold walls $(=\frac{\overline{Nu_o}+\overline{Nu_i}}{2})$
P	Non-dimensional pressure
p	Pressure [Pa]
Pr	Prandtl number $(=\frac{\nu}{\alpha})$
R_i	Radius of the inner cylinder [m]
R_o	Radius of the outer cylinder [m]
Ra	Rayleigh number $(=\frac{g\beta L^3(T_h-T_c)}{\nu\alpha})$
SF	Slip factor
T	Temperature [K]
T_c	Temperature of cold wall [K]
T_h	Temperature of hot wall [K]
U	Non-dimensional horizontal velocity
u	Horizontal velocity $[m/s]$
V	Non-dimensional vertical velocity
v	Vertical velocity $[m/s]$
X	Non-dimensional horizontal coordinate
x	Horizontal coordinate
Y	Non-dimensional vertical coordinate
y	Vertical coordinate

Greek Symbols

α	Thermal diffusivity $[m^2/s]$
β	Coeffcient of thermal expansion $[K^{-1}]$
λ	Slip length
ν	Kinematic viscosity $[m^2/s]$
ρ	Density $[kg/m^3]$
Θ	Non-dimensional temperature $(=\frac{T-T_c}{T_h-T_c})$
θ	Angular coordinate measured from upper vertical axis [°]

of Prandtl number on natural convection in eccentric and concentric annuli. Shahraki[6] employed the modified Boussinesq approximation for simulating the heat transfer characteristics in the annulus of two horizontal cylinders with a radius ratio of 2.6 and for various Rayleigh numbers from 10^3 to 10^5 at various eccentricities. Guj & Stella[7] numerically visualized the fluid flow and thermal fields in a two-dimensional annulus of eccentric cylinders for Pr=0.71, $0.53 \times 10^4 \leq Ra \leq 8.27 \times 10^4$ with eccentricity from 0 to 1 and concluded that heat transfer reduced at lower gaps. Shi et al.[8] proposed a finite difference-based lattice BGK model to obtain velocity and temperature distribution in a horizontal annuli. Waheed[9] followed a heat line approach to understand how heat transfer took place in a horizontal annulus.

The previously reported experimental and numerical analyses on natural convection inside the annulus have considered no-slip boundary condition and the influence of slip boundary condition applied on the walls has not been studied so far. This paper is hence dedicated to finding out the effect of wall-slip on natural convection inside a cylindrical annulus at Ra (based on gap width)=5×10^4. The results of our studies will be helpful for industrial and academic research relating to wettability of surfaces especially in heat pipes where the presence of hydrophobic surfaces can limit the formation of residuals and thus improving heat transfer characteristics.

The structure of the presentation of our findings in the rest of the article is as follows. Solution methodology in Section 2 describes the numerical method, computational domain, and boundary conditions. Results and discussion in Section 3 include the grid sensitivity studies and validation of the code followed by visual inspection of flow and thermal fields using streamlines and isotherms. Heat transfer characteristics for various slip configurations have been explained using the variation in local Nusselt number along the hot and cold walls and average Nusselt

IOP Conf. Series: Materials Science and Engineering 1132 (2021) 012042 doi:10.1088/1757-899X/1132/1/012042

number. Conclusions in Section 4 summarises the overall numerical results.

2. Solution methodology

2.1. Numerical method

Numerical simulations are performed using the open-source computational fluid dynamics toolbox OpenFOAM. The generic solver *buoyantBoussinesqPimpleFoam* is used to solve the steady, laminar and incompressible governing equations in two-dimensional form as given below. A second-order central differencing *Gauss linear* scheme is used for the spatial discretization of convective and diffusive terms.

$$Continuity : \frac{\partial U}{\partial X} + \frac{\partial V}{\partial Y} = 0 \tag{1}$$

$$X - momentum : U\frac{\partial U}{\partial X} + V\frac{\partial U}{\partial Y} = -\frac{\partial P}{\partial X} + \mathrm{Pr}\left(\frac{\partial^2 U}{\partial X^2} + \frac{\partial^2 U}{\partial Y^2}\right) \tag{2}$$

$$Y - momentum : U\frac{\partial V}{\partial X} + V\frac{\partial V}{\partial Y} = -\frac{\partial P}{\partial Y} + \mathrm{Pr}\left(\frac{\partial^2 V}{\partial X^2} + \frac{\partial^2 V}{\partial Y^2}\right) + RaPr\Theta \tag{3}$$

$$Energy : U\frac{\partial \Theta}{\partial X} + V\frac{\partial \Theta}{\partial Y} = \frac{\partial^2 \Theta}{\partial X^2} + \frac{\partial^2 \Theta}{\partial Y^2} \tag{4}$$

The non-dimensional variables in the governing equations are defined as follows

$$X = \frac{x}{L}, \quad Y = \frac{y}{L}, \quad U = \frac{uL}{\alpha}, \quad V = \frac{vL}{\alpha}, \quad \Theta = \frac{T - T_{\mathrm{c}}}{T_{\mathrm{h}} - T_{\mathrm{c}}}, \quad P = \frac{pL^2}{\rho\alpha^2} \tag{5}$$

Rayleigh Number (Ra) based on the gap width L is given by $Ra = \frac{g\beta L^3(T_{\mathrm{h}}-T_{\mathrm{c}})}{\nu\alpha}$ and Prandtl number (Pr) is defined as $Pr = \frac{\nu}{\alpha}$. ρ and ν are density and kinematic viscosity of the fluid and β is the coefficient of thermal expansion.

The fluid properties are assumed to be constant, except for the variation in density which is obtained by the Boussinesq approximation. The radiation heat transfer is considered negligible and the acceleration due to gravity (g) acts downwards in the negative y-direction.

The heat transfer from the walls is quantified in terms of Nu_l and Nu_{avg} along the walls using

$$Nu_l = \left.\frac{\partial \Theta}{\partial n}\right|_{\mathrm{wall}}, \quad \overline{Nu} = \frac{1}{W}\int_0^W Nu_l \, \mathrm{d}S \tag{6}$$

where n is the unit normal to the wall, W is the surface area of the wall, dS is the infinitesimal surface area of the wall. The term Nu_{avg} is the surface averaged Nusselt number along both the walls given by

$$Nu_{avg} = \frac{\overline{Nu_o} + \overline{Nu_i}}{2} \tag{7}$$

where $\overline{Nu_o}$ and $\overline{Nu_i}$ are $\overline{Nu}$ for outer and inner walls respectively.

IOP Conf. Series: Materials Science and Engineering 1132 (2021) 012042 doi:10.1088/1757-899X/1132/1/012042

2.2. Computational domain and boundary conditions

A schematic of the computational domain is shown in Figure 1(a). The gap width is chosen as 1 unit and the ratio of gap width to the inner radius (L/R_i) is 1.6. The inner and outer cylinder walls are kept at uniform hot ($\Theta = 1$) and cold ($\Theta = 0$) temperatures respectively.

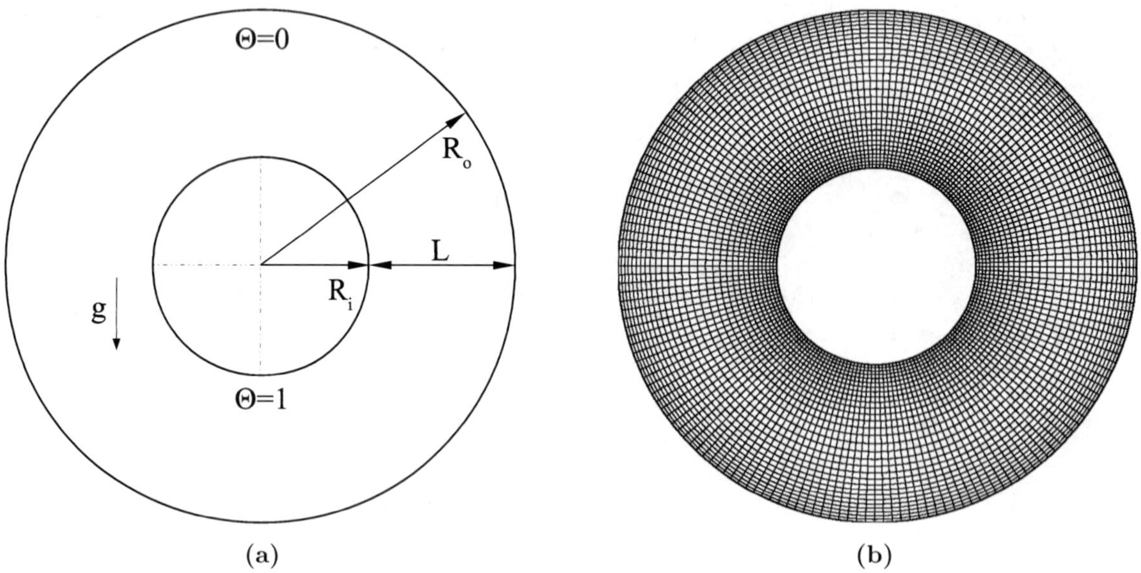

Figure 1: Schematic of (a) computational domain (b) grid structure used for the present study

Slip conditions on the walls are varied using the *partialSlip* boundary condition available in OpenFOAM. *partialSlip* represents a mixed boundary condition. For a 2-D flow, the first-order approximation for the slip-velocity is given by

$$U + \lambda \frac{dU}{dY} = 0 \tag{8}$$

where λ is the slip length [10].

Let U_0 be the tangential velocity on the wall face boundary and U_1 the cell centre velocity on boundary layer. Equation (8) can be re-written in terms of U_0, U_1 and D as follows

$$U_0 + \lambda \frac{(U_0 - U_1)}{D/2} = 0 \tag{9}$$

where D is the size of the adjacent cell. The above equation after re-arranging becomes

$$U_0 = \frac{2\lambda}{D + 2\lambda} U_1 \tag{10}$$

Here, we introduce a new term $Slip\,factor(SF)$ defined as

$$SF = \frac{2\lambda}{D + 2\lambda} \tag{11}$$

Hence equation (10) becomes

$$U_0 = (SF)U_1 \tag{12}$$

For $\lambda \to 0$, we have $SF = 0$, which indicates no-slip condition.
For $\lambda \to \infty$, we have $SF = 1$, which indicates full-slip condition.

Conditions imposed at the walls other than the slip boundary conditions are as follows:

Inner cylinder wall : Dirichlet boundary condition, Θ=1
Outer cylinder wall : Dirichlet boundary condition, Θ=0

3. Results and discussion

3.1. Grid sensitivity study

Grid sensitivity studies are conducted to minimise the influence of grid size on the computational results and to optimise the computational time. To accurately capture the boundary layer phenomenon, a finer mesh is adopted near the walls and a coarser mesh for other regions. Table 1 shows the variation in Nu_{avg} while increasing the grid size. The percentage deviation of the successive values of Nu_{avg} is indicated in the bracket. It can be observed that the % difference keeps decreasing. However, since the variation between grid sizes 5400 and 9600 is less than 0.3%, the latter grid size is selected for the present study.

Table 1: Grid sensitivity study: Ra=10^5, Pr=0.71, SF=0

Number of cells	Nu_{avg}
2400	1.387
3750	3.71 (167.48%)
5400	3.688 (0.592%)
9600	3.677 (0.297%)
12150	3.673 (0.111%)
15000	3.672 (0.026%)

3.2. Validation

The computational code used for the present study is compared and validated with the previous literature. Comparison is done by plotting the local Nusselt number (Nu_l) values along the hot and cold walls for Ra= 5×10^4 (Figures 2(a) and(b)). The curve representing the present study is seen to follow other curves of the previous studies very closely.

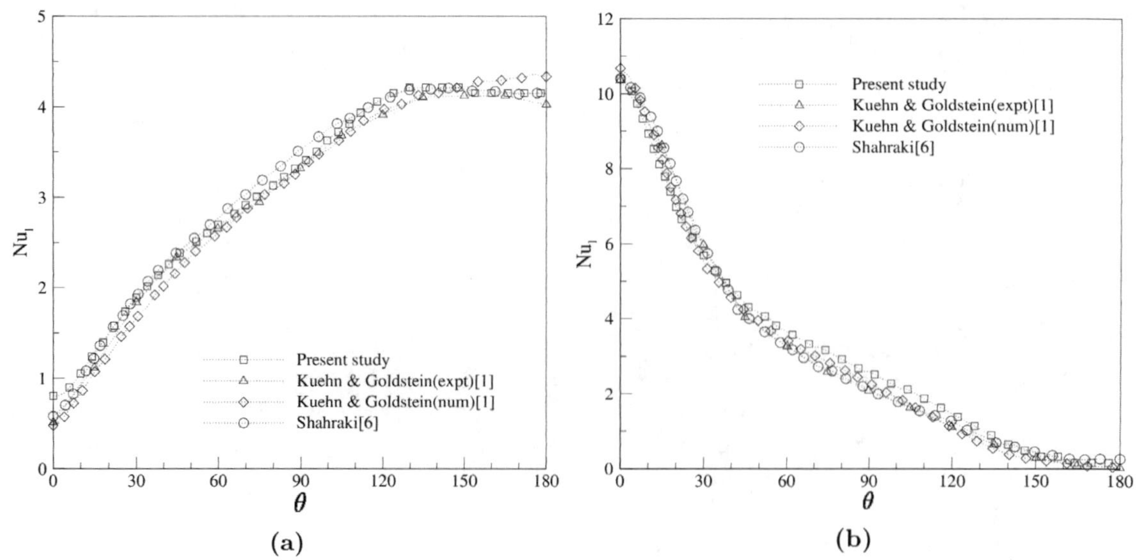

Figure 2: Comparison with previous studies : Nu_l along (a) hot wall (b) cold wall

Table 2 shows the comparison of the average Nusselt number (Nu_{avg}) for different Rayleigh numbers. The percentage deviation of the calculated value of Nu_{avg} from the data in the literature is indicated in brackets (absolute values) and the results are found to be in accordance with the existing data. The above observations conclude that the present values are in good agreement with the previously reported results.

Table 2: Comparison of Nu_{avg} : present study vs previous studies

Ra	Present study	Shahraki[6]	Kuehn & Goldstein[1]	Shi et al.[8]
1.02×10^5	3.677	3.647 (0.808%)	3.66 (0.464%)	3.531 (4.134%)
6.19×10^4	3.276	3.309 (1%)	3.32 (1.325%)	3.361 (2.53%)
9.5×10^3	2.052	1.990 (3.11%)	2.01 (2.089%)	1.999 (2.651%)

3.3. Streamlines
Figure 3 shows the streamlines for various values of SF at Ra=5×10^4. In the region around the hot inner cylinder, the fluid gets heated up, becomes less dense, and begins to rise. As it reaches the vicinity of the cold outer cylinder, it interacts with the cold surface, becomes heavier, and moves down. While descending, some of the cold fluid again warms up due to the interaction with the hot surface and its neighbourhood thus giving rise to a pair of recirculating convection currents manifested as vortices.

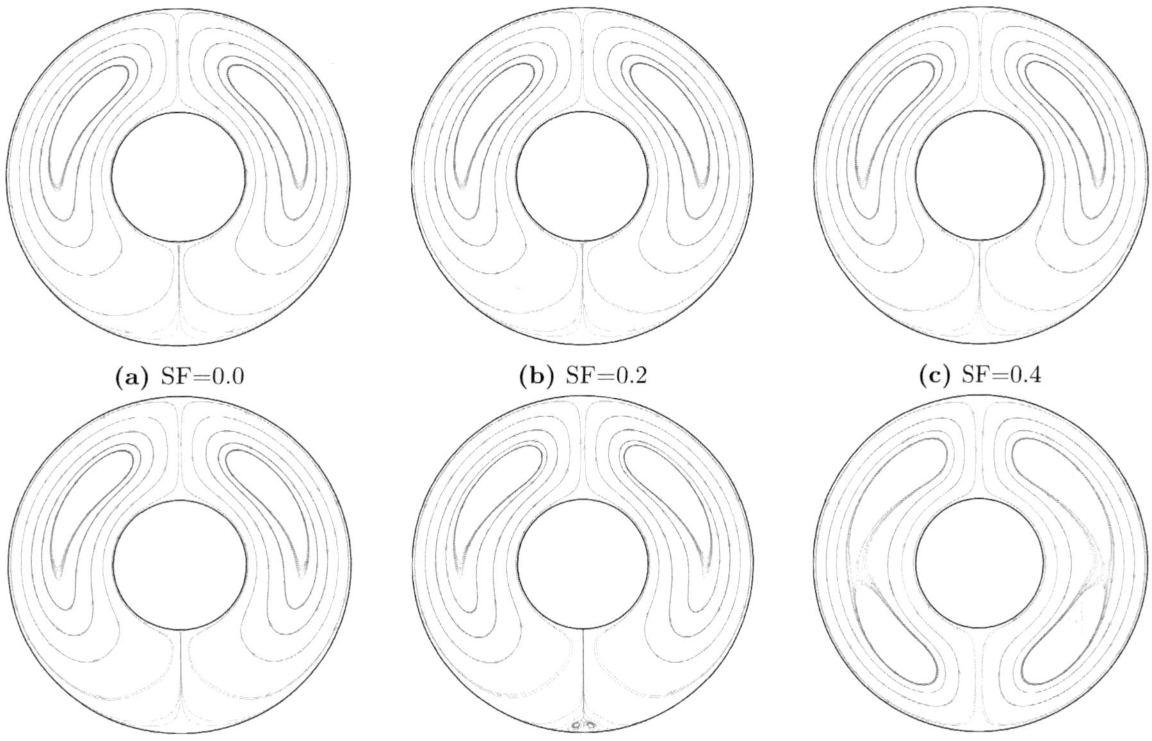

(a) SF=0.0 (b) SF=0.2 (c) SF=0.4

(d) SF=0.6 (e) SF=0.8 (f) SF=1.0

Figure 3: Streamlines for various slip configurations at $Ra = 5 \times 10^4$

For no-slip condition, SF=0 (Figure 3(a)), counter-rotating crescent-shaped vortices can be seen on either side of the vertical axis of symmetry, similar to the ones observed by Kuehn & Goldstein[1] and Date[4].

When the slip factor is increased from SF=0 to SF=0.6, no noticeable changes are observed, except that the center of rotation of the vortices moves slightly upward. Visual inspections, however, reveals that these cores remain in the upper half, which points to strong convection currents in this region. For SF=0.8 (Figure 3(e)), a new pair of secondary vortices appear near to the bottom of the annulus which indicates boundary layer separation from the cold outer cylinder. Significant transitions are seen in the flow structures when the boundary condition changes to full-slip (Figure 3(f)). The centres of the counter-rotating primary vortices on either side of the vertical axis split into two, creating two additional inner vortices. However, the outer recirculating envelope remains in position. It is also observed that the pair of secondary vortices near the cold wall which appears at the bottom of the annulus for SF=0.8 totally disappears for the full-slip condition.

3.4. Isotherms

The isotherms for various slip configurations are shown in Figure 4. It is observed from previous literature that as Rayleigh number nears 10^4, radial temperature inversion appears, indicating the separation of thermal boundary layers.

IOP Conf. Series: Materials Science and Engineering 1132 (2021) 012042 doi:10.1088/1757-899X/1132/1/012042

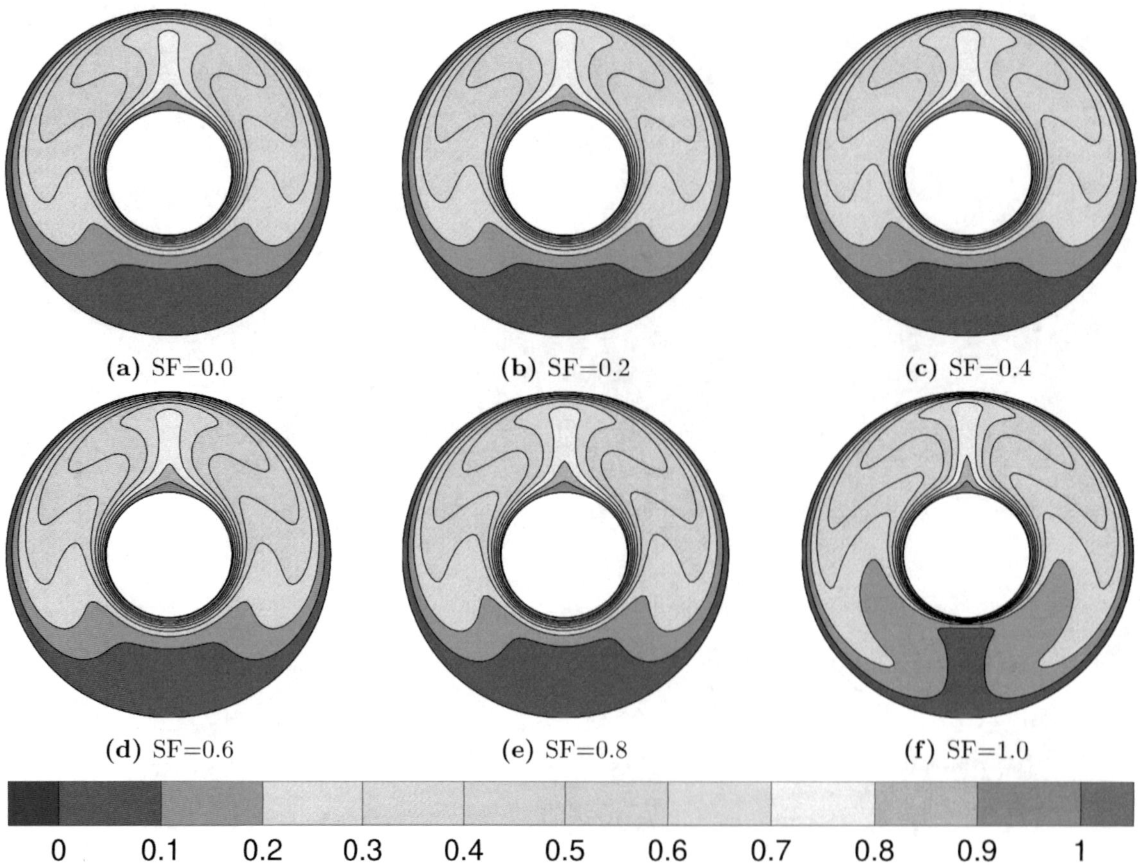

Figure 4: Isotherms for various slip configurations plotted at Ra $=10^5$

Due to the domination of convection currents, the thermal boundary layer separation occurs on the upper half of the hot cylinder which develops into a plume in the upward direction before the flow attains a steady state. The isotherms observed for the no-slip condition (Figure 4(a)) are in good agreement with previously published literature (Kuehn & Goldstein[1] and Date[4]). The isotherms are seen to be distorted as a result of the migration of the plumes after impinging the cold surface. Owing to the formation of plumes in the upper half, the thermal gradients in this region are very high compared to the other half. This could be the reason why the core of the primary vortices are present in the upper half as found in the streamlines (Figures 3(a)-(e)).

The isotherms for SF values ranging from 0 to 0.6 are similar however, for SF=0.8 (Figure4(e)), the hot plume originating from the top half of the inner cylinder slightly extends upwards which contributes to the further distortion of isotherms. This upward movement influences the thermal distribution throughout the annulus and close visual inspection reveals the wings of the bottom-most isotherm becoming slightly closer to each other which subsequently reduces the thermal stagnation at the bottom of the annulus. As full-slip condition is imposed, the plumes above the hot wall shrink and slightly extend vertically upwards towards the cold wall. The influence of this change is also visible in the lower half as the isotherms expand more towards the bottom wall of the outer cylinder. The thermally stagnant region shrinks considerably and is seen to inhabit a narrow space between the lowermost sections of the inner and outer cylinders.

3.5. Variation in Nusselt number

Figures 5(a) and (b) show variation in local Nusselt number (Nu_l) along the right half of hot and cold walls for various configurations of SF. Due to the symmetric nature of flow and thermal fields, the heat transfer characteristics too are the same on either side of the vertical axis hence, the local Nusselt number variation along the left wall (of both inner and outer cylinders) is the same as that of the right side of which the latter is considered for analysis.

Moving from θ=0° to θ=180° in the clockwise direction, for no-slip (SF=0) condition, the Nu_l along the hot surface increases monotonically up to θ=130° after which the local Nusselt number does not vary appreciably. The boundary layer thickness reduces while moving down the surface which enhances the heat transfer rate. However, between θ=130° and θ=180°, the growing influence of the thermally stagnant region at the bottom of the annulus actively opposes the plume dynamics thus maintaining uniformity in the local Nusselt number. A similar trend is observed for all slip configurations up to SF=0.4. For SF=0.6 and 0.8, after the initial monotonic increase up to θ=130°, the influence of partial-slip is seen to dominate the effect of plume which could be the reason for the slightly decreasing trend (negative slope) in the Nu_l-θ curve between θ=130° and θ=180°. Significant transitions occur when full-slip (SF=1) is imposed on the walls. The shrinkage of the hot plume decreases the boundary layer thickness in its vicinity thereby producing an enhancement in heat transfer rate between θ=0° and θ=30°. Between θ=30° and θ=120°, the increase in Nu_l is monotonic however beyond θ=120° the combined effect of full-slip and the shrinkage in the thermally stagnant region produces a steep rise in the Nu_l-θ curve.

The variation in the Nu_l-θ curve for the cold wall is monotonically decreasing up to θ=150° for all slip configurations. Between SF=0 and SF=0.8, the variation in Nu_l between θ=150° and θ=180° is negligible as this zone falls in the thermally stagnant region located at the bottom of the annulus. However, for full-slip configuration, since the thermally inactive region shrinks, the local Nusselt number keeps decreasing up to θ=180°. It is worth noting that with an increase in SF, local Nusselt number keeps increasing at all points on the hot and cold walls.

Figure 5(c) illustrates the variation in the average Nusselt number for the entire range of slip configurations. As evidenced, the Nu_{avg} variation is gradual and monotonically increasing up to SF=0.8 however, between SF=0.8 and SF=1, the curve witnesses a steep increase in the heat transfer rate which could be the effect of nullified viscous retardation in combination with the narrow stagnant region. The presence of twin vortices on either side of the vertical axis is also seen to enhance the heat transfer characteristics for the full-slip condition.

4. Conclusions

Numerical analysis is conducted to study the effect of slip boundary condition on natural convection in the annulus between concentric isothermal cylinders for Ra=5×10^4 and Pr=0.71. It is observed that an increase in slip factor from SF=0 (no-slip) to SF=1 (full-slip) improves the overall heat transfer characteristics. The other inferences are as follows:

- For SF values between 0 and 0.6, a pair of primary vortices are seen on either side of the vertical axis whose cores are found in the upper half of the annulus. When SF reaches 0.8, a pair of secondary vortices appears at the bottom of the annulus in addition to the primary vortices. For full-slip configuration, the secondary vortices cease to exist and the cores of primary vortices break into two inner vortices enclosed by a single recirculating envelope on either side of the vertical axis.
- For all values of SF, the thermal boundary layer on the upper half of the annular region develops into a plume projecting upwards. An increase in SF from 0.6 to 0.8 is accompanied by a slight contraction of the thermally stagnant zone located at the bottom of the annulus. When full-slip (SF=1) is applied, the plume originating from the top of the hot wall shrinks and rises more, while the stagnant zone occupies the narrow region at the bottom of the annulus.

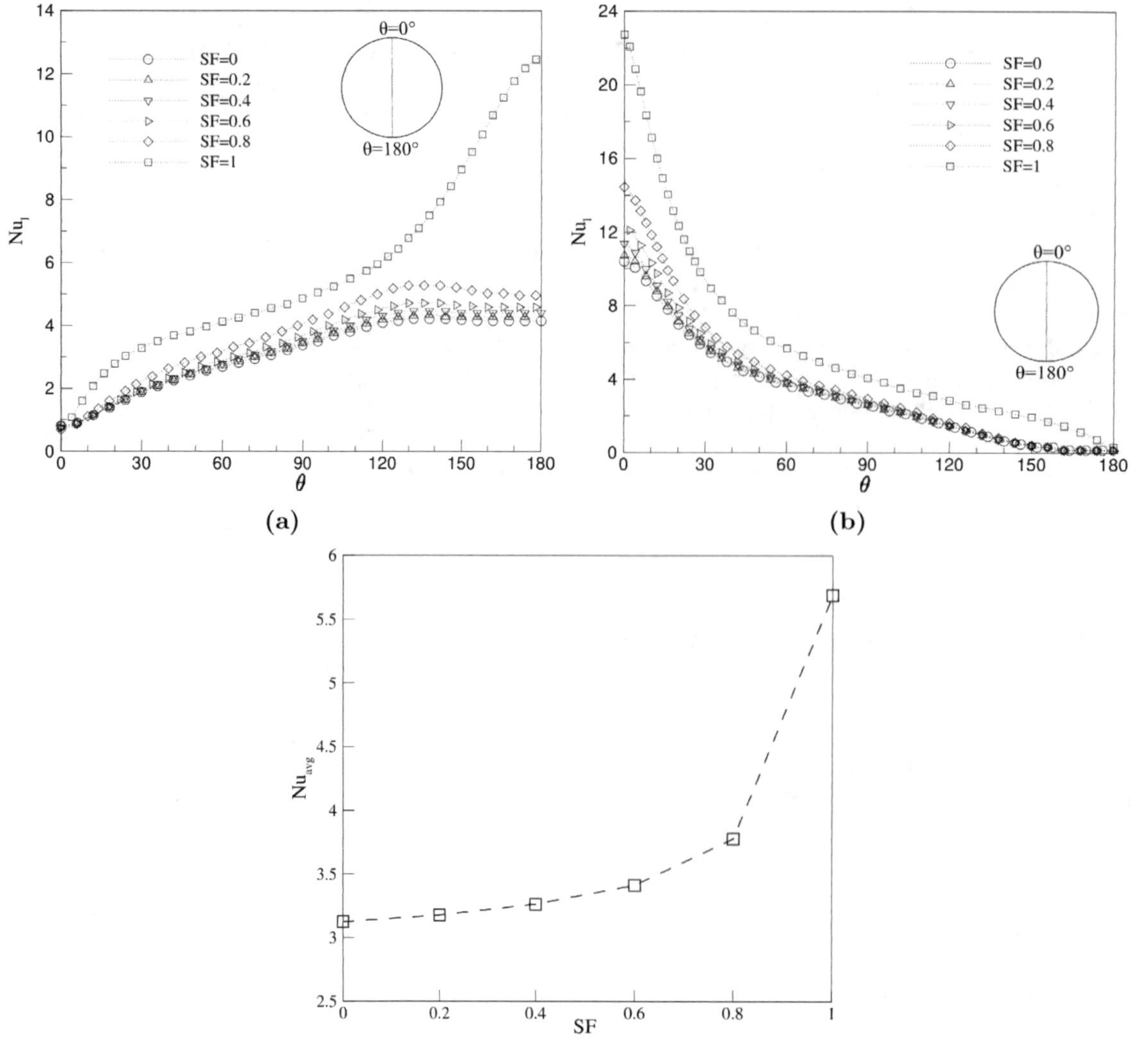

Figure 5: Nusselt number variation : (a) Nu_l along hot wall (b) Nu_l along cold wall and (c) Nu_{avg} with SF

- Between SF=0 and SF=0.4, local Nusselt number along the hot wall monotonically increases up to $\theta = 130°$ beyond which Nu_l remains constant. For SF=0.6 and 0.8, after the monotonic increase up to $\theta = 130°$, Nu_l reduces which could be due to the increased influence of partial-slip. The combined effect of full-slip and the shrinkage in the thermally stagnant region produces a steep rise in the Nu_l-θ curve beyond $\theta = 120°$. Along the cold wall, Nu_l continuously decreases up to θ=150° for all slip configurations. Beyond θ=150°, the local Nusselt number remains constant up to SF=0.8. For full-slip configuration, the local Nusselt number monotonically decreases throughout the cold wall.
- Nu_{avg} variation is gradual and monotonically increasing up to SF=0.8 however, between SF=0.8 and SF=1, the curve witnesses a steep increase in the average Nusselt number.

IOP Conf. Series: Materials Science and Engineering 1132 (2021) 012042 doi:10.1088/1757-899X/1132/1/012042

References

[1] Kuehn T H and Goldstein R J 1976 *Journal of Fluid Mechanics* **74**(4)
[2] Boyd R D 1981 *International Journal of Heat and Mass Transfer* **24**(9)
[3] Rao Y F, Miki Y, Fukuda K, Takata Y and Hasegawa S 1985 *International Journal of Heat and Mass Transfer* **28**(3)
[4] Date A W 1986 *International Journal of Heat and Mass Transfer* **29**(10)
[5] Projahn U and Beer H 1985 *Wärme-und Stoffübertragung* **19**(4)
[6] Shahraki F 2002 *Numerical Heat Transfer, Part A: Applications* **42**(6)
[7] Guj G and Stella F 1995 *Numerical Heat Transfer, Part A: Applications* **27**(1)
[8] Shi Y, Zhao T S and Guo Z L 2006 *Computers & Fluids* **35**(1)
[9] Waheed M A 2007 *Numerical Heat Transfer, Part A: Applications* **53**(3)
[10] Legendre D, Lauga E and Magnaudet J 2009 *Journal of Fluid Mechanics* **633** 437–447

IOP Conf. Series: Materials Science and Engineering 1132 (2021) 012043 doi:10.1088/1757-899X/1132/1/012043

Effect of wall-slip on natural convection in the annulus between a circular cylinder and square enclosure

A S Ananthapadmanabhan, Sabin Santhosh, Damu Murali and S Ajith Kumar

Department of Mechanical Engineering, Amrita Vishwa Vidyapeetham, Amritapuri, India.

E-mail: ananthugravet@gmail.com

Abstract. Two-dimensional natural convection in the annulus between a hot circular cylinder and cold square enclosure is numerically investigated for Rayleigh number (Ra)=10^6 using OpenFOAM. The Prandtl number (Pr) and radius to length (R/L) ratio are fixed at 0.71 and 0.2 respectively and slip condition is imposed by varying the $Slipfactor$ (SF) from 0 (no-slip) to 1 (full-slip) in steps of 0.2. For each value of SF, the flow patterns are visualised using streamlines and heat transfer behaviour is examined using isotherms. The full-slip boundary condition (SF=1) witnesses significant phenomena like vortex splitting, shrinkage in thermally stagnant region and reduction in thermal boundary layer thickness. The heat transfer rate is found to increase with an increase in SF and results are quantified using local Nusselt number (Nu_l) and average Nusselt number (Nu_{avg}) along the hot wall. It is noticed that for full-slip (SF=1) configuration, the natural convection patterns and heat transfer characteristics undergo notable changes.

1. Introduction

Natural convection heat transfer in an annulus deserves considerable attention owing to various industrial applications in electronic cooling, nuclear engineering and safety, heat exchanging equipments etc; to cite a few. One of the earliest researches in this area was reported by Kuehn & Goldstein[1] wherein they studied natural convection in the horizontal annulus between two coaxial circular cylinders numerically and experimentally.

Convection analysis in rectangular enclosures with an internal heat source was primarily reported by Nasrine & Ghaddar[2]. The authors used a spectral element method to simulate natural convection phenomena in the annulus and established a correlation between Rayleigh number (Ra) and Nusselt number (Nu) that could explain the overall heat transfer characteristics. Moukalled & Acharya[3] investigated heat transfer inside a square enclosure containing a cylindrical heater for Rayleigh numbers 10^4 to 10^7 at three different R/L ratios 0.1, 0.2 and 0.3. They reported that the overall heat transfer increased with an increase in Rayleigh number at a constant R/L ratio however, the contribution of convection heat transfer to the overall heat transfer decreased as the R/L ratio increased. Cesini et al.[4] conducted numerical and experimental study of natural convection from a horizontal cylinder embedded in a rectangular cavity and studied the influence of the aspect ratio of the cavity and Rayleigh number (Ra) on the thermal fields and the Nusselt number (Nu). The authors also reported an enhancement in the

IOP Conf. Series: Materials Science and Engineering 1132 (2021) 012043 doi:10.1088/1757-899X/1132/1/012043

Nomenclature

g	Acceleration due to gravity $[m/s^2]$
L	Enclosure length [m]
Nu_l	Local Nusselt number along circular cylinder
Nu_{avg}	Surface-Averaged Nusselt number along circular cylinder
P	Non-dimensional pressure $(=\frac{pL^2}{\rho\alpha^2})$
p	Pressure [Pa]
Pr	Prandtl number $(=\frac{\nu}{\alpha})$
Ra	Rayleigh number $(=\frac{g\beta L^3(T_h-T_c)}{\nu\alpha})$
SF	Slip factor
T	Temperature [K]
T_c	Cold wall temperature [K]
T_h	Hot wall temperature [K]
U	Non-dimensional horizontal velocity
u	Horizontal velocity $[m/s]$
V	Non-dimensional vertical velocity
v	Vertical velocity $[m/s]$
X	Non-dimensional horizontal coordinate
x	Horizontal coordinate
Y	Non-dimensional vertical coordinate
y	Vertical coordinate

Greek Symbols

α	Thermal diffusivity $[m^2/s]$
β	Coeffcient of thermal expansion $[K^{-1}]$
λ	Slip length
ν	Kinematic viscosity $[m^2/s]$
ρ	Density $[kg/m^3]$
Θ	Non-dimensional temperature $(=\frac{T-T_c}{T_h-T_c})$
θ	Angle from top of circular cylinder

average heat transfer coefficients with increase in Ra.

The effect of natural convection in the annulus for arbitrary positions of the inner cylinder was conducted by Jami et al.[5]. They reported that the heat transfer was maximum when the cylinder was kept at the centre of the enclosure with no heat generation from it. Kim et al.[6] investigated the influence of location of the inner circular cylinder relative to the outer square enclosure on the heat transfer characteristics for different Rayleigh numbers. Similar to Kim et al.[6] a study conducted by Jaddoa et al.[7] considered the effect of aspect ratio of the system along with the location of the cylinder. The results obtained in their study showed that the maximum Nusselt number was observed on top of the cylinder when it was moved diagonally and vertically. It was also noted that the surface averaged Nusselt number increased with increase in the aspect ratio of the geometry. Hussain and Hussein[8] conducted a study on natural convection which was an extension to the study done by Kim et al.[6], but in this investigation a constant heat flux per unit surface area on the inner cylinder was considered. It was reported that the flow field did not have much impact at low Rayleigh numbers but had considerable effect at high Rayleigh numbers. A nonlinear behaviour for surface averaged Nusselt number which was a function of position was also observed. The influence of Prandtl number (Pr) on natural convection inside an annulus was investigated by Liao & Lin[9] with the help of flow structure map in the Ra-Pr plane. They reported that by decreasing Pr, the flow underwent a shift from steady single to steady double thermal plume and then after a further decrease in Pr, it became unsteady.

Rabani & Talebi[10] numerically simulated natural convection heat transfer in a square enclosure containing a partially heated circular cylinder and the effect of increase in heating angle, displacement heating, and Rayleigh number on the flow and temperature contours were reported. A three dimensional natural convection in the annulus between an inner hot cylinder and an outer cold cubic enclosure was studied by Souayeh et al.[11] for various tilted angles of the enclosure and Rayleigh numbers. The results obtained showed that for an inclination of 90°and Rayleigh number 10^6, optimum average heat transfer was attained. Lee et al.[12] studied the natural convection in the annulus which emphasized the effects on flow and thermal fields in the system. It was found that the alteration in the localized heated zone at the bottom of the enclosure wall affected the flow and thermal fields. The authors reported a minor variation in the thermal and flow fields due to the changes in the size of heated section. The effect on natural convection by four different polygonal enclosures with an inner cylinder was studied by Saleh et.al[13] and reported that the profile of the Nusselt number was not influenced by the polygonal geometry. Mun et al.[14] investigated the natural convection in the annulus of four inner cylinders which were arranged in a diamond array within a square enclosure and reported the effect of distance between the cylinders and Rayleigh number on the time dependence of flow fields and isotherms. A three dimensional natural convection study for a sinusoidal bottom wall temperature variation in the annulus of an outer square enclosure and inner circular cylinder was conducted by Lee et al.[15].The thermal plume was observed to be influenced by bottom wall temperature fluctuation and Rayleigh number. It was also reported that the thermal and flow fields depended on three factors such as size, count and rotational direction of the thermal plumes.

Numerous investigations have been carried out in the past which discuss the heat transfer between an eccentric cylinder and a square enclosure[16], a square enclosure with two inner cylinders at different vertical positions[17], a cylindrical heat source kept in a square enclosure with adiabatic vertical walls and varying temperature on the bottom wall[18],the effect of tilted angle of enclosure on the natural convection[19]. All the reported experimental and numerical analyses on the annular convection phenomena have used no-slip boundary condition on the hot and cold surfaces and to the best of our understanding there is hardly any report on how the natural convection phenomena is affected by the slip condition applied on the boundary walls. This paper is hence devoted to find out the effect of slip on natural convection in the annulus of a horizontal circular cylinder and a square enclosure. The results of our investigations can be insightful in industrial and academic research pertaining to wettability of surfaces especially in heat pipes where the presence of hydrophobic surfaces can limit the formation of residuals thereby improving anti-freezing characteristics.

The structure of the presentation of our investigations in the rest of the article is as follows. Numerical methodology in Section 2 describes the solver details, governing equations and the boundary conditions imposed. Results and discussion in Section 3 includes the grid independency and code validation followed by the presentation of results. The flow and thermal fields visualised using streamlines and isotherms have been illustrated followed by a thorough discussion on the heat transfer characteristics quantified using the local and surface averaged Nusselt number. Conclusions in Section 4 summarises the overall numerical findings.

2. Numerical methodology

Numerical simulations have been performed using the finite volume open source computational fluid dynamics (CFD) toolbox OpenFOAM. The generic solver *buoyantBoussinesqPimpleFoam* is used to solve the two-dimensional steady, laminar and incompressible governing equations in cartesian co-ordinates as given below. For the spatial discretization of diffusive and convective terms, a second order central differencing *Gauss linear* scheme is used.

$$Continuity: \frac{\partial U}{\partial X} + \frac{\partial V}{\partial Y} = 0 \tag{1}$$

$$X - momentum: U\frac{\partial U}{\partial X} + V\frac{\partial U}{\partial Y} = -\frac{\partial P}{\partial X} + \Pr\left(\frac{\partial^2 U}{\partial X^2} + \frac{\partial^2 U}{\partial Y^2}\right) \tag{2}$$

$$Y - momentum: U\frac{\partial V}{\partial X} + V\frac{\partial V}{\partial Y} = -\frac{\partial P}{\partial Y} + \Pr\left(\frac{\partial^2 V}{\partial X^2} + \frac{\partial^2 V}{\partial Y^2}\right) + RaPr\Theta \tag{3}$$

$$Energy : U\frac{\partial \Theta}{\partial X} + V\frac{\partial \Theta}{\partial Y} = \frac{\partial^2 \Theta}{\partial X^2} + \frac{\partial^2 \Theta}{\partial Y^2} \quad (4)$$

The non-dimensional variables in the governing equations are defined as follows

$$X = \frac{x}{L}, \quad Y = \frac{y}{L}, \quad U = \frac{uL}{\alpha}, \quad V = \frac{vL}{\alpha}, \quad \Theta = \frac{T - T_c}{T_h - T_c}, \quad P = \frac{pL^2}{\rho\alpha^2} \quad (5)$$

where u and v are the components of velocity in x and y directions, ρ is the density of the fluid, α is the thermal diffusivity, T_c and T_h are temperatures at cold and hot walls and p is the pressure.

Rayleigh number, which is defined on the basis of the length of the enclosure is $Ra = \frac{g\beta L^3(T_h - T_c)}{\nu\alpha}$ and Prandtl number as $Pr = \frac{\nu}{\alpha}$ where, ν is the kinematic viscosity of the fluid, g is the acceleration due to gravity acting downwards in the negative y direction and β is the coefficient of thermal expansion.

A schematic of the computational domain is given in Figure 1(a) which consists of a square enclosure of length L with an inner cylinder (R=0.2L) located at its centre. The inner cylinder is kept at a constant high temperature Θ=1 whereas the outer square enclosure is maintained at a relatively low temperature Θ=0.

Slip conditions on the walls are varied using the *partialSlip* boundary condition (BC) available in OpenFOAM. *partialSlip* represents a mixed boundary condition which is an implementation of the Navier slip law as expressed below. For a 2-D flow, the first order approximation for the velocity slip boundary is given by

$$U + \lambda\frac{dU}{dY} = 0 \quad (6)$$

here λ is the slip length as defined in Legendre et al.[20], Silvia and Bodgan[21]

Let U_0 be the tangential velocity on the wall face boundary and U_1 the tangential cell center velocity at the boundary cell. Equation (6) can be re-written in terms of U_0, U_1 and D as follows

$$U_0 + \lambda\frac{(U_0 - U_1)}{D/2} = 0 \quad (7)$$

where D is the size of the boundary cell. The above equation after re-arranging becomes

$$U_0 = \frac{2\lambda}{D + 2\lambda}U_1 \quad (8)$$

Here, we introduce a new term $Slip\, factor(SF)$ defined as

$$SF = \frac{2\lambda}{D + 2\lambda} \quad (9)$$

Hence equation (8) becomes

$$U_0 = (SF)U_1 \quad (10)$$

For $\lambda \to 0$, we have $SF = 0$, which indicates no-slip condition.
For $\lambda \to \infty$, we have $SF = 1$, which indicates full-slip condition.

IOP Conf. Series: Materials Science and Engineering 1132 (2021) 012043 doi:10.1088/1757-899X/1132/1/012043

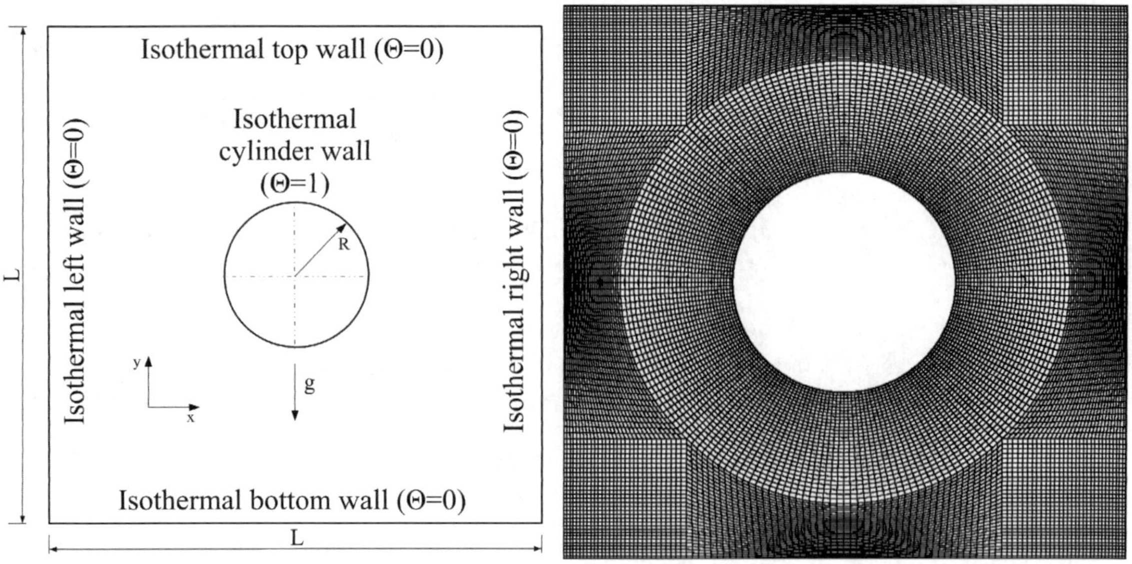

(a) (b)

Figure 1: Schematic of (a) computational domain (b) grid structure used in the problem

The rate of heat transfer is computed at each wall and is expressed in terms of local Nusselt number (Nu_l) and average Nusselt number (Nu_{avg}).

$$Nu_l = \left.\frac{\partial \Theta}{\partial n}\right|_{\text{wall}}, \quad Nu_{avg} = \frac{1}{W}\int_0^W Nu_l \, \mathrm{d}S \tag{11}$$

here n is the unit normal vector to the wall and W is the surface area of the wall.

Conditions imposed at the walls other than the slip boundary conditions are as follows:
Left, right, top and bottom walls : Dirichlet boundary condition, Θ=0
Inner cylinder wall : Dirichlet boundary condition, Θ=1

The radiation heat transfer is considered negligible and the fluid properties are assumed to be constant, except for the variation in density which is obtained by the Boussinesq approximation.

3. Results and discussion

Computations have been carried out for Ra=10^6 and Pr=0.71 by changing the slip conditions, $0 \leq$ SF ≤ 1. Section 3.1 discusses the grid sensitivity studies and details of validation of the present numerical code is explained in Section 3.2. Section 3.3 and Section 3.4 are dedicated to the illustration of flow and thermal fields and the heat transfer characteristics are examined in Section 3.5 by plotting the local Nusselt number (Nu_l) distribution and average Nusselt number (Nu_{avg}) along the walls.

3.1. Grid independency

Grid independency tests are conducted to minimise the influence of number of cells on the computational results and to optimise the computational cost. In order to accurately capture the boundary layer phenomenon, finer mesh is adopted near the walls and coarser mesh for other regions.

Table 1: Grid independency data for Ra=10^5, Pr=0.71, SF=0

Number of cells	Nu_{avg}
5100	7.977
8000	7.937(-0.501%)
11500	7.930(-0.088%)
15600	7.928(0.025%)
20300	7.899(-0.365%)
25600	7.905(0.076%)

Table 1 shows the variation in Nu_{avg} while increasing the number of cells. It is evident that, the percentage difference is minimum (0.025%) when the cell count is 15600 beyond which an increase in grid count is accompanied by accumulation of numerical errors hence an optimum grid structure comprising of 15600 cells is employed throughout our computations.

3.2. Validation

Validation of the solver for the present computations is performed by comparing the Nusselt number data available in the literature. Table 2 compares the Nu_{avg} values with no-slip condition at the walls obtained using the present code with the existing data. The percentage deviation of the computed Nu_{avg} values with those reported in the literature are indicated in brackets. The results obtained are found to be in good agreement and the maximum deviation is less than 3%.

Table 2: Validation: Nu_{avg} comparison with previous studies for Pr=0.71, SF=0

Ra	Present study	Park et al.[16]	Kim et al.[6]	Kim et al.[22]	Moukalled & Acharya[3]
10^4	5.147	5.128(-0.374%)	5.108(-0.767%)	5.113(-0.669%)	5.286(-2.625%)
10^5	7.93	7.836(-1.203%)	7.767(-2.102%)	7.75(-2.326%)	—
10^6	14.377	14.462(-0.589%)	14.11(-1.89%)	14.2(-1.245%)	14.228(-1.242%)

3.3. Streamlines

Figure 2 illustrates the flow fields for various slip configurations visualised using str eamlines. The fluid in the vicinity of the hot wall gets heated up, becomes less dense and begins to a scend. As soon it reaches the top wall, it interacts with the cold surface, migrates outwards, becomes more dense and moves down. While descending, some of the cold fluid a gain w arms u p d ue t o the interaction with the hot surface and its neighbourhood thus giving rise to a pair of recirculating convection currents manifested as vortices. The physical mechanism of the formation of natural convection currents is explained in Moukalled & Acharya[3] and Kim et al.[6].

The no-slip condition represented by SF=0 witnesses two recirculating primary vortices symmetric about the vertical axis. The cores of these vortices are found to occupy the upper half of the domain indicating strong convection currents. In addition to the occurrence of counter-rotating primary vortices on either side of the vertical axis, small recirculating secondary vortices are also visible near to the bottom wall.

IOP Conf. Series: Materials Science and Engineering 1132 (2021) 012043 doi:10.1088/1757-899X/1132/1/012043

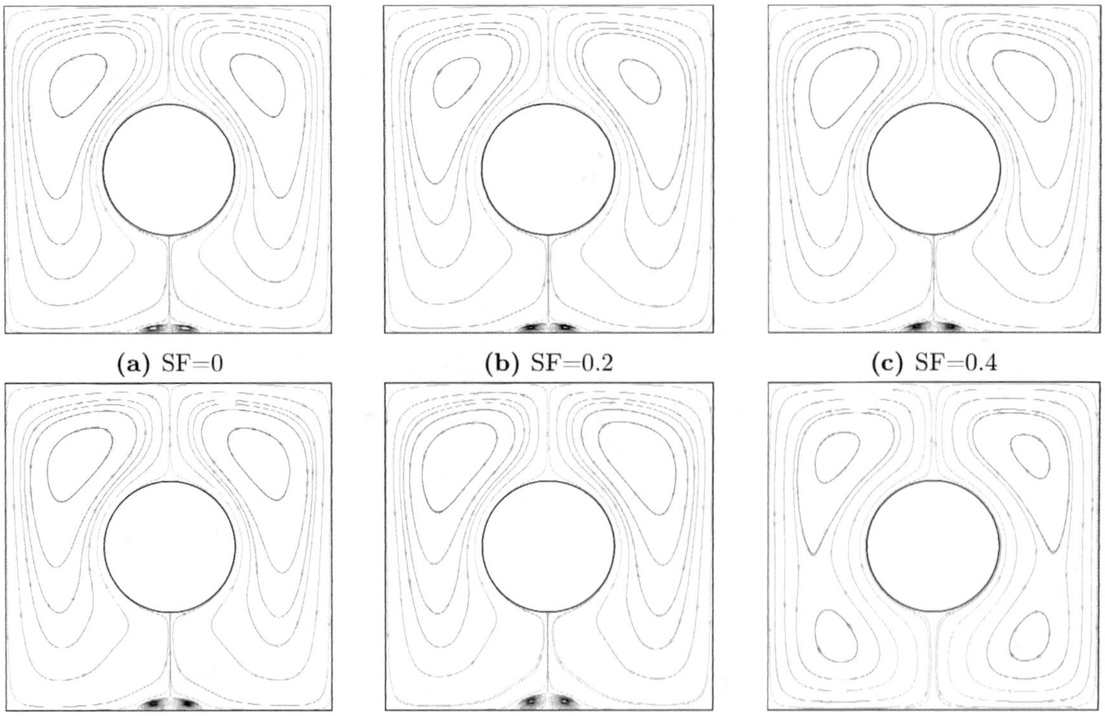

(a) SF=0 **(b)** SF=0.2 **(c)** SF=0.4

(d) SF=0.6 **(e)** SF=0.8 **(f)** SF=1

Figure 2: Streamlines for various slip configurations at Ra=10^6

With increase in slip factor, the counter-rotating secondary vortices are seen to enlarge gradually and consistently upto SF=0.8. In all these cases, the pair of primary vortices continue to exist with their core at the upper half. Introduction of full-slip (SF=1) on the surface of hot and cold walls brings notable changes in the convection currents. The core of the counter-rotating primary vortices on either side of the vertical axis splits into two thereby giving rise to two additional inner vortices however, the circulating outer envelope remains in position. Also, the tiny recirculating secondary vortex pair near to the bottom wall vanishes. This drastic variation in convection has significant contribution to the thermal fields as discussed below.

3.4. Isotherms

Previous literature have observed the dominance of convection currents in the annular region at Ra=10^5 and above. Due to this, the thermal boundary layer on the top half of the hot cylinder gets separated and develops into a plume in the upward direction before the flow becomes steady. The effect of the plume formation is felt in the adjacent layers finally giving way to distorted isotherms. The thermal gradients at the upper half of the domain is very high compared to the other half due to the formation and progression of the thermal plume. This is exactly the same reason why the cores of the primary vortices belong to the upper half as evidenced in the streamlines. Figure 3 illustrates the temperature contours (isotherms) for various slip configurations. The isotherms at no-slip condition (SF=0) is in excellent match with previously recorded literature (Kim et al.[6], Sheikholeslami et al.[23]). The increased thermal stagnation at Ra=10^6, which is predominant in the lower half of the domain, acts as a hindrance to the penetration of flow to the bottom of the annulus. This is another reason for the occurrence of weak thermal gradients and convection currents at the bottom.

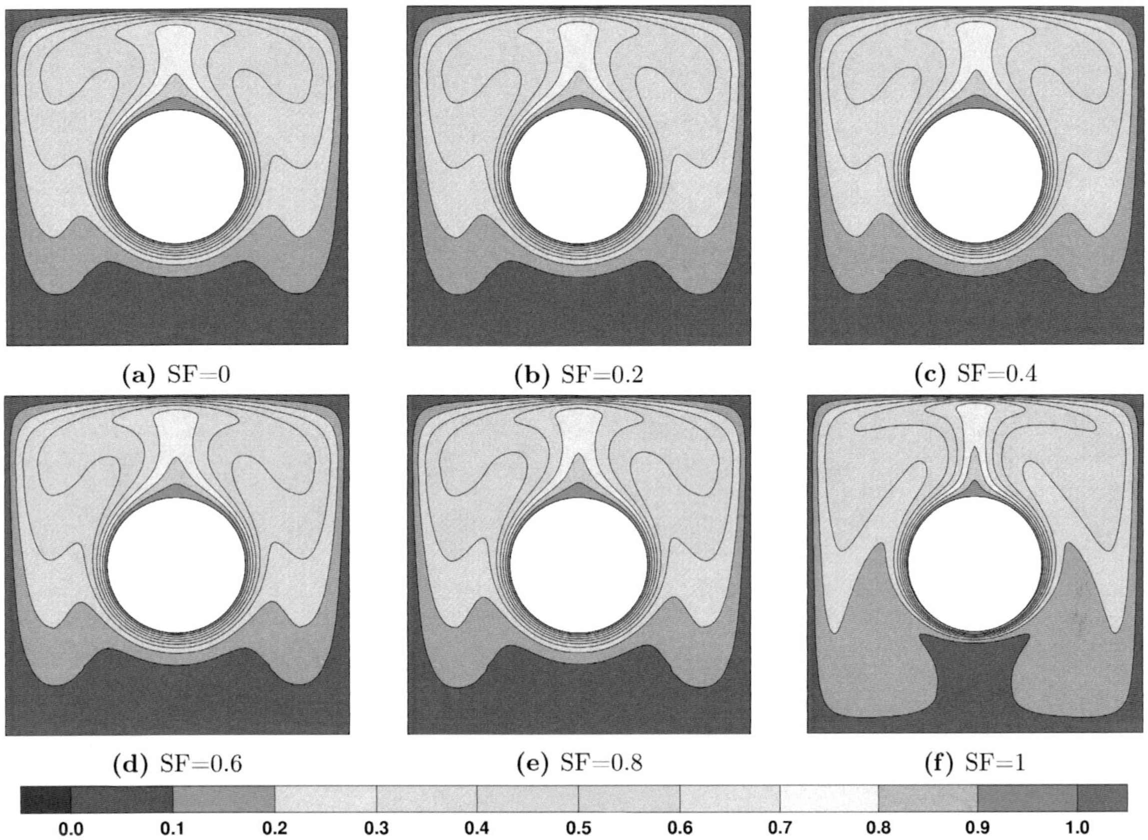

(a) SF=0 (b) SF=0.2 (c) SF=0.4

(d) SF=0.6 (e) SF=0.8 (f) SF=1

Figure 3: Isotherms for various slip configurations at Ra=10^6

With increase in slip factor, close observations reveal the plume shifting slightly vertically upward towards the top wall upto SF=0.8. Visual investigations in this range, however, reveal little variation in the thermally stagnant region near to the bottom wall. Introduction of full-slip conditions on the hot and cold walls brings changes everywhere. The hot plume shrinks and extends further upward whose impact reaches the lower half. The stagnant region suddenly shrinks and occupies the space between the bottom wall and lower half of the inner cylinder. The boundary layer on the lower hot surface becomes thinner influencing t he h eat transfer characteristics which is discussed in the next section.

3.5. Variation in Nusselt number

Figure 4(a) shows the distribution of local Nusselt number (Nu_l) along the right half of the hot wall. Due to the symmetric nature of flow a nd t hermal fi elds, th e he at tr ansfer characteristics too are same on either side of the vertical axis hence, the Nu_l variation along the left surface is same as that of the right side of which the later is considered for analysis.

IOP Conf. Series: Materials Science and Engineering 1132 (2021) 012043 doi:10.1088/1757-899X/1132/1/012043

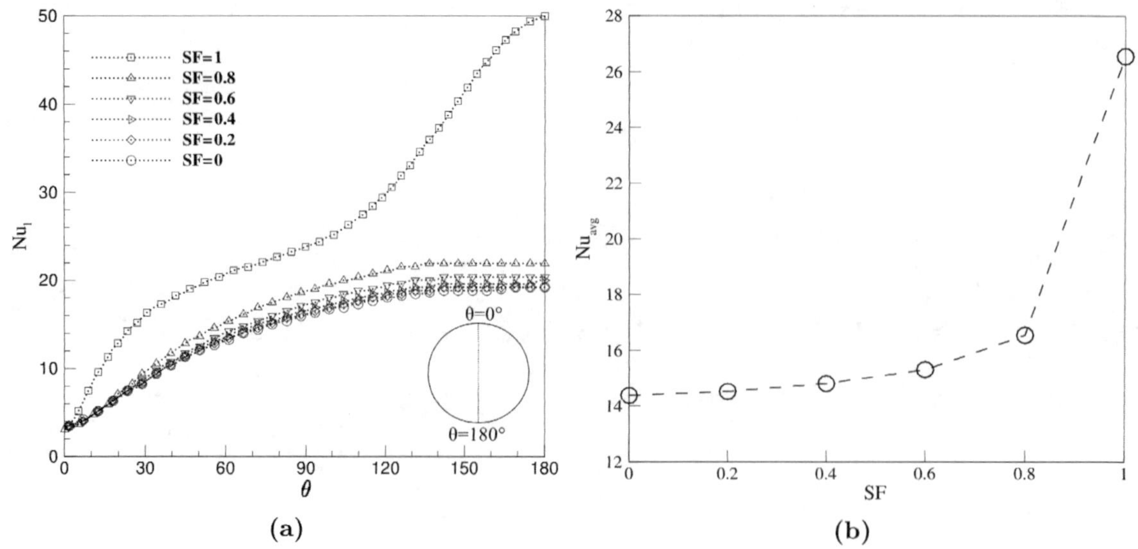

Figure 4: Nusselt number variation: (a) Nu_l variation along the hot wall and (b) Nu_{avg} variation with slip factor

Moving from θ=0° to θ=180° in the clockwise direction, the development of plume increases the boundary layer thickness in its vicinity thereby reducing the rate of heat transfer. The influence of plume is less felt on the lower half and the comparatively thinner boundary layer enhances the heat transfer characteristics. The local Nusselt number (Nu_l) monotonically increases between θ=0° and θ=150° irrespective of the slip conditions. Between θ=150° and θ=180°, the local Nusselt number is nearly constant (except for SF=1) as the heat transfer takes place mainly due to conduction since the flow is restricted from penetrating to the bottom of the enclosure. Nu_l profile for SF=0 matches with previous studies such as Kim et al.[6] and Aithal[24]. The trend shown in the Nu_l-θ plot is same for slip values in the range SF=0 to SF=0.8, however, the heat transfer characteristics for the full-slip boundary condition receives special attention. For SF=1, the slope of Nu_l-θ curve between θ=0° to θ=30° is large due to greater heat transfer rate. Due to the splitting up of inner core of the vortex, the convection currents are evenly distributed which enhances the overall heat transfer characteristics. This is reflected in the higher Nu_l values throughout the hot surface. Between θ=120° and θ=180°, the Nu_l-θ curve is more steeper owing to the shrinkage in thermal stagnation and decrease in thermal boundary layer thickness. This can also be visualised in the corresponding temperature profile (Figure 3(f)).

Figure 4(b) shows the average Nusselt number (Nu_{avg}) for the entire range of slip conditions. The Nu_{avg} value at no-slip condition matches well with the previously reported values by Kim et al.[6]. It is noticed that the heat transfer rate increases with increase in slip factor. Between SF=0 and SF=0.8, the introduction of slip conditions, even though marginal, enhances the heat transfer rate. However, there is steep rise in Nu_{avg} with the introduction of full-slip on the hot and cold walls. The shrinkage of thermal plume and thermal stagnation together with the distribution of convection currents to the lower side contribute to the enhancement in overall heat transfer characteristics.

4. Conclusions

Numerical analysis is conducted to investigate the effect of wall-slip on natural convection in the annulus between a square enclosure and circular cylinder. It is observed that increase in slip factor from SF=0 to SF=0.8 brings marginal variation in flow and thermal fields and overall heat transfer rate however, introduction of full-slip (SF=1) induces stronger convection currents. The other inferences are given below

- Between SF=0 and 0.8, the flow field witnesses two counter-rotating primary vortices symmetric about the vertical axis with the vortex cores occupying the upper half. The boundary layer near to the bottom wall separates giving rise to a pair of tiny secondary vortices.
- At full-slip (SF=1), the cores of the primary vortices on either side breaks into 2 vortices thereby evenly distributing the convection currents.
- For all cases investigated, the thermal boundary layer on the top half of the hot circular cylinder develops into a plume projecting upwards.
- Thermal stagnation is predominant for slip values between SF=0 and SF=0.8. For full-slip (SF=1) condition, the stagnation shrinks and occupies the space between the bottom wall and lower half of the circular cylinder.
- The increase in slip factor from SF=0 to 0.8 is accompanied by a monotonic increase in local Nusselt number (Nu_l) upto θ=150° beyond which the variation is nearly constant. For full-slip condition, the distribution of convection currents and shrinkage in thermal stagnation bring about significant enhancement in the overall heat transfer rate.
- The Nu_{avg} marginally increases between SF=0 and 0.8 and the introduction of full-slip brings a steep increase in the heat transfer rate.

References

[1] Kuehn T H and Goldstein R J 1976 *Journal of Fluid Mechanics* 695–719
[2] Nasrine K and Ghaddar 1992 *International Journal of Heat and Mass Transfer* 2327–2334
[3] Moukalled F and Acharya S 1996 *Journal of Thermophysics and Heat Transfer* 524–531
[4] Cesini G, Paroncini M, Cortella G and Manzan M 1998 *International Journal of Heat and Mass Transfer* 1801–1811
[5] Jami M, Mezrhab A and Naji H 2008 *Engineering Computations: International Journal for Computer-Aided Engineering and Software* 480–489
[6] Kim B S, Lee D S, Ha M Y and Yoon H S 2008 *International Journal of Heat and Mass Transfer* 1888–1906
[7] Jaddoa A A, Ekaid A L and Al-Sadawi L 2020 *Journal of Engineering Science and Technology* 2472 – 2491
[8] Hussain S H and Hussein A K 2010 *International Communications in Heat and Mass Transfer* 1115–1126
[9] Liao C C and Lin C A 2014 *Applied Thermal Engineering* 1–7
[10] Rabani R and Talebi S 2015 *Progress in Computational Fluid Dynamics* 372–387
[11] Souayeh B, Ben-Cheikh N and Ben-Beya B 2016 *International Journal of Thermal Sciences* 325–339
[12] Lee H, Doo J, Ha M and Yoon H 2013 *International Journal of Heat and Mass Transfer* 435–450
[13] Saleh H, Alsabery A I and Hashim I 2015 *Advances in Mechanical Engineering* 1–10
[14] Mun G S, Seo Y M, Park Y G and Ha M Y 2017 *International Journal of Heat and Mass Transfer* 755–770
[15] Lee S H, Seo Y M, Yoon H S and Ha M Y 2016 *International Journal of Heat and Mass Transfer* 807–823
[16] Park Y G, Yoon H S and Ha M Y 2012 *International Journal of Heat and Mass Transfer* 7911–7925
[17] Park Y G, Ha M Y, Choi C and Park J 2014 *International Journal of Heat and Mass Transfer* 501–518
[18] Park J, Kim M, Mun G S, Park Y G and Ha M Y 2018 *Journal of Mechanical Science and Technology* 3201–3211
[19] Park H K, Ha M Y, Yoon H S, Park Y G and Son C 2013 *International Journal of Heat and Mass Transfer* 295–314
[20] Legendre D, Lauga E and Magnaudet J 2009 *Journal of Fluid Mechanics* 437–447
[21] Silvia J B and Bogdan V 2015 *International Journal of Heat and Mass Transfer* 1–10
[22] Kim M, Doo J H, Park Y G, Yoon H S and Ha M Y 2014 *Journal of Mechanical Science and Technology* 5013–5025
[23] Sheikholeslami M, Gorji-Bandpy M, Pop I and Soleimani S 2013 *International Journal of Thermal Sciences* 147–158
[24] Aithal S 2018 *International Journal of Heat and Mass Transfer* 961–975

IOP Conf. Series: Materials Science and Engineering 1132 (2021) 012044 doi:10.1088/1757-899X/1132/1/012044

Chemical and Enzymatic transesterification of rubber seed oil – A comparative study

Eldhose K Joy[1], Jilse Sebastian[2], Eldose K K[3], Jithesh S R[4]

[1,3,4] Department of mechanical Engineering, Adi Shankara Institute of Engineering and Technology, Kalady.
[2] Department of mechanical Engineering, St. Joseph College of Engineering and technology, Pala.

Email:eldhosekj.me@adishankara.ac.in,jilsesebastian@gmail.com,eldhosekk.me@adishankara.ac.in, jithesh.me@adishankara.ac.in,

Abstract. Biodiesel produced from a wide variety of vegetable oils is expected to replace diesel fuel. Conventional diesel engines are the major sources of expelling the particulate matter, soot, smoke and oxides of nitrogen to the environment which are harmful. Biodiesel is an attractive alternative fuel to replace petroleum derived diesel because it is environment friendly and can be produced from edible or non-edible vegetable oils. The conflict between food and fuel led to the identification of more than fifty non-edible oil sources for biodiesel production. Utilization of locally available non-edible oil sources for production of biodiesel reduces the cost of fuel and increases production interest. Rubber seed oil is a non-edible type vegetable oil, identified as a potential feedstock for biodiesel production. This paper comprises the synthesis of biodiesel from high free fatty acid rubber seed oil through transesterification by chemical and enzymatic methods. The transesterification of rubber seed oil with chemical catalyst and enzyme catalyst were investigated in the work. The variables affecting the reaction such as molar ratio, catalyst concentration, time and temperature were analysed. Kinetic study was carried out to find the reaction constants and activation energy of both chemical and enzymatic transesterification reactions. The properties of fuel such as kinematic viscosity, calorific value and specific gravity were found out and compared with the properties of diesel. Base catalyst concentration was varied from 0.5wt% to 2wt% for alkaline transesterification process and the result revealed that maximum yield of 90% was obtained with a base catalyst concentration of 1.5wt% with 1:9 molar ratio at 55± 3 °C. The best condition arise in enzymatic reaction was 10wt% enzyme and 1:4 methanol ratio with 5wt% addition of water as the solvent.

1. Introduction

Energy security is the constant availability and supply of affordable energy for consumers and industry. Risks to energy security include disruptions to the supply of imported fossil fuels, limited availability of fuel, and energy price hikes. Besides these, the greenhouse gas emissions get along with petroleum fuels creates many environmental problems. These issues look forward for a better replacement of fossil fuels. Many countries have already implemented deriving biofuels from locally available sources and using them as alternatives to petroleum products. Vegetable oils are promising alternatives to diesel since their properties are close to diesel. Since edible oils are used as food, non-edible oil sources (more than fifty are identified) are preferred for the application. In India, there are a variety of non-edible oils that can be potential sources for biodiesel production [1].

The high demand for diesel in the industrialized world and pollution problems caused by its wide use make it necessary to develop safe renewable energy sources with a smaller environmental impact than the conventional fuel sources. Biologically derived fuels are renewable, sustainable and expandable to meet the growing demand. They have the potential to be domestically and globally available for energy security [2]. Biodiesel is free from sulphur and its combustion will reduce the number of serious pollutants. Thus, the need for development of biodiesel as an alternative fuel for the diesel engines could reduce the harmful emissions without compromising on the performance. A major economic challenge for the biodiesel industry is the high cost of refined vegetable oils which constitutes between 70 and 85% of the overall production cost. In order to reduce the cost of biodiesel and the social impact, alternative feedstock that are readily available in large quantities and at low cost must be considered. One of these ways is to use non-edible vegetable oils for biodiesel production. An important technological challenge in biodiesel production is the development of heterogeneous catalysts and enzyme catalysts. The use of heterogeneous catalysts would result in simpler and cheaper separation processes, a reduced water effluent load, and reduced capital and energy costs. Enzyme catalysis proceeds without the generation of by-products, with easy recovery of biodiesel product, mild reaction conditions, and is insensitive to FFAs. The catalyst can be reused, thus contributing to reduced energy consumption and waste generation compared to the conventional base-catalyzed chemical process. Rubber seed oil is a promising non-edible oil being explored as an alternative fuel resource for diesel engine. The large availability of rubber seed from the existing rubber plantations in India and South East Asia region can potentially reduce the price of biodiesel production [3].

The high viscosity and free fatty acid of rubber seed oil are incompatible to use in diesel engines. The commonly employed methods, to reduce the viscosity of the vegetable oils are blending with diesel, emulsification, pyrolysis, cracking and transesterification. Among this transesterification is the most common method for biodiesel production [4]. High acid value feedstock's follows acid pretreatment and alkaline esterification process [5]. A two-step transesterification process is developed to convert the high FFA oils to its mono-esters. The first step, acid catalyzed esterification reduces the FFA content of the oil from 17% to less than 2%. The factors affecting the process, the conversion efficiency and FFA reduction were studied. The maximum conversion efficiency reported at 0.5wt% acid catalyst and 9:1 molar ratio with temperature range of 45±5 ^{o} C. Alkaline transesterification process was carried out at 1:6 molar ratio giving a maximum yield 98% with 0.5wt% NaOH as base catalyst. The oil used in this study was the filtered rubber seed oil of 17% FFA [6]. Rubber seed oil has high FFA content of 45% was investigated by Junaid Ahammed et al. [7]. A methanol ratio of 50%v/v of rubber seed oil and 10%v/v of sulphuric acid was found as the optimum conditions for the reduction of acid value. The maximum conversion of methyl esters (biodiesel) was reported as 96.8% at 6:1 molar ratio, 1% catalyst concentration, 55 ^{o} C reaction temperature and 67.5 minutes reaction time. Satyanarayana and Muraleedharan employed two step pretreatment processes for the production of methyl ester from rubber seed oil having acid value of 48 (24% FFA). The methanol ratio of 75%v/v of rubber seed oil with 1%v/v concentrated sulphuric acid in two steps with one hour reaction reduced an acid value to 1.72mgKOH/g. The base catalyst used in this study was KOH and a yield of 98% was reported after 40 minutes reaction time at 55±5 ^{o} C. A two-step pretreatment process (acid esterification) was developed in this study for the reduction of acid value in an effective range [8]. The production of biodiesel using a biocatalyst eliminates the disadvantages of the alkali process by producing product of very high purity and less or no downstream processing [9]. Several literatures are available on the enzymatic biodiesel production by Candida Antarctica lipase [10-13]. A number of studies are reported on various enzymes and their reaction condition for biodiesel production. However, only few studies are related with rubber seed oil.

2. Characteristics of Rubber seed oil

Selection of vegetable oil used for biodiesel production is one major task to overcome the high cost of refined oils, problems related to food and fuel, treatment of feed stocks etc. Nowadays, researchers are looking into the locally available non edible sources for the fuel production in order to overcome the cost and other issues. In Kerala, rubber tree is a major plantation crop which accounts about 92% of India's rubber production. Use of rubber seed oil for biodiesel production gives an added benefit to the farmers apart from the natural rubber latex as the main product.

Commercially available rubber seed oil has dark brown in colour [8], and high acid value. The fatty acid composition of rubber seed oil consists of 18.9% saturation comprising of palmitic and stearic acids and 80.5% unsaturation comprising mainly of oleic, linoleic and linolenic acids. Saturated fatty acid methyl esters increase the cloud point, cetane number and improve stability whereas more poly-unsaturation reduces the cloud point and cetane number and stability [6]. The fatty acid composition of rubber seed oil was analysed by GC MS technique in Care Keralam Ltd., Koratty. The molecular formula of rubber seed oil is determined as $C_{18}H_{32}O_{32}$ and molecular weight is about 278. Table 2.1 shows the results of GCMS analysis

Table 2.1. Fatty acid composition of rubber seed oil

Fatty acid	Composition (%)
Saturated	
1. Palmitic acid $C_{16:0}$	10.2
2. Stearic acid $C_{18:0}$	8.7
Unsaturated	
3. Oleic acid $C_{18:1}$	24.6
4. Linoleic acid $C_{18:2}$	39.6
5. Linolenic acid $C_{18:3}$	16.3
Others	0.6

The GC MS detailed chromatogram of rubber seed oil was as shown in Fig. 2.1. The various properties of rubber seed oil was found out by different test methods. The major properties of RSO for the present work are compared with the previous studies and described in Table 2.2.

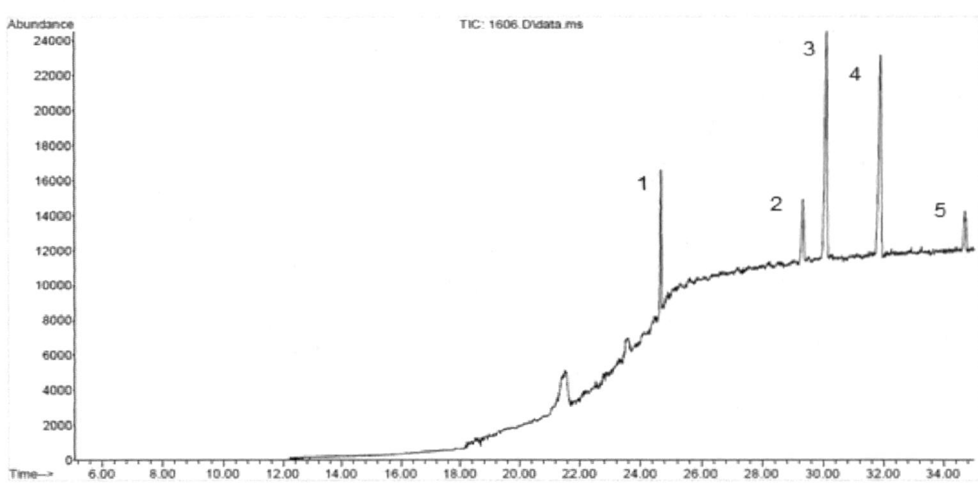

Fig. 2.1. GC MS chromatogram of rubber seed oil (1) Linolenic acid (2) Palmitic acid (3) Linoleic acid (4) Oleic acid (5) Stearic acid and 1, 2-Cyclononadiene

Table 2.2. Property comparison of rubber seed oil with previous studies

Property	Ramadhas et al. [6]	Satyanarayana and Muraleedharan [8]	Present work
Specific gravity	0.91	0.883	0.91
Viscosity (mm^2/s) at 40 °C	66.2	34.69	59.10
Flash point (°C)	198	237	210
Calorific Value (MJ/kg)	37.5	38.44	37.8
Acid value (mgKOH/g)	34	48	82

3. Experimental section

3.1. Materials and Equipment

The feed stock used in this work is rubber seed oil (RSO). The chemicals and other reagents used in the preparation of biodiesel are sulphuric acid, KOH, methanol (99.8% pure), toluene, ethanol, n-hexane, t butanol, de-ionised water for enzymatic method. The enzyme used in the present work is Candida Antarctica lipase B 10000 immobilized.

The equipment and instruments used for the production process and analysis of biodiesel from rubber seed oil are Magnetic stirrer with hot plate, Incubated shaker, Centrifuge, Weighing balance, Micro pipette, Brookfield viscometer, Bomb calorimeter and Pensky Martin apparatus.

3.2. Acid value estimation of rubber seed oil

The primary part of biodiesel preparation using vegetable oils is determination of acid value or FFA content in the oil. It is very essential to estimate the base catalyst amount or to decide whether the feed stock requires any type of pre-treatment. The acid value of the collected rubber seed oil was found out by standard titration (AOCS) method [14]. 5 g of rubber seed oil was well mixed with 125 ml solution (50% isopropanol and 50% toluene) titrated against 0.1N KOH solution using Phenolphthalein as the colour indicator.

Acid value can be calculated using the equation

$$AV = [(A-B) \times 56.1 \times N] \div W \quad (1)$$

Where, A= number of ml of KOH to neutralize sample
B= number of ml of KOH to neutralize blank
N= normality of KOH (0.1N in this case)
W= weight of RSO sample.

The acid value (eqn. 1) of rubber seed oil was found to be 82 (FFA of 41%) and the direct alkali transesterification will not convert the high FFA into methyl ester. Rubber seed oil was directly used in this work without any type of pre-treatment process because the enzymes are capable of transesterify any range of FFA and the comparison will be easier with chemical and enzymatic method.

3.3 Acid Esterification (Two step pretreatment process) of rubber seed oil

If the acid value of the oils or fats is very high, one-step esterification pre-treatment is insufficient to reduce the FFA content effectively because of the large quantity of water produced during the reaction. The acid catalyst generally used for esterification reaction is Sulphuric acid (H2SO4). For effective alkaline esterification the acid value of the untreated feed stock oil is expected to be below 3%. The initial step was to reduce the fatty acid content of RSO. The acid catalysed esterification was considered in this work for the FFA reduction of the feed stock. Five sets of experiments were conducted with varying oil to alcohol molar ratio and acid catalyst concentration for the reduction of FFA content. In each run 200 ml of rubber seed oil was filled in the flask of 500 ml capacity and preheated to 50 ° C. Pre-determined amount of methanol was then added to the oil and stirring was started. After a few minutes acid catalyst was added to the mixture and stirring continued. Different methanol to oil molar ratios of 0.20, 0.40, 0.50, 0.60, and 0.65 were used in the first step for a reaction duration of 1 h keeping temperature 45±5 ° C throughout and the acid catalyst was 1% (v/v). The mixture was then poured into the separating funnel and kept for separation. Water and methanol mixture along with sulphuric acid content moved to top. There was a layer of residue settled at the bottom containing colouring pigments, gum forming agents, wax etc.

The second step was carried out with molar ratios 0.20, 0.35, 0.30, 0.20, and 0.20, respectively, keeping other parameters as same as in the first step. Among the results obtained, an overall molar value of 80% v/v methanol to oil and 2% v/v of H_2SO_4 are found to give the best results for 2 h reaction duration and at constant temperature 45±5 ° C. A three-step reaction was also employed with lower methanol ratio, and it was observed that the increase in methanol ratio caused a significant reduction in acid value. The residue and methanol water mixture were separated after the reaction. The residue was observed as settled at the bottom and methanol water mixture as top layer for 20%, 40%, 60% molar ratios and both layers at the top for 50% and 65% molar ratios. Investigations were also carried out with varying acid catalyst concentration with optimum alcohol to oil molar ratio. It was found that there is no significant reduction of FFA value by increasing the acid catalyst concentration. Table 3.1 describes the experimental data obtained in acid esterification stage and the final FFA reduction. A favourable molar ratio of 80% and acid catalyst amount of 2% were observed with two step acid esterification.

Table 3.1. Experimental data of acid esterification of rubber seed oil

Set no.	No of reaction steps	Methanol ratio (% v/v)	Acid Qty H_2SO_4 (% v/v)	Reaction time (min)	Temper-ature (°C)	Titration End point value (ml)	% FFA
1	1st step	20	1	60	45±5	29	16.29
	2nd step	20	1	60	45±5	15	8.41
	3rd step	20	1	60	45±5	5	2.8
2	1st step	40	1	60	45±5	20	11.22
	2nd step	30	1	60	45±5	9	5.04
	3rd step	20	1	60	45±5	3	1.68
3	1st step	50	1	60	45±5	14	7.85
	2nd step	30	1	60	45±5	3.65	2.03
4	1st step	60	1	60	45±5	16	8.97
	2nd step	20	1	60	45±5	4	2.44
5	1st step	65	1	60	45±5	15	8.41
	2nd step	20	1	60	45±5	4.5	2.52

IOP Conf. Series: Materials Science and Engineering 1132 (2021) 012044 doi:10.1088/1757-899X/1132/1/012044

3.4 Alkaline transesterification of rubber seed oil

Alkaline transesterification was carried out after the two step pretreatment process. The base catalyst used was KOH (85% assay) and molar ratio of oil to ethanol was always kept as 1:9 in every experiment. The yield obtained was investigated by varying the base catalyst concentration from 0.5wt% to 2wt%. The results showed an increase in FAME yield with increase in catalyst concentration. Properly mixed solution of KOH and methanol of required quantity was added to 150 ml of treated and preheated oil (50 ° C) along with stirring. Maximum yield of 90% was obtained with 45 ml methanol and 2.04 g KOH (1.5wt %) catalyst. The temperature was maintained at 55±3 ° C throughout the experiment for 40 minutes reaction time. After the completion of the reaction, products were allowed to settle for 24 hours. Glycerin was removed when it was settled at the bottom. The top layer (biodiesel) was purified by washing 3-5 times with demineralised water at room temperature.

4. Enzymatic transesterification

Enzymes are biological catalysts which accelerate the rate of a chemical reaction without undergoing a permanent change in the structure. Enzymes have been used in many industrial applications like reported in the production of detergents, textiles, starch hydrolysis and, genetic engineering, rare sugars, leather, pulp and lipase-based reactions. Enzyme catalyzed transesterification reaction or hydrolysis is used in biofuel production, enzyme based esterification etc. The microbes that are suggested for biodiesel production include *Aspergillusniger Candida cylindracea, Candida rugosa, Chromo-bacteriumviscosum, Pseudomonas cepacia, Pseudomonas fluorescens, Rhizopusoryzae* etc. Among these Candida antarctica displayed high activity in methanolysis and ethanolysis.

4.1 Selection of enzyme

Most of the literatures recommended the use of Candida antarctica lipase B (Novozym 435) for the transesterification of vegetable oils like soybean, rapeseed, sunflower etc. Most of the literatures recommended the use of Candida antarctica lipase B (Novozym 435) for the transesterification of vegetable oils like soybean, rapeseed, sunflower etc. Vorgelegt Von [15] studied different lipase abilities to transesterify the rubber seed oil with dimethyl carbonate and identified that Candida antarctica lipase B (Novozym 435) is the best member for rubber seed oil transesterification.

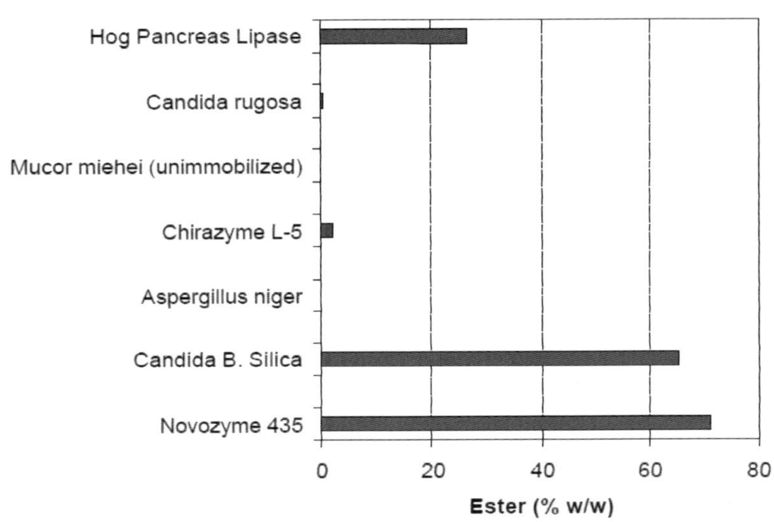

Fig. 4.1. Lipase abilities to transesterify RSO with Dimethyl carbonate

Based on the literature, the enzyme used in this work was selected as Candida antarctica lipase B 10000 immobilized and the enzyme sample supplied from Fermeta Biotec Ltd., Thane, India.

4.2 Experimental Methodology

Five sets of experiments were carried out to study the yield variation of biodiesel with different molar ratio, enzyme weight, solvent addition and time. A 10 ml of rubber seed oil was used in every experimental runs. The oil was poured into the clean and dried conical flask of 100 ml working volume. The enzyme was weighed according to the wt% of oil in each set and added to the rubber seed oil. Methanol was measured by micropipette and added to the oil in stepwise at equal time interval. The reaction mixture with methanol, oil and enzyme were shaken in incubated shaker at 170-200 rpm and 37 ° C. The methanol and solvents were added in three stages at equal time interval. The experimental runs for each set of enzymatic transesterifications of rubber seed oil with different conditions are shown in Table 4.2

Table 4.2. Experimental details of enzymatic transesterification of RSO

Set No.	Oil Qty (ml)	Enzyme weight to oil (wt%)	Molar ratio (%v/v)	Solvent Addition (wt %)	Steps for addition	Time in hour	Tempe-rature (°C)	Shaker Speed in rpm
I	10	10	2	No	3	24	37	200
	10	5	3	No	3	24	37	170
II	10	5	4	No	3	24	37	170
	10	5	5	No	3	24	37	170
	10	8	4	No	3	24	37	170
III	10	10	4	No	3	24	37	170
	10	12	4	No	3	24	37	170
	10	10	6	5 (water)	3	24	37	170
IV	10	10	5	0.75 (t butanol)	3	24	37	170
	10	10	5	No	3	24	37	170
	10	10	2	5 (Water)	3	48	37	170
V	10	10	3	5 (Water)	3	48	37	170
	10	10	2	10 (Water)	3	48	37	170
	10	10	3	10 (Water)	3	48	37	170
VI	10	10	4	5 (Water)	3	30	37	170
	10	10	5	5 (Water)	3	30	37	170

The conical flask was taken out from the shaker after the completion of reaction. Enzyme was separated out from the reaction mixture by filtration method. The remaining mixture was poured into centrifuge tube and separated by centrifuging the mixture at 5000 rpm for 15 minutes. Glycerol content settled at the bottom of the tube was separated by separating glass funnel. The separated mixture was heated to above 90 o C to remove unreacted methanol, water contents and any other volatile impurities. Viscosity of final mixture was measured by Brookfield viscometer. Enzyme after reaction was washed by distilled water and t butanol, dried and stored under the required temperature.

5. Results and discussion

5.1. Two-step acid pre-treatment of rubber seed oil

The reaction parameters of acid esterification have been optimized by performing five sets of experiments. Single step esterification reaction was not able to reduce the acid value. A best suited methanol ratio was obtained after two step pre-treatment process with 80% v/v methanol. There were a three-step reaction employed for lower methanol ratio, and the increase in methanol ratio caused a significant reduction in acid value.

5.1.1. Effect of molar ratio. The alcohol to oil molar ratio is an important parameter to determine the conversion efficiency and acid value reduction. The alcohol used in the present work is methanol which is a low-cost chemical. The amount of alcohol used in both acid esterification and alkaline transesterification plays a significant role in production cost of biodiesel. Experiments started from 20%v/v methanol and a better reduction in acid value was observed by increasing the percentage of oil-alcohol ratio. Use of excess alcohol was not given any significant reduction of acid value. The result obtained from pre-treatment process shows less oil-methanol ratio in comparison with the work done by Satyanarayana (Rubber seed oil having 24% FFA used 75% methanol in two step pre-treatment) [8].

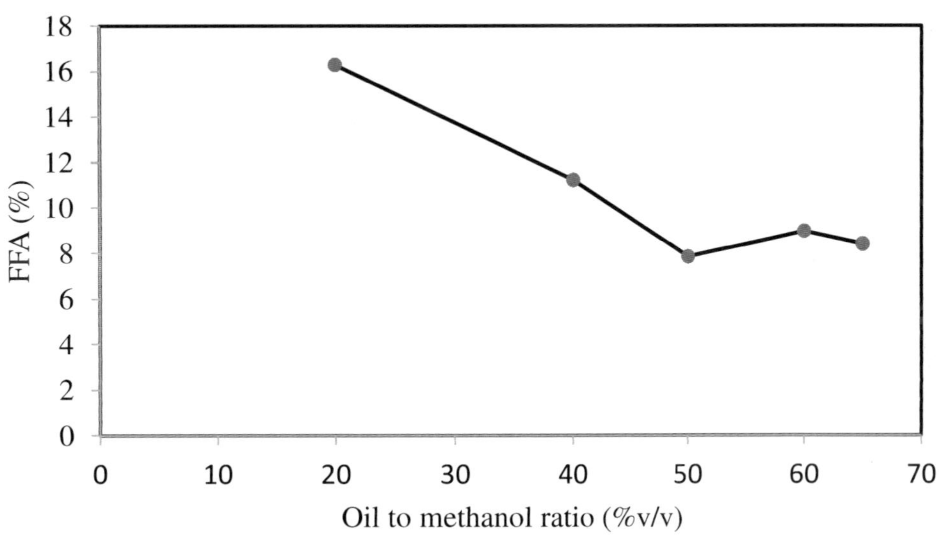

Fig. 5.1. FFA reduction with methanol ratio (After first step pre-treatment)

A better acid value reduction (FFA of 2.03%) reached with total 2 % (v/v) acid catalyst in two step pre-treatment. The residue and methanol water mixture were separated after the reaction was observed on bottom and top, respectively at 20%, 40%, 60% and both were at the top in 50% and 65% molar ratios.

5.1.2. Effect of acid catalyst. Experiments were conducted with varying acid catalyst amount from 1% to 4% with favourable methanol ratio. Table 3.1 shows that the maximum conversion efficiency is obtained at 2 % (v/v) H_2SO_4. Junaid Ahammed et al. [7] conducted experiments with 10wt% H_2SO_4 in which 45% FFA rubber seed oil was reduced to 0.82% in single step acid treatment. A better acid value reduction (Fig. 6.2) of 4.06 (FFA of 2.03%) was obtained in the present work with total 2 % (v/v) acid catalyst in two step pre-treatments.

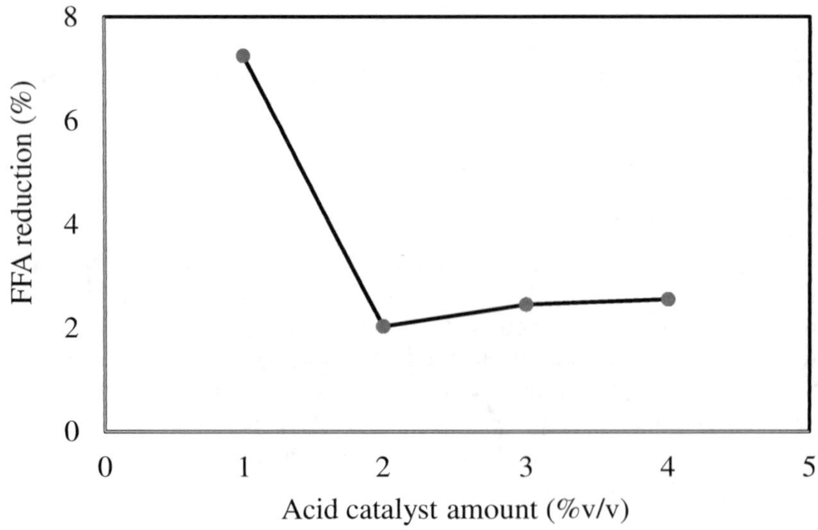

Fig. 5.2. FFA variation with acid catalyst amount

5.1.3. Conversion efficiency of treated oil. The maximum conversion efficiency was obtained at 1:12 molar ratio. The conversion efficiency with various methanol ratios and acid catalyst amount are plotted as shown in Figs. 6.3 and 6.4. The yield/conversion efficiency is calculated as:

$$\text{Treated oil conversion efficiency} = \frac{\text{Amount of treated oil collected}}{\text{Amount of rubber seed oil taken}} \text{x}100 \quad (2)$$

The reaction time in the investigation was 30 and 60 minutes, and observed that the latter was sufficient for completing the reaction in each step of acid esterification. Reaction temperature was always maintained at 45±5 °C. The increase in temperature above 55 °C led the product to become dark in colour and not suitable for further treatment. The mild temperature condition is observed as advantage to reduce the heat input for the reaction.

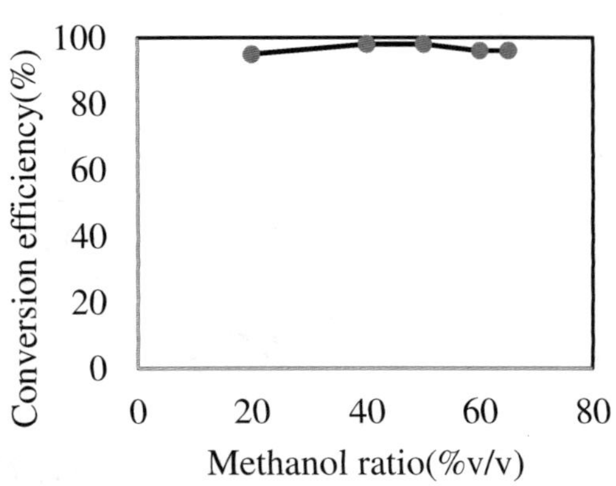

Fig. 5.3. Treated oil conversion efficiency with methanol ratio

(After first step pretreatment)

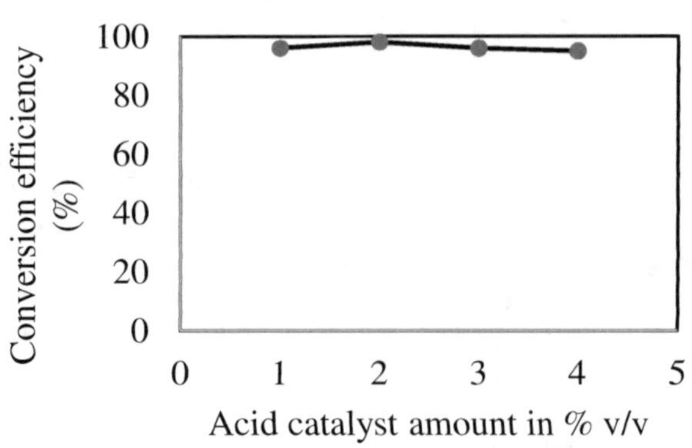

Fig. 5.4. Treated oil conversion efficiency with acid catalyst amount

IOP Conf. Series: Materials Science and Engineering 1132 (2021) 012044 doi:10.1088/1757-899X/1132/1/012044

5.1.4. Comparison of acid esterification with previous studies. Comparison of variables (methanol ratio, acid catalyst amount, reaction time and reaction temperature) in acid esterification reaction with similar works available in literature [9, 18] is given in Table 5.1. The results show a lower acid catalyst amount and lower methanol ratio as compared with the previous studies in two step acid pre-treatment of high acid value rubber seed oil. The results obtained from the present study show the use of comparatively lower methanol ratio and significant reduction in acid catalyst amount.

Table 5.1. Comparison of variables in acid esterification with existing literatures

Variable	Satyanarayana and Muraleedharan (2009)	Junaid Ahammed et al. (2012)	Present study
FFA value	24%	42%	41%
Treatment steps	Two steps	Single step	Two steps
Methanol ratio	75%	50%	80%
Acid catalyst amount	1%v/v	10wt%	2%v/v
Reaction time	1 h	1.5 h	2 h
Optimum acid value reduction	< 2 mgKOH/g	0.820 mgKOH/g	2.03 mgKOH/g
Reaction temperature	63±2 °C	45 °C	45±5 °C

5.2. Alkaline transesterification

The experiments were conducted to verify the base catalyst amount for maximum yield of biodiesel along with temperature and reaction time. In this study it was found that 55±3 °C was sufficient for completion of the reaction and the same temperature was maintained until the reaction was completed. The lower temperature increases the reaction time and higher temperature reduces the FAME yield. It has been observed that 40 minutes reaction duration was sufficient for alkaline transesterification of rubber seed oil. Results obtained from the reactions show that increase in reaction duration will not give high biodiesel yield.

5.2.1. Effect of base catalyst. Base catalyst concentration was varied from 0.5wt% to 2wt% in the present study and the result revealed that maximum yield of 90% was obtained with a base catalyst concentration of 1.5wt% (refer Fig. 6.5). Emulsion was observed with lower catalyst concentrations (0.5 - 1wt %) due to insufficient catalyst amount to completely neutralize the entire FFA resulting in saponification reaction. The soap content in the biodiesel will form emulsion during water washing. The water reacted with soap and unreacted base catalyst form the emulsion. An excellent medium to break the emulsion is NaCl solution. Sodium chloride solution of 50% by volume was added to the emulsion and heated to 60 oC and stirred. It has been observed that the biodiesel was separated from the emulsion solution and coming out at the top.

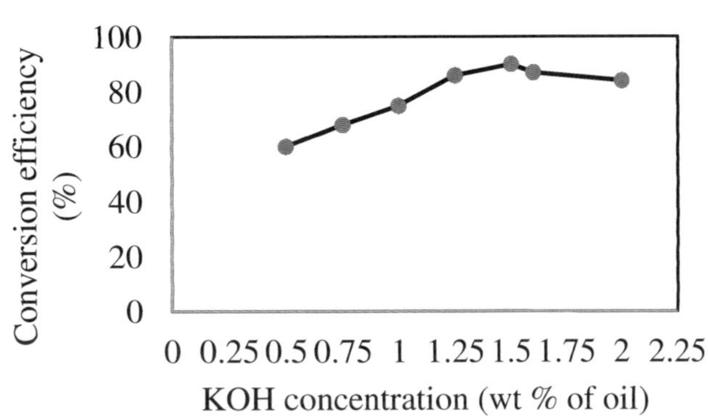

Fig. 5.5. Effect of catalyst concentration on conversion efficiency

IOP Conf. Series: Materials Science and Engineering 1132 (2021) 012044 doi:10.1088/1757-899X/1132/1/012044

5.2.2. GC MS Analysis of RSO methyl ester. The GC MS analysis of the rubber seed oil methyl ester sample was done by using Flame Ionization Detector (FID) method. The detailed chromatogram of the analysis is shown in Fig. 5.6 The results show the content in the sample are all methyl esters of corresponding fatty acids of rubber seed oil. The complete ester conversion was achieved in chemical method. The maximum methyl ester conversion of Oleic acid and Linoleic acid is observed from the chromatogram peaks.

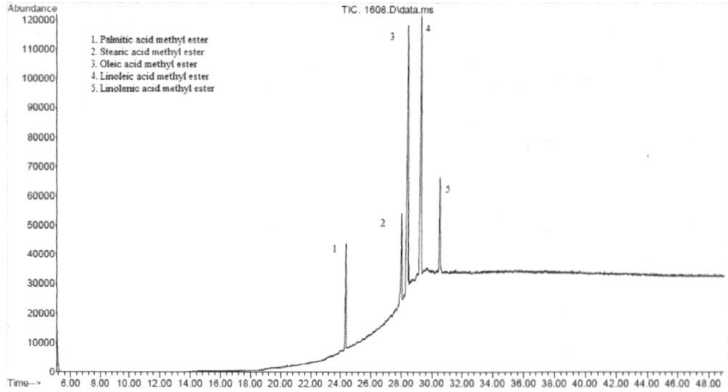

Fig. 5.6. GC MS results of rubber seed oil methyl ester by chemical method

5.3. Enzymatic transesterification

Rubber seed oil was transesterified by Candida antarctica lipase B 10000 immobilized and the results obtained from the various experimental conditions are discussed below. The effect of molar ratio, enzyme weight, and solvent addition is discussed.

5.3.1 Variation of Kinematic viscosity. Six set of experiments were carried out with different enzyme weight and methanol ratio. Stepwise addition of methanol used in reaction helps to decrease the inhibition rate of reaction. The kinematic viscosity of final mixture separated by centrifuging and separating funnel is measured by Brookfield viscometer. The ASTM standard for the Kinematic viscosity at 40°C of biodiesel is 6 cSt.

Table 5.2. Experimental details and kinematic viscosity variation of enzymatic transestirficat-ion

Set No.	Oil Qty (ml)	Enzyme weight to oil (%)	Molar ratio (%v/v)	Solvent Addition (wt%)	Steps for addition	Time in hour	Temp-erature (°C)	Shaker Speed in rpm	Kinematic viscosity cSt (30°C)
I	10	10	2	No	3	24	37	200	14.33
	10	5	3	No	3	24	37	170	25.56
II	10	5	4	No	3	24	37	170	24.33
	10	5	5	No	3	24	37	170	25.44
	10	8	4	No	3	24	37	170	14.78
III	10	10	4	No	3	24	37	170	13.89
	10	12	4	No	3	24	37	170	17.11
	10	10	6	5 (water)	3	24	37	170	12.89
IV	10	10	5	0.75 (t butanol)	3	24	37	170	15.33
	10	10	5	No	3	24	37	170	16.00
	10	10	2	5 (Water)	3	48	37	170	14.78
V	10	10	3	5 (Water)	3	48	37	170	11.22
	10	10	2	10 (Water)	3	48	37	170	13.33
	10	10	3	10 (Water)	3	48	37	170	15.22

VI	10	10	4	5 (Water)	3	30	37	170	10.87
	10	10	5	5 (Water)	3	30	37	170	12.78

Kinematic viscosity value of the separated mixture reduced to 9.09 cSt after 48 h reaction with water as the solvent medium. Lower enzyme concentration was not sufficient for the viscosity reduction and 10wt% of enzyme shows a better condition of reaction. Table 5.2 gives the details of kinematic viscosity value after each set of experiment with different reaction condition.

5.3.2. Effect of methanol ratio. The methanol ratio of the reaction was varying from 2 to 6 with solvent free system and with solvent addition. The maximum viscosity reduction achieved is 10.89 cSt with solvent free system and 9.09 cSt with water as the solvent system.

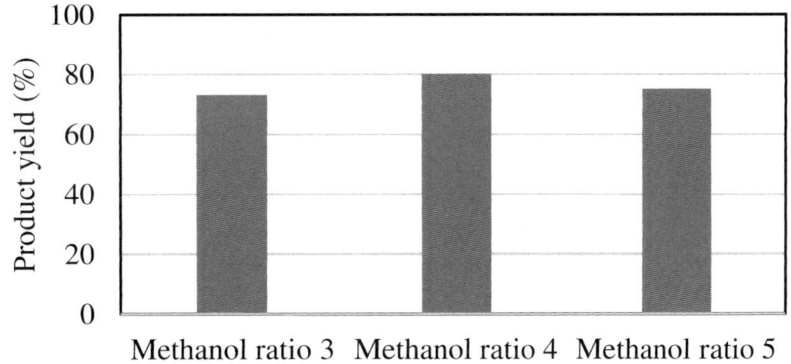

Fig. 5.7. Product yield variation with methanol ratio with 5wt% enzyme

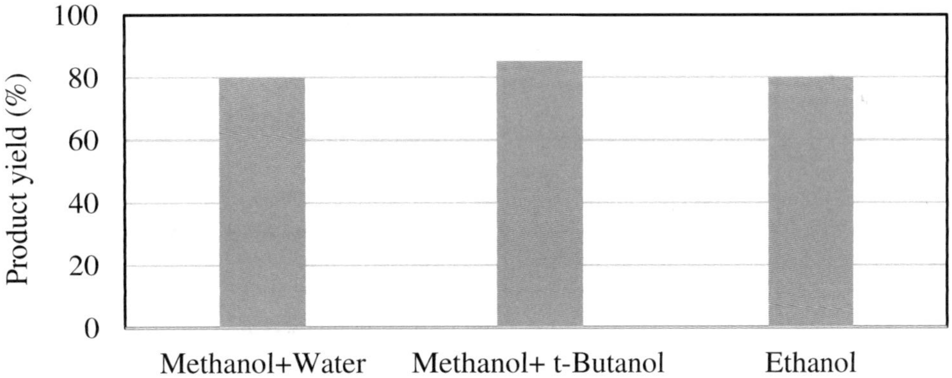

Fig. 5.8. Variation of Product yield with solvent medium and ethanol

The best solvent medium for enzyme transesterification is identified as water with methanol. The viscosity value obtained with t butanol mediated reaction is more than that of water-based reaction. The maximum yield of the final mixture obtained is 80% with methanol ratio of 3 in solvent free system and 85% with t butanol solvent system.

5.3.3. Effect of enzyme concentration. The variation of product yield and kinematic viscosity was analyzed with different enzyme concentrations. Enzyme weights were varying from 5 to 12 and a suitable wt% of enzyme is found as 10wt%. The maximum viscosity value was reduced to 9.09 cSt with 5wt% water and 10wt% enzyme. The kinematic viscosity value variation with different enzyme concentration is shown in Fig. 5.9

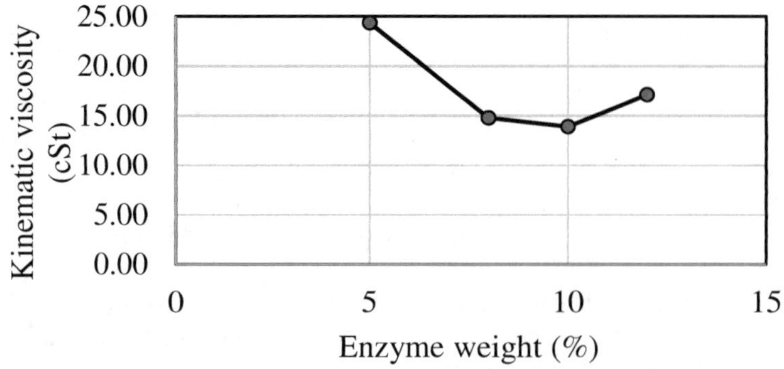

Fig. 5.9. Variation of Kinematic viscosity with enzyme weight

Maximum product yield was 80% with 5wt% of enzyme to the oil with the viscosity value of 19.71 cSt measured in that reaction. Figure 6.22 shows the product yield with different enzyme concentrations.

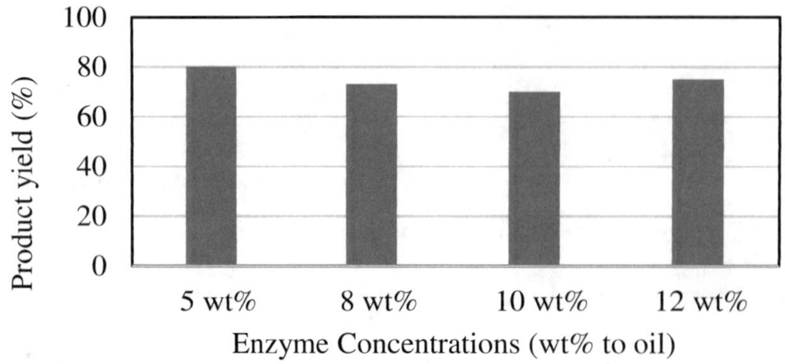

Fig. 5.10. Product yield variation with enzyme concentration without solvent

5.3.4. Effect of solvent addition. The addition of solvent to the reaction mixture will help the enzyme activity. The solvent prevents the inhibition rate by dissolving the alcohol and gives the flexibility to the enzyme.

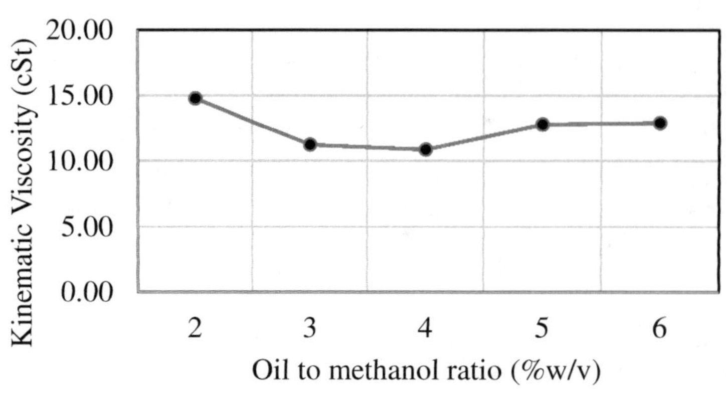

Fig. 5.11. Kinematic viscosity variation with methanol ratio in water-based reaction

Water and t butanol are the solvents used in this work, the viscosity variation and product yield was measured. Water based reaction gives more viscosity reduction than t butanol. Water concentrations of 5wt% and 10wt% were analysed and the 5wt% water was found to be the optimum condition. Figure

6.23 shows the variation of kinematic viscosity at 5wt% water mediated reaction with different methanol ratio.

The maximum viscosity reduction of 8.80 cSt was achieved at 5wt% water with 4 molar ratio of methanol to oil. The product yield with water based solvent reaction reached to 80% at 1:6 molar ratio. The excess alcohol ratio was not supporting to achieve a viscosity reduction during the reaction; it might inhibit the reaction rate.

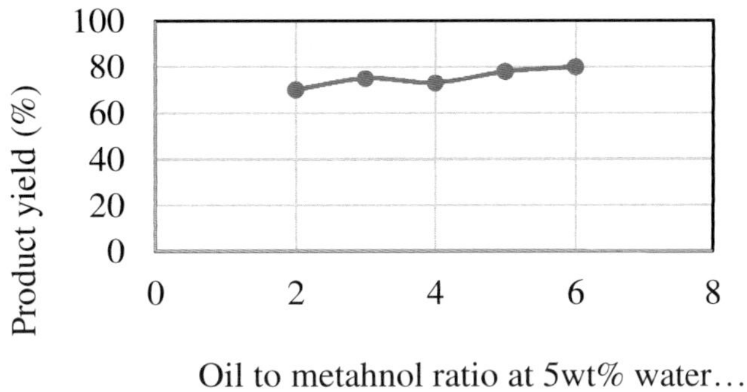

Fig. 5.12. Product yield variation with 5wt% water

5.4 GC MS analysis of biodiesel by enzymatic method

GC MS analysis of rubber seed methyl ester by enzymatic method was done. The sample has lowest viscosity value obtained after the enzymatic transesterification with water as the solvent medium. The chromatogram of GC MS analysis is as shown in Fig. 5.13. Result shows the presence of Tetraoxatetradecan-1-ol and Isopropyl linoleate in the final mixture. The complete ester conversion was not taken place during enzymatic transesterification.

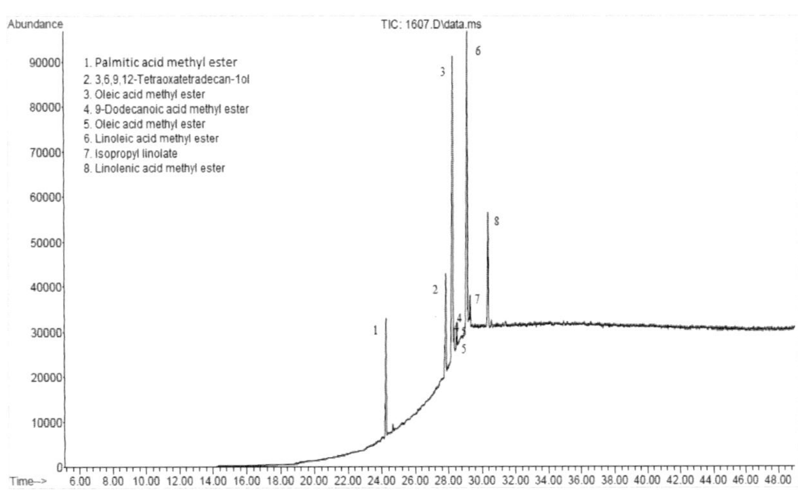

Fig. 5.13. GC MS analysis of RSO methyl ester by enzymatic method

5.5 Separation of biodiesel in enzymatic method.

The mixture of biodiesel and oil are obtained after enzymatic transesterification. The biodiesel is separated out from the mixture by 3/27 conversion process. It is a quantitative and qualitative test used in biodiesel industry. The test procedure is: 3 ml of sample is cooled to 20 °C and mixed with 27 ml methanol and allow settling. The basic chemistry behind this test is that biodiesel is fully dissolved in methanol, and oil and other unreacted compounds settle at the bottom of reaction funnel. The mixture after centrifuging is cooled to 20 °C and mixed with required amount of methanol and allowed to settle. Figure 6.25 shows the separation process, in which the oil and impurities are settled at the bottom separated by a separating funnel. The mixture of methanol and biodiesel are separated by the distillation of methanol. The biodiesel collected after complete distillation is used to compare the properties with chemical method.

6. Comparison of properties of rso methyl ester

The various properties of rubber seed oil methyl ester produced by chemical and enzymatic transesterification were compared. Table 6.6 gives the property comparison of rubber seed methyl ester obtained from both the production method with ASTM biodiesel standards.

Table 6.1. Comparison of properties of RSO methyl ester

Property	Rubber seed oil	Biodiesel standard ASTM 6751	Biodiesel (Chemical method)	Biodiesel (Enzymatic method)	Diesel
Specific gravity	0.91	0.82-0.90	0.87	0.86	0.835
Kinematic viscosity at 40 ^oC (cSt)	59.10	1.9-6.0	4.78	5.16	4.5
Calorific value (MJ/kg)	37.8	35	37.3	36.2	42
Acid value (mgKOH/g)	82	0.5	0.228	0.32	< 0.5

7. Conclusion

In this study, a comparison of biodiesel production by chemical and enzymatic method has been carried out. High free fatty acid rubber seed oil was used in this study without any kind of treatment and filtration. Experimental investigation was carried out for chemical transesterification, kinetic study and enzyme-based biodiesel production. The acid value reduction of rubber seed oil with high FFA and methyl ester production were investigated and found that direct alkaline transesterification cannot be used for high FFA feed stocks. Two stage acid esterification process was required to reduce FFA from 41% to 2.03% with less amount of acid catalyst. An overall molar value of 80% v/v methanol to oil and 2% v/v of H_2SO_4 are found to give the best results for 2 h reaction duration and at constant temperature 45±5 ^{o} C. Base catalyst concentration was varied from 0.5wt% to 2wt% for alkaline transesterification process and the result revealed that maximum yield of 90% was obtained with a base catalyst concentration of 1.5wt% with 1:9 molar ratio at 55± 3 ^{o}C for 40 minutes. The enzyme-based transesterification did not give a complete conversion of methyl ester. The best condition arise in enzymatic reaction was 10wt% enzyme and 1:4 methanol ratio with 5wt% addition of water as the solvent. The kinematic viscosity was reduced to 10.87 cSt after 30 h reaction with 37 ^{o} C in an incubated shaker at 170 rpm. The calorific value of methyl ester produced by enzymatic method (36.2 MJ/kg) is found to be less compared with chemical method (37.3 MJ/kg) and the acid value and viscosity have higher values in the product from enzymatic method.

References

[1] Kaul S, Kumar A, Bhatnagar A K, Goyal H B and Gupta A K. Biodiesel: A clean and sustainable fuel for future. Scientific strategies biofuels. All India seminar on national policy on nonedible oils as biofuels. SUTRA, IISc Bangalore, India

[2] Linghong Zhang, Chunbao Xu, Pascale Champagne. Overview of recent advances in thermochemical conversion of biomass. Energy conservation and management,(2010) Vol-51: 969-982

[3] Wendy Pei Qin Ng, Mook Tzeng Lim, Sohibatul Muizzah bt Mohamad Izhar, Hon Loong Lam, Suzana Yusup. Overview on economics and technology development of rubber seed utilisation in Southeast Asia. Clean Techn Environ Policy (2014) 16:439–453. DOI10.1007/s10098-013-0667-6More references

[4] A.S. Ramadhas, S. Jayaraj, C. Muraleedharan, Biodiesel production from high FFA rubber seed oil, Fuel 84 (2005) 335–340.
[5] Amit Sarin. Biodiesel Production and Properties. ISBN: 978-1-84973-470-7
[6] A.S. Ramadhas, S. Jayaraj, C. Muraleedharan, Biodiesel production from high FFA rubber seed oil, Fuel 84 (2005) 335–340
[7] Junaid Ahmad, Suzana Yusup, Awais Bokhari, Ruzaimah Nik, Mohammad Kamil. Study of fuel properties of rubber seed oil based biodiesel. Energy Conversion and Management 78 (2014) 266–275
[8] Satyanarayana M, Muraleedharan C. Biodiesel Production from Vegetable Oils: A Comparative Optimization Study, Journal of Bio based Materials and Bioenergy Vol. 3, 1–7, 2009. doi:10.1016/j.energy.2010.09.050
[9] Hideki Fukuda, Akihiko Kond, Hide Noda. Biodiesel Fuel Production by Transesterification of Oils. Journal of Bioscience and Bioengineering. Vol. 92, No. 5,405%416.2001
[10] Mukesh Kumar Modi, J.R.C. Reddy, B.V.S.K. Rao, R.B.N. Prasad. Lipase-mediated conversion of vegetable oils into biodiesel using ethyl acetate as acyl acceptor. Bioresource Technology 98 (2007) 1260–1264
[11] Nie K, Xie F, Wang F, Tan T., 2006. Lipase catalyzed methanolysis to produce biodiesel: Optimization of the biodiesel production. J. Mol. Catal. B: Enzym. 43, 142–147
[12] Royon D, Daz M, Ellenrieder G, Locatelli S., 2007. Enzymatic production of Biodiesel from cotton seed oil using t-butanol as a solvent. Bioresour. Technol. 98, 648–653.
[13] Nevena Ognjanovic, Dejan Bezbradica, Zorica Knezevic Jugovic. Enzymatic conversion of sunflower oil to biodiesel in a solvent-free system: Process optimization and the immobilized system stability. Bioresource Technology 100 (2009) 5146–5154.
[14] J. Van Gerpen, B. Shanks, and R. Pruszko, D. Clements, G. Knothe. Biodiesel Production Technology August 2002–January 2004. July 2004 NREL/SR-510-36244.
[15] Vorgelegt Von. Lipase catalysed transesterification of plant oils with Dialkyl carbonates. Master of Science. Tjahjono Herawan, 29 November 2004

IOP Conf. Series: Materials Science and Engineering 1132 (2021) 012045 doi:10.1088/1757-899X/1132/1/012045

Design of a sustainable flood carrier system in Cuddalore of Tamilnadu

[1]P K Suresh, [1]S Sakthival, [1]R Sundaravadivelu and [2]P C Aneesh

[1] Department of Ocean Engineering,IIT Madras, India
[2] Department of Civil engineering, Adishankara Institute of Engineering and Technology, Kalady, Kerala, India

[1] sureshpk2000@gmail.com, rsun@iitm.ac.in, [2] hod.ce@adishankara.ac.in

Abstract. Cuddalore district is located along the south west coast of Indian in the state of Tamilnadu bordering bay of Bengal.. The district is blessed with presence of wetlands, minor port and beach. The annual rainfall of the place is about 1300 mm. The district receives 70% of rainfall during north east monsoon from October to December. Agriculture is the main occupation. The rainfall along the coast is higher compared to upper regions. The high intensity rainfall occurrence is a common one. The district has anciently constructed hydraulic structures for conserving water called tanks. Perumal tank is the last source. The entire flood water during rainy season flows through a drain called Paravanar . This runs almost parallel to the coast for about 18km and finally merges with Bay of Bengal. Because of length and shallow slope, it takes more time for the water to recede. Hence a new canal was designed to drain some water directly in to the sea. HECRAS numerical model was adopted to study canal performance. The details of hydraulic studies and coastal sediment transport computations are highlighted.

Introduction

The district of Cuddalore is located along 78°45'E and 11°15'N coordinates along the south west coast of India in Tamilnadu state. The sea of Bay of Bengal is bordering the district on its east side. The river Pennar is draining in the sea along the north side and river Vellar is confluencing at the southside. Apart from this another drain called Paravanar river is also draining in the sea in between Pennar (Fig 1).Cuddalore is a low lying area and highly flood prone as described by Evany Nithya and Priyanka(2019).

Hydrology of Paravanar

The river is drain carrying the rainwater during floods. The drain commences from surplus portion of the existing Perumal tank. The river takes a crescent like plan shape at a place called Aruvamukku and then flows for a distance of about 18km and finally confluences with Bay of Bengal at Cuddalore port. The average slope of the river is 1 in 3000. The annual rainfall of the district is about 1300mm.Number of effective rainfall days are around 30 in a year. Nearly 60% of the rainfall occur during north east monsoon period from October to December. Out of this it is common to have high intensity rainfall based on studies of Sandeep and Jegankumar (2018) which results in drainage problem. Moreover, due to very shallow slope the drainage process is getting more delayed (Fig 2 to 4).

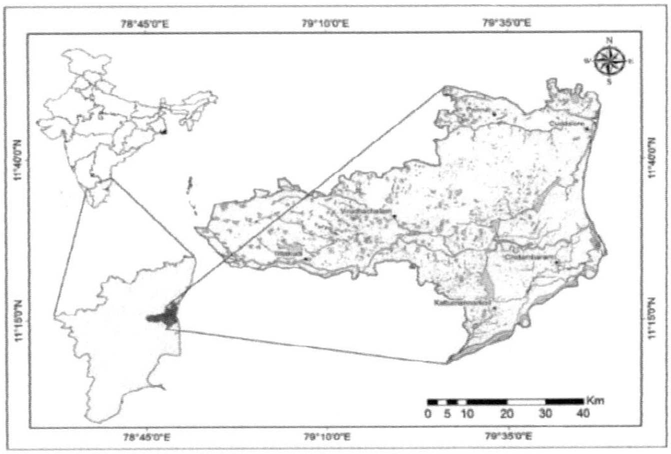

Figure 1. Index map

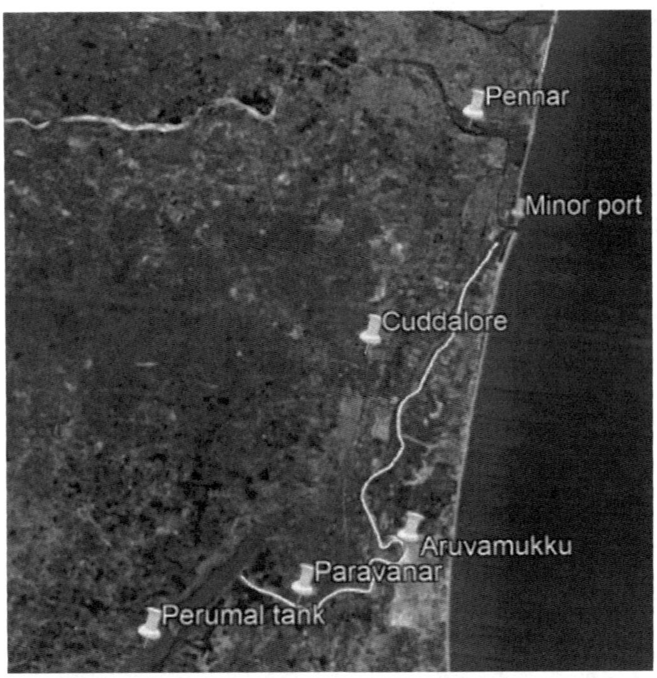

Figure 2. Details of Paravanar

Figure 3. Aruvamukku location

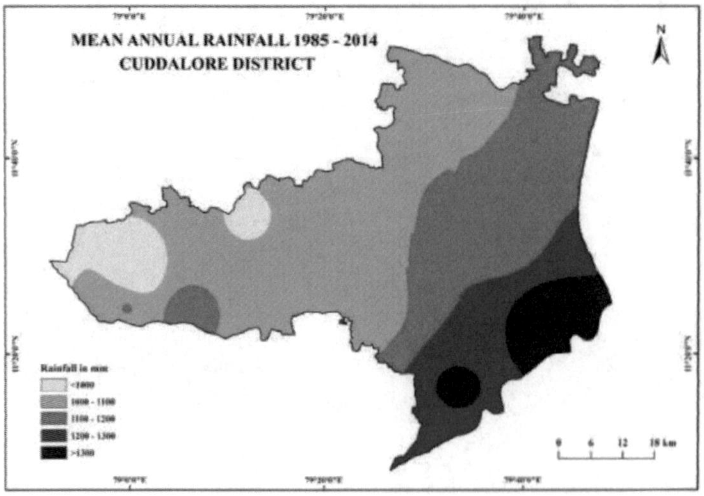

Figure 4. Rainfall pattern

The district of Cuddalore is generally a low lying one .The slope of river Paravanar is very shallow. The place is highly flood prone especially during north east monsoon. So this drain because of its geometry takes a long time for complete drainage of flood water. This in turn results in heavy water logging as observed by Yovan Felix and Sasipraba (2020). Because of waterlogging the people living here are in high flood threat for more days even if the rainfall stops. It also results in unhygienic situation. Based on their representation the PWD officials have proposed a flood carrier directly connecting Paravanar river to sea by a flood carrier canal.(Fig 5). The existing river Paravanar can carry 1190 m^3/s. It is proposed to diver 600 m^3/s to new canal.

Figure 5. Proposal

This study was taken up to examine the feasibility of the proposal. The study consists of
(1) Desk studies (2) Model studies (3) Design components if needed

IOP Conf. Series: Materials Science and Engineering 1132 (2021) 012045 doi:10.1088/1757-899X/1132/1/012045

1.1. Desk studies

This includes study of shoreline changes and rainfall data. The satellite imageries will give the historical coastline changes over a period of time. In the present case the imageries during 2003, 2005, 2006, 2008.2011 and 2015 were downloaded from google earth and analysed. The pattern of changes that takes place along the estuary was studied. The river mouth is already trained by constructing a pair of breakwater. Hence it was not all creating any block. Accretion of beach was noticed on the south side and erosion on the north side of breakwater indicating that littoral drift is towards north. (Fig 6).

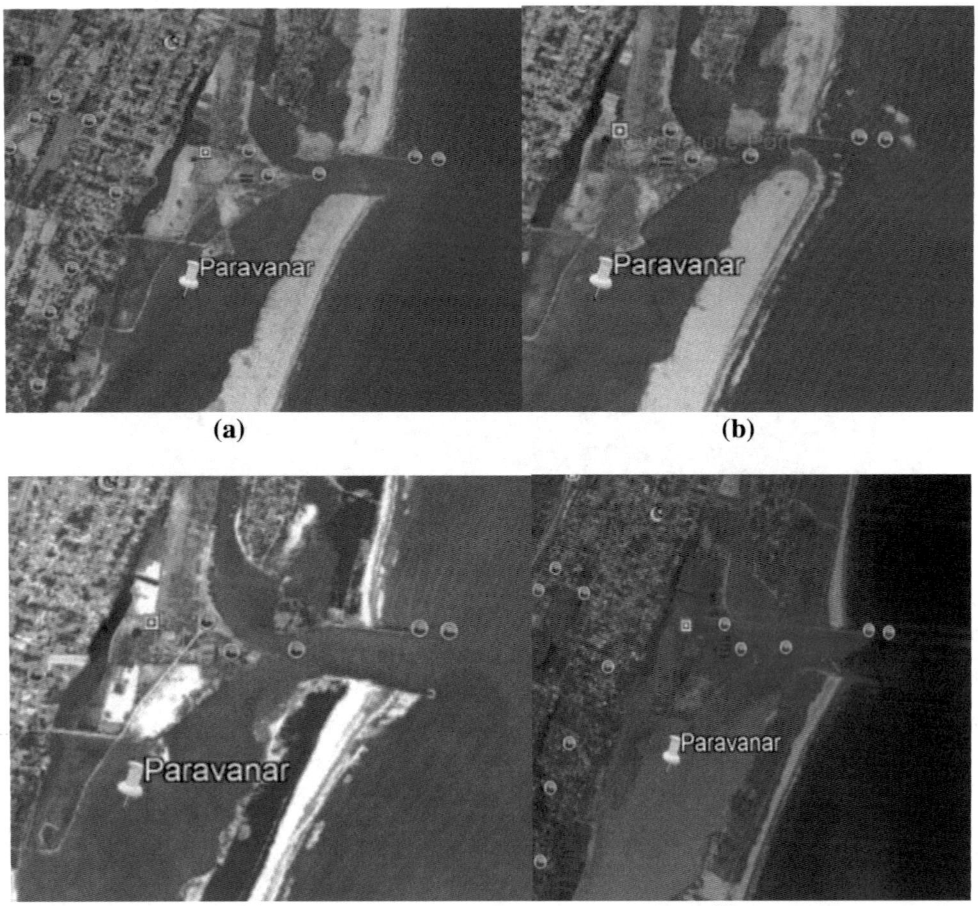

Figure 6. Historical changes along estuary: (a) Paravanar estuary 2003 / May. (b) Paravanar estuary 2006 / August. (c) Paravanar estuary 2011/ July. (d) Paravanar estuary 2015/December

2.1.2. Rainfall. The district has a hot tropical climate. The summer season, which is very oppressive, is from March to May. The southwest monsoon, which follows, lasts till September. October to December constitutes northeast monsoon season. January to February is the comparatively cooler period. The annual normal rainfall for the period (1901- 2000) ranges from 1050 – 1400 mm. The normal annual rainfall over the district varies from about 1050 mm. The contributions of individual seasons are as follows: NE-57%, SW-31%, Summer- 7% and winter 5%. But the rainfall pattern in third decade shows high magnitudes of north east monsoon. (Fig 7 &8).

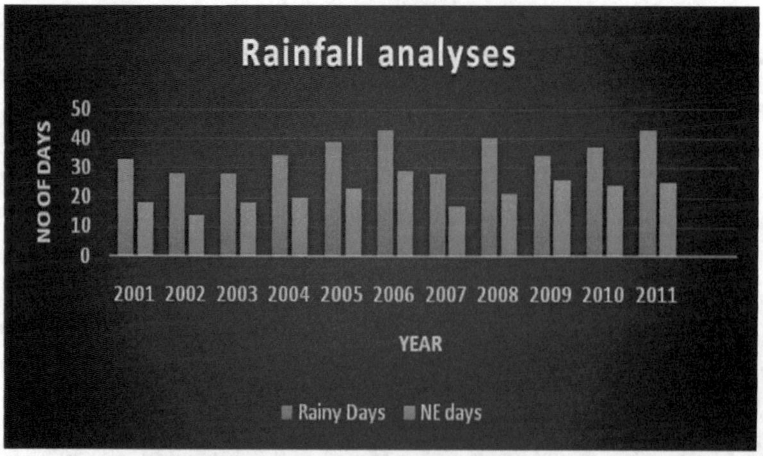

Figure 7. Rainfall in last 10 years

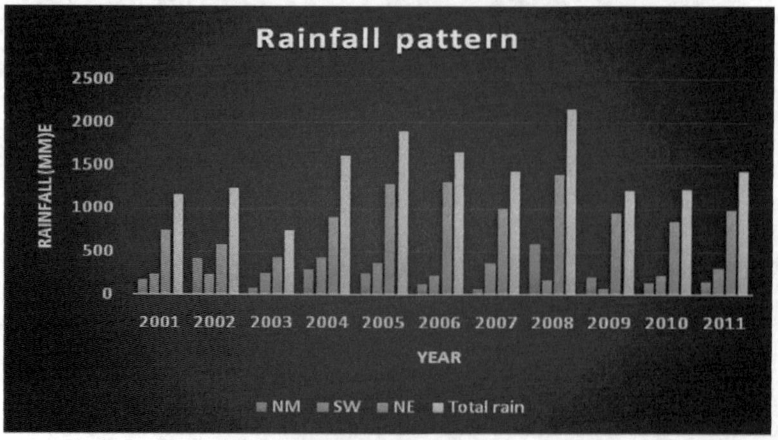

Fig 8. Rainfall pattern as per season

The study of rainfall pattern indicates that north east monsoon is dominating. The number of rainy days is in the range of about 25. But during high intensity rains the flooding process is the creating inundation

Table1. Proposal

The hydraulic particulars	scheme
High tide level:	(+) 1.30m
Low tide level:	(+) 0.70m
Width of Paravanar at proposed site:	200 m
Sill level of proposed regulator at Aruvalmukku:	(+) 0.70m
Length of proposed canal:	1600 m
Top width of proposed canal:	80 m
Bottom width of proposed canal:	60 m
Bed fall proposed:	1 in 3000
Maximum design discharge:	21200 cusecs/600 m3/s
Top bund level:	1.50m above high water level
Maximum flow depth :	4.5m

Based on the field particulars hydraulic computations were made using the model HEC-RAS developed by US army engineers (Table 1). The computations were performed for following conditions

1) Maximum discharge of 21200 cusecs
2) High tide level (+) 1.30m
3) Low tide level (+) 0.30m
4) Storm surge of 1.50m

1.2. Numerical model study

In order to assess the existing hydraulic conditions of the field, numerical model studies were made. For the present study the U.S. Army Corps of Engineers' River Analysis System (HEC-RAS) software is used. This software is developed by the Hydrologic Engineering Center (HEC-2008), which is a division of the Institute for Water Resources (IWR), U.S. Army Corps of Engineers. HEC-RAS allows users to perform one-dimensional steady and unsteady flow calculations (HEC, 2008). In a HEC-RAS steady state simulation, water surface profiles are computed from one cross-section to the next by solving the standard step iterative procedure to solve the energy equation. The energy equation is intended to calculate water surface profiles for steady gradually varied flow. The inputs are geometric, flow and boundary data. The geometric data consisting of cross section and chainages were keyed in after making a base map of the study area.
The model run was performed for different tidal heights such as low tide, high tide and storm surge. In all the case no spill over was noticed. The average velocity was about 2m/s (Fig. 11 & 12)

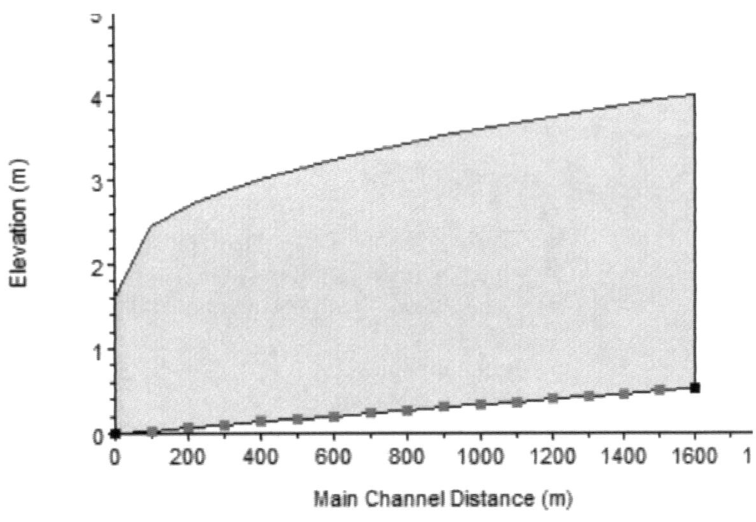

Figure 9. Backwater profile

1.3. Wave climate

A quantitative understanding of wave characteristics in the near shore is essential for the estimation of sediment transport and morphological changes along the coastal areas .The coast has three seasons namely non monsoon (January-May), south west monsoon (June-September) and north east monsoon (October-December). The procedure to generate wave climate was as per Suresh & Sundar (2009) The wave directions are indicated vide fig 13 to 15 and table as taken from Suresh (2010).Dominant wave direction is from south east (Tab 1).

IOP Conf. Series: Materials Science and Engineering 1132 (2021) 012045 doi:10.1088/1757-899X/1132/1/012045

Table 2. Average monthly wave climate

Month	Wave Height (m)	Period (sec)	Angle deep water (deg) w r t north
January	0.78	7	96.4
February	0.66	6	101.84
March	0.66	10	118.33
April	0.97	11	132.07
May	1.07	8	140.06
June	1.26	10	140.54
July	1.06	8	138.06
August	1.22	9	143.96
September	1.02	8	145.14
October	0.96	8.5	120.98
November	0.93	7	93.6
December	0.86	8	67.22

1.4. Sediment transport

In order to understand the shoreline dynamics, it is very much essential to have an estimation of littoral drift that takes place along the coast. This will be useful to arrive any training works at river entrance. Here the littoral drift is estimated adopting Van Rijn (2001) approach (Table 2). It is to be mentioned that the net littoral drift along east coast during Jan to Sept is towards north, while during Oct to Dec it is directed toward south as pointed out by NarasimhaRao (1983).

Formula of Van Rijn (2001)

$Q = 40\ K_{swell}*K_{grain}*K_{slope}*(H_{sb})^{3} \sin(2\alpha_b)$

Q = alongshore sediment transport (kg/s)

T_P = Peak period, Swell correction factor

K_{swell} = $T_P/6$

α_b = Breaker angle

D_{50} = Particle size (mm)

K_{grain} = Particle size correction factor = $0.20 / D_{50}$

K_{slope} = Slope correction factor = $(\tan\beta/0.01)0.5$

H_{sb} = Breaker height

Table 2. Average monthly Sediment transport

Month	Sediment transport (Cum)
January	3431.433
February	650.712
March	-5273.74
April	-30968.3
May	-30291.3
June	-63990.5

July	-28824.7
August	-51817.5
September	-26906.9
October	-14331.6
November	6377.89
December	*16437.04*

3. Conclusions

The proposed carrier canal designed for 600cumecs or 12200 cusecs was modelled for different boundary conditions without any spill

- The channel is hydraulically capable of carrying the designed discharges
- The sea bed bathymetry and coast line changes using the imageries wereanalysed. The estimation of alongshore sediment transport is directed towards north direction
- Even though it was initially felt that there is a need to construct training wall for about 100m length to keep the outlet of proposed channel open, the littoral drift pattern indicate that it will create high erosion on its north side which is populated. Hence it is recommended not to adopt any such hard measures
- The present site is located at about 10km south side of existing Cuddalore minor port. It was assessed that present opening of new inlet to sea will not be permanently open as the net drift is towards north. Hence the confluence point is to be periodically dredged and the dredge material should be dumped on the coast on the north side so as to maintain the sediment transport equilibrium. Hence no artificial interferences were made at exit and it is designed as a sustainable one.
- The present canal will definitely reduce the receding of flood water because the distance is reduced by the construction of flood carrier
- The project in general will help the social population in clearing the flood.

References

[1]. Evany Nithya and S.Priyanka (2019 "Vulnerable Area Assesment Due to Flood in Cuddalore District by Morphometric Analysis Method Special Issue - 2019International Journal of Engineering Research & Technology (IJERT) ISSN: 2278-0181 Published by, www.ijert.org ICONEEEA - 2k19 Conference Proceedings

[2]. Sandeep. P and R. Jegankumar (2018) "Assessment of Spatial Variability of Rainfall Pattern in Cuddalore district of Tamil Nadu using GIS techniques"2018International Journal of Creative Research Thoughts IJCRT | Volume 6, Issue 1 March 2018 | ISSN: 2320-2882

[3]. Yovan Felix & T. Sasipraba (2020) "Spatial and temporal analysis of flood hazard assessment of Cuddalore District, Tamil Nadu, India. Using geospatial techniques", Journal of Ambient Intelligence and Humanized Computing (2020)

[4]. US Army corps of engineers Hydrologic Engineering Center HEC-RAS River Analysis System Application guide March 2008

[5]. Suresh,P.K and Sundar,V (2009): "Simulation of Shoreline Changes and Estimation of Littoral Drift"HYDRO-2009,Central Water Power Research Station,Pune pp 303-308

IOP Conf. Series: Materials Science and Engineering 1132 (2021) 012045 doi:10.1088/1757-899X/1132/1/012045

[6]. P K Suresh 2010 Nearshore *sediment dynamics along the coast of Tamilnadu* Doctoral theses IIT Madras.

[7]. Van Rijn, L.C (2001) "Longshore Sediment Transport" Delft Hydraulics

[8]. Narasimha Rao, T.V.S.(1983)"Studies on Wave Induced Long shoreSediment Transport on beaches and Related Phenomena" M.S.Thesis,I.I.T.Madras.

IOP Conf. Series: Materials Science and Engineering 1132 (2021) 012046 doi:10.1088/1757-899X/1132/1/012046

Design and analysis of a solar drier with a parabolic shaped dish type collector for drying peanut

Jobin Varghese[1], Rupesh S[2], Jithu Augustine[1], Adithya Nair[1], Prajith[1]

[1]Adishankara Institute of Engineeirng and Technology, Kalady.
[2]PES College of Engineering, Mandya

jobinvargheset@gmail.com

Abstract

Abstract

The present work deals with the performance evaluation of an active indirect solar dryer integrated with a parabolic dish collector to dry peanuts. Peanut is a widely available edible seed in India, obtained from a legume crop (Arachis hypogaea) which is used either directly in food or for extracting cooking oil. The system mainly consists of a parabolic collector which is provided with an insulator made of plywood with air gaps and aluminium foil wrapping, and a drier consisting of four trays where the peanuts are dried. The flow of hot air through the tray is maintained using an exhaust fan positioned at the top of the drier. The fan exhausts the moisture extracted from the product to make the flow of air without any localization. From the analysis it is found that the parabolic dish collector generates a maximum temperature of 79 °C at 12.30 PM and dries 1.5 kilograms of peanut in 5 hours of a day. Thus, hot air from the parabolic dish concentrator reduces moisture content of peanut by 21% more than obtained from open solar drying.

Keywords: Peanut, parabolic collector, solar dryer, trays

1. Introduction

The renewable energy has a vital role in the life of farmers in improvising the technology in farming which will increase the productivity. Lack of adequate infrastructure for processing of food and its storage in Asia-Pacific region leads to average wastage of 10 to 40% [1].

As a result of lack of processing technology availability in India, large quantity of food produced in the country is going to garbage, even when a numerous people are in need of their daily food. This leads to the formation of gap between the gross food production and its availability. Post-harvest loses should be minimized to improve the distribution of food to needy people [2]. Some of the conventional food preservation methods are bottling, canning, freezing, sugaring, salting (curing), drying, smoking, refrigerating, pickling and freezing. Among these methods, drying is one of the practical methods to preserve food at a very low cost and by means of renewable energy. Drying method not only an effective method, but also an ecofriendly to reduce postharvest losses and thus a solution to supply shortages. Drying process dehydrates the food products which enhances the shelf life by resisting the growth of bacteria [3]. Drying is a widely employed technology to dry crops, fruits, vegetables, herbs, woods, agricultural products and food stuffs that significantly contributes to the revenue developed by a country like India through agricultural technology [4]–[6].

Although drying of food under direct sunlight is economical, the dried products lack quality due to the contamination caused by dust, insects, birds and rain [7]. Moreover, direct sunlight drying is not only time consuming but also leads to loss of vitamins, nutrients and unpleasant colour change due to its exposure to direct ultraviolet rays. These problems pertaining to direct sunlight drying is eliminated through dehydrating using solar dryers. In addition, solar dryers generate a higher temperature, minimum relative humidity, lesser moisture content of the product and lower damage of drying components compared to direct solar drying [8].

IOP Conf. Series: Materials Science and Engineering 1132 (2021) 012046 doi:10.1088/1757-899X/1132/1/012046

Solar dryers remove the 20% more moisture than open air drying to produce better quality of dried products [9]. Type of solar dryers include tunnel dryers, hybrid dryers, horizontal and vertical type dryers, multi-pass dryers and active and passive dryers [10]–[17].

On the basis of heating modes and utility of solar dryers, the solar dryers are classified as forced air circulation or active solar dryers and natural air circulation or passive solar dryers. Active or passive solar drying system are classified based on design or working principle of the dryer, mode of drying and product type to be dried as direct mode or integral type, distributed or indirect mode and mixed mode solar dryers, respectively. The direct mode of drying may affect the quality and property of a product to be dried, as an example, the chlorophyll may be damaged easily due to solar direct drying method. Indirect mode of drying is more preferred when there is less available space for drying large quantities of product. In this case, colour and the nutrients of the product undergoing drying is preserved. Moreover, forced convection dryers dries the product faster than natural convection solar dryer and those dried with forced convection has a higher quality than direct solar dryers [18], [19]. The quality of dried products using forced convection method can be further increased by the right choice of solar collector with appropriate solar tracking facility. Depending upon the concentration ratio of the collector, the solar radiation is amplified to generate more heat flux. When more heat is generated in the collector due to concentration ratio, the air can carry enough heat to the drying chamber to dry the product that requires higher temperature. This will considerably reduce the time of drying and hence efficiency of the dryer is improved.

Through there are many works carried out on the efficiency of parabolic solar dish collector, no studies are found to be reported in the moisture removal of the peanuts in a dryer with an efficient parabolic solar dish collector. In the present work, a solar dryer is designed with a parabolic dish collector to increase the concentration ratio of the solar radiation and hence carrying more heat in the air to remove the moisture of peanuts at a faster rate. Moreover, this works also demonstrates with the further improvement in solar radiation absorption rate with manual solar tracking collector which can rotate in one direction.

2. Materials and Method

Various steps are followed in designing a solar dryer. The first step involved is the study of climatic data and geographical nature of the location. Here, the study is conducted at Kalady, Ernakulam district. Second step is the study of insolation and calculation based on the size and configuration of the collector used in the dryer. Performance analysis of the drier is carried out by considering major design parameters pertaining to solar drier design [20]–[23]. When all the dimensions are determined for the design of dryer, an exhaust fan is chosen to mount at the top of the drying chamber to make the exhaust of air containing the moisture.

2.1. Climatic and geographical data

Kalady is a place in Ernakulam district of Kerala state, India. It is located at a latitude of 10.1710° N and longitude of 76.4468° E. Solar radiation (Insolation) on the surface of Kalady is 855.3876 W/m^2 [23]. The angle at which the solar collector should be tilted is determined from the declination angle (Ψ).

2.2. Declination (Ψ)

Elevated angle between equatorial plane and sun's direction (Ψ) [21] is formulated using the day 'i' in the year, where i varies from 1 to 365.

$$\Psi = 23.45 \text{ Sin } [0.9863(284 + i)] \tag{1}$$

General equation for optimum slope of collector of the solar collector (β) with respect to horizontal is defined with the declination of (Ψ), and latitude(θ) of Kalady [21]. The solar collector is tilted at an angle (β) of 11.8686 South.

$$\beta = \Psi + \text{latitude angle } (\theta) \tag{2}$$

The solar radiation on 12^0 tilted surfaces is calculated as 867.43 W/m^2.

2.3. Design consideration

As solar drying method has a higher drying rate than open sun drying, solar drier with the collector having manual solar tracking system is implemented for drying of the peanut. The dryer has a parabolic dish shaped collector made of galvanized iron sheet with a clear glass cover of 5 mm thickness [5] and an exhaust fan at the exit of the drying chamber. The inner portion of parabolic dish shaped collector is first coated with a copper and then it is painted black over it [24], [25] to provide maximum absorption of heat from the sunlight. The recommended air gap is 5 cm [26] for a tropical climate is also maintained. Collector surface is insulated by providing an air gap using a plywood material cut in the circular shape to reduce thermal losses.

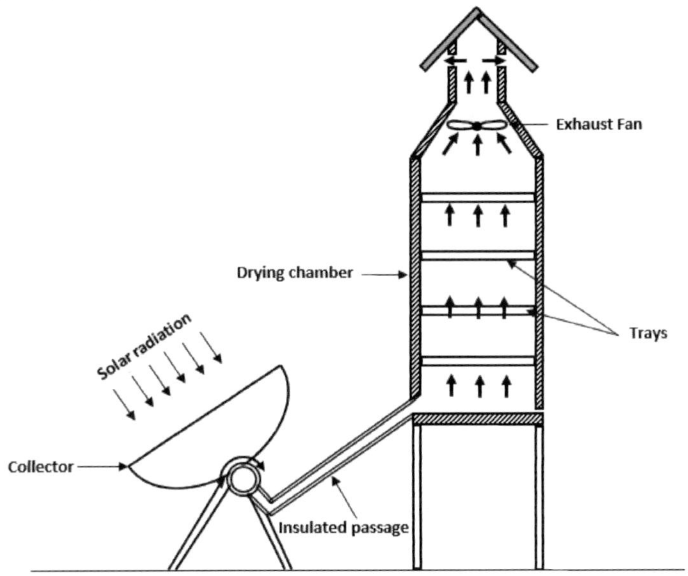

Figure 1. Schematic diagram of solar dryer with dish shaped parabolic collector and drying chamber

The reason for choosing plywood is that it is an ecofriendly material. Heat loss by radiation and damaging of plywood due to moisture absorption is prevented by wrapping an aluminum foil over it [19]. The axial flow exhaust fan assures the constant air flow through the drying chamber at a flow rate of 2 m/s. For this study, 500 grams of peanut is spread over the trays in the drying chamber. The hot air from the outlet of the solar collector is fed to trays in the drier chamber through the inlet holes drilled just below it. In drying chamber, slots are made in the drying chamber with a clearance between them to hold the trays. The clearance between the slots in the chamber helps to keep the trays separated from one another, so that the peanut placed for drying will be exposed to adequate hot air. This arrangement will speed up the drying process. Heat from the solar collector is carried to trays, where the crops are placed for drying, through a connecting pipe that connects the solar collector and the drier chamber.

A parabolic dish shaped collector has a diameter of 1.5 meter and area of 1.77m^2 is selected. The major function of collector is to convert the solar radiation into useful heat energy. A thermal analysis is performed for calculating heat gain and losses when the air is flowing between glass cover and absorber plate, which is called as a top flow, while air flow between absorber plate and bottom insulation which is called as bottom flow [3]. It is observed that more heat is lost through upper portion of the collector than the bottom flow portion as the bottom portion has less temperature difference with ambient air.

2.4. Efficiency of the drier

Efficiency of a solar drier (η_d) is calculated as the ratio of product of mass of moisture in a crop (m) and the latent heat of vaporization (h_{fg}) to the product of area of collector (A_c), and insolation (I_c) falling on the collector and time duration of drying (t).

$$\eta_d = \frac{mh_{fg}}{A_c I_c t} \quad (3)$$

IOP Conf. Series: Materials Science and Engineering 1132 (2021) 012046 doi:10.1088/1757-899X/1132/1/012046

The parameter that controls the efficiency of a drier is the time period of drying.

$$Q = m \times h_{fg} = \rho C_p V (T_c - T_a) \begin{cases} T_c - T_a = T_b \\ T_c - T_a > T_b \end{cases} \left(\begin{array}{c} \textit{Ideal case} \\ \textit{When losses are considerable} \end{array} \right. \quad (4)$$

The amount of moisture removal can be related to the hot air flowing from the collector outlet is by equating mass of the moisture to be removed from the crop (m) times the latent heat of vapourization of water (h_{fg}) to the heat required (Q) to remove the crop moisture (eqn.4). However, in case of air, the amount of heat carried by the air (Q) can be formulated (eqn. 4) by multiplying density of air (ρ) with Specific heat capacity of air, Volume flow rate of the air (V) and the difference between the temperature carried by the air from the collector (T_c) to the ambient temperature (T_a). There are two cases (eqn. 4) to be dealt with: (1) ideal case and (2) when losses are considerable. In the ideal case, it is assumed that there is no loss of heat happens when the air carries heat from collector to the drying chamber. In this ideal case, the difference between the air at the outlet of collector and the ambient temperature should be equal to the required temperature (T_b) in the drying chamber. The required temperature in the drying chamber is the temperature required for drying a particular crop, here it is peanut.

However, if the heat loss when the air carries the air from the collector to the drying chamber is considered, then the difference between the temperature of collector (T_c) and ambient temperature (T_a) should be greater than the temperature in the drying chamber. If only this constraint is met, the temperature loss while carrying the hot air from collector to the drying chamber can be compensated, and required air can be delivered to the drying chamber.

The percentage of moisture removal from the peanut (eqn. 5) is calculated as the ratio of the difference between the initial mass of the crop (M_i) and the final mass (M_f) of the peanut after drying to the initial mass of the peanut (M_i).

$$\text{Moisture (\%)} = \textit{Moisture drop in the crop (\%)} = \left(\frac{M_i - M_f}{M_f}\right) \times 100 \quad (5)$$

3. Results and Discussion

The solar radiation, insolation (I_c), falling on the exposed surface of the dish type solar collector is in the range 540 to 760 W/m^2. As a result of this insolation falling over the surface of the solar collector, the temperature of the air (T_{ci}), ambient air temperature, entering the inlet of the solar collector is in the range from 31^0 C to 37.5 ^{0} C and the temperature of the air exiting through the outlet of the collector (T_{co}) is in the range of 59 ^{0} C to 79 ^{0} C.

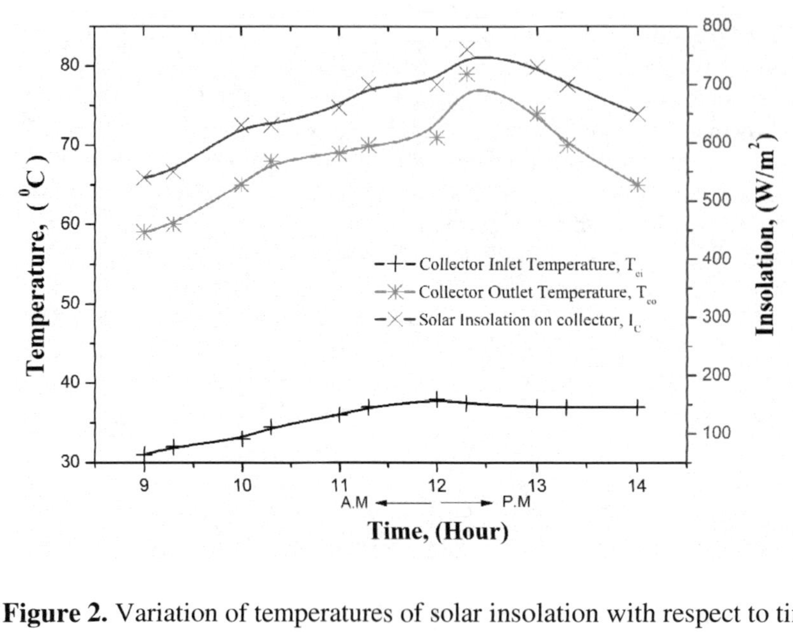

Figure 2. Variation of temperatures of solar insolation with respect to time

Collector inlet temperature, collector outlet temperature and the solar insolation falling on the collector is plotted against time of exposure of collector from 9 AM to 2 PM on a hot day in figure 2. The maximum solar radiation is incident on the solar collector during the noon12 hours 30 minutes, which is 760 W/m^2 and the minimum solar radiation falling on the collector is at 9 A.M. morning, which is 540 W/m^2. The ambient air drawn into the solar collector is heated by the radiation of sunlight falling on the absorber to a maximum temperature of 79 ^{0} C, for 760 W/m^2, and the hot air is taken out of the solar collector to the drying chamber.

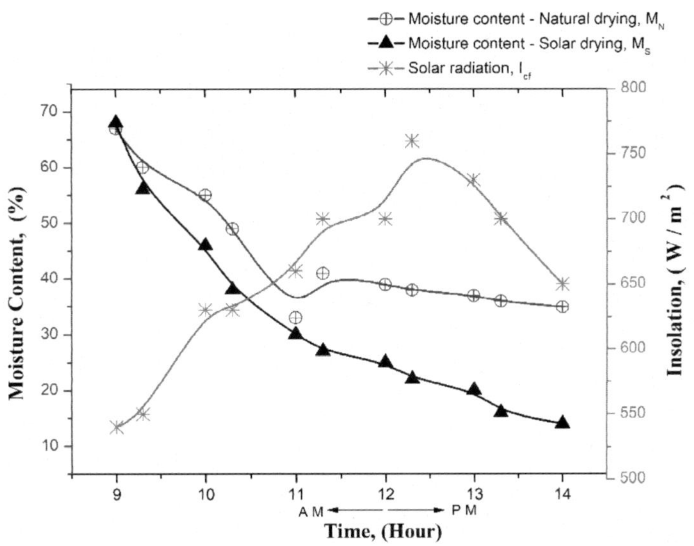

Figure 3. Reduction in the moisture content and solar insolation with time

Figure 3 shows a comparison of natural and solar drying. The crop is openly exposed to sunlight for natural drying, whereas dish collector solar drying system is utilized for drying the crop under solar drying method. For comparison between natural and solar drying method, solar dryer is exposed to same insolation or solar radiation as in figure 3. It is evident from the figure 3 that the crop having 68% of moisture is reduced to 14% when dried for 5 hours, where 35% of moisture is still exist in the crop after natural drying. Hence, the solar drying with solar dish collector is more effective and capable of removing 21% of extra moisture when compared to natural drying method.

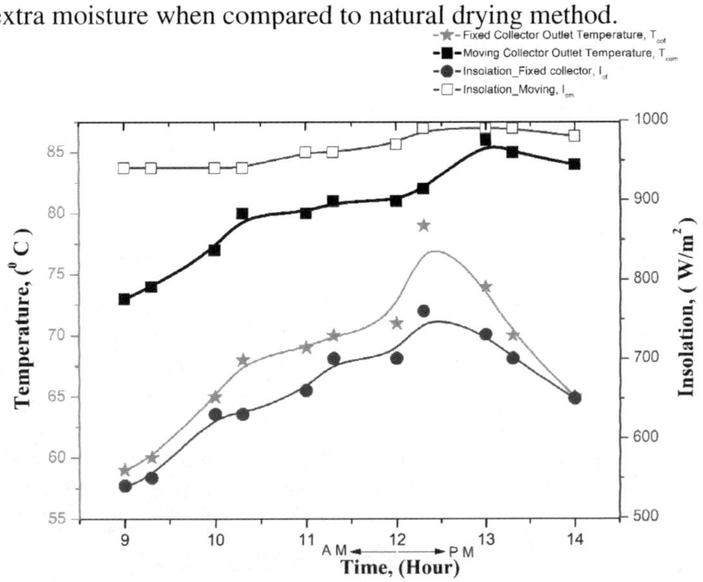

Figure 4. Collector exit temperature (T_{co}), tray temperatures (T_1, T_2, T_3 and T_4) and moisture removal rate (%) with time

In the solar drying system, the hot air from the collector is delivered to the trays through the vents just below the trays. As hot air passes through these four trays, Tray1, Tray 2, Tray 3 and Tray 4 have temperature T_1, T_2, T_3 and T_4 respectively. In figure 4, tray 1 has a temperature of 61 °C, whereas tray 2 has 60.5 °C, tray 3 has 59.5 °C and tray 4 has 67.5 °C. Tray 4 is exposed to highest temperature because tray 4 is placed in the bottom most position, and as a result the hot air first reaches tray 4 and in addition to the supplied heat to the tray 4, the hot air coming through the vents of other trays will also add heat to the tray 4. Though the insolation is 660 W/ m^2 at 11 AM on the day of testing, tray 4 is subjected to highest temperature in this time period. However, for an insolation of 760 W/ m^2 and collector outlet temperature of 79 °C at 12.30 PM, the hot air reaching the tray 4 has only a temperature of 63.5 °C which is attributed to heat loss between collector and the tray. As the tray 4 is subjected to enough temperature, the moisture level of the crop placed for drying on this tray is drastically reduced from 68

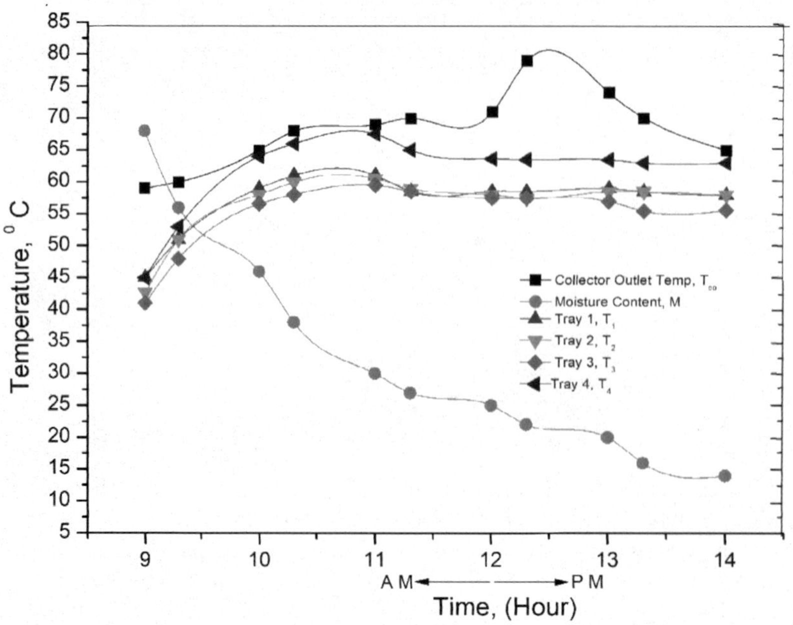

% gh
solar drying method is efficient than natural drying method, movable type solar collectors can absorb more temperature from the sunlight than fixed type collectors (figure 5) for the same time period of sunlight exposure as shown in figure 5. However, the moving type solar collectors align itself to the incident sunlight to make the angle of incidence perpendicular to surface of the collector. This enable the collector to be exposed to a more insolation than the fixed type collectors by reducing the reflection of incident sunlight and as a result, the temperature of a moving type collector is increased to a higher level than the fixed type as in figure 5. Thus, dryers with moving type collectors are having higher heat absorption capability than fixed collector type dryers.

4. Conclusions

Peanut is dried using a solar dryer with a parabolic dish type collector and the moisture reduction over a time period of 5 hours in a day is analysed. The study reveals that the lowest tray in a dryer chamber, for which an air hole is provided, dries the peanut at a faster rate than other trays because of the highest air temperature to which it is exposed. Moreover, solar drying method extracted a higher percentage of moisture from the peanut than the open sun drying method. Although the slope of the drying curve is steep at peak hours of temperature in a day, the increase in temperature does not reduce the drying time drastically. The parabolic dish collector generated a maximum temperature of 79 °C at 12.30 PM and dries 1.5 kilograms of peanut in 5 hours of a day. Application of parabolic collector in solar dryer improved the moisture removal by 21% when compared to direct solar drying.

References

[1] S. Adhikari, Postharvest Management of Fruit and Vegetables in the Asia-Pacific Region. 2006.

[2] Y. Abdullahi, M. Momoh, M. M. Garba, and M. Musa, “Design and Construction of an Adjustable and Collapsible Natural Convection Solar Food Dryer,” International Journal of Computational Engineering Research, vol. 3, no. 6, pp. 1–8, 2013.

[3] G. Prakash and H. P. Garg, Solar Energy: Fundamentals and Applications. Tata McGraw-Hill Publishing Company, 2000.

[4] E. A. Almuhanna, “Utilization of a Solar Greenhouse as a Solar Dryer for Drying Dates under the Climatic Conditions of the Eastern Province of Saudi Arabia,” Journal of Agricultural Science, vol. 4, no. 3, p. p237, 2012.

[5] C. L. Hii, S. V.Jangam S V, Ong S P, Mujumdar A S (eds) (2012) Solardrying:Fundamentals, applications and innovations. TPR Group Publication, Singapore.

[6] V. I. Umogbai and H. Iorter, “Design, construction and performance evaluation of a passive solar dryer for maize cobs,” African Journal of Food Science and Technology, vol. 4(5), no. 5, pp. 110–115, 2013.

[7] N. Rajeshwari and A. Ramalingam, “Low cost material used to construct Effective box type solar dryer,” Archives of Applied Science Research, vol. 4, pp. 1476-1482,

[8] G. Tiwari, V. K. Katiyar, V. Dwivedi, A. K. Katiyar, and C. K. Pandey, “A Comparative study of commonly used Solar Dryers in India,” International Journal of Current Engineering and Technology, vol. 3, no. 3, pp. 994–999, 2013.

[9] A. Madhlopa, S. A. Jones, and J. D. Kalenga Saka, “A solar air heater with composite-absorber systems for food dehydration,” Renewable Energy, vol. 27, no. 1, pp. 27–37, 2002.

[10] A. Brett, D. R. Cox, R. Simmons, G. Anstee, “A solar tunnel dryer for natural convection drying of vegetables and other commodities in Cameroon”. J Am Med Assoc, vol. 35, pp. 31–35, 1996.

[11] M. Isiaka, A. M. I. El-Okene, and U. S. Muhammed, “Effect of selected factors on drying process of tomato in forced convection solar energy dryer,” Research Journal of Applied Sciences, Engineering and Technology, vol. 4, no. 19, pp. 3637–3640, 2012.

[12] E. A.Arinze, S. Sokhansanj, G. J. Schoenau, and F. G. Trauttmansdorff, “Experimental evaluation, simulation and optimization of a commercial heated-air batch hay drier: Part 1, drier functional performance, product quality, and economic analysis of drying,” Journal of Agricultural and Engineering Research, vol. 63, no. 4, pp. 301–314, 1996.

[13] J. K. Afriyie, H. Rajakaruna, M. A. A. Nazha, and F. K. Forson, “Simulation and optimisation of the ventilation in a chimney-dependent solar crop dryer,” Solar Energy, vol. 85, no. 7, pp. 1560–1573, 2011.

[14] S. Karale, “A review paper on Solar Dryer 1 * UmeshToshniwal and 2,” 2013.

[15] B. M. A. Amer, M. A. Hossain, and K. Gottschalk, “Design and performance evaluation of a new hybrid solar dryer for banana,” Energy Conversion and Management, vol. 51, n o. 4, pp. 813–820, 2010.

[16] M. Mohanraj and P. Chandrasekar, “Performance of a forced convection solar drier integrated with gravel as heat storage material for chili drying,” Journal of Engineering Science and Technology, vol. 4, no. 3, pp. 305–314, 2009.

[17] J. Banout, I. Kucerova, and S. Marek, “Using a double-pass solar drier for jerky drying,” Energy Procedia, vol. 30, no. 2, pp. 738–744, 2010.

[18] N. A. Vlachos, T. D. Karapantsios, A. I. Balouktsis, and D. Chassapis, “Design and Testing of a New Solar Tray Dryer,” Drying Technology, vol. 20, no. 6, pp. 1243–1271, 2002.

[19] D. Jain and R. K. Jain, “Performance evaluation of an inclined multi-pass solar air heater with in-built thermal storage on deep-bed drying application,” Journal of Food Engineering, vol. 65, no. 4, pp. 497–509, 2004.

[20] D. R. Pangavhane and R. L. Sawhney, "Review of research and development work on solar dryers for grape drying," Energy Conversion and Management, vol. 43, no. 1, pp. 45– 61, 2002.
[21] K. Sukhatme and S. P. Sukhatme, Solar Energy: Principles of Thermal Collection and Storage. Tata McGraw-Hill, 1996.
[22] F. K. Forson, M. A. A. Nazha, F. O. Akuffo, and H. Rajakaruna, "Design of mixed-mode natural convection solar crop dryers: Application of principles and rules of thumb," Renewable Energy, vol. 32, no. 14, pp. 2306–2319, 2007.
[23] Sudhanshu, "Solar Radiation Hand Book," Solar Energy, pp. 1986–2000, 2008.
[24] A. J. D. M. Queiroz, H. J. Dantas, R. M. F. De Figueirêdo, and K. D. S. Melo, "Solar drying of jack fruit almonds," Engenharia Agricola, vol. 31, no. 6, pp. 1150–1161, 2011.
[25] A. A. Gatea, "Performance evaluation of a mixed-mode solar dryer for evaporating moisture in beans," Journal of Agricultural Biotechnology and Sustainable Development, vol. 3, no. 4, pp. 65–71, 2011.
[26] F. M. Mohamed, A. S. Jassim, Y. H. Mahmood, and M. A. K. Ahmed, "Design and Study of Portable Solar Dish Concentrator," International Journal of Recent Research and Review, vol. III, no. September, pp. 52–59, 2012.

AUTHOR INDEX

Institute of Physics
Dirac House, Temple Back
Bristol BS1 6BE UK

ISSN: 1757-8981
ISBN 978-1-7138-3410-6

CURRAN ASSOCIATES INC.
proceedings.com